Fossil Fuel Emissions Control Technologies

Fossil Fuel Emissions Control Technologies

Stationary Heat and Power Systems

Bruce Miller
Senior Scientist and Associate Director
EMS Energy Institute
Penn State University

AMSTERDAM • BOSTON • HEIDELBERG • LONDON
NEW YORK • OXFORD • PARIS • SAN DIEGO
SAN FRANCISCO • SINGAPORE • SYDNEY • TOKYO
Butterworth-Heinemann is an imprint of Elsevier

Butterworth-Heinemann is an imprint of Elsevier
225 Wyman Street, Waltham, MA 02451, USA
The Boulevard, Langford Lane, Kidlington, Oxford OX5 1GB, UK

ISBN: 978-0-12-801566-7

Library of Congress Cataloging-in-Publication Data
A catalog record for this book is available from the Library of Congress.

British Library Cataloguing-in-Publication Data
A catalogue record for this book is available from the British Library.

For Information on all Butterworth-Heinemann publications
visit our website at http://store.elsevier.com/

Dedication

For my wife Sharon for her patience
and support during the writing of this book.

Contents

Preface

Fossil fuels – including coal, oil, and natural gas – are currently the world's primary energy source. Historically they have supplied most of the world's energy and will continue to do so for many decades. Despite concerns regarding climate change, experts predict that petroleum, coal, and natural gas will remain the primary energy sources because they are abundant, inexpensive relative to renewables, and familiar to the user. The implementation of technologies to extract petroleum and natural gas from unconventional sources has resulted in a dramatic increase in their production, which will alter the percentages of coal, oil, and natural gas used but fossil fuels in general will still be the dominate primary energy used. Today's share of fossil fuels in the global mix, approximately 82% on an equivalent energy basis, is the same level as it was 25 years ago. According to the International Energy Agency, the rise in renewable energy sources only reduces this to around 75% in 2035. Similar predictions by the U.S. Energy Information Administration indicate that liquid fuels, natural gas, and coal will still supply more than 75% of total world consumption in 2040.

Fossil fuels are a major energy source for stationary heat and power generation, nearly all of which require some form of pollution control. Of these fuels, coal is the primary source for power generation, followed by natural gas with a very small amount of petroleum-based fuels. If fossil fuels are to be extensively used, then we must use them in an environmentally-sound manner. That is the intent of this book. This book provides information on emissions control technologies for stationary heat and power systems. Background is provided on energy usage in the United States, fossil fuel usage in boilers and combustion turbines for heat and power production, characteristics and composition of fossil fuels, specifically coal, liquid fuels (petroleum products), and gaseous fuels (primarily natural gas), and the types of emissions from fossil fuel-fired systems. Fossil fuel-fired emission regulations are presented. Discussions of particulate formation and reduction technologies, sulfur oxides formation and capture, and nitrogen oxides formation and control are provided. Mercury emissions and reduction is discussed and information on acid gases, and organic and inorganic hazardous pollutants formation and control is provided. Greenhouse gases and carbon dioxide emissions and control are presented. The technology chapters provide information on pollutant formation, various control options and efficiencies, fundamentals of pollutant capture/reduction, basic design principles of control devices, and simplified comparative economics of control options.

I am grateful to the staff of Elsevier for their support in various phases of this project, namely the senior acquisition editor Kenneth McCombs and the project manager Anusha Sambamoorthy, as well as other members of the production team, for their experience and professional support. No book is flawless, and this one is certainly no different from others.

Bruce G. Miller
Senior Scientist and Associate Director,
EMS Energy Institute,
The Pennsylvania State University,
University Park, Pennsylvania
April 2015

Introduction

1

1.1 Organization of this book

Fossil fuels historically have supplied most of the world's energy and will continue to do so for many decades. Experts predict that petroleum, coal, and natural gas will remain the primary energy sources because they are abundant, inexpensive relative to renewables, and familiar to the user. Today's share of fossil fuels in the global mix, approximately 82% on an equivalent energy basis, is the same level as it was 25 years ago. According to the International Energy Agency (IEA), the rise in renewable energy sources only reduces this to around 75% in 2035 [1]. Similar predictions by the U.S. Energy Information Administration (EIA) indicate that liquid fuels, natural gas, and coal still supply more than 75% of total world consumption in 2040 as shown in Figure 1.1 [2].

According to EIA, petroleum and other liquid fuels remain the largest source of energy but their share of the world marketed energy consumption declines from 34% in 2010 to 28% in 2040. Liquids consumption increases from 175 quadrillion (10^{15}) Btu in 2010 to $\approx$230 quadrillion Btu in 2040. On a worldwide basis, liquids consumption increases only in the industrial and transportation sectors while it declines in the commercial and electric power sectors.

Worldwide, natural gas consumption increases from 113 trillion cubic feet ($\approx$115 quadrillion Btu) in 2010 to 185 trillion cubic feet ($\approx$190 quadrillion Btu) in 2040. Over the projection period, natural gas demand increases in all end-use sectors, with the largest increments in the electric power and industrial sectors. These two sectors account for 77% of the net increase in global natural gas use over the projection period.

Coal continues to be important in world energy markets, especially in non-Organization for Economic Cooperation and Development (non-OECD) Asia (see Appendix A for regional definitions of countries) [2]. Coal consumption increases from 150 quadrillion Btu in 2010 to $\approx$215 quadrillion Btu in 2040 with the major growth in China, India, and other non-OECD Asia countries. Coal provides the largest share of world electricity generation throughout the projection period although its share declines from 40% of total generation in 2010 to 36% in 2040 [2].

Similarly in the U.S., fossil fuels currently provide approximately 82% of the total primary energy consumption, which is projected to remain at 80% in 2040 as shown in Figure 1.2. [3]. Total primary energy consumption increases from 95 quadrillion Btu in 2012 to 106 quadrillion Btu in 2040. Energy usage in the United States is discussed in more detail in the next section.

Total U.S. consumption of petroleum and other liquids, which was 35.4 quadrillion Btu in 2012, will decline to 35.4 quadrillion Btu in 2040. The decrease is

Fossil Fuel Emissions Control Technologies. DOI: http://dx.doi.org/10.1016/B978-0-12-801566-7.00001-4

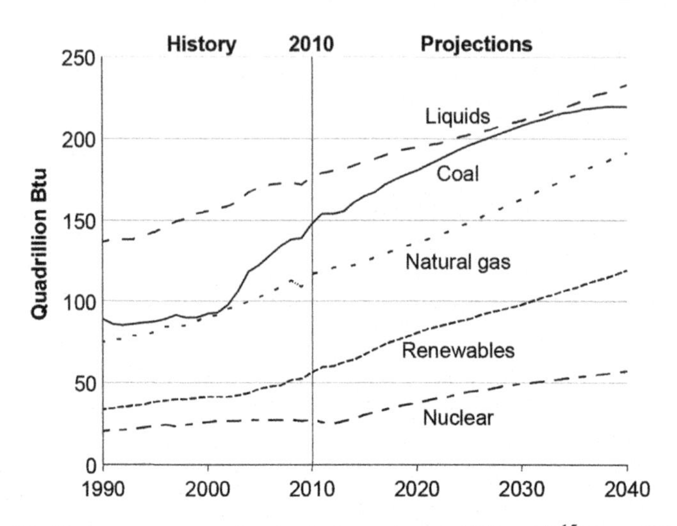

Figure 1.1 World energy consumption by fuel type in quadrillion (10^{15}) Btu, 1990–2040.

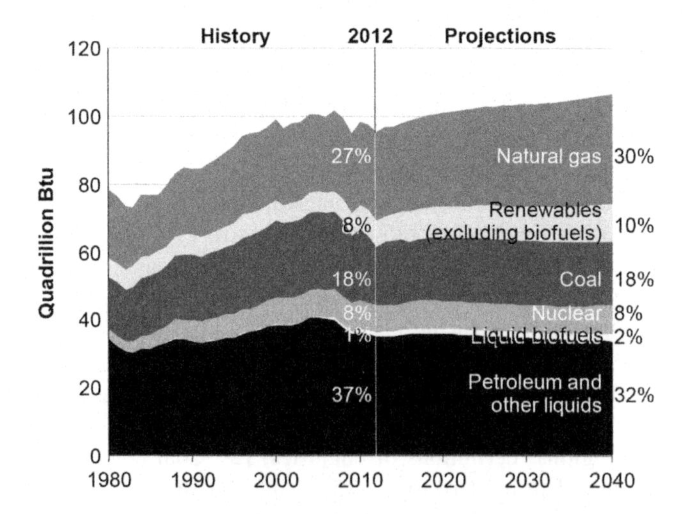

Figure 1.2 U.S. primary energy consumption by fuel (quadrillion Btu), 1980–2040.

attributed to a shift from motor gasoline to distillate and the increased use of natural gas and LNG (liquefied natural gas) in vehicles.

Natural gas consumption rises from 25.6 trillion cubic feet (Tcf) in 2012 to 31.6 Tcf in 2040 with the largest share of its growth due to electricity generation. A portion of this growth is attributed to the retirement of 50 GW of coal-fired capacity by 2021. Natural gas consumption also increases in the industrial sector due to surging shale gas production that is accompanied by slower growth in natural gas prices.

Total coal consumption increases from 17.3 quadrillion Btu (891 million short tons) to 18.7 quadrillion Btu (979 million short tons) in 2040. Coal consumption remains relatively steady over the projection period even though there are retirements of coal-fired power plants and little capacity is added because the remaining coal-fired capacity is used more intensively.

It has been shown that fossil fuels are a major source of energy worldwide and in the U.S., a trend that will continue for many years. A major portion of fossil fuels usage is for stationary heat and power generation, nearly all of which require some form of pollution control. Of these fuels, coal is the primary source for power generation, followed by natural gas with a very small amount of petroleum-based fuels. This book provides information on emissions control technologies for stationary heat and power systems. Continuing in Chapter 1, background will be provided on energy usage in the United States, fossil fuel usage in boilers and combustion turbines for heat and power production, characteristics and composition of fossil fuels, specifically coal, liquid fuels (petroleum products), and gaseous fuels (primarily natural gas), and the types of emissions from fossil fuel-fired systems. Fossil fuel-fired emission regulations are presented in Chapter 2. Discussions of particulate formation and reduction technologies, sulfur oxides formation and capture, and nitrogen oxides formation and control are provided in Chapters, 3, 4, and 5, respectively. Chapter 6 discusses mercury emissions and reduction, while Chapter 7 provides information on acid gases, and organic and inorganic hazardous pollutants formation and control. In Chapter 8, greenhouse gases and carbon dioxide emissions and control are presented. The technology chapters will provide information on pollutant formation, various control options and efficiencies, fundamentals of pollutant capture/reduction, basic design principles of control devices, and simplified comparative economics of control options.

1.2 Overview of energy usage in the United States

The United States has always been a resource-rich nation but in the colonial days nearly all energy was supplied by muscle power from both man and animal, waterpower, wind, and wood. The history of energy use in the United States begins with wood being the dominant energy source from the founding of the earliest colonies until late last century as shown in Figure 1.3 [4]. Consumption is illustrated in quadrillion Btu in Figure 1.3. Although wood use continued to expand along with the nation's economic growth, energy shortages led to a search for other energy sources. Hence, coal began to be used in blast furnaces for coke production and in the making of coal-gas for illumination in the early 1800s. Natural gas found limited application in lighting. It was still not until well after mid-century that the total work output from engines exceeded that of work animals.

Westward expansion from the seacoast was a major factor in increasing the use of coal. As railroads drove west to the plains and mountains, they left behind the plentiful wood resources along the east coast. Coal became more attractive as deposits were found along the railroad right-of-way and it had a higher energy content than wood. This meant more train-miles traveled per pound of fuel. Demand for coal in

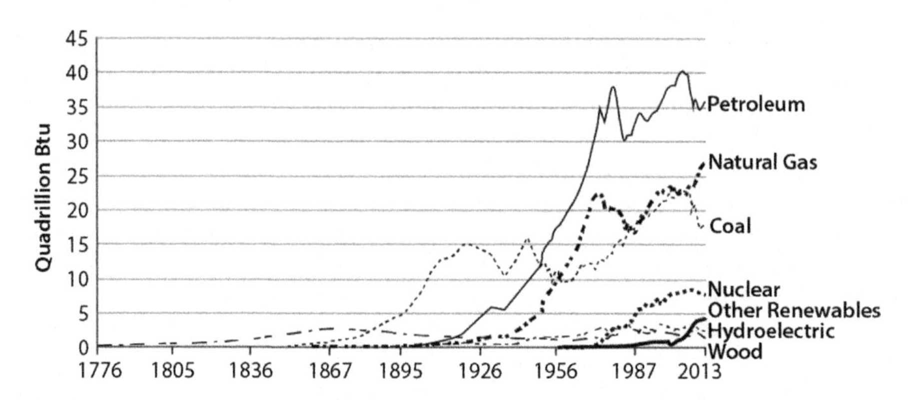

Figure 1.3 United States energy consumption by source, 1775–2013.

coke production also rose because the railroads were laying thousands of miles of new track and iron and steel were needed for the rails and spikes. The rapid growth of the transportation and industrial sectors was fueled by coal.

Coal ended the long dominance of wood in the United States about 1885, only to be surpassed in 1951 by petroleum and then by natural gas a few years later. Hydroelectric power and nuclear electric power appeared about 1890 and 1957, respectively. Solar photovoltaic, advanced solar thermal, and geothermal technologies represent further recent developments in energy sources. Figure 1.4 shows the trends of these fuels from 1949 to 2013 [modified from 5].

Petroleum was initially used as an illuminant and an ingredient in medicines but was not used as a fuel for many years. At the end of World War I, coal still accounted for approximately 75% of United States' total energy use. This changed, however, after World War II. Coal relinquished its place as the premier fuel in the United States as railroads lost business to trucks that operated on gasoline and diesel fuel. The railroads themselves began switching to diesel locomotives. Natural gas also started replacing coal in home stoves and furnaces. The coal industry survived, however, mainly because nationwide electrification created new demand for coal among electric utilities.

Most of the energy produced today in the United States comes from fossil fuels – coal, natural gas, crude oil, and natural gas plants liquids (see Figures 1.5 and 1.6) [modified from 5]. Although United States energy production takes many forms, fossil fuels together far exceed all other forms of energy. In 2013, fossil fuels accounted for 78% of total energy production [4].

For most of its history, the United States was self-sufficient in energy although small amounts of coal were imported from Britain and Nova Scotia during colonial times. Through the late 1950s, production and consumption of energy were nearly in balance. However, beginning in the 1960s and continuing today, consumption outpaces domestic production (see Figure 1.7) [4]. This is further shown in Figure 1.8 where, in 2013, the United States produced approximately 82 quadrillion Btu but consumed more than 97 quadrillion Btu with crude oil imports totaling nearly 25 quadrillion Btu [4]. In October 1973, the Arab members of the Organization of

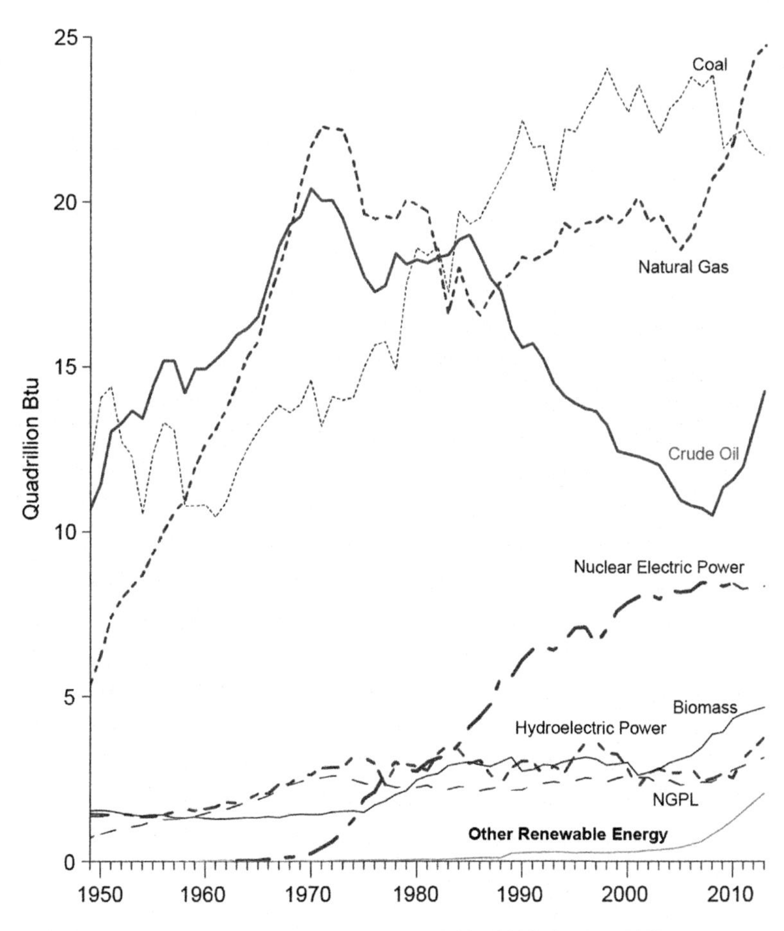

Figure 1.4 Primary energy production by source, 1949–2013 (in Quadrillion Btu).

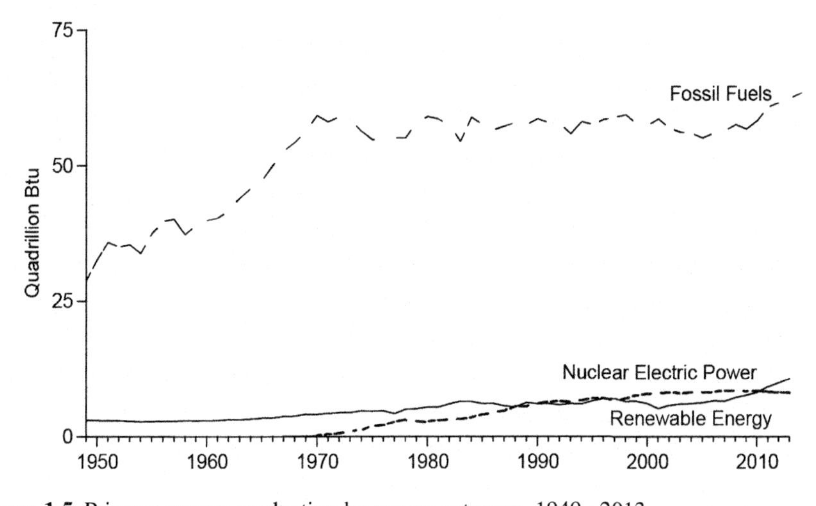

Figure 1.5 Primary energy production by source category, 1949–2013.

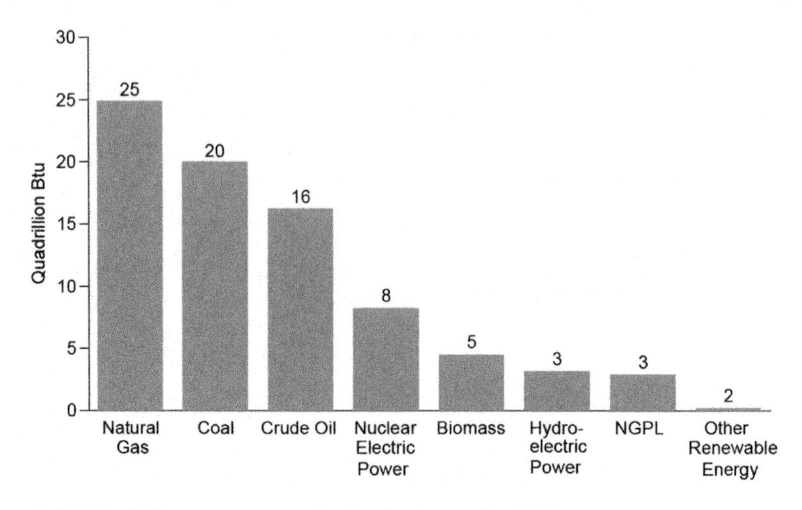

Figure 1.6 United States energy production by source for 2013.

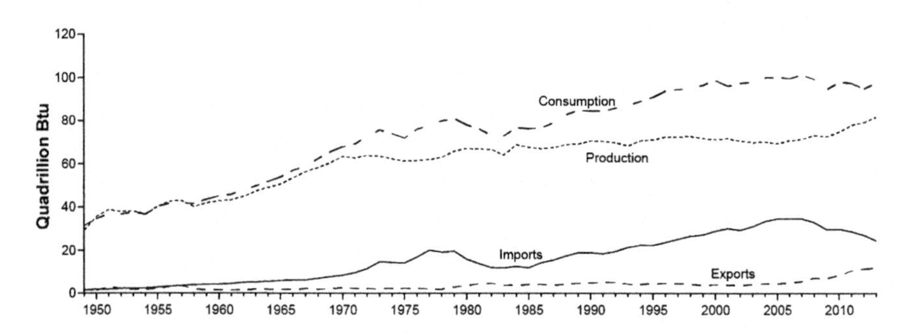

Figure 1.7 United States' primary energy overview.

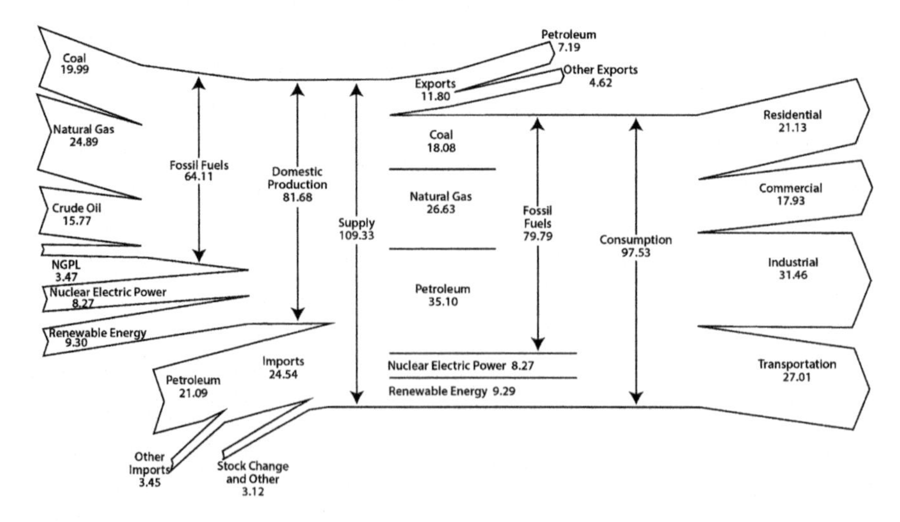

Figure 1.8 United States' energy flow in 2013 (quadrillion Btu).

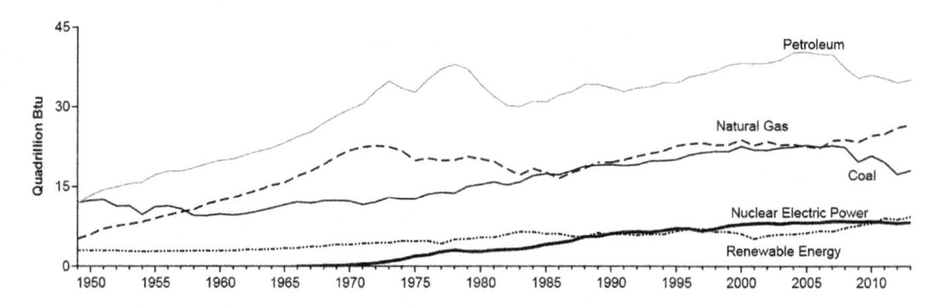

Figure 1.9 Overview of primary energy consumption in the United States from 1949 to 2013.

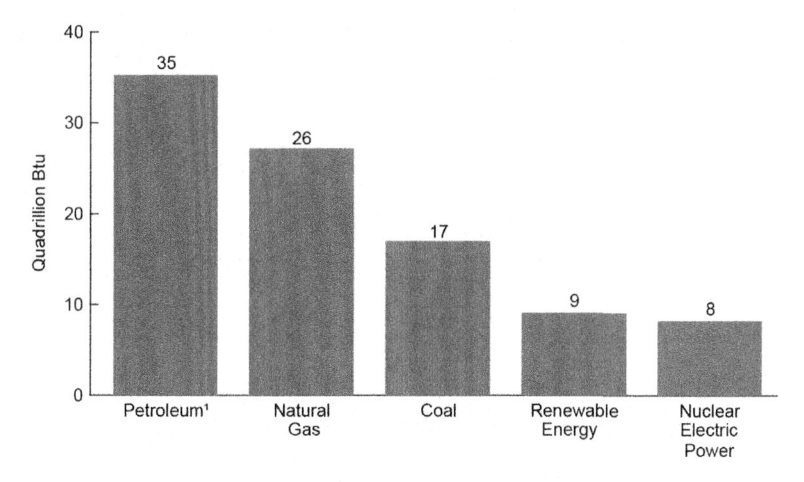

Figure 1.10 United States' primary energy consumption by major fuel source in 2013.

Petroleum Exporting Countries (OPEC) embargoed the sale of oil to the United States, causing prices to rise sharply and sending the country into a recession. A listing of OPEC countries is provided in Appendix A. Although petroleum imports declined for two years, they increased again until prices rose dramatically from about 1979 through 1981, which suppressed imports. The increasing import trend resumed in 1986 and has continued ever since. However, with new technology developments to recover unconventional oil resources, the United States recently is experiencing increased petroleum production which can be observed in Figure 1.7.

Energy is crucial in the operation of the industrialized United States economy, and energy usage and spending is high. Figure 1.9 illustrates the amount of primary energy consumption in the United States from 1949 to 2013 [4]. Primary energy consumption by major source for 2013 is highlighted in Figure 1.10 [4]. Since 2005, American consumers are spending over a trillion dollars a year on energy [6]. Energy is consumed in four major sectors – residential, commercial, industrial, and transportation – as illustrated in Figure 1.11, which shows the energy consumed by each of the four sectors from 1949 to 2013 [4].

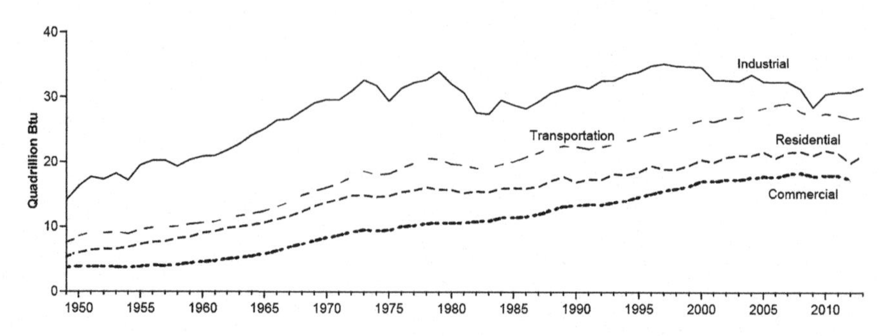

Figure 1.11 Energy consumption by sector from 1949 to 2013 (in Quadrillion Btu).

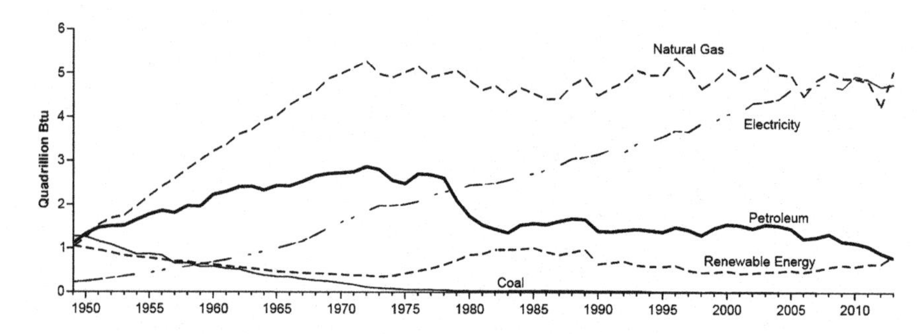

Figure 1.12 Residential energy consumption in the United States.

Industry is historically the largest user of energy and the most vulnerable to fluctuating prices and consequently shows the greatest volatility. This is observed in Figure 1.11. In particular, steep drops occurred in 1975, 1980–83, and again in 2001 and 2007 in response to high oil prices and economic slowdowns. Transportation was the next largest energy-consuming sector, followed by residential and commercial use.

Energy sources have changed over time for the various sectors. In the residential and commercial sectors, coal was the leading source as late as 1951 but then decreased, as observed in Figures 1.12 and 1.13 [4]. Coal was replaced by other forms of energy. Meanwhile electricity's use and related losses during generation, transmission, and distribution increased dramatically. The expansion of electricity reflects the increased electrification of American households, which typically rely on a wide range of electrical appliances and systems. Home heating in the United States also underwent a big change. Over a third of all housing units were heated by coal in 1950 but less than 0.5% was coal heated in 2005. Similarly, distillate fuel oil lost a significant share of the home heating market, dropping from 22% to 9% over the same period. Home heating by natural gas and electricity on the other hand, rose from 26% to 52% and from 1% to 31%, respectively [5].

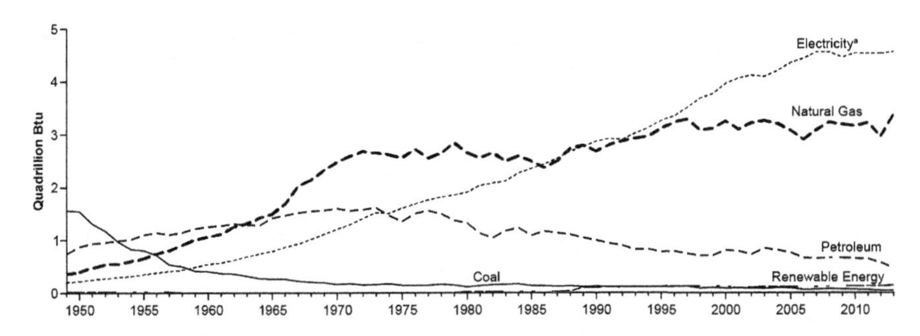

Figure 1.13 Commercial sector energy consumption in the United States.

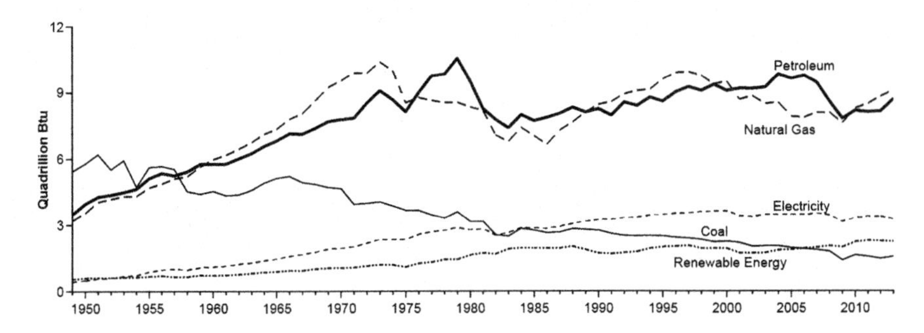

Figure 1.14 Industrial energy consumption in the United States.

In the industrial sector, the use of coal, once the leading energy source, decreased as the consumption of both natural gas and petroleum rose (see Figure 1.14) [4]. Electricity and its associated losses also grew steadily. Approximately 71% of the energy consumed in the industrial sector is used for manufacturing. The remainder goes to mining, construction, agriculture, fisheries, and forestry. The large consumers of energy in the manufacturing industries, for which the fuel of choice is primarily natural gas, include petroleum and coal products, chemicals and associated products, paper and associated products, and metal industries. Less than 5% of all energy consumed in the United States is used for nonfuel purposes such as asphalt and road oil for roofing products, road construction and road conditioning, liquefied petroleum gases for feedstocks at petrochemical plants, waxes for packaging, cosmetics, pharmaceuticals, inks, and adhesives, and gases for chemical and rubber manufacture [3].

The transportation sector's use of energy, which is mainly petroleum (petroleum accounted for 97% of the sector's energy in 2013 [3]), has more than tripled over the last 50 years as shown in Figure 1.15 [4]. Motor gasoline accounts for about 62% of the petroleum consumed in this sector with distillate fuel oil and jet fuel the main other petroleum products used in this sector.

Electric power sector energy consumption is shown in Figure 1.16 [4]. Coal has been the dominant energy source for electric power generation until recently.

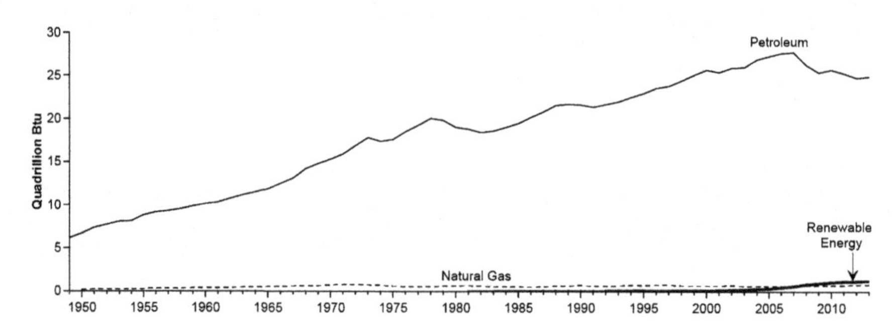

Figure 1.15 Transportation energy consumption in the United States.

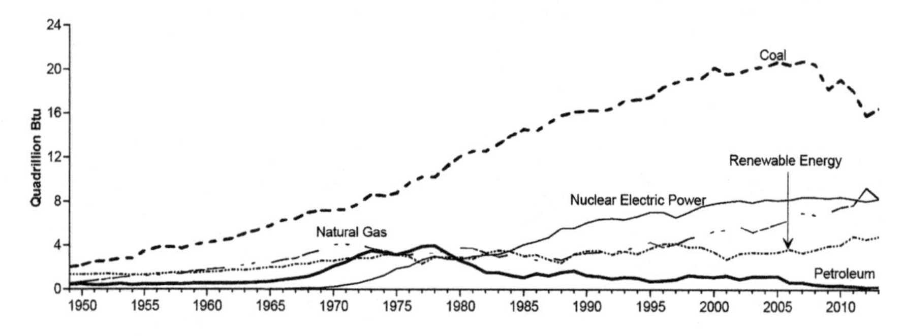

Figure 1.16 Electric power sector energy consumption in the United States.

While coal is still the leading fuel source for electric power generation, the use of natural gas has been steadily increasing since the late 1980s as more stringent emissions legislation has been passed and older coal-fired units have been retired. In addition, the recent technology advances in recovering unconventional natural gas (from tight gas shales) have resulted in a larger supply of natural gas and lower natural gas prices, which can be observed in Figure 1.16. The use of petroleum in electric power generation has decreased to about 1% of the total fuel used in this sector. Renewable energy use in electric power generation has been slowly increasing over the last 10 years and provides about 12% of the energy for this sector.

1.3 Fossil fuel usage in boilers for heat and power production

Figure 1.17 provides a summary of primary energy consumption by source and sector in the U.S. [4]. Approximately 66% of the fuels consumed in the U.S. in 2012 for electric power generation were coal, natural gas, or petroleum. In 2012, 38.1 quadrillion Btu of energy, or 40% of the U.S. total, were consumed for electric power generation of which 25.2 quadrillion Btu or approximately 27% of all energy consumed in the U.S. were derived from fossil fuels. The fossil fuels were

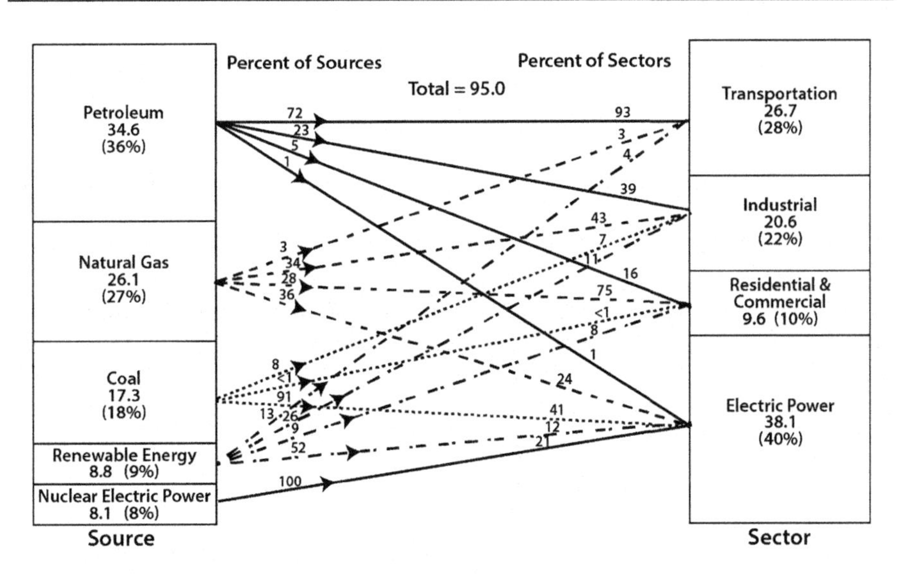

Figure 1.17 Primary energy consumption by source and sector, 2012.

consumed in a sector that requires pollution control measures. Approximately 89% of fuels consumed in the industrial sector were fossil fuels. These fuels are used for a combination of heat and power generation and feedstocks for chemicals production. Industrial boilers also require pollution control devices depending upon the boiler size (i.e., thermal input).

From Figure 1.17 and data in [3], it can be determined that 35.0 quadrillion Btu of energy from fossil fuels are consumed for heat and power applications. This is out of a total 78.0 quadrillion Btu of fossil fuels consumed in all sectors and a total of 95.0 quadrillion Btu of all energy sources. Petroleum, natural gas, and coal are all used as fuels for stationary heat and power units. All electric power-generating units require some form of pollution control and most industrial boilers, above a given thermal input, also must meet emissions requirements (this is discussed in detail in Chapter 2). Only a small portion of the petroleum used for nontransportation purposes ($\approx$0.75 quadrillion Btu) is used as chemical feedstocks indicating that about 7.55 quadrillion Btu (0.35 for electric power generation and 7.21 for the industrial sector) are used for heat and power applications. Conversely, with natural gas a small portion of the natural gas consumed in the industrial sector is used for heat and power generation (1.45 quadrillion Btu out of about 8.87 quadrillion Btu) indicating that approximately 10.85 quadrillion Btu (1.45 for the industrial sector and 9.40 for the electric power sector) are used for heat and power applications. For coal on the other hand, nearly all of it or 16.64 quadrillion Btu (15.74 for electric power and 0.9 for industrial boilers) are used for heat and power applications. A total of 35.0 quadrillion Btu of petroleum, natural gas, and coal are used for heat and power applications where emissions control is required. This quantity is out of a total of 78.0 quadrillion Btu fossil fuels consumed for all applications or approximately 45% of all fossil fuels used for heat and power applications require some form of emissions control. The following chapters will focus on emissions produced, legislative actions requiring their capture or reduction, and the technologies used and under development to control them.

1.4 Characteristics and composition of fossil fuels

This section presents an introductory overview of the chemical and physical characteristics of fossil fuels that are used in boilers for stationary heat and power production. The purpose of this section is to provide the reader with basic coal, liquid fuels, and gaseous fuels information as a prelude to the subsequent chapters, because emissions are a primary function of fuel composition and a secondary function of combustion/boiler equipment and operation.

1.4.1 Coal

1.4.1.1 Introduction to coal

Coal is a sedimentary rock composed of both organic and inorganic material. Coal is composed of macerals, discrete minerals, inorganic elements held molecularly by the organic matter, and water and gases contained in submicroscopic pores. Organically, coal consists primarily of carbon, hydrogen, and oxygen, and lesser amounts of sulfur and nitrogen. Inorganically, coal consists of a diverse range of ash-forming compounds distributed throughout the coal. The inorganic constituents can vary in concentrations from several percentage points to parts per billion of the coal.

Coal is formed from the accumulation of vegetative debris that has undergone physical and chemical changes over millions of years. These changes include decaying of the vegetation, deposition and burying by sedimentation, compaction, and transformation of the plant remains into the organic rock found today. The geochemical process that transformed plant debris to coal and its effect on coal composition (which is described in much detail in [7–12]) is called coalification and can be simply expressed as:

$$\text{Peat} \rightarrow \text{Lignite} \rightarrow \text{Subbituminous Coal} \rightarrow \text{Bituminous Coal} \rightarrow \text{Anthracite}$$

Chemically there is a decrease in moisture and volatile matter (i.e. methane, carbon dioxide) content, as well as an increase in the percentage of carbon, a gradual decrease in the percentage of oxygen, and ultimately, as the anthracitic stage is approached, a marked decrease in the content of hydrogen [9,12]. For example, carbon content (on a dry, mineral-matter free basis) increases from approximately 50% in herbaceous plants and wood, to 60% in peat, 70% in lignite, 75% in subbituminous coal, 80–90% in bituminous coal, and >90% in anthracite [9,13–15]. This change in carbon content is known as carbonification.

1.4.1.2 Coal classification

Coals differ throughout the world in the degree of metamorphism or coalification (rank of coal), in the kinds of plant materials deposited (type of coal), and in the range of impurities included (grade of coal). Efforts to classify coals began nearly 200 years ago and were prompted by the need to establish some order in the confusion of different coals. Two types of classification systems arose: some schemes are intended to aid scientific studies and other systems are designed to assist coal producers and users. The scientific systems of classification are concerned with origin, composition, and fundamental properties of coals while the commercial systems address trade and market issues, utilization, technological properties, and suitability for certain end uses. It is the latter classification systems that will be discussed in this section.

1.4.1.2.1 Basic coal analysis

Prior to discussing rank, type, grade, and classification systems of coal, a brief description of basic coal analyses, which classification schemes are based upon, will be provided. These analyses provide important information on coal behavior and are used in the marketing of coals.

There are three analyses that are used in classifying coal, two of which are chemical analyses and the other a calorific determination. The chemical analyses include proximate and ultimate analysis. The proximate analysis gives the relative amounts of moisture, volatile matter, and ash, i.e. inorganic material left after all the combustible matter has been burned off, and, indirectly, the fixed carbon content of the coal. The ultimate analysis gives the amounts of carbon, hydrogen, nitrogen, sulfur, and oxygen comprising the coal. Oxygen is typically determined by difference, i.e., subtracting the total percentages of carbon, hydrogen, nitrogen, and sulfur from 100 because of the complexity in determining oxygen directly. However, this technique does accumulate all the errors in determining the other elements into the calculated value for oxygen. The third important analysis, the calorific value, also known as heating value, is a measure of the amount of energy that a given quantity of coal will produce when burned.

Since moisture and mineral matter (or ash) are extraneous to the coal substance, analytical data can be expressed on several different bases to reflect the composition of as-received, air-dried, or fully water-saturated coal or the composition of dry, ash-free (daf), or dry, mineral-matter free (dmmf) coal. The most commonly used bases in the various classification schemes are shown in Figure 1.18. Descriptions of the most commonly used bases are [16]:

- as received – data are expressed as percentages of the coal with the moisture. This is also sometimes referred to as as-fired and is commonly used by the combustion engineer to monitor operations and for performing calculations since it is the whole coal that is being utilized;
- dry basis (db) – data are expressed as percentages of the coal after the moisture has been removed;
- dry, ash-free (daf) basis – data are expressed as percentages of the coal with the moisture and ash removed;

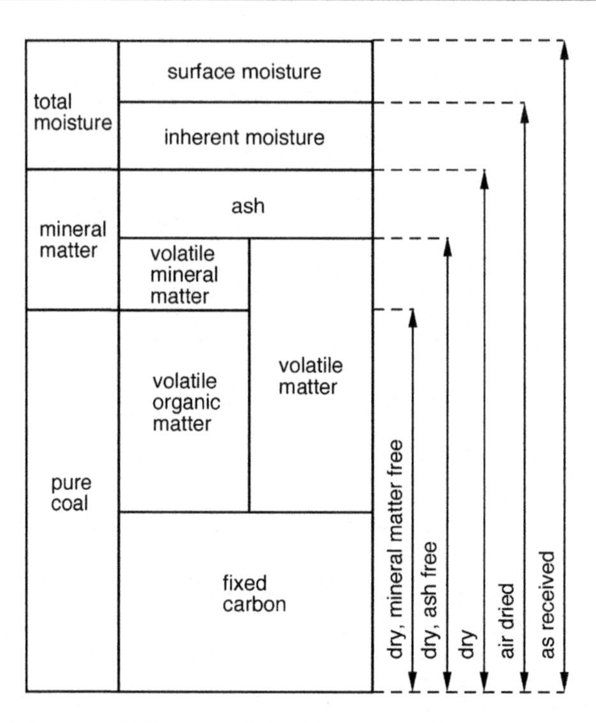

Figure 1.18 Relationship of different analytical bases to coal components.

- dry, mineral matter-free (dmmf) basis – the coal is assumed to be free of both moisture and mineral matter and the data are a measure of only the organic portion of the coal;
- moist, ash-free (maf) basis – the coal is assumed to be free of ash but still contains moisture; and
- moist, mineral matter-free (mmmf) basis – the coal is assumed to be free of mineral matter but still contains moisture.

1.4.1.2.2 Coal rank

The degree of coal maturation is known as rank of coal and is an indication of the extent of metamorphism the coal has undergone. Because vitrinite precursors (see Section 1.4.1.2.3 for a brief discussion on maceral types), such as humates and humic acids, were the major constituents in peat, the extent of metamorphism is often determined by noting the changes in the properties of vitrinite [12]. Some of the more important properties of vitrinite that change with metamorphism are listed in Box 1.1 [12]. Metamorphism did affect the other macerals, but the relationship between the severity of metamorphism and the magnitude of change are different from those of vitrinite.

Several useful rank-defining properties are elemental carbon content, volatile matter content, moisture-holding capacity, heating value, and microscopic reflectance of vitrinite [12]. Figure 1.19 illustrates the relationship between rank and fixed carbon content [17], the one chemical property most used to express coal rank. In the United States, lignites and subbituminous coals are referred to as being

Box 1.1 Properties of Vitrinite Affected Progressively by Metamorphism

Increase with increasing rank

- Reflectance (microscopic) and optical anisotropy
- Carbon content
- Aromaticity, $f_a = C_{aromatic}/C_{total}$
- Condensed-ring fusion
- Parallelization of molecular moieties
- Heating value (a slight decrease at very high rank)

Decrease with increasing rank

- Volatile matter (especially oxygenated compounds)
- Oxygen content (especially as functional groups)
- Oxidizability
- Solubility (especially in aqueous alkalis and polar hydrocarbons)

Increase initially to a maximum, then decrease

- Hardness (minor increase at very high rank)
- Plastic properties
- Hydrogen content

Decrease initially to a minimum, then increase

- Surface area
- Porosity (and moisture holding capacity)
- Density (in helium)

low in rank while bituminous coals and anthracites are classified as high-rank coals. In Europe, low-rank coals are referred to as brown coal or lignite and hard coals are anthracite and bituminous coals, and in some cases, subbituminous coals are included as well. Figure 1.19 illustrates this relationship between rank and fixed carbon content [17]. Fixed carbon content in Figure 1.19 is calculated on a dry, mineral matter-free basis. Figure 1.19 also shows the comparison between heating value and rank with the heating value calculated on a moist, mineral matter-free basis. Note that the heating value increases with increasing rank but begins to decrease with semianthractic and higher rank coals. This decrease in heating value is due to the significant decrease in volatile matter, which is shown in Figure 1.19 [17].

Vitrinite reflectance is another important rank-measuring parameter. The advantage of this technique is that it measures a rank-sensitive property on only one petrographic constituent; therefore, it is applicable even where the coal type is atypical [12].

1.4.1.2.3 Coal type

Coal is composed of macerals, discrete minerals, inorganic elements held molecularly by the organic matter, and water and gases contained in submicroscopic pores. Macerals are organic substances derived from plant tissues that have been

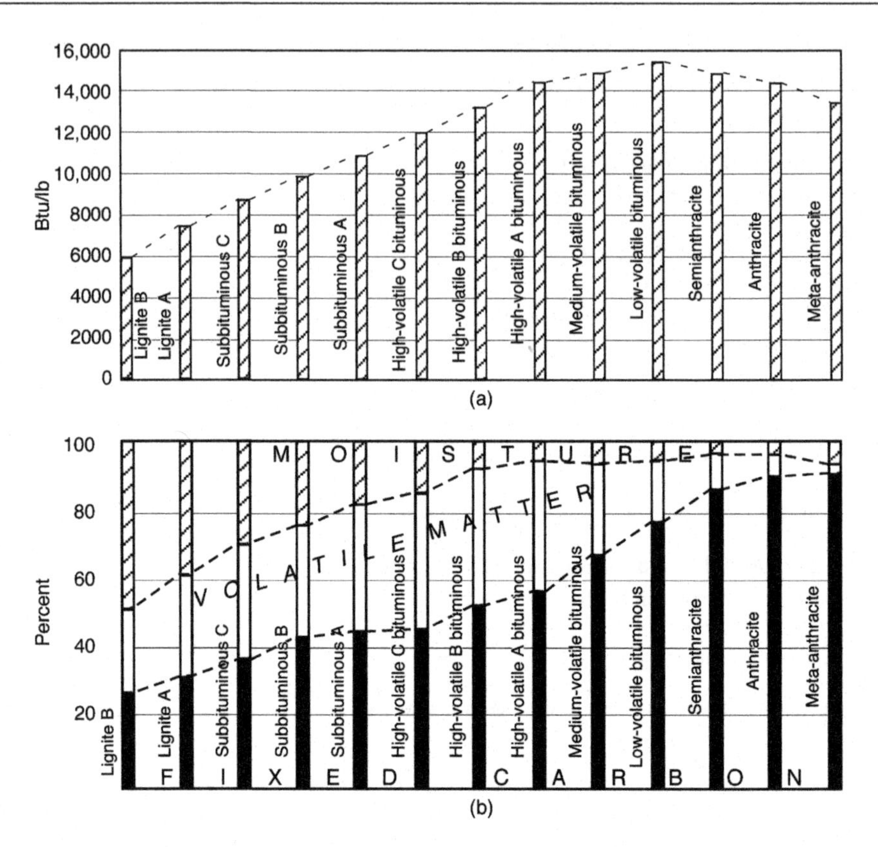

Figure 1.19 Comparison of heating values (on a moist, mineral matter-free basis) (a) and proximate analyses of coals of different ranks (b).

incorporated into the sedimentary strata, subjected to decay, compacted, and chemically altered by geological processes. This organic matter is extremely heterogeneous and a classification system has been developed to characterize it [12,18,19]. Classifying the coal, known as petrography, was primarily used to characterize and correlate coal seams and resolve questions about coal diagenesis and metamorphism but now is an important tool in assessing coals for industrial applications [20].

Vitrinite group macerals are derived from the humification of woody tissues and can either possess remnant cell structures or be structureless [11]. Vitrinite contains more oxygen than the other macerals at any given rank level, and are characterized by a higher aromatic fraction.

Liptinite group macerals are not derived from humifiable materials but rather from relatively hydrogen-rich plant remains such as resins, spores, cuticles, waxes, fats, and algal remains, which are fairly resistant to bacterial and fungal decay [13,20]. Liptinites are distinguishable by a higher aliphatic (i.e. paraffin) fraction and a correspondingly higher hydrogen content, especially at lower rank [18].

The inertinite group macerals were derived mostly from woody tissues, plant degradation products, or fungal remains. While they were derived from the same original plant substances as vitrinite and liptinite, they have experienced a different primary transformation [18]. Inertinite group macerals are characterized by a high carbon content that resulted from thermal or biological oxidation, as well as low hydrogen content and an increased level of aromatization [16,18].

All macerals are classified into three maceral groups – vitrinite, liptinite (sometimes also referred to as exinite), and intertinite – and they are characterized by their appearance, chemical composition, and optical properties. Each maceral group includes a number of macerals and other subcategories. A detailed discussion of macerals and petrography will not be presented for the purpose of this book; however, extensive discussions of petrography can be found elsewhere [11–13,18,19].

Petrographic analysis has many uses. Initially it was primarily used to characterize and correlate seams and resolve questions about coal diagenesis and metamorphism but later it influenced developments in coal preparation (i.e. crushing, grinding, and removal of mineral constituents) and conversion technologies [13]. Industrially, petrographic analysis can provide insight into the hardness of a coal (i.e. its mechanical strength) as well as a coal's thermoplastic properties, which is of significant importance in the coking industry.

1.4.1.2.4 Coal grade

Grade of coal refers to the amount of mineral matter that is present in the coal and is a measure of coal quality. Sulfur content, ash fusion temperatures (i.e. measurement of the behavior of ash at high temperatures) and quantity of trace elements in coal are also used to grade coal. Formal classification systems have not been developed around grade of coal; however, grade is important to the coal user.

Mineral matter may occur finely dispersed or in discrete partings in the coal. Some of the inorganic matter and trace elements are derived from the original vegetation, but the majority is introduced during coalification by wind or water to the peat swamp or through movement of solutions in cracks, fissures, and cavities [21]. Coal mineralogy can affect: the ability to remove minerals during coal preparation/cleaning; coal combustion and conversion (i.e., production of liquid fuels or chemicals) characteristics; and metallurgical coke properties.

1.4.1.2.5 Classification systems

Since the rank of the coal is most important for the coal industry, almost every coal-producing country has its own economic coal classification, which is based mainly on rank parameters. An excellent discussion of the many classification systems, scientific as well as commercial, is provided by van Krevelen [16]. There are two primary commercial classification systems in use – the ASTM (American Society of Testing Materials) system used in the United States/North America and an international ECE (Economic Commission for Europe) Codification system developed in Europe. The classification systems used commercially are primarily based on the content of volatile matter [16]. In some countries, a second parameter is also used and in the United States, for example, this is the heating value (see Figure 1.19). For many European countries this parameter is either the caking or the coking properties.

1.4.1.2.5.1 The ASTM classification system The ASTM classification system (ASTM D388) distinguishes between four coal classes, each of which is subdivided into several groups and is shown in Table 1.1. As previously mentioned, high-rank coals, i.e. medium volatile bituminous coals or those of higher rank, are classified based on their fixed carbon and volatile matter contents (expressed on a dmmf basis), while low-rank coals are classified in terms of their heating value (expressed on an mmmf basis).

This classification system was developed for commercial applications but has proved to be satisfactory for certain scientific uses as well [16]. For example, if a given coal is described as being a certain rank, then an estimate of some properties can be made and, if the coal is classified as subbituminous/lignitic or anthracitic, then it would not be considered for certain applications such as for coke production.

1.4.1.2.5.2 International classification/codification system Because of the significant amount of coal trade in the world, the ECE Coal Committee developed a new classification system in 1988 for higher rank coals [16]. The original international system had deficiencies in that it was primarily developed for trading

Table 1.1 ASTM coal classification by rank

Class/Group	Fixed carbon[a] (%)	Volatile matter[b] (%)	Heating value[b] (Btu/lb)
Anthracitic			
Metaanthracite	>98	<2	
Anthracite	92–98	2–8	
Semianthracite	86–92	8–14	
Bituminous			
Low volatile	78–86	14–22	
Medium volatile	69–78	22–31	
High volatile A	<69	>31	>14,000
High volatile B			13,000–14,000
High volatile C			10,500–13,000[c]
Subbituminous			
Subbituminous A			10,500–11,500[c]
Subbituminous B			9,500–10,500
Subbituminous C			8,300–9,500
Lignitic			
Lignite A			6,300–8,300
Lignite B			<6,300

[a]Calculated on dry, mineral matter-free coal; correction from ash to mineral matter is made by means of the Parr formula: mineral matter = 1.08[percent ash + 0.55(percent sulfur)]. Ash and sulfur are on a dry basis.
[b]Calculated on mineral matter-free coal with bed moisture content.
[c]Coals with heating values between 10,500 and 11,500 Btu/lb are classified as high volatile C bituminous if they possess caking properties or as subbituminous A if they do not.

Northern Hemisphere coals, which have had distinctly different characteristics than those from the Southern Hemisphere, e.g. Australia and South Africa. As trade of Southern Hemisphere coals increased, it became apparent that a new classification system was needed. This new system, which in reality is a system of codes, is better known as a Codification System. The Codification System for hard coals, combined with the ISO (International Organization for Standardization) Codification of Brown Coals and Lignites (which was established in 1974), provides a complete codification for coals in the international trade.

The ISO Codification of Brown Coals and Lignites is given in Table 1.2 [16]. Total moisture content of run-of-mine coal and tar yield (i.e., determination of the yields of tar, water, gas, and coke residue by low-temperature distillation) are the two parameters coded.

The ECE International Codification of Higher Rank Coals is much more complicated and is listed in Table 1.3. Eight basic parameters define the main properties of the coal, which are represented by a 14-digit code number. The codification: is commercial; includes petrographic, rank, grade, and environmental information; is for medium- and high-rank coals only; is for blends and single coals; is for raw and washed coals; and is for all end use applications [16]. The major drawback of this system is that it is complicated.

1.4.1.3 Analyses of selected U.S. coals

This section contains examples of analyses for the various ranks of coals to illustrate the differences in coal composition as a function of rank. Coal analyses exhibit variability both within and between coal seams but the following tables show some of the major differences between the ranks of coal. Properties of selected United States' anthracites, bituminous coal, and low-rank coals are given in Table 1.4 [22], Table 1.5 [22], and Table 1.6 [22–24], respectively. These tables show the variability in various constituents that can lead to emissions such as sulfur, nitrogen, and mineral matter, i.e. fly ash and bottom ash. Also, where available, chlorine and mercury concentrations in the coal are also provided.

Table 1.2 Codification of brown coals and lignites

Parameter	Total moisture content (run-of-mine coal)		Tar yield (dry, ash free)	
Digit:	1		2	
Coding:	Code	Weight %	Code	Weight %
	1	≤20	0	≤10
	2	>20 to 30	1	>10 to 15
	3	>30 to 40		
	4	>40 to 50	2	>15 to 20
	5	>50 to 60	3	>20 to 25
	6	>60	4	>25

Table 1.3 International codification of higher rank coals[a]

Parameter	Vitrinite reflectance (mean random)		Characteristics of reflectogram[b]				Maceral group composition (mmf)				Petrographic tests
							Inertinite[c]		Liptinite		
Digit	1, 2		3				4		5		
Coding	code	R_{random}%	code	standard dev		type	code	vol %	code	vol %	
	02	0.2–0.29	0	≤1	no gap	Seam coal	0	0–<10	1	0 to <5	
	03	0.3–0.39	1	>0.1 ≤0.2	no gap	Simple blend	1	10–<20	2	5 to <10	
	04	0.4–0.49	2	>0.2	no gap	Complex blend	2	20–<30	3	10 to <15	
	–	–	3		1 gap	Blend with 1 gap	–	–	–	–	
	48	4.8–4.89	4		2 gaps	Blend with 2 gaps	7	70–<80	7	30 to <35	
	49	4.9–4.99	5		>2 gaps	Blend with >2 gaps	8	80–<90	8	35 to <40	
	50	≥5.0					9	≥90	9	≥40	

Parameter	Crucible swelling no.		Volatile matter,[d] daf		Ash, dry		Total sulfur, dry		Gross calorific value, daf		Technological tests
Digit	6		7, 8		9, 10		11, 12		13, 14		
Coding	code	number	code	wt.%	Code	wt.%	code	wt.%	code	MJ/kg	
	0	0–0.5	48	≥48	00	0 to <1	00	0–<0.1	21	<22	
	1	1–1.5	46	46 to <48	01	1 to <2	01	0.1–<0.2	22	22 to <23	
	2	2–2.5	44	44 to <46	02	2 to <3	02	0.2–<0.3	23	23 to <24	
	–	–	–	–	–	–	–	–	–	–	
	7	7–7.5	12	12 to <14	20	20 to <21	29	2.9–<3.0	37	37 to <38	
	8	8–8.5	10	10 to <12	–	–	30	3.0–<3.1	38	38 to <39	
	9	9–9.5	09	9 to <10			–	–	39	≥39	
			–	–							
			02	2 to <3							
			01	1 to <2							

[a]Higher rank coals are coals with gross calorific value (maf) ≥24 MJ/kg and those with gross calorific value (maf) <24 MJ/kg provided mean random vitrinitic reflectance ≥0.6%. To convert from MJ/kg to Btu/lb, multiply by 429.23.
[b]A reflectogram as characterized by code number 2 can also result from a high rank seam coal.
[c]It should be noted that some of the intertinite may be reactive.
[d]When the ash content of the coal is more than 10%, it must be reduced, before analysis, to below 10% by dense medium separation. In these cases, the cutting density and resulting ash content should be noted.

Table 1.4 Properties of selected United States anthracites

Rank	**Anthracite**	**Anthracite**	**Anthracite**	**Semi-Anthracite**	**Semi-Anthracite**	**Semi-Anthracite**
Sample description	**Mammoth Seam, PA**	**Buck Mtn. Seam, PA**	**#8 Leader Seam, PA**	**#8 Seam, PA**	**Gunnison Co., CO**	**L. Spadra Seam, AR**
Prox. anal. (%, as-rec.)						
Moisture	3.1	2.5	1.3	1.0	1.1	0.8
Ash	23.4	13.6	8.2	25.1	7.0	6.3
Volatile matter	3.9	5.6	7.0	8.2	10.2	12.1
Fixed carbon	69.6	78.3	83.5	65.7	81.7	80.8
Ult. anal. (%, dry, ash free)						
Carbon	93.5	90.8	91.3	89.6	91.5	92.2
Hydrogen	1.9	2.6	3.9	3.8	3.4	3.9
Nitrogen	1.2	0.8	0.6	1.4	1.6	2.1
Sulfur	1.0	0.6	1.2	1.6	0.8	0.8
Oxygen	2.4	5.2	3.0	3.6	2.7	1.0
HHV[a] (Btu/lb)	10,860	12,603	14,125	11,194	14,098	14,372
Maceral group anal. (Vol % min. free[b])						
Vitrinite	89.5	96.3	97.5	70.8	92.4	87.7
Inertinite	10.5	3.7	2.5	29.2	7.6	12.3
Liptinite	0.0	0.0	0.0	0.0	0.0	0.0
Vitrinite Reflectance (mean-maximum % in oil)	5.72	4.17	2.82	2.15	3.10	2.47
Ash fusion temps. (Red, °F)						
Initial deform.	2700+	–[c]	–	–	–	2150
Softening	2700+	–	–	–	–	2250
Hemispherical	2700+	–	–	–	–	2485
Fluid	2700+	–	–	–	–	2560

(Continued)

Table 1.4 (Continued)

Rank	Anthracite	Anthracite	Anthracite	Semi-Anthracite	Semi-Anthracite	Semi-Anthracite
Sample description	**Mammoth Seam, PA**	**Buck Mtn. Seam, PA**	**#8 Leader Seam, PA**	**#8 Seam, PA**	**Gunnison Co., CO**	**L. Spadra Seam, AR**
Ash comp. (%)						
SiO_2	56.5	54.9	56.7	58.8	37.4	34.4
Al_2O_3	28.4	32.4	32.5	32.3	25.3	26.2
TiO_2	1.77	3.82	1.81	1.27	0.80	1.01
Fe_2O_3	6.10	3.29	4.55	4.18	12.8	13.8
MgO	1.32	1.09	0.41	0.64	2.02	4.96
CaO	0.84	1.53	0.58	0.29	10.8	7.19
Na_2O	0.29	0.12	0.16	0.28	1.25	1.80
K_2O	2.70	1.22	2.30	2.99	1.20	1.50
P_2O_5	0.25	0.06	0.09	0.06	1.69	0.66
SO_3	0.70	2.10	0.20	0.10	7.69	9.00
Trace elements (ppm whole coal, dry)						
Chlorine	1300	100	100	400	100	100
Chromium	41	44	27	71	4	16
Mercury	–	0.05	0.08	–	–	0.03
Nickel	21	27	15	53	4	29
Lead	–	11	<3	19	–	10
Strontium	82	17	11	5	178	116
Vanadium	58	24	22	76	11	20

[a]Higher heating value.
[b]White light.
[c]Not available.

Table 1.5 **Properties of selected United States bituminous coals**

Rank	**Low Vol.**	**Low Vol.**	**Med Vol.**	**Med. Vol.**	**Med. Vol.**	**High Vol.A**
Sample description	**L.Kittanning Seam, PA**	**Elk Lick Seam, WV**	**U.Freeport Seam, PA**	**Dutch Creek Seam, CO**	**Sewell Seam, WV**	**Pittsburgh #8, PA**
Prox. anal. (%, as-rec.)						
Moisture	2.0	1.3	6.1	1.1	1.5	2.4
Ash	10.1	18.3	10.1	5.3	4.2	10.0
Volatile matter	17.4	16.0	22.9	26.9	24.6	35.2
Fixed carbon	70.5	64.4	60.9	66.7	69.7	52.4
Ult. anal. (%, dry, ash free)						
Carbon	88.8	87.2	87.0	87.4	88.2	83.3
Hydrogen	4.7	4.7	5.5	5.9	5.0	5.7
Nitrogen	1.6	1.5	1.7	1.7	1.5	1.4
Sulfur	1.6	3.3	2.9	0.7	0.7	1.3
Oxygen	3.2	3.3	2.9	4.3	4.6	8.3
HHV[a] (Btu/lb)	14,025	12,535	14,006	14,889	14,871	13,532
Maceral group anal. (Vol % min. free[b])						
Vitrinite	90.0	92.7	88.5	94.4	77.2	83.0
Inertinite	9.9	7.3	11.5	5.6	18.0	8.9
Liptinite	0.1	0.0	0.0	0.0	4.8	8.1
Vitrinite reflectance (mean-maximum % in oil)	1.73	1.63	1.24	1.28	1.35	0.87

(*Continued*)

Table 1.5 **(Continued)**

Rank	**Low Vol.**	**Low Vol.**	**Med Vol.**	**Med. Vol.**	**Med. Vol.**	**High Vol.A**
Sample description	**L.Kittanning Seam, PA**	**Elk Lick Seam, WV**	**U.Freeport Seam, PA**	**Dutch Creek Seam, CO**	**Sewell Seam, WV**	**Pittsburgh #8, PA**
Ash fusion temps. (Red, °F)						
Initial deform.	2700+	2530	–	2020	2750	2225
Softening	2700+	2550	–	2130	2800+	2330
Hemispherical	2700+	2600	–	2200	2800+	2390
Fluid	2700+	2670	–	2250	2800+	2445
Ash comp. (%)						
SiO_2	42.7	50.8	39.2	24.8	52.8	55.8
Al_2O_3	39.0	27.5	27.5	22.7	33.3	25.8
TiO_2	1.86	1.55	1.05	0.82	1.02	1.21
Fe_2O_3	10.7	14.5	18.7	10.8	4.59	6.37
MgO	0.57	0.80	0.34	3.52	0.97	0.91
CaO	1.71	1.15	4.06	9.40	1.78	3.20
Na_2O	0.43	0.49	0.11	2.84	0.70	0.49
K_2O	1.61	2.40	0.91	0.33	2.29	2.19
P_2O_5	0.94	0.39	0.84	0.90	1.16	0.56
SO_3	0.40	0.82	5.00	13.7	0.20	2.10
Trace elements (ppm whole coal, dry)						
Chlorine	1300	700	100	700	2600	2300
Chromium	26	32	19	3	11	21
Mercury	–[c]	–	0.16	–	–	–
Nickel	10	15	20	<1	11	11
Lead	–	8	13	–	–	–
Strontium	203	152	172	200	104	111
Vanadium	37	38	26	6	11	7

Rank	**High Vol A**	**High Vol A**	**High Vol A**	**High Vol A**	**High Vol B**	**High Vol B**
Sample Description	**L.Kittanning Seam, WV**	**U.Kittanning Seam, WV**	**Blind Canyon Seam, UT**	**L. Elkhorn Seam, KY**	**Hazard #5 Seam, KY**	**Hiawatha Seam, UT**
Prox. anal. (%, as-rec.)						
Moisture	1.8	1.5	4.7	1.3	3.8	4.9
Ash	11.8	10.4	5.6	9.7	9.1	7.1
Volatile matter	33.6	32.1	42.4	30.9	34.6	39.5
Fixed Carbon	52.8	56.0	47.3	58.1	52.5	48.5
Ult. anal. (%, dry, ash free)						
Carbon	84.5	85.2	81.3	86.5	82.8	80.6
Hydrogen	5.6	5.5	6.2	5.5	5.5	5.7
Nitrogen	1.4	1.5	1.6	1.4	1.6	1.4
Sulfur	1.0	2.0	0.4	0.6	0.8	0.9
Oxygen	7.5	5.8	10.5	6.0	9.3	11.4
HHV[a] (Btu/lb)	13,272	13,678	13,923	13,897	13,369	13,264
Maceral group anal. (Vol % min. free[b])						
Vitrinite	52.5	89.0	69.1	71.2	60.3	89.2
Inertinite	30.3	7.6	13.6	16.8	21.1	9.6
Liptinite	17.2	3.4	17.3	11.9	18.7	1.2
Vitrinite reflectance (mean-maximum % in oil)	0.91	1.07	0.66	1.03	0.80	0.58

(Continued)

Table 1.5 **(Continued)**

Rank	**High Vol A**	**High Vol A**	**High Vol A**	**High Vol A**	**High Vol B**	**High Vol B**
Sample Description	**L.Kittanning Seam, WV**	**U.Kittanning Seam, WV**	**Blind Canyon Seam, UT**	**L. Elkhorn Seam, KY**	**Hazard #5 Seam, KY**	**Hiawatha Seam, UT**
Ash fusion temps. (Red, °F)						
Initial deform.	2700 +	2310	1900	2700 +	2700 +	2110
Softening	2700 +	2400	2020	2700 +	2700 +	2210
Hemispherical	2700 +	2460	2090	2700 +	2700 +	2260
Fluid	2700 +	2510	2240	2700 +	2700 +	2360
Ash comp. (%)						
SiO_2	53.0	46.1	50.3	58.2	54.1	49.5
Al_2O_3	34.9	29.5	15.3	28.3	31.1	21.2
TiO_2	2.28	1.28	0.96	1.58	1.81	1.12
Fe_2O_3	5.49	15.0	6.88	3.37	3.86	4.19
MgO	0.70	0.82	1.26	1.10	0.83	0.97
CaO	0.89	2.59	12.0	0.69	2.09	8.09
Na_2O	0.16	0.24	6.91	0.56	0.32	4.91
K_2O	2.14	1.87	0.59	3.74	2.08	0.21
P_2O_5	0.08	0.21	0.38	0.08	0.42	0.88
SO_3	1.00	1.30	4.80	0.50	1.20	8.00
Trace elements (ppm whole coal, dry)						
Chlorine	600	3700	1200	900	1100	300
Chromium	31	18	6	21	31	19
Mercury	–[c]	–	–	0.08	–	–
Nickel	18	6	<0.4	15	24	24
Lead	–	–	–	7	10	8
Strontium	39	62	85	73	297	262
Vanadium	28	7	9	25	34	8

Rank	**High Vol B**	**High Vol B**	**High Vol C**	**High Vol C**	**High Vol C**
Sample description	**Kentucky #9 Seam, KY**	**Illinois #9 Seam, IL**	**Wadge Seam, CO**	**Illinois #5 Seam, IL**	**Indiana #6 Seam, IN**
Prox. anal. (%, as-rec.)					
Moisture	6.8	5.3	9.5	9.5	11.5
Ash	11.4	10.3	6.4	9.5	13.5
Volatile matter	38.4	37.0	38.0	36.4	33.5
Fixed carbon	43.4	47.4	46.1	44.6	41.5
Ult.aanal. (%, dry, ash free)					
Carbon	79.1	79.9	77.5	78.4	79.7
Hydrogen	5.8	5.6	5.5	5.2	4.9
Nitrogen	1.4	1.6	1.8	1.4	1.6
Sulfur	4.8	4.6	0.6	4.9	3.4
Oxygen	8.9	8.3	14.6	10.1	10.4
HHV[a] (Btu/lb)	12,687	12,633	12,762	12,722	12,292
Maceral group anal. (Vol % min. free[b])					
Vitrinite	85.7	87.9	89.2	87.5	89.0
Inertinite	6.6	12.1	8.7	10.6	8.7
Liptinite	7.7	0.0	2.1	1.9	2.3
Vitrinite reflectance (mean-maximum % in oil)	0.56	0.33	0.60	0.62	0.43
Ash fusion temps. (Red, °F)					
Initial deform.	1755	2090	2700 +	1945	2035
Softening	1900	2125	2700 +	2065	2080
Hemispherical	2000	2185	2700 +	2160	2365
Fluid	2040	2285	2700 +	2245	2425

(*Continued*)

Table 1.5 (Continued)

Rank	High Vol B	High Vol B	High Vol C	High Vol C	High Vol C
Sample description	**Kentucky #9 Seam, KY**	**Illinois #9 Seam, IL**	**Wadge Seam, CO**	**Illinois #5 Seam, IL**	**Indiana #6 Seam, IN**
Ash comp. (%)					
SiO_2	41.2	–	48.7	43.1	50.1
Al_2O_3	15.6	–	31.1	16.0	18.8
TiO_2	0.75	–	0.92	0.78	0.89
Fe_2O_3	22.0	–	3.60	24.2	20.2
MgO	0.70	–	1.47	0.81	0.86
CaO	8.08	–	6.22	5.66	2.16
Na_2O	0.66	–	0.92	0.45	0.57
K_2O	1.89	–	0.84	1.76	2.57
P_2O_5	0.04	–	1.03	0.10	0.15
SO_3	9.40	–	3.90	6.10	2.30
Trace elements (ppm whole coal, dry)					
Chlorine	1600	–	100	800	500
Chromium	23	–	4	10	14
Mercury	–[c]	–	–	–	–
Nickel	8	–	<2	13	6
Lead	–	–	–	–	–
Strontium	21	–	175	17	24
Vanadium	19	–	7	14	19

[a]Higher heating value.
[b]White light.
[c]Not available.

Table 1.6 Properties of selected United States low-rank coals

Rank	**Subbit. B**	**Subbit. B**	**Subbit. B**	**Subbit. B**	**Lignite A**	**Lignite**
Sample description	**Wyodak Seam, WY**	**Rosebud Seam, MT**	**Black Thunder, WY**	**Antelope WY**	**Beulah Seam, ND**	**L.Wilcox Seam, TX**
Prox. anal. (%, as-rec.)						
Moisture	26.3	19.8	26.3	11.0	33.4	28.5
Ash	5.6	7.3	5.6	5.6	6.4	15.3
Volatile Matter	33.1	32.3	33.1	38.4	37.4	44.2
Fixed Carbon	35.0	41.6	35.0	45.0	22.8	12.0
Ult. anal. (%, dry, ash free)						
Carbon	75.5	75.3	75.5	75.1	73.1	72.3
Hydrogen	6.1	5.1	6.2	5.2	4.5	5.2
Nitrogen	1.0	1.1	1.0	0.9	1.0	1.4
Sulfur	0.5	0.9	0.4	0.3	0.8	0.9
Oxygen	16.9	17.6	16.9	18.5	20.6	20.2
HHV[a] (Btu/lb)	12,145	11,684	12,220	11,758	11,062	9882
Maceral group anal. (Vol % min. free[b])						
Huminite	85.7	79.8	–	–	73.8	90.0
Inertinite	9.7	18.8	–	–	19.5	6.9
Liptinite	4.6	1.4	–	–	6.7	3.1
Huminite Rreflectance (mean-maximum % in oil)	0.29	0.51	–	–	0.35	0.44
Ash fusion temps. (Red, °F)						
Initial deform.	2100	2145	–	–	2000	2310
Softening	2130	2170	–	–	2070	2360
Hemispherical	2160	2205	–	–	2100	2500
Fluid	2185	2310	–	–	2140	2540

(Continued)

Table 1.6 **(Continued)**

Rank	Subbit. B	Subbit. B	Subbit. B	Subbit. B	Lignite A	Lignite
Sample description	**Wyodak Seam, WY**	**Rosebud Seam, MT**	**Black Thunder, WY**	**Antelope WY**	**Beulah Seam, ND**	**L.Wilcox Seam, TX**
Ash comp. (%)						
SiO_2	31.7	31.3	31.4	35.5	11.5	47.9
Al_2O_3	16.1	16.8	14.0	17.1	11.1	22.2
TiO_2	1.27	0.79	1.6	1.02	0.57	1.75
Fe_2O_3	4.84	4.78	6.4	5.41	9.29	3.51
MgO	4.64	5.21	4.5	4.45	9.94	2.79
CaO	23.5	17.1	21.2	18.1	27.4	14.0
Na_2O	1.80	0.20	1.0	1.39	9.39	0.37
K_2O	0.40	0.10	0.5	0.40	0.59	0.55
P_2O_5	0.89	0.24	1.2	2.22	0.49	0.11
SO_3	12.7	19.3	18.2	9.8	18.2	7.00
Trace elements (ppm whole coal, dry)						
Arsenic	1	–	–	–	–	–
Cadmium	<0.2	–	–	–	–	–
Chlorine	–[c]	200	–	–	800	1100
Chromium	3	7	–	–	5	17
Mercury	0.12	–	–	–	–	–
Nickel	2	7	–	–	<0.4	5
Lead	5	5	–	–	–	–
Selenium	2	–	–	–	–	–
Strontium	164	225	–	–	772	158
Vanadium	11	5	–	–	14	34

[a]Higher heating value.
[b]White light.
[c]Not available.

1.4.2 Liquid fuels

This section discusses various liquid fuels from petroleum/crude oil. There are small quantities of liquid fuels produced from coal, i.e. coal-to-liquids (CTL), and most of these are produced in China. Because the quantities of CTL produced are very small relative to petroleum-derived liquids and their composition is similar to that of petroleum-derived liquids, which results in them being subjected to the same emissions regulations and using similar emissions control technologies as petroleum-derived fuels, they are not discussed separately in this section.

1.4.2.1 Introduction to petroleum

Petroleum is typically defined as a flammable liquid that is a complex mixture of hydrocarbons present in certain rock strata that was derived from the decomposition and transformation of plants and animals. Others define it as a complex mixture of organic liquids called crude oil and gases, which is predominately natural gas. Operationally, petroleum is referred to as a mixture of gases, liquids, and condensates upon production. Theories of the origin of petroleum are divided into two groups according to their view of the primary source material as either organic or inorganic [25]. Early theories leaned towards inorganic sources; however, the most widely accepted theory is that the chief primary source material of petroleum was organic. There are three reasons for this. First, vast amounts of organic matter and hydrocarbons are found in the earth's sediments. Carbon and hydrogen are predominate in the remains of organic material, both plant and animal. Second, almost all crude oils contain nitrogen and many have been found to contain molecular structures, i.e. porphyrins, that are biomarkers [25]. All organic matter contains both nitrogen (a component of amino acids/proteins) and porphyrins. Two of the most important chemical compounds in the world are porphyrins – chlorophyll and heme [26]. Chlorophyll is the green pigment of plants that catalyzes the photosynthesis process; heme is a constituent of hemoglobin, which is responsible for the transport of oxygen in the bloodstream. Figure 1.20 shows the similarity between a vanadium-containing porphyrin found in petroleum to that of chlorophyll and heme. Third, petroleum is optically active in that it contains constituents that can rotate polarized light. This trait is only known in nature for compounds that derive from living organisms.

1.4.2.2 Formation of petroleum

Petroleum was formed through the compression and heating of organic material buried in sea or lake bottoms over geologic time. Marine organisms, mainly zooplankton and algae, mixed with sediments and were buried under conditions of low oxygen. Maturation is the process by which organic matter is converted into petroleum. This is illustrated in Figure 1.21 where petroleum is formed through diagenesis and catagenesis [modified from 27].

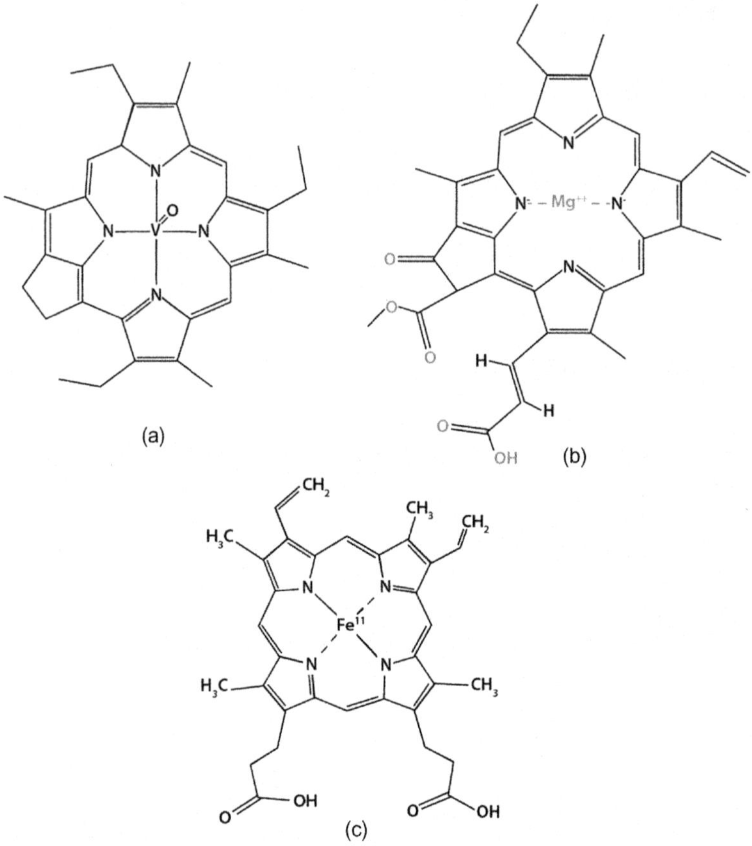

(a)

(b)

(c)

Figure 1.20 Examples of porphyrins: (a) vanadium porphyrin compound; (b) chlorophyll; (c) heme.

1.4.2.2.1 Diagenesis

Diagenesis is a process of compaction under mild conditions of temperature (less than 122°F; 50°C) and pressure. The reactions occur in the shallow subsurface where many of the reactions are facilitated by bacteria and consequently diagenesis is sometimes called the biochemical stage of fuel formation [26]. The first reaction zone occurs at depths of 0–3 feet (0–1 meter) below the surface where oxygen can still diffuse through the covering sediments to react with the organic matter [26]. The moldering reaction zone is about 3–6 feet (1–2 meters) below the surface. In this region the oxygen concentration is low enough to allow reactions other than oxidation to occur. Hydrolysis of proteins and carbohydrates occur in this zone. Putrefaction occurs in the region 6–30 feet (2–10 meters) below the surface where the oxygen concentration is so low that the conditions are effectively anaerobic [26]. Proteins, carbohydrates, lipids, and lignin undergo decomposition in this zone. Diagenesis results in a decrease of oxygen and a correlative increase in carbon.

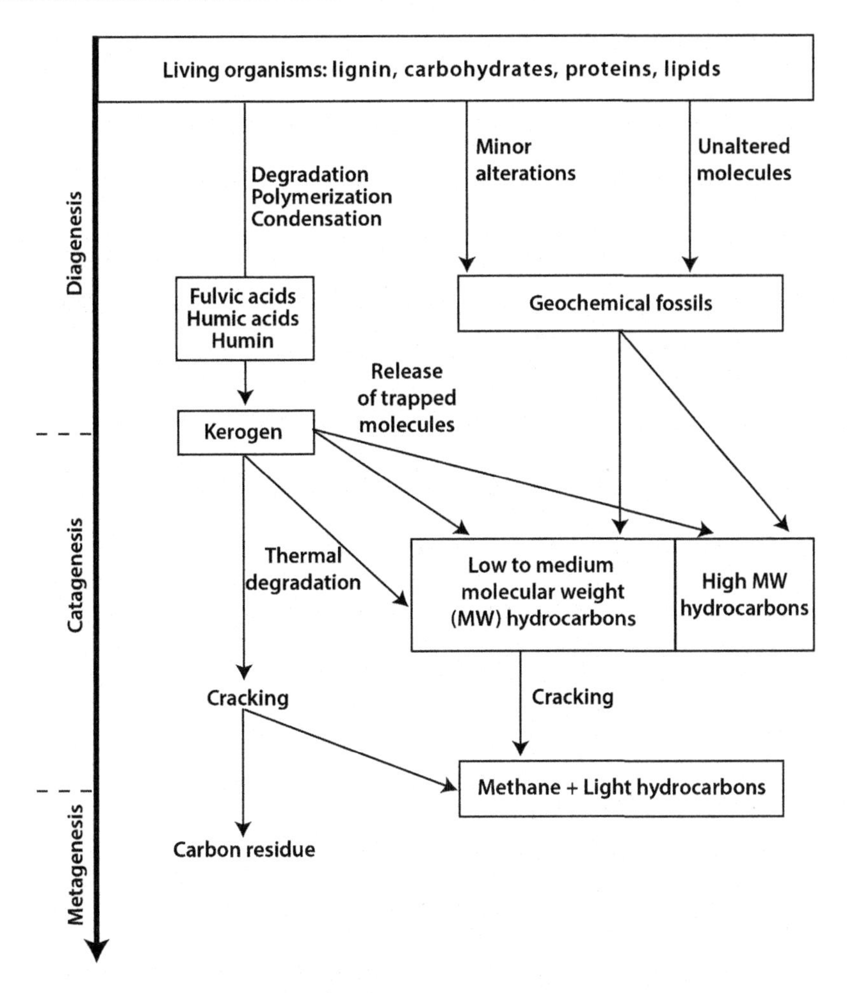

Figure 1.21 The petroleum maturation process.

At the end of putrefaction, the original organic matter has been converted into a waxy material known as kerogen. Kerogen is defined as disseminated organic matter in sediments that is insoluble in normal petroleum solvents [28]. In petroleum studies the kerogen is classified into three basic types (I, II, and III) based on the nature of the original organic source material. Table 1.7 summarizes some of the characteristics of the three basic types of kerogen [modified from 26,27].

Kerogen types I and II, also referred to as sapropelic kerogens, are hydrogen-rich, with predominately straight chain hydrocarbon compounds. Type II kerogen is also highly aliphatic, containing material derived from both algae and plankton. Type III kerogen is oxygen rich and the H/C ratio is lower than the other two. Type III kerogen tends to be rich in aromatic compounds and derives mainly from the lignin of the higher (woody) plants.

Table 1.7 Characteristics of kerogens

Type	Name	Source	H/C	O/C
I	Algal kerogen	Mainly algae	1.65	0.06
II	Liptinic kerogen	Mainly plankton, some contribution from algae	1.28	0.10
III	Humic kerogen	Mainly higher plants	0.84	0.13

1.4.2.2.2 Catagenesis

The conversion of kerogen to petroleum occurs in the catagenesis phase, also known as the geochemical stage of fuel formation. Petroleum is predominately generated from Types I and II kerogen. Catagenesis reactions are driven by temperature and pressure conditions in deeper subsurface environments. The deeply buried sediments undergo chemical reactions at elevated temperatures (140–400°F; 60–200°C) and pressures. During this phase, there is a redistribution of hydrogen available in the kerogen to form a hydrogen-rich fraction, which would ultimately become methane and a hydrogen-poor fraction, which would become graphite. During catagenesis, the organic compounds undergo a variety of thermal degradation reactions including reduction of double bonds by incorporating hydrogen or sulfur atoms, cracking reactions, and condensation reactions. Oil formation occurs during the initial phase of catagenesis at temperatures between 140 and 250°F (60 and 120°C). At higher temperatures the kerogen is cracked into lighter and smaller hydrocarbons such as methane.

1.4.2.3 Refining of crude oil into fuels

Analyses of crude oil from around the world shows that the elemental composition varies over a narrow range: 82–87% carbon, 12–15% hydrogen, with the balance being oxygen, nitrogen, and sulfur. However, the specific compounds present in these crude oils contain hundreds of individual compounds. These compounds fall into several classes of hydrocarbons: alkanes (alkanes are also called paraffins), which can be straight chained (normal or n-alkanes with stoichiometry of C_nH_{2n+2}), branched chain (also called isoparaffins with stoichiometry of C_nH_{2n+2}), or cyclic (also called cycloalkanes, cycloparaffins, or naphthenes with ring structures containing 5, 6, or 7 carbons with the general stoichiometry of C_nH_{2n}); and aromatics (compounds that contain at least one benzene ring (C_6H_6)). Individual compounds have a specific boiling point but because crude oil is a mixture of many compounds, the crude oil fractions have a wide range of boiling points.

Crude oil is refined to produce useful products. An oil refinery consists of several processes designed to produce physical and chemical changes in the crude oil to convert it to many products including fuels. Crude oil is passed through several operations to separate it into fractions, to purify the products, and, in some cases, to

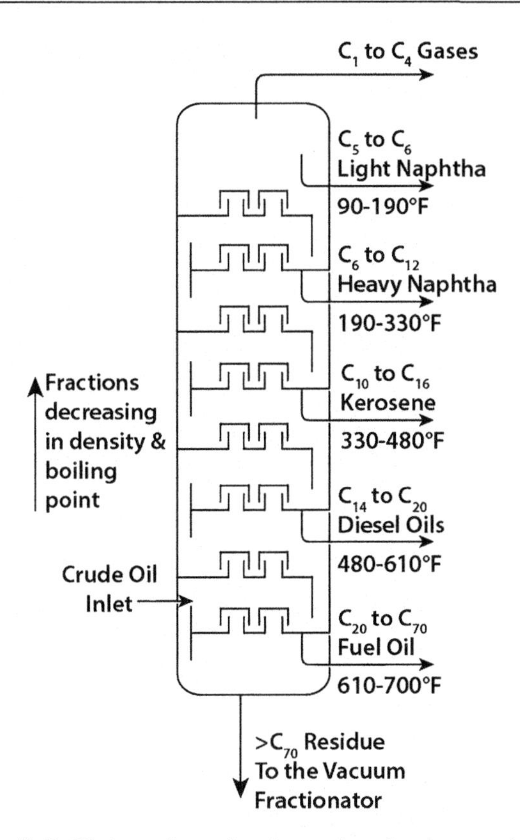

Figure 1.22 Crude oil distillation column showing various fractions produced.

increase the yield of the more valuable products relative to the others. One of the first major processes in the refinery is distillation, which is the key operation in petroleum refining. Distillation takes advantage of the boiling point ranges of the various crude oil components to separate the crude oil into fractions or cuts. Figure 1.22 is a generalized schematic diagram of an atmospheric crude fractionator showing the major fractions that are produced [modified from 26,29]. The fractions or cuts from the distillation unit undergo further treatment (i.e. hydrotreating, isomerization, reforming, cracking, blending, etc.) to produce the final fuel products. The final liquid products used in stationary systems for producing heat and power are discussed in more detail in the following section. The light hydrocarbon gases that come off of the top of the atmospheric crude unit are typically separated into wet gases (propane and butane, or liquefied petroleum gas (LPG)) and dry gases (methane and ethane). The dry gases are primarily used as fuel throughout the refinery for the heaters and boilers. Gases are discussed in Section 1.4.4. The liquid products from the distillation step are shown in Figure 1.22 and their typical uses are given in Table 1.8 [modified from 26,29]. The bottoms from the distillation

Table 1.8 Examples of petroleum fractions produced by distillation of crude oil that are used as fuels

Petroleum fraction (other names)	Number of carbon atoms	Typical uses	Boiling range, °F
Light gas			
Fuel gas	C_1–C_2	Fuel	<85°F
LPG	C_3–C_4	Fuel	<85°F
Light naphtha (LSR-light straight run naphtha)	C_5–C_6	Gasoline	85–185°F
Heavy naphtha (HSR-heavy straight run naphtha)	C_6–C_{12}	Gasoline/jet fuel	185–350°F
Kerosene	C_{10}–C_{16}	Jet fuel/kerosene	350–480°F
Light gas oil (LAGO-light atmospheric gas oil; distillate; diesel; heating oil)	C_{14}–C_{20}	Heating fuel, diesel fuel	480–610°F
Heavy gas oil (HAGO-heavy atmospheric gas oil)	C_{20}–C_{70}	Fuel oil, feed to cracking units	610–800°F
Resid	>C_{70}	Heavy fuel oil, asphalt	>1,050°F

column undergo further processing to produce oils and coke. Petroleum coke, which is used as a boiler fuel, is discussed in Section 1.4.3.

Table 1.8 contains the petroleum fraction name (in some cases alternative names are provided), the number of carbon atoms in the fraction, typical uses, and boiling point ranges. Note that boiling point ranges in the literature can vary for the various fractions because of the variability in composition of the fractions.

1.4.2.4 Fuel oil properties

Fuel oil burned in furnaces for home heating, for process heating in industry to raise steam or produce electricity, or in electric utility generation plants may be a distilled fraction of crude oil, a residuum from refinery operations, or a blend of these. Fuel oils are graded on the basis of viscosity. To achieve a desired viscosity of oil, different fuel grades may be blended. Table 1.9 describes the various grades of oils and the petroleum fractions that they are comprised of. Two major categories of fuel oils are burned by combustion sources – distillate oils and residual oils. These oils are further distinguished by grade numbers with Nos. 1 and 2 being distillate oils, Nos. 5 and 6 being residual oils, and No. 4 either distillate oil or a mixture of distillate and residual oils. Although viscosity is a primary

Table 1.9 Grades of fuel oil

ASTM grade	Description
No. 1	This grade is a middle distillate oil intended for vaporizing pot-type burners and other burners requiring this grade. It is the kerosene refinery cut that boils off right after the heavy naphtha cut used for gasoline.
No. 2	This grade is a heavier middle distillate oil than No. 1 fuel oil that is intended for use in atomizing type burners, which spray the oil into a combustion chamber. No. 2 fuel oil is used in most domestic burners and in many medium capacity commercial and industrial boilers. This is typically obtained from the light gas oil cut.
No. 3	This grade is no longer used. ASTM merged this grade into the No. 2 fuel oil specification.
No. 4 (Light)	This grade is a heavy distillate oil or distillate/residual fuel blend (e.g. a blend of No. 2 and No. 6 fuel oils) meeting the specification viscosity range. It is intended for use both in pressure-atomizing commercial and industrial burners equipped to atomize oils of higher viscosity. Its viscosity range allows it to be pumped and atomized at relatively low temperatures.
No. 4	This grade is usually a heavy distillate/residual fuel blend (e.g. a blend of No. 2 and No. 6 fuels) but it can be a heavy distillate fuel meeting the specification viscosity range. It is intended for use in burners equipped with devices that atomize oils of higher viscosity than domestic burners can handle. Its viscosity range allows it to be pumped and atomized at relatively low temperatures except in extremely cold weather where preheating is required. It may be obtained from the heavy gas oil cut or it may be a blend of No. 6 and No. 2 fuel oils to adjust the viscosity.
No. 5 (Light)	This grade is a residual fuel of intermediate viscosity for burners capable of handling fuel more viscous than No. 4 fuel oil without preheating except in colder climates. It may be obtained from the heavy gas oil cut or it may be a blend of No. 6 and No. 2 fuel oils to adjust the viscosity.
No. 5 (Heavy)	This grade is residual fuel more viscous than No. 5 (Light). Preheating may be required for burning and, in cold climates, may be required for handling. It may be obtained from the heavy gas oil cut or it may be a blend of No. 6 and No. 2 fuel oils to adjust the viscosity.
No. 6	This grade is sometimes referred to as Bunker C oil and is a high viscosity residual oil after the more valuable cuts of crude oil have been boiled off. It is used mostly in commercial, industrial, and utility units. It requires preheating for burning and handling.

indication of fuel oil grade, the American Society for Testing Materials (ASTM) has detailed specifications for each grade of fuel oil (i.e. ASTM D396). The ASTM specifications for fuel oil properties are given in Table 1.10 [modified from 14,30,31].

Table 1.10 Detailed requirements for fuel oils[a]

Property	Grade of fuel oil						
	No. 1	No. 2	No. 4 (Light)	No. 4 (Heavy)	No. 5 (Light)	No. 5 (Heavy)	No. 6
Flash point, °F (°C), min	100 (38)	100 (38)	100 (38)	130 (55)	130 (55)	130 (55)	150 (65)
Pour point[b], °F (°C), max	0 (-18)	20 (-7)	20 (-7)	20 (-7)	—	—	—
Water & sediment[c], vol.%, max	0.05	0.05	0.05	0.05	1.00	1.00	2.00
Carbon residue on 10% distillation residue, %, max	0.15	0.35	—	—	—	—	—
Ash, wt.%, max	—	—	0.05	0.10	0.15	0.15	—
Sulfur[d], wt. %, max	0.50	0.50	—	—	—	—	—
Distillation temperature, °F (°C)							
10% volume recovered, max	419 (215)	—	—	—	—	—	—
90% volume recovered, min	—	540 (282)	—	—	—	—	—
90% volume recovered, max	550 (288)	640 (338)	—	—	—	—	—
Saybolt viscosity[e], s							
Universal at 100°F (38°C) min/max	-/-	(32.6)/(37.9)	(32.6)/(45)	45/125	150/300	350/750	(900)/(9000)
Furol at 122°F (50°C) min/max	-/-	-/-	-/-	-/-	-/-	(23)/(40)	45/300
Kinematic viscosity[e], centistokes							
100°F (38°C), min/max	1.4/2.2	2.0/3.6	2.0/5.8	(5.8)/(26.4)	(32)/(65)	(75)/(162)	-/-
122°F (50°C), min/max	-/-	-/-	-/-	-/-	-/-	(42)/(81)	(92)/(638)
Gravity, °API, min	35	30	—	—	—	—	—
Copper strip corrosion rating, 3h at 122°F (50°C), max	No. 3	No. 3	—	—	—	—	—

[a]It is the intent of these classifications that failure to meet any requirement of a given grade does not automatically place an oil in the next lower grade unless, in fact, it meets all requirements of the lower grade.

[b]Lower or higher pour points may be specified whenever required by conditions of storage or use. When a pour point less than 0°F (-18°C) is specified, the minimum viscosity shall be 1.8cs (32.0s, Saybolt Universal) and the minimum 90% recovered temperature shall be waived. Where low sulfur fuel oil is required, No. 6 fuel oil will be classified as Low Pour (+60°F (15°C) max) or High Pour (no max). Low Pour fuel should be used unless tanks and lines are heated.

[c]The amount of water by distillation plus the sediment by extraction shall not exceed the value shown in the table. For No. 6 fuel oil, the amount of sediment by extraction shall not exceed 0.50%. A deduction in quantity shall be made for all water and sediment in excess of 1.0%.

[d]Other sulfur limits may apply in selected areas in the United States and in other countries.

[e]Viscosity values in parentheses are for information only and not necessarily limiting.

1.4.2.5 Fuel oil analyses

A typical analysis of fuel oil contains the following information:

- Ultimate analysis;
- API gravity;
- Heating value;
- Viscosity;
- Pour point;
- Flash point; and
- Water and sediment concentration.

The ultimate analysis for an oil is similar to that of coal and the results indicate the quantities of carbon, hydrogen, nitrogen, sulfur, oxygen, and ash. The API gravity is used by the petroleum industry to determine the relative density of oil and heavier (denser) liquid fuels are denoted by lower API gravities. API gravity is defined as:

$$\text{API gravity} = (141.5/\text{Specific Gravity}) - 131.5 \quad (1.1)$$

where specific gravity is the specific gravity of crude oil at 60°F.

The heating value indicates the heat released by the combustion of the oil. Viscosity is the measure of its internal resistance to flow and the lower the value the easier the oil is to pump and atomize. The pour point is the lowest temperature at which a liquid fuel flows under standardized conditions. The flash point is the temperature to which a liquid fuel must be heated to produce vapors that flash but do not burn continuously when ignited. The water and sediment concentration is a measure of the contaminants in the liquid fuel. Examples of fuel oil analyses are given in Table 1.11 [14,26,31]. The analyses provided in Table 1.11 were drawn from several sources so values are reported as ranges for some properties and as single numbers for others. The importance of viscosity is further illustrated in Figure 1.23, which shows the effect of temperature on viscosity for various grades of fuel oil [modified from 14].

Some general trends to be observed from Table 1.11 and Figure 1.23 are that viscosity, sulfur and ash content, and heating value increase with increasing grade of oil. Fuel oils Nos. 4–6, and in some cases No. 2 fuel oil, are the fuels that are most commonly used in the commercial, industrial, and utility applications, which are those sectors that will have some form of emissions control requirements.

1.4.3 Petroleum coke

Petroleum coke is a byproduct of petroleum refining, useful in the production of electrodes used as carbon anodes for the aluminum industry, graphite electrodes for steel making, as fuel in the firing of solid fuel boilers used to generate electricity, and as a fuel for cement kilns [32]. Fuel-grade coke is coke that is high in sulfur content and its metals content is higher than what is acceptable to make carbon anodes, hence

Table 1.11 **Typical analyses and properties of fuel oils**

	Grade of fuel oil				
	No. 1 Distillate (Kerosene)	**No. 2 Distillate**	**No. 4 Very Light Residual**	**No. 5 Light Residual**	**No. 6 Residual**
Color	Light	Amber	Black	Black	Black
Percent by weight					
Carbon	85.9–86.7	86.1–88.2	86.5–89.2	86.5–89.2	86.5–90.2
Hydrogen	13.3–14.1	11.8–13.9	10.6–13.0	10.5–12.0	9.5–12.0
Sulfur	0.01–0.5	0.05–1.0	0.2–2.0	0.5–3.0	0.7–3.5
Oxygen & Nitrogen	0.2	0.2	0.48	0.70	0.92
Ash	Trace	Trace	0–0.1	0–0.1	0.01–0.5
Sediment & Water, vol.%	Trace	Trace	0.5 max	1.0 max	2.0 max
Gravity					
°API, 60°F	40–44	28–40	15–30	14–22	7–22
Specific, 60/60°F	0.825–0.826	0.887–0.825	0.966–0.876	0.972–0.922	1.022–0.922
lb/gal	6.71–6.87	6.87–7.39	7.30–8.04	7.68–8.10	7.65–8.51
Pour Point, °F	-50 to 0	-40 to 0	-10 to 50	-10 to 80	15 to 85
Temperature for pumping, °F	Atmospheric	Atmospheric	15 min	35 min	100
Temperature for atomizing, °F	Atmospheric	Atmospheric	25 min	130	200
Viscosity					
Kinematic @ 100°F, centistokes	1.4–2.2	1.9–3.0	10.5–65	65–200	260–750
Saybolt Univ. @ 100°F, SSU	31	32–38	60–300	232	–
Saybolt Furol @122°F, SFS	–	–	–	20–40	45–300
Heating Value					
Btu/lb	19,670–19,860	19,170–19,750	18,280–19,400	18,100–19,020	17,410–18,990
Btu/gal	137,000	141,000	146,000	148,000	150,000

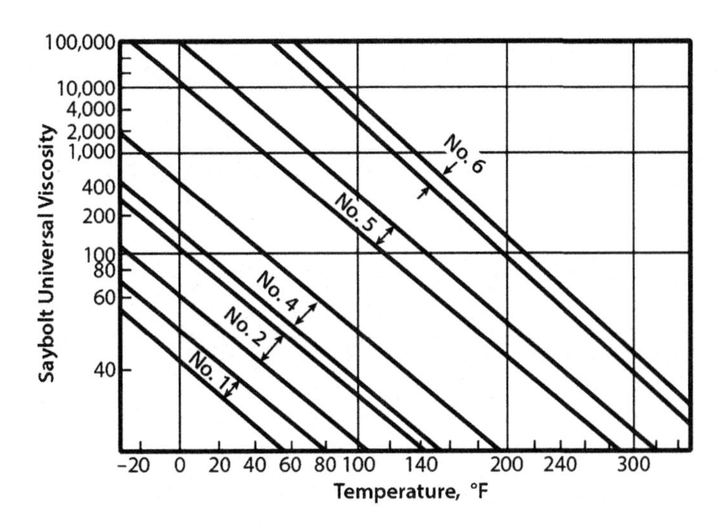

Figure 1.23 Viscosity ranges for fuel oils.

fuel-grade coke is less expensive than coke used for carbon anodes or graphite electrodes. Fuel-grade coke is high in heat content, typically >14,000 Btu/lb, and low in ash content, typically <1% by weight ash, as compared to coals [32]. The high sulfur concentrations and, for many petroleum cokes, high vanadium and nickel concentrations are the less desirable characteristics of this boiler fuel. Examples of typical analyses of different fuel-grade petroleum cokes are provided in Table 1.12 [modified from 32].

1.4.4 Gaseous fuels

This section focuses on natural gas and other gases obtained during the production of natural gas. Natural gas is defined as a mixture of hydrocarbons with varying quantities of nonhydrocarbons, which exists in either the gas phase or in solution with crude oil. It typically consists of methane, CH_4 (a high percentage), ethane (C_2H_6) together with other hydrocarbons up through the nonanes (C_9H_{20}), and nonhydrocarbons, such as carbon dioxide, hydrogen, nitrogen, argon, helium, and hydrogen sulfide. Natural gas is considered a premium fuel because of its ease to handle and burn and, of all the fossil fuels, it produces the least amount of pollutants that require some form of control.

1.4.4.1 Formation of natural gas

Methane is formed in two general ways: diagenesis of accumulated organic matter, and catagenesis of kerogen via hydrogen abstraction or hydrogen capping methyl radicals produced in thermal cracking [26]. Wet gas, i.e. a condensate consisting of liquid hydrocarbons dissolved in the gas, is essentially formed during the catagenetic stage.

Table 1.12 Typical fuel characteristics of petroleum coke

Analysis	Fuel (coke) type			
	Delay (Sponge)	**Shot**	**Fluid**	**Flexicoke**
Proximate analysis (wt.%)				
Fixed Carbon	87.20	89.58	91.50	94.90
Volatile Matter	4.48	3.07	4.94	1.25
Ash	0.72	1.06	1.32	0.99
Moisture	7.60	6.29	2.24	2.86
Ultimate analysis (wt.%)				
Carbon	81.12	81.29	84.41	92.00
Hydrogen	3.60	3.17	2.12	0.30
Oxygen	0.04	0.63	0.82	0.00
Nitrogen	2.55	1.60	2.35	1.11
Sulfur	4.37	5.96	6.74	2.74
Ash	0.72	1.06	1.32	0.99
Moisture	7.60	6.29	2.24	2.86
Higher Heating Value (Btu/lb)	14,298	14,363	14,017	13,972
Hardgrove Grindability Index	54	39	35	55
Ash elemental composition (wt.%)				
SiO_2	10.1	13.8	23.6	1.6
Al_2O_3	6.9	5.9	9.4	0.5
TiO_2	0.2	0.3	0.4	0.1
Fe_2O_3	5.3	4.5	31.6	2.4
CaO	2.2	3.6	8.9	2.4
MgO	0.3	0.6	0.4	0.2
Na_2O	1.8	0.4	0.1	0.3
K_2O	0.3	0.3	1.2	0.3
SO_3	0.8	1.6	2.0	3.0
NiO	12.0	10.2	2.9	11.4
V_2O_5	58.2	57.0	19.7	74.5

Methane can be formed bacterially during early diagenesis, but most of the dry gas in the subsurface was formed thermally during catagenesis and metagenesis.

1.4.4.2 Natural gas characteristics and properties

The ability to condense hydrocarbons other than methane from natural gas provides a classification of the gas as dry gas, which by definition contains less than

Table 1.13 Typical analyses of selected samples of natural gas from U.S. fields

Sample number:	1	2	3	4	5
Constituents, % by volume					
Methane, CH_4	83.40	84.00	93.33	90.00	84.10
Ethane, C_2H_6	15.80	14.80	–	5.00	6.70
Ethylene, C_2H_4	–	–	0.25	–	–
Hydrogen, H_2	–	–	1.82	–	–
Carbon monoxide, CO	–	–	0.45	–	–
Carbon dioxide, CO_2	–	0.70	0.22	–	0.80
Nitrogen, N_2	0.80	0.50	3.40	5.00	8.40
Oxygen, O_2	–	–	0.35	–	–
Hydrogen sulfide, H_2S	–	–	0.18	–	–
Ultimate Analysis, wt.%					
Carbon	75.25	74.72	69.12	69.26	64.84
Hydrogen	23.53	23.30	23.20	22.68	20.85
Nitrogen	1.22	0.76	5.76	8.06	12.90
Sulfur	–	–	0.34	–	–
Oxygen	–	1.22	1.58	–	1.41
Density, lb/ft^3	0.0486	0.0486	0.0433	0.0458	0.0481
Heating value					
Btu/lb	23,170	22,900	22,080	21,820	20,160
Btu/ft^3 at 60°F & 30 inches Hg	1,130	1,120	960	1,020	970

0.013 L/m^3 condensable hydrocarbons, and wet gas, which contains greater than 0.040 L/m^3 of the condensable hydrocarbons [26]. Common natural gas liquids (NGLs) include ethane, propane (C_3H_8), normal butane (n-C_4H_{10}), isobutane (i-C_4H_{10}), and pentane (C_5H_{12}). NGLs are used for a wide range of commercial and industrial purposes including feedstocks for petrochemical plants, burned for space heating and cooking, and blended with vehicle fuel. NGLs are high value products and have limited use in stationary heat and power applications. The composition of natural gas is variable and depends on the location where it is produced. Table 1.13 provides typical analyses of natural gas from several different U.S. oil fields [modified from 31].

1.5 Air emissions from fossil fuel usage

Coal, fuel oils, and natural gas are the primary fossil fuels used in boilers for producing process steam and/or electricity in commercial, industrial, and utility stationary power systems, and in combustion turbines for producing electricity.

Table 1.14 Emissions of interest in fossil fuel-fired stationary power systems

Emissions of interest by fossil fuel category		
Natural Gas	**Fuel Oil**	**Coal**
Nitrogen oxides (NO_x)	NO_x	NO_x
PIC[a]- CO; VOCs[b]	PIC-CO; VOCs	PIC-CO; VOCs
PM[c]	Sulfur oxides (SO_2/SO_3)	SO_2/SO_3
CO_2	PM-soot/ash	PM-ash
	Acid gases-HCl[d]	Acid gases-HCl[d]
	Mercury	Mercury
	CO_2	CO_2

[a]Products of incomplete combustion.
[b]Carbon monoxide; volatile organic compounds.
[c]Particulate matter.
[d]Hydrogen chloride.

Subsequent chapters will go into more depth on emissions legislation for boiler systems as well as stationary combustion turbines, pollutant formation mechanisms, and types of pollution control technologies used for the various fossil fuel-fired systems. In summary, the primary emissions produced when firing fossil fuels are listed in Table 1.14. All of these are currently regulated as air emissions with the exception of carbon dioxide (CO_2) but future regulations for CO_2 emissions are anticipated. CO_2 emissions are also discussed in upcoming chapters on regulations and CO_2 control options.

References

[1] IEA (International Energy Agency). World energy outlook 2013. International Energy Agency; 2013.
[2] EIA (Energy Information Administration). International energy outlook 2013. U.S. Department of Energy; 2013.
[3] EIA. Annual energy outlook 2014 early release. U.S. Department of Energy; 2014.
[4] EIA. April 2014 Monthly energy review. U.S. Department of Energy; 2014
[5] EIA. Annual energy review 2011. U.S. Department of Energy; 2012.
[6] EIA. Annual energy review 2008. U.S. Department of Energy; 2009.
[7] Miller BG. Clean coal engineering technology. Oxford, United Kingdom: Butterworth-Heinemann; 2011.
[8] Miller BG, Tillman DA, editors. Combustion engineering issues for solid fuel systems. Oxford, United Kingdom: Elsevier; 2008.
[9] Schobert HH. Coal: the energy source of the past and future. Washington, DC: American Chemical Society; 1987.
[10] Van Krevelen DW, Schuyer J. Coal science: aspects of coal constitution. Amsterdam: Elsevier Science Publishers; 1957.

[11] Mitchell G, Basics of coal and coal characteristics, Iron & Steel Society, Selecting Coals for Quality Coke Short Course, 1997.
[12] Elliott MA, editor. Chemistry of coal utilization second supplementary volume. New York: John Wiley & Sons; 1981.
[13] Berkowitz N. An introduction to coal technology. New York: Academic Press; 1979.
[14] Singer JG, editor. Combustion: fossil power systems. Windsor, Connecticut: Combustion Engineering, Inc; 1981.
[15] Miller BG, Falcone Miller S, Cooper R, Gaudlip J, Lapinsky M, McLaren R, et al. Feasibility Analysis for Installing a Circulating Fluidized Bed Boiler for Cofiring Multiple Biofuels and Other Wastes with Coal at Penn State University, U.S. Department of Energy, National Energy Technology Laboratory, DE-FG26-00NT40809, 2003, Appendix J.
[16] Van Krevelen DW. Coal: typology – physics – chemistry – constitution. 3rd ed. Amsterdam: Elsevier Science Publishers; 1993.
[17] Averitt P, Coal Resources of the U.S., January 1, 1974, U.S. Geological Survey Bulletin No. 1412, 1975 (reprinted 1976).
[18] Taylor GH, M. Teichmüller A. Davis, C.F.K. Diessel, R. Littke, and P. Robert, Organic Petrology, Bebrüder Borntraeger, Berlin, 1998.
[19] Bustin RM, Cameron AR, Grieve DA, Kalkreuth WD. Coal Petrology: Its Principles, Methods, and Applications. Geological Association of Canada; 1983.
[20] Suárez-Ruiz I, Crelling JC, editors. Applied coal petrology: the role of petrology in coal utilization. Oxford: Elsevier; 2008.
[21] Mackowsky MT. Mineral Matter in Coal. In: Murchson D, Westoll TS, editors. Coal and Coal-Bearing Strata. London: Oliver & Boyd, Ltd; 1968.
[22] Penn State Earth and Mineral Sciences Energy Institute, 2015, Coal Sample Bank, Available from: www.energy.psu.edu/copl/index.html.
[23] Tillman DA, Miller BG, Johnson DK, and Clifford DJ. Structure, Reactivity, and Nitrogen Evolution Characteristics of a Suite of Solid Fuels. In *Proc. of the 29th International Technical Conference on Coal Utilization & Fuel Systems*, Coal Technology Association, Gaithersburg, M.D., April 2004.
[24] Miller, B.G., unpublished data, 2005.
[25] Levorsen AI. Geology of petroleum. New York: W.H. Freeman and Company; 1967.
[26] Schobert HH. The chemistry of hydrocarbon fuels. London: Butterworths; 1990.
[27] Tissot BP, Welte DH. Petroleum formation and occurrence – a new approach to oil and gas exploration. Springer-Verlag; 1978.
[28] Selley RC. Elements of petroleum geology. New York: W.H. Freeman and Company; 1985.
[29] Olsen T. An oil refinery walk-through, Chemical Engineering Progress, Vol. 110, No. 5, 2014.
[30] American Society for Testing Materials. ASTM Standards for Industrial Fuel Applications Including Burners, Diesel Engines, Gas Turbines, and Marine Applications, ASTM, 1999.
[31] Kitto JB, Stultz SC, editors. Steam: its generation and use. 41st ed. The Babcock & Wilcox Company; 2005.
[32] Tillman DA, Harding NS. Fuels of opportunity: characteristics and uses in combustion systems. Elsevier; 2004.

Federal regulations and impact on emissions

2

The emphasis of this chapter is on federal legislation and regulatory trends in the United States and their impact on air quality and emissions. A history of legislative action in the United States as it pertains to stationary heat and power plants utilizing fossil fuels is presented. Impending legislation of emissions currently not regulated is also discussed. The types and quantities of emissions are presented along with their trends over time as a consequence of legislative action.

2.1 History of legislative action for fossil fuel-fired stationary heat and power plants

In the United States, industrial-based cities, such as Pittsburgh, Cincinnati, St. Louis, Cleveland, Detroit, Chicago, and Louisville, had significant air pollution problems by the late 1800s because of the amount and types of fuels used in each area [1]. Cities in the eastern U.S., such as New York, Philadelphia, and Boston, primarily used anthracite from eastern Pennsylvania while cities in the Midwest predominately used high sulfur bituminous coal. Early efforts to control air pollution led to court challenges before judges who were not sympathetic to efforts to abate air pollution because the judiciary embraced the concept that the fledgling economy could expand only if the business community was protected from ordinances and lawsuits by those who were injured. During World War I, air pollution control efforts were inconsequential and the war-generated economic expansion led to increased air pollution emissions. World War I was followed by the Great Depression, which prevented the placement of additional economic burdens on the private sector. During World War II, the national need for war-oriented production and the lack of money for domestic programs prevented implementation of air pollution control programs. During the period following the Second World War, many cities started enacting air pollution ordinances, which began to be upheld by the courts [1]. One of the most successful efforts at air pollution abatement was the clean-up of Pittsburgh air in the late 1940s. A major air pollution incident in Donora, Pennsylvania in 1948 was also a catalyst in cleaning up the Pittsburgh-area air and in the passage of air pollution legislation.

Donora, which is located south of Pittsburgh in the Monongahela Valley, was a town of approximately 26,000 people in the 1940s and home to a steel mill and zinc works [2]. The inhabitants were accustomed to dreary days, dirty buildings, and barren ground where no vegetation would grow, but they ignored the effect that

Fossil Fuel Emissions Control Technologies. DOI: http://dx.doi.org/10.1016/B978-0-12-801566-7.00002-6

the steel and zinc mills had on the population and environment, as about two-thirds of the workers were employed in the steel and zinc mills. This continued until the "killer smog" of October 1948. Nearly half of the town's inhabitants were sick by the end of the second day of the thick smog, 20 people died in three days, 50 more deaths than would be expected from other causes occurred in the month following the episode, and many people experienced breathing difficulties for the rest of their lives [2]. It was discovered later that the cause for many of the fatalities was not sulfur dioxide from firing coal in the steel and zinc mills, as initially thought and often reported [3], but was fluoride poisoning from the fluorspar used in the zinc mill [2]. Regardless of the cause, this episode brought increased public awareness towards air pollution and helped in the passage of air pollution laws.

The major development of air pollution legislative and regulatory acts occurred from 1955 to 1970; however, the early acts were narrow in scope as the U.S. Congress was hesitant to grant the federal government a high degree of control since air pollution problems were viewed as local or regional. This was found to be impractical, since some states were hesitant to regulate industry, and atmospheric transport of pollutants is not bounded by geographic lines. By the mid-1970s, the basis for national regulation of air pollution was developed and the actual regulations are continually changing. Regulations on emissions from power plants essentially started in 1970 with the passing of the Clean Air Act Amendments of 1970. There were a few regulatory changes in the 1980s but the Clean Air Act Amendments of 1990 resulted in significant regulatory changes. From 1992 through 2008 various NO_x regulations and trading programs were established. In 2004 stationary combustion turbine NESHAP (National Emissions Standards for Hazardous Air Pollutants) regulations went into effect. Interstate transport of emissions was addressed through the Clean Air Interstate Rule (CAIR) in 2005 and the Cross-State Air Pollution Rule (CSAPR) in 2011/2014. Boiler MACT (maximum achievable control technology) rules were published in 2011 and later amended in 2013 to specifically address smaller boilers and commercial and industrial solid waste incinerators that were considered major sources of air toxics. In 2012, $PM_{2.5}$ (i.e. fine particulate) emissions (National Ambient Air Quality Standards) standards became more stringent and the Utility MACT rule, i.e. MATS (Mercury & Air Toxics Standards), was finalized. Many of the regulations focus on utility power plants because they are a concentrated source of emissions and, for some pollutants, a significant source. In addition, legislation for controlling carbon dioxide is currently (as of 2014) being proposed. These are discussed in detail in the following sections. Figure 2.1 is a summarized timeline of recent key regulatory EPA (U.S. Environmental Protection Agency) actions addressing air emissions [modified from 4]. Of these, the GHG emissions actions are the most unknown at the present time.

2.1.1 Pre-1970 federal legislation

The history of federally enacted air pollution legislation begins in 1955 with the Air Pollution Control Act of 1955. The act was narrow in scope because of the federal government's hesitation to encroach on states' rights; however, it was the first

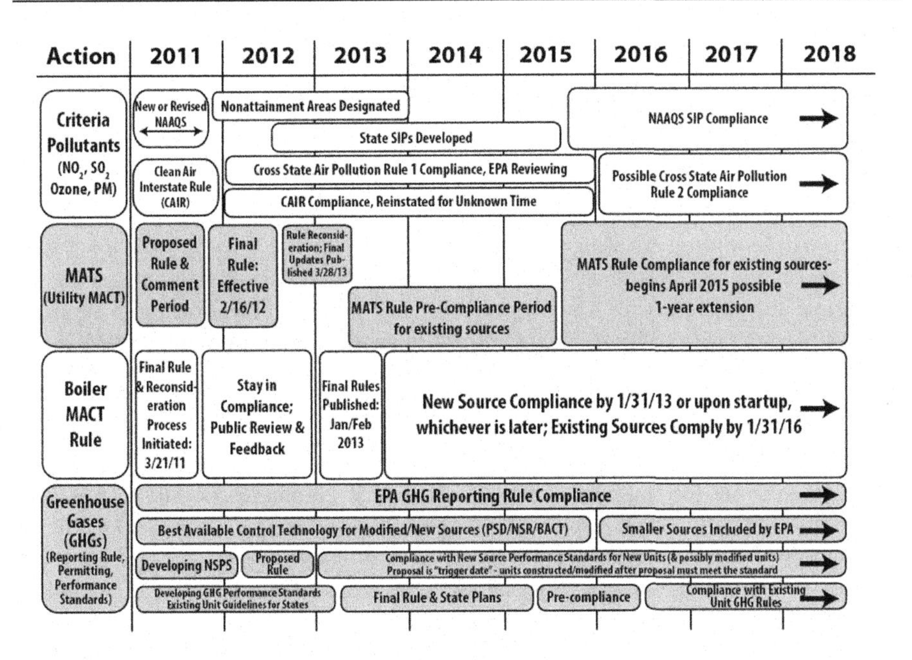

Figure 2.1 Summary of key EPA air emissions regulatory actions.

step toward identifying air pollution sources and its effects, and setting the groundwork for effective legislation and enforcement by regulatory agencies that was developed over the next 15 years. The act initiated the following [5]:

- Research by the United States Public Health Service on the effects of air pollution.
- Provision for technical assistance to the states by the federal government.
- Training of individuals in the area of air pollution.
- Research on air pollution control.

The Air Pollution Control Act of 1955 was amended in 1960 and 1962 (i.e. Air Pollution Control Act Amendments of 1960 and 1962) because of worsening conditions in urban areas due to mobile sources. Through these acts, Congress directed the Surgeon General to study the effect of motor vehicle exhausts on human health. A more formal process for the continual review of the motor vehicle pollution problem was included in the Clean Air Act of 1963.

The Clean Air Act of 1963 provided, for the first time, federal financial aid for air pollution research and technical assistance [5]. The act encouraged state, regional, and local programs for the control and abatement of air pollution, while reserving federal authority to intervene in interstate conflicts, thereby preserving the classical three-tier system of government. The act provided for [5]:

- Acceleration in the research and training program.
- Matching grants to state and local agencies for air pollution regulatory control programs.
- Developing air quality criteria to be used as guides in setting air quality standards and emissions standards.

- Initiating efforts to control air pollution from all federal facilities.
- Federal authority to abate interstate air pollution.
- Encouraging efforts by automotive companies and the fuel industries to prevent pollution.

The Clean Air Act of 1963 also provided for research authority to develop standards for sulfur removal from fuels, and a formal process for reviewing the status of the motor vehicle pollution problem. This in turn led to the Motor Vehicle Air Pollution Control Act of 1965, which formally recognized the technical and economic feasibility of setting automotive emission standards. The act also gave the secretary of the Department of Health, Education, and Welfare (HEW) the authority to intervene in intrastate air pollution problems of "substantial significance."

2.1.1.1 National air quality control act of 1967

The first federal legislation to impact stationary combustion sources was the National Air Quality Control Act of 1967. The act provided for a two-year study on the concept of national emissions standards for stationary sources and was the basis for the 1970 legislative action. Among the provisions of the National Air Quality Control Act of 1967 were [5]:

- Establishment of eight specific areas in the United States on the basis of common meteorology, topography, and climate.
- Designation of air quality control regions (AQCR) within the United States where evaluations were to be conducted to determine the nature and extent of the air pollution problem.
- Development and issuance of air quality criteria (AQC) for specific pollutants that have identifiable effects on human health and welfare.
- Development and issuance of information on recommended air pollution control techniques, which would lead to recommended technologies to achieve the levels of air quality suggested in the AQC reports.
- Requirement of a fixed time schedule for state and local agencies to establish air quality standards consistent with air quality criteria. The states were allowed to set higher standards than recommended in the AQC reports; however, if a state did not act, the secretary of HEW had the authority to establish air quality standards for each air quality region. The states were given primary responsibility for action, but a very strong federal fallback authority was provided.

The federal program was not implemented according to the required time schedule because federal surveillance of the overall program was understaffed and the process to set up the AQCRs proved to be too complex. Consequently, both President Nixon and Congress proposed new legislation in 1970.

2.1.2 Clean air act amendments of 1970

The Clean Air Act Amendments of 1970 extended the geographical coverage of the federal program aimed at the prevention, control, and abatement of air pollution from stationary and mobile sources. The act transferred administrative functions assigned to the secretary of HEW to the newly created Environmental Protection Agency (EPA). The act provided for the first time national ambient air quality standards and national emission standards for new stationary sources. It initiated the study of aircraft emissions and imposed carbon monoxide, hydrocarbons, and nitrogen oxide emissions control on automobiles.

The major goal of the act was the achievement of clean air throughout the United States by the middle of the decade. Two types of pollutants were to be regulated – criteria air pollutants and hazardous air pollutants (HAPs). The criteria air pollutants were to be regulated to achieve attainment of National Ambient Air Quality Standards by establishing emission standards (developed by state and local agencies) for existing sources and national emission standards for new sources through promulgation of New Source Performance Standards (NSPS). HAPs were to be regulated under National Emission Standards for Hazardous Air Pollutants (NESHAP). The major provisions of the act were [5]:

- EPA was to establish National Ambient Air Quality Standards (NAAQS) including primary standards for the protection of public health and secondary standards for the protection of public welfare.
- New Source Performance Standards (NSPS) were to be required with each state implementing and enforcing the standard of performance. Before a new stationary source could begin operation, state or federal inspectors were required to certify that the controls would function and the new stationary sources had to remain in compliance throughout the lifetime of the plant.
- National Emission Standards for Hazardous Air Pollutants (NESHAP) were to be established and would apply to existing as well as new plants.
- Funding was provided for fundamental air pollution studies, research on health and welfare effects of air pollutants, research on cause and effects of noise pollution, and research on fuels at stationary sources, including methods of cleaning fuels prior to combustion rather than flue gas cleaning techniques, improved combustion techniques, and methods for producing new or synthetic fuels with lower potential for creating polluted emissions.
- State and regional grant programs were authorized, and matching grants established for implementing standards.
- The designation of air quality control regions was to be completed.
- Establishment of statewide plans for implementation, i.e. state implementation plans (SIPs) for implementation designed to achieve primary or public health standards within three years was required, and the overall plan had to be convincing as to its ability to meet and maintain the standards.
- Industry was required to monitor and maintain emission records, make these records available to EPA officials, and EPA was given the right of entry to examine records.
- Fines and criminal penalties were imposed for violation of implementation plans, emission standards, and performance standards that were stricter than those under earlier law.
- New automobile emission standards were set.
- Aircraft emission standards were to be developed by EPA.
- Citizen's suits were permitted against those alleged to be in violation of emission standards including the United States, and suits could be brought against the EPA Administrator if he failed to act in cases where the law specified that he must.

2.1.2.1 Air quality criteria and national ambient air quality standards

The Air Quality Act of 1967 addressed the development and issuance of air quality criteria (AQC), and the need for such criteria was reaffirmed in the 1970 amendments. AQC indicate qualitatively and quantitatively the relationship between various levels of exposure to pollutants and the short- and long-term effects on health and welfare [5]. AQC describe effects that can be expected to occur when pollutant levels reach or exceed specific values over a given time period, and delineate the effects from combinations of contaminants as well as from individual pollutants. Economic and technical considerations are not relevant to the establishment of AQC.

The development of AQC is essential in providing a quantitative basis for air quality standards. Standards prescribe the pollutant levels that cannot be legally exceeded during a specific time period in a specific geographical region. The Clean Air Act Amendments of 1970 required federal promulgation of national primary and secondary standards that are to be established equitably in terms of the social, political, technological, and economic aspects of the problem. Standards are subject to revision as aspects change over time.

The purpose of the primary standards is immediate protection of the public health, including the health of sensitive populations such as asthmatics, children, and the elderly. Primary standards are to be achieved regardless of cost and within a specified time limit. Secondary standards are to protect the public welfare from known or anticipated adverse effects, including protection against decreased visibility, damage to animals, crops, vegetation, and buildings. Both standards have to be consistent with AQC and, in addition, the standards have to prevent the continuing deterioration of air quality in any portion of an air quality control region.

The Clean Air Act Amendments of 1970 defined the first six criteria pollutants as carbon monoxide (CO), nitrogen dioxide (NO_2), sulfur dioxide (SO_2), total particulate matter (PM), hydrocarbons, and photochemical oxidants, and NAAQS were established for these. Subsequently, the list has been revised with the following major actions and the standards have undergone periodic updates, specifically to make them more stringent:

- Lead was added to the list in 1976.
- The photochemical oxidant standard was revised and restated as ozone (O_3) in 1979.
- The hydrocarbon standard was withdrawn in 1983.
- The total suspended particulate matter standard was revised in 1987 to include only particles with an aerodynamic particle size of less than or equal to 10 μm and referred to as the PM_{10} standard.
- The $PM_{2.5}$ (i.e. particles with an aerodynamic particle size of less than or equal to 2.5 μm) standard was added in 1997.

The current list of NAAQS is shown in Table 2.1 and includes carbon monoxide, nitrogen dioxide, ozone, lead, PM_{10}, $PM_{2.5}$, and sulfur dioxide [6].

Table 2.1 National ambient air quality standards

Pollutant	Standard value	Standard type
Carbon monoxide (CO)		
8-hour average 1-hour average	9 ppm 35 ppm	Primary (No secondary) Primary (No secondary)
Nitrogen dioxide (NO_2)		
1-hour average Annual Arithmetic Mean	0.100 ppm 0.053 ppm	Primary Primary & Secondary
Ozone (O_3)		
8-hour average	0.075 ppm	Primary & Secondary
Lead (Pb)		
Rolling 3-month average	0.15 $\mu g/m^3$	Primary & Secondary
Particulate (PM_{10})		
24-hour average	150 $\mu g/m^3$	Primary & Secondary
Particulate ($PM_{2.5}$)		
Annual Arithmetic Mean Annual Arithmetic Mean 24-hour average	12 $\mu g/m^3$ 15 $\mu g/m^3$ 35 $\mu g/m^3$	Primary Secondary Primary & Secondary
Sulfur dioxide (SO_2)		
1-hour average 3-hour average	0.075 ppm 0.50 ppm	Primary Secondary

2.1.2.2 National emission standards

Emission standards place a limit on the amount or concentration of a pollutant that may be emitted from a source. It is often necessary for certain industries to be regulated by emission standards promulgated by the federal government or state governments in order to maintain or improve ambient air quality within a region to comply with national or state air quality standards. There are a number of factors that must be considered when establishing emission standards [5]:

- The availability of technology that is appropriate for the clean-up of a given type of industry must be determined.
- Monitoring stations must be available to measure the actual industrial emissions for which control is considered, as well as the ambient air quality so that the effectiveness of the standards can be determined.
- Regulatory agencies must be organized to cope with the measurement and enforcement of the standards.

- The synergistic effects of various pollutants must be determined.
- Models must be developed that reasonably predict the effects of reducing various emissions on the ambient air quality.
- Reasonable estimates of future emissions must be made based on the growth or decline of industry and population within a region.

Emission standards can be categorized into the following general types of standards [5]:

- Visible emissions standards – the opacity of the plume from a stack or a point of fugitive emissions is not to equal or exceed a specified opacity.
- Particulate concentration standards – the maximum allowable emission rate is specified in mass/volume: grams per dry standard cubic meter (g/dscm) or grains per dry standard cubic foot (gr/dscf), and for combustion processes it is common to specify the concentration at a fixed oxygen (O_2) or carbon dioxide (CO_2) level so as to prevent dilution, thereby lowering the concentration.
- Particulate process weight (or mass) standards – the maximum allowable particulate emissions are tied to the actual mass of material being processed or used, and in combustion systems the standards are commonly reported in pounds of particulate matter per million Btu of fuel burned (lb/10^6 Btu or lb/MM Btu).
- Gas concentration standards – gas standards are typically reported in mass per volume or volume per volume (g/dscm or parts per million (ppm)).
- Prohibition of emissions – processes are banned outright.
- Fuel regulations – fuel standards may be specified for various fuel-burning equipment such as limiting sulfur concentration in a fuel.
- Zoning restrictions – emissions may be limited by passing zoning ordinances that dictate facilities that can be constructed.
- Dispersion-based standards – these standards limit the allowable emission of pollutants based on their contribution to the ambient air quality.

National emissions or performance standards have been set for a number of industries, including fossil fuel-fired electric utility steam generating units, stationary gas turbines, and stationary combustion turbines. National standards are necessary for industries that are spread geographically across the country and provide a basic commodity essential for the development of the country. Unfair economic advantages might be gained by the states if the standards were set by the states with one state relaxing the standards to attract industries. These national emissions standards are referred to as New Source Performance Standards (NSPS) and apply to construction of new sources as well as sources that undergo operational and physical changes, which either increase emission rates or initiate new emissions from the plant [5].

2.1.2.2.1 40 CFR part 60, subpart D – standards of performance for fossil fuel-fired steam generators

The first five final standards were published on December 23, 1971. The complete text for each NSPS is available in the Code of Federal Register in Title 40 (Protection of the Environment), Part 60 (Standards of Performance for New Stationary Sources) with steam electric plants found in Subpart D [7,8].

The 1971 regulations addressed fossil fuel (or a combination of fossil fuel/wood residue) usage in utility and industrial steam generation units and, the regulations were amended for lignite in March 7, 1978. Affected generating facilities are those capable of firing fossil (or fossil/wood residue) fuels at a heat input rate of 73 MW (Megawatts) or 250 million (MM) Btu/h and were under construction or modification after August 17, 1971.

The maximum SO_2 emissions allowable from electric utility steam generating units are 1.2 lb SO_2/MM Btu for solid fossil fuels or a mixture of solid fossil fuel and wood residue. The SO_2 emissions limit for liquid fossil fuel or the simultaneous use of liquid fossil fuel and wood residue is 0.80 lb SO_2/MM Btu. When different fossil fuels are burned simultaneously in any combination, the applicable standard (lb/MM Btu) is determined using the following formula:

$$PS_{SO_2} = [y(0.80) + z(1.20)]/(y + z) \quad (2.1)$$

where

PS_{SO_2} = Prorated standard for SO_2, lb SO_2/MM Btu
y = Percentage of total heat input derived from liquid fossil fuel, MM Btu/h
z = Percentage of total heat input from all fossil fuel, MM Btu/h

Particulate matter is limited to 0.10 lb/MM Btu heat input and opacity is not to exceed 20% except for one 6-minute period per hour of not more than 27%. An affected facility that combusts only gaseous or liquid fossil fuel (excluding residual fuel oil) with a potential SO_2 emission rate of 0.06 lb/MM Btu or less and that does not use post-combustion technology to reduce SO_2 or particulate matter emissions are exempt from the particulate matter standards.

NO_x emissions, expressed as NO_2, cannot exceed 0.20 lb/MM Btu heat input for gaseous fossil fuels or 0.30 lb/MM Btu heat input for liquid fossil fuel, liquid fossil fuel and wood residue, or gaseous fossil fuel and wood residue. NO_x standards are 0.70 lb NO_x/MM Btu heat input for solid fossil fuel or solid fossil fuel and wood residue (except for lignite); 0.60 lb NO_x/MM Btu heat input for lignite or lignite and wood residue; and 0.80 lb NO_x/MM Btu for lignite that is mined in North Dakota (ND), South Dakota (SD), or Montana (MT) and burned in a cyclone-fired unit. When different fossil fuels are burned simultaneously in any combination, the applicable standard (lb/MM Btu) is determined using the following formula:

$$PS_{NO_x} = [w(0.60) + x(0.20) + y(0.80) + z(1.20)]/(w + x + y + z) \quad (2.2)$$

where

PS_{NO_x} = Prorated standard for NO_x, lb NO_x/MM Btu
w = Percentage of total heat input derived from lignite, MM Btu/h
x = Percentage of total heat input derived from gaseous fossil fuel, MM Btu/h
y = Percentage of total heat input derived from liquid fossil fuel, MM Btu/h
z = Percentage of total heat input from all fossil fuel (except lignite), MM Btu/h

2.1.2.2.2 40 CFR part 60, subpart Da – standards of performance for electric utility steam generating units

The original regulations were significantly revised as of February 6, 1980 and have undergone several revisions since. EPA promulgated new regulations for the control of SO_2, NO_x, and particulate matter from steam generating units with more than 73 MW or 250 MM Btu/h of heat input (40 CFR Part 60 Subpart Da) and under construction, modification, or reconstruction after September 18, 1978. IGCC (Integrated gasification combined cycle) electric utility steam generating units (both the stationary combustion turbine and any associated duct burners) under construction, modification, or reconstruction after February 28, 2005 were included as well [9].

The standards for SO_2 emissions have undergone several revisions and the current standards, as of May 27, 2014 are summarized in Table 2.2. In Table 2.2, the emission limits are listed for affected facilities, which are identified by date of construction, reconstruction, or modification, and fuel type.

The standards for NO_x emissions have also undergone several revisions and the current standards, as of May 27, 2014, are summarized in Table 2.3. The emission limits in Table 2.3 are listed for affected facilities, which are identified by date of construction, reconstruction, or modification, and fuel type.

The standards for particulate matter, as of May 27, 2014, are summarized in Table 2.4. In Table 2.4, the emission limits are listed for affected facilities, which are identified by date of construction, reconstruction, or modification, and fuel type.

2.1.2.2.3 40 CFR part 60, subpart Db – standards of performance for industrial-commercial-institutional steam generating units

Smaller sources have also been addressed over time, including industrial/ commercial/ institutional steam generating units that have heat inputs greater than 29 MW (100 MM Btu/h) where construction, modification, or reconstruction commenced after June 19, 1984 but on or before June 19, 1986, which are subject to the following standards [10]:

- Coal-fired facilities having heat input capacity between 29 and 73 MW (100 and 250 million Btu/h) are subject to the particulate matter and NO_x standards under this subpart. Coal-fired facilities having a heat input capacity greater than 73 MW and meeting the applicability requirements under Subpart D (Standards of performance for fossil-fuel-fired steam generators) are subject to particulate matter and NO_x standards under this subpart and to the SO_2 standards under Subpart D.
- Oil-fired facilities having a heat input capacity between 29 and 73 MW are subject to the NO_x standards under this subpart. Oil-fired facilities having a heat input capacity greater than 73 MW and meeting the applicability requirements under Subpart D are also subject to the NO_x standards under this subpart and the particulate matter and SO_2 standards under Subpart D.
- Steam generating units meeting the applicability requirements under Subpart Da (Standards of performance for electric utility steam generating units) are not subject to subpart Db.

Table 2.2 Subpart Da standards for SO_2

Affected facility	Fuel type	Emission limits
One for which construction, reconstruction, or modification commenced before or on 02/28/2005	Solid fuel or solid-derived fuel	1.20 lb/MM Btu heat input and 90% reduction; 0.60 lb/MM Btu heat input and 70% reduction; 1.4 lb/MWh gross energy input; or 0.15 lb/MM Btu heat input
	Liquid or gaseous fuels	0.80 lb/MM Btu heat input and 90% reduction; or 0% reduction when emissions <0.20 lb/MM Btu heat input
	Solvent refined coal	1.20 lb/MM Btu heat input and 85% reduction
	100% anthracite	1.20 lb/MM Btu heat input
One which is classified as a resource recovery unit (i.e., combusts >75% nonfossil fuel		1.20 lb/MM Btu heat input
One which is located in a noncontinental area	Solid fuel or solid-derived fuel	1.20 lb/MM Btu heat input
	Liquid or gaseous fuels	0.80 lb/MM Btu heat input
One firing different fuels simultaneously		Standard is determined by Equations (2.3) and (2.4) if SO_2 emissions >0.60 lb/MM Btu heat input[a] Standard is determined by Equations (2.5) and (2.6) if SO_2 emissions ≤0.60 lb/MM Btu heat input[a]
One for which construction, reconstruction, or modification commenced after 02/28/2005 but before 05/04/2011		
Construction		1.4 lb/MWh gross energy output or 95% reduction
Reconstruction		1.4 lb/MWh gross energy output; 0.15 lb/MM Btu heat input; or 95% reduction
Modification		1.4 lb/MWh gross energy output; 0.15 lb/MM Btu heat input; or 90% reduction

(*Continued*)

Table 2.2 (Continued)

Affected facility	Fuel type	Emission limits
One for which construction, reconstruction, or modification commenced after 02/28/2005	≥75% (heat input) coal refuse	
Construction		1.4 lb/MWh gross energy output; or 94% reduction
Reconstruction		1.4 lb/MWh gross energy output; 0.15 lb/MM Btu heat input; or 94% reduction
Modification		1.4 lb/MWh gross energy output; 0.15 lb/MM Btu heat input; or 90% reduction
One for which construction, reconstruction, or modification commenced after 02/28/2005 but before 05/04/2011	Solid or solid-derived fuel	1.20 lb/MM Btu heat input
	Other than solid or solid-derived fuel	0.54 lb/MM Btu heat input
One for which construction, reconstruction, or modification commenced after 05/04/2011		
Construction or reconstruction		1.0 lb/MWh gross energy output; or 1.2 lb/MWh net energy output; or 97% reduction
Modification		1.4 lb/MWh gross energy output; or 90% reduction
One in a noncontinental area for which construction, reconstruction, or modification commenced after 05/04/2011	Solid or solid-derived fuel	1.20 lb/MM Btu heat input
	Other than solid or solid-derived fuel	0.54 lb/MM Btu heat input

[a] $E_s = (0.80x + 1.20y)/100$ and $\%Ps = 10$ [Eqs. (2.3) and (2.4)]

$E_s = (0.80x + 1.20y)/100$ and $\%Ps = (10x + 30y)/100$ [Eqs. (2.5) and (2.6)]

where:

E_s = Prorated SO_2 emissions limit (lb/MM Btu heat input).

$\%P_s$ = Percentage of potential SO_2 emissions allowed (Note that $100 - \%P_s = \%$ reduction).

x = Percentage of total heat input derived from the combustion of liquid or gaseous fuels (excluding solid-derived fuels).

y = Percentage of total heat input derived from the combustion of solid fuel (including solid-derived fuels).

Table 2.3 **Subpart Da standards for NO_x**

Affected facility	Fuel type	Emission limits
One for which construction, reconstruction, or modification commenced before 07/10/1997	Gaseous fuels:	
	Coal-derived fuels	0.50 lb/MM Btu heat input
	All other fuels	0.20 lb/MM Btu heat input
	Liquid fuels:	
	Coal-derived fuels	0.50 lb/MM Btu heat input
	Shale oil	0.50 lb/MM Btu heat input
	All other fuels	0.30 lb/MM Btu heat input
	Solid fuels:	
	Coal-derived fuels	0.50 lb/MM Btu heat input
	Any fuel containing >25%, by weight, coal refuse	Exempt from NO_x standards
	Any fuel containing >25%, by weight, lignite if the lignite is mined in ND, SD, or MT, and is combusted in a slag tap furnace	0.80 lb/MM Btu heat input
	Any fuel containing >25%, by weight, lignite not subject to the 0.80 lb/MM Btu heat input limit	0.60 lb/MM Btu heat input
	Subbituminous coal	0.50 lb/MM Btu heat input
	Bituminous coal	0.60 lb/MM Btu heat input
	Anthracite	0.60 lb/MM Btu heat input
	All other fuels	0.60 lb/MM Btu heat input
One firing two or more fuels simultaneously		Standard is determined by Equation (2.7)[a]
One for which construction, reconstruction, or modification commenced after 07/09/1997 but before 03/01/2005		
Construction		1.6 lb/MWh gross energy output
Reconstruction		0.15 lb/MM Btu heat input
One for which construction, reconstruction, or modification commenced after 02/28/2005 but before 05/04/2011		

(*Continued*)

Table 2.3 (Continued)

Affected facility	Fuel type	Emission limits
Construction		1.0 lb/MWh gross energy output
Reconstruction		1.0 lb/MWh gross energy output; or 0.11 lb/MM Btu heat input
Modification		1.4 lb/MWh gross energy output; or 0.15 lb/MM Btu heat input;
IGCC electric utility plant for which construction, reconstruction, or modification commenced after 02/28/2005 but before 05/04/2011		1.0 lb/MWh gross energy output
	Liquid fuel exclusively or in combination with solid-derived fuel where liquid fuel contributes >50% if total heat input	1.5 lb/MWh gross energy output
One for which construction, reconstruction, or modification commenced after 05/03/2011		
Construction or reconstruction		0.70 lb/MWh gross energy output; or 0.76 lb/MWh net energy output
	≥ 75% coal refuse by heat input	0.85 lb/MWh gross energy output; or 0.92 lb/MWh net energy output
Modification		1.1 lb/MWh gross energy output

[a]$E_n = (0.20w + 0.30x + 0.50y + 0.60z + 0.80v)/100$ [Eqs. (2.7)]
where:
E_n = Applicable NO_x emissions limit (lb/MM Btu).
w = Percentage of total heat input derived from the combustion of fuels subject to the 0.20 lb/MM Btu input standard.
x = Percentage of total heat input derived from the combustion of fuels subject to the 0.30 lb/MM Btu input standard.
y = Percentage of total heat input derived from the combustion of fuels subject to the 0.50 lb/MM Btu input standard.
z = Percentage of total heat input derived from the combustion of fuels subject to the 0.60 lb/MM Btu input standard.
v = Percentage of total heat input derived from the combustion of fuels subject to the 0.80 lb/MM Btu input standard.

The standards for SO_2 emissions have undergone several revisions and the current standards, as of May 27, 2014, are summarized as follows:

- For a facility that has commenced construction, reconstruction, or modification on or before February 28, 2005 and fires coal or oil, the SO_2 emissions cannot exceed 0.20 lb/MM Btu

Table 2.4 Subpart Da standards for particulate matter

Affected facility	Fuel type	Emission limits
A-One for which construction, reconstruction, or modification commenced before 03/01/2005		0.03 lb/MM Btu heat input; <20% opacity (6-minute average) except for one 6-minute period per hour of not more than 27% opacity
	Only natural gas and/or synthetic natural gas	Exempt from opacity standard
B-One for which construction, reconstruction, or modification commenced after 02/28/2005 but before 05/04/2011	Solid, liquid, or gaseous fuel	0.14 lb/MWh gross energy output; or 0.015 lb/MM Btu heat input
C-Alternative to B; one for which construction, reconstruction, or modification commenced after 02/28/2005 but before 05/04/2011	Solid, liquid, or gaseous fuel	0.030 lb/MM Btu heat input AND
Construction or reconstruction		99.9% reduction
Modification		99.8% reduction
D-One for which construction, reconstruction, or modification commenced after 05/03/2011	Solid, liquid, or gaseous fuel	
Construction or reconstruction		0.090 lb/MWh gross energy output; or 0.097 lb/MWh net energy output
Modification		Emissions limits specified in B or C
E-Produces ≤0.060 lb/MM Btu	Gaseous or liquid fuels (excluding residual fuel oil)	Exempt from PM emissions limits

heat input, or 10% of the potential emissions rate (90% reduction) and the emissions limit is determined by the equation

$$E_s = (K_a H_a + K_b H_b)/(H_a + H_b) \tag{2.8}$$

where

E_s = SO_2 emissions limit, lb/MM Btu
K_a = 1.2 lb/MM Btu
K_b = 0.80 lb/MM Btu
H_a = Heat input from the combustion of coal
H_b = Heat input from the combustion of oil

- For a facility that commenced construction, reconstruction, or modification on or before February 28, 2005, the SO_2 emissions are limited to 1.2 lb/MM Btu heat input when firing coal or 0.50 lb/MM Btu heat input if firing oil other than low sulfur oil for: 1) facilities that have an annual capacity factor for coal and oil of 30% or less; 2) or the facilities are located in a noncontinental area; 3) or facilities firing coal or oil, alone or in combination with any other fuel, in a duct burner as part of a combined cycle system where 30% or less of the heat entering the steam generating unit is from the combustion of coal and oil in the duct burner and 70% or more of the heat entering the steam generating unit is from the exhaust gases entering the duct; or 4) facilities that burn coke oven gas alone or in combination with natural gas or low sulfur distillate oil.
- For a facility that commenced construction, reconstruction, or modification on or before February 28, 2005 that combusts coal refuse alone in a fluidized-bed combustion steam generation unit, the SO_2 emissions are limited to 0.20 lb/MM Btu heat input, or 20% of the potential emissions rate (i.e. 80% reduction) and 1.2 lb/MM Btu heat input. If coal or oil is fired with coal refuse, the facility is subject to the standards in the previous two bullets.
- When firing coal or oil, either alone or in combination with any other fuel and using an emerging SO_2 control technology, SO_2 emissions are limited to no more than 50% of the potential SO_2 emissions rate (i.e. 50% reduction) determined by the following equation

$$E_s = (K_c H_c + K_d H_d)/(H_c + H_d) \tag{2.9}$$

where
E_s = SO_2 emissions limit, lb/MM Btu
K_c = 0.6 lb/MM Btu
K_d = 0.40 lb/MM Btu
H_c = Heat input from the combustion of coal
H_d = Heat input from the combustion of oil

- Percent reduction requirements are not applicable to facilities combusting only very low sulfur oil.
- For a facility that commenced construction, reconstruction, or modification after February 28, 2005 that combusts coal, oil, natural gas, a mixture of these fuels, or a mixture of these fuels with any other fuels, the SO_2 emissions cannot exceed 0.20 lb/MM Btu heat input, or 8% of the potential emissions rate (i.e. 92% reduction) and 1.2 lb/MM Btu heat input. Exceptions are: 1) units firing very low sulfur fuel oil, gaseous fuel, a mixture of these fuels, or a mixture of these fuels with any other fuels with a potential SO_2 emission rate of 0.32 lb/MM Btu heat input or less are exempt; 2) units located in a noncontinental area that combust coal, oil, or natural gas have SO_2 emissions limits of 1.2 lb/MM Btu heat input when firing coal, 0.50 lb/MM Btu heat input when firing oil or natural gas; 3) and modified facilities that combust coal or a mixture of coal with other fuels, the SO_2 emissions cannot exceed 0.20 lb/MM Btu heat input, or 10% of the potential emission rate (i.e. 90% reduction) and 1.2 lb/MM Btu heat input.

The standards for NO_x emissions have undergone several revisions and the current standards, as of May 27, 2014, are contained in Table 2.5 and summarized below.

Table 2.5 NO_x emission limits for fuel/steam generating units

Fuel/steam generating unit type	Emissions limits (lb NO_2 /MM Btu heat input)
(1) Natural gas and distillate oil, except (4)	
Low heat release rate	0.10
High heat release rate	0.20
(2) Residual oil:	
Low heat release rate	0.30
High heat release rate	0.40
(3) Coal:	
(i) Mass-feed stoker	0.50
(ii) Spreader stoker and fluidized-bed combustion	0.60
(iii) Pulverized coal	0.70
(iv) Lignite, except (v)	0.60
(v) Lignite mined in North Dakota, South Dakota, or Montana and combusted in a slag tap furnace	0.80
(vi) Coal-derived synthetic fuels	0.50
(4) Duct burner used in a combined cycle system:	
Natural gas and distillate oil	0.20
Residual oil	0.40

Additional NO_x standards include:

- When firing simultaneously mixtures of only coal, oil, or natural gas, the NO_x emissions limits are determined by the equation

$$E_n = [(EL_{go}H_{go}) + (EL_{ro}H_{ro}) + (EL_cH_c)]/(H_{go} + H_{ro} + H_c) \tag{2.10}$$

where
E_n = NO_x emissions limit (expressed as NO_2), MM Btu/h
EL_{go} = Appropriate emissions limit from Table 2.5 for combustion of natural gas or distillate oil, MM Btu/h
H_{go} = Heat input from combustion of natural gas or distillate oil, MM Btu
EL_{ro} = Appropriate emissions limit from Table 2.5 for combustion of residual oil, MM Btu/h
H_{ro} = Heat input from combustion of residual oil, MM Btu
EL_c = Appropriate emissions limit from Table 2.5 for combustion of coal, MM Btu/h
H_c = Heat input from combustion of coal, MM Btu

- If a facility simultaneously combusts coal, oil, natural gas or any combination of the three, and wood or any other fuel, the NO_x emissions limits will be that for coal, oil, or natural gas (or the combination of the three) except for those facilities that are limited to a capacity factor of 10% or less for coal, oil, natural gas, or any combination of the three.

- The NO_x emissions limit for a facility that simultaneously combusts natural gas and/or distillate oil with a potential SO_2 emission rate of 0.60 lb/MM Btu or less with wood, municipal-type solid waste, or other solid fuel except for coal is 0.30 lb NO_x/MM Btu heat input unless the facility is limited to an annual capacity factor of 10% or less for natural gas, distillate oil, or a mixture of these fuels.
- The NO_x emissions limits for a facility that commenced construction after July 9, 1997 are: 1) 0.20 lb/MM Btu heat input if the facility fires coal, oil, natural gas, or any combination of the three, alone or with any other fuels except for facilities that are limited to annual capacity factor of 10% or less for coal, oil, natural gas, or any combination of the three; or 2) if the facility has a low heat release rate and combusts natural gas or distillate oil in excess of 30% of the heat input on a 30-day rolling average from the combustion of all fuels, the limit is determined by the equation

$$E_n = [(0.10 \times H_{go}) + (0.20 \times H_r)]/(H_{go} \times H_r) \tag{2.11}$$

where

E_n = NO_x emission limit, lb/MM Btu
H_{go} = 30-day heat input from combustion of natural gas or distillate oil
H_r = 30-day heat input from combustion of any other fuel;

or 3) after February 27, 2007, units where more than 10% of total annual output is electrical or mechanical may comply with an optional limit of 2.1 lb/MWh gross energy output, based on a 30-day rolling average.

The standards for particulate matter emissions have undergone several revisions and the current standards, as of May 27, 2014, are summarized as follows:

- The particulate matter emissions limit for facilities that commenced construction, reconstruction, or modification on or before February 28, 2005 is: 1) 0.051 lb/MM Btu heat input for facilities that combust only coal, or those that combust coal and other fuels that have an annual capacity factor for the other fuels of 10% or less; 2) 0.10 lb/MM Btu heat input for facilities that combust coal and other fuels and have an annual capacity factor for the other fuels greater than 10%; and 3) 0.20 lb/MM Btu heat input if the facilities combusts coal or coal and other fuels and have an annual capacity factor for coal or coal and other fuels of 30% or less, has a maximum heat input capacity of 73 MW or less, has a federally enforceable requirement limiting operation to an annual capacity factor of 30% or less for coal or coal and other fuels, and construction of the facility commenced after June 19, 1984.
- Facilities that combust coal, oil, wood, or mixtures of these fuels must not have opacity exceed 20% over a 6-minute average except for one 6-minute period per hour where it can exceed 27%.
- The particulate matter emissions limit for units where construction, reconstruction, or modification commenced after February 28, 2005 and combusts coal, oil, wood, a mixture of these fuels is 0.030 lb/MM Btu heat input. As an alternative, the emissions limit is 0.051 lb/MM Btu heat input with 99.8% particulate matter reduction.
- A facility not located in a noncontinental area where construction, reconstruction, or modification commenced after February 28, 2005 that combusts only oil that contains no more than 0.30 weight % sulfur, coke oven gas, or a mixture of these fuels in combination with other fuels and not using post-combustion technology to reduce SO_2 or particulate matter emissions (except a wet scrubber) is not subject to a particulate matter limit.

- A facility located in a noncontinental area where construction, reconstruction, or modification commenced after February 28, 2005 that combusts only oil that contains no more than 0.50 weight % sulfur, coke oven gas, or a mixture of these fuels and not using post-combustion technology to reduce SO_2 or particulate matter emissions (except a wet scrubber) is not subject to a particulate matter limit.

2.1.2.2.4 40 CFR part 60, subpart Dc – standards of performance for small industrial-commercial-institutional steam generating units

This subpart applies to industrial/ commercial/ institutional steam generating units for which construction, modification, or reconstruction commenced after June 9, 1989 and which have a maximum design heat input capacity of 29 MW (100 MM Btu/h) or less, but greater than or equal to 2.9 MW (10 MM Btu/h) [11]. Under this subpart, steam generating units are not subject to NO_x emissions nor to SO_2 or particulate matter emission limits during periods of combustion research. Also, heat recovery steam generators and fuel heaters that are associated with stationary combustion turbines that meet the requirements of Subpart KKKK (discussed later in this section) are not subject to Subpart Dc. However, this subpart does apply to all other heat recovery steam generators, fuel heaters, and other facilities that are capable of combusting more than or equal to 2.9 MW heat input of fossil fuels but less than or equal to 29 MW heat input of fossil fuels.

The standards for SO_2 emissions have undergone several revisions and the current standards, as of May 27, 2014, are summarized as follows:

- When firing coal only, the SO_2 emissions cannot exceed 0.20 lb/MM Btu heat input, or 10% of the potential emissions rate (i.e. 90% reduction) and 1.2 lb/MM Btu heat input. If coal is combusted with other fuels, the SO_2 emissions cannot exceed 0.20 lb/MM Btu heat input, or 10% of the potential emissions rate (i.e. 90% reduction) and the emission limit is determined according to the following equation

$$E_s = (K_a H_a + K_b H_b + K_c H_c)/(H_a + H_b + H_c) \tag{2.12}$$

where

E_s = SO_2 emissions limit, lb/MM Btu
K_a = 1.2 lb/MM Btu
K_b = 0.60 lb/MM Btu
K_c = 0.50 lb/MM Btu
H_a = Heat input from the combustion of coal, except coal combusted in a facility using an emerging technology for SO_2 control, MM Btu/h
H_b = Heat input from the combustion of coal in a facility that uses an emerging technology for SO_2 control, MM Btu/h
H_c = Heat input from the combustion of oil, MM Btu/h

- When firing only coal refuse in a fluidized-bed combustion steam generating unit, the SO_2 emissions cannot exceed 0.20 lb/MM Btu heat input, or 20% of the potential emission rate (i.e. 80% reduction) and 1.2 lb/MM Btu heat input. If coal is fired with refuse, the SO_2 emissions cannot exceed 0.20 lb/MM Btu heat input, or 10% of the potential emissions rate (i.e. 90% reduction) and the emissions limit is determined using Eq. (2.12).
- When firing only coal in a facility that uses an emerging technology for SO_2 control, SO_2 emissions are limited to no more than 50% of the potential SO_2 emissions rate (i.e. 50% reduction) and less than 0.60 lb SO_2/MM Btu. If coal is combusted with other fuels, the facility is subject to 50% SO_2 reduction and the emissions limit is determined using Eq. (2.12).

- When firing coal alone or in combination with any other fuel, SO_2 emission limits are set by Eq. (2.12) but % reductions are not required for: 1) facilities that have a heat input capacity of 22 MW (75 MM Btu/h) or less; 2) facilities that have annual capacity for coal of 55% or less and are subject to a federally enforceable requirement limiting the operation of the facility to an annual capacity factor of 55% or less; 3) facilities are located in a noncontinental area; or 4) facilities that combust coal in a duct burner as part of a combined cycle system where 30% or less of the heat entering the steam generating unit is from the combustion of coal in the duct burner and 70% or more of the heat entering the steam generating unit is from exhaust gases entering the duct burner.
- When firing oil, the SO_2 emissions are limited to 0.50 lb/MM Btu, or, as an alternative, the concentration of the sulfur in the fuel cannot exceed 0.5 weight %. In either case, sulfur reduction requirements are not applicable.
- When combusting coal, oil, or coal and oil with any other fuel, the SO_2 emissions cannot exceed 0.20 lb/MM Btu heat input, or 10% of the potential emissions rate (i.e. 90% reduction) and the emission limit is determined according to Eq. (2.12) for facilities that have a heat input capacity greater than 22 MW and have an annual capacity factor for coal greater than 55%.
- Reduction of the potential SO_2 emission rate through fuel pretreatment is not credited toward the % reduction requirement unless fuel pretreatment results in a 50% or greater reduction in the potential SO_2 emission rate and emissions from the pretreated fuel (without either combustion or post-combustion SO_2 control) are equal or less than 0.60 lb/MM Btu.
- Compliance with the % reduction requirements, fuel oil sulfur limits, and emission limits are determined on a 30-day rolling average.
- Compliance with the emission limits or fuel oil sulfur limits may be determined from the fuel supplier for: 1) distillate oil-fired facilities with heat input capacities between 2.9 and 29 MW (10 and 100 MM Btu/h); 2) residual fuel oil facilities with heat input capacities between 2.9 and 8.7 MW (10 and 30 MM Btu/h); and 3) coal-fired facilities with heat input capacities between 2.9 and 8.7 MW (10 and 30 MM Btu/h).

The particulate matter (PM) standards in Subpart Dc, effective as of May 27, 2014, state that:

- A facility that commenced construction, reconstruction, or modification on or before February 28, 2005 and combusts coal or mixtures of coal with other fuels and has a heat input capacity of 8.7 MW (30 MM Btu/h) or greater is subject to the following particulate matter emission limits: 1) 0.051 lb/MM Btu if the facility combusts only coal or coal with others fuels and has an annual capacity factor for the other fuels of 10% or less; and 2) 0.10 lb/MM Btu if the facility combusts coal with other fuels, has an annual capacity factor for the other fuels greater than 10%, and is subject to a federally enforceable requirement limiting operation of the facility to an annual capacity factor greater than 10% for fuels other than coal.
- Facilities that combust coal, oil, or wood must not have opacity exceed 20% over a 6-minute average except for one 6-minute period per hour where it can exceed 27%.
- A facility that commenced construction, reconstruction, or modification after February 28, 2005 and combusts coal or mixtures of coal, oil, wood, a mixture of these fuels, or a mixture of these fuels with any other fuels and has a heat input capacity of 8.7 MW (30 MM Btu/h) or greater is limited to 0.030 lb/MM Btu except for: 1) an alternative requirement that is available for facilities where modifications commenced after February 28, 2005 where particulate matter limits are 0.051 lb/MM Btu heat input derived from the combustion of coal, oil, wood, a mixture of these fuels, or a mixture of these fuels with any other

fuels and 99.8% reduction in PM; 2) a facility that commences modification after February 28, 2005 and combusts over 30% wood by heat input on an annual basis and has a heat input capacity of 8.7 MW or greater is limited to 0.10 lb particulate matter/MM Btu; and 3) a facility that commences construction, reconstruction, or modification after February 28, 2005 and combusts only oil that contains no more than 0.50 weight % sulfur or a mixture of 0.50 weight % sulfur oil with other fuels not subject to a particulate matter standard and not using a post-combustion technology (except a wet scrubber) to reduce particulate matter or SO_2 emissions is not subject to particulate matter limits.

2.1.2.2.5 40 CFR part 60, subpart GG – standards of performance for stationary gas turbines

Standards for NO_x and SO_2 emissions have been promulgated for stationary gas turbines with a heat input at peak load equal to greater than 10.7 gigajoules per hour (10 million Btu/h), based on the lower heating value of the fuel fired [12]. This is for facilities that commenced construction, modification, or reconstruction after October 3, 1977.

The standards for nitrogen oxides, as of May 27, 2014, are:

- The NO_x emissions limit for electric utility stationary gas turbines with a heat input at peak load greater than 107.2 gigajoules per hour (100 MM Btu/h) based on the lower heating value of the fuel fired is determined by the following equation

$$STD = 0.0075\ (14.4/Y) + F \tag{2.13}$$

where

STD = allowable ISO corrected NO_x emission concentration (% by volume at 15% oxygen on a dry basis)

Y = manufacturer's rated heat rate at manufacturer's rated load (kilojoules per watt hour), or actual measured heat rate based on lower heating value of the fuel as measured at actual peak load for the facility. The value of Y cannot exceed 14.4 kilojoules per watt hour.

F = NO_x emission allowance for fuel-bound nitrogen as defined in Table 2.6. Note that the use of F is optional. The F-value can either be determined from Table 2.6 or a value of zero can be used.

The NO_x emission limit for stationary gas turbines with a heat input at peak load equal to or greater than 10.7 gigajoules per hour (10 million Btu/h) but less than or equal to 107.2 gigajoules per hour (100 million Btu/h) based on the lower heating value of the fuel fired is determined by the following equation

Table 2.6 NO_x emission allowance for fuel-bound nitrogen, F-value

Fuel-bound nitrogen (Percent by weight)	F-value (NO_x % by volume)
$N < 0.015$	0
$0.015 < N \leq 0.1$	0.04 (N)
$0.1 < N \leq 0.25$	0.004 + 0.0067 (N − 0.1)
$N > 0.25$	0.005

$$STD = 0.0150\ (14.4/Y) + F \quad (2.14)$$

where

STD = allowable ISO corrected NO_x emission concentration (% by volume at 15% oxygen on a dry basis)

Y = manufacturer's rated heat rate at manufacturer's rated load (kilojoules per watt hour), or actual measured heat rate based on lower heating value of the fuel as measured at actual peak load for the facility. The value of Y cannot exceed 14.4 kilojoules per watt hour.

F = NO_x emission allowance for fuel-bound nitrogen as defined in Table 2.6. Note that the use of F is optional. The F-value can either be determined from Table 2.6 or a value of zero can be used.

- The NO_x emission limit for stationary gas turbines with a manufacturer's rated base load at ISO conditions of 30 MW or less is determined by Eq. (2.14).

Exemptions to the NO_x standard include:

- Stationary gas turbines with a heat input at peak load equal to or greater than 10.7 gigajoules per hour (10 million Btu/h) but less than or equal to 107.2 gigajoules per hour (100 million Btu/h) based on the lower heating value of the fuel fired and that have commercial construction prior to October 3, 1982.
- Stationary gas turbines using water or steam injection for control of NO_x emissions when ice fog is deemed a traffic hazard.
- Emergency gas turbines, military gas turbines for use in other than a garrison facility, military gas turbines installed for use as military training facilities, and fire-fighting gas turbines.
- Stationary gas turbines engaged by manufacturers in research and development of equipment for both gas turbine emission control techniques and gas turbine efficiency improvements.
- When mandatory water restrictions are required by governmental agencies because of drought conditions (on a case-by-case basis).
- Stationary gas turbines (except electric utility stationary gas turbines) with a heat input at peak load greater than 107.2 gigajoules per hour that commenced construction, modification, or reconstruction between the dates of October 3, 1977 and January 27, 1982 and were required in the September 10, 1979 Federal Register to comply.
- Stationary gas turbines with a heat input greater than or equal to 10.7 gigajoules per hour (10 million Btu/h) when fired with natural gas.
- Regenerative cycle gas turbines with a heat input less than or equal to 107.2 gigajoules per hour (100 million Btu/h).

The standards for SO_2 are less complex than the NO_x standards. The SO_2 emission limit is 0.015% SO_2 by volume at 15% oxygen on a dry basis. No fuel can be burned in any stationary gas turbine that contains total sulfur in excess of 0.8% by weight.

2.1.2.2.6 40 CFR part 60, subpart KKKK – standards of performance for stationary combustion turbines

Standards for NO_x and SO_2 emissions have been promulgated for stationary combustion turbines with a heat input at peak load equal to greater than 10.7 gigajoules per hour (10 million Btu/h), based on the higher heating value of the fuel fired [13]. This is for facilities that commenced construction, modification, or reconstruction after February 18, 2005. Only heat input to the combustion turbine is included. Any additional heat input to associated heat recovery steam generators (HRSG) or duct burners is not included. However, this subpart does apply to emissions from any associated HRSG and duct burners. Stationary combustion turbines regulated under this subpart are exempt from the requirements of Subpart GG. Heat recovery steam generators and duct burners regulated under this subpart are exempt from the requirements of Subparts Da, Db, and Dc.

The emission limits for nitrogen oxides, as of May 27, 2014, are provided in Table 2.7. If there are two or more turbines connected to a single generator, each turbine must meet the emission limits for NO_x.

The SO_2 emissions regulations, as of May 27, 2014, are:

- If the turbine is located in a continental area, compliance with one of the following is required: 1) an SO_2 emissions limit of 110 ng/J (0.90 lb/MWh) gross output; 2) the fuel cannot contain total sulfur with potential sulfur emissions in excess of 0.060 lb SO_2/MM Btu heat input; or 3) for each stationary combustion turbine burning at least 50% biogas on a calendar month basis, as determined based on total heat input, SO_2 emissions are limited to 0.15 lb SO_2/MM Btu.
- If the turbine is located in a noncontinental area or a continental area that the EPA Administrator determines doe not have access to natural gas and where the removal of the sulfur compounds would cause more environmental harm than benefit, compliance with one of the following is required: 1) SO_2 emissions are limited to 6.2 lb SO_2/MWh gross out; or 2) the fuel cannot contain total sulfur with potential sulfur emissions in excess of 0.42 lb SO_2/MM Btu heat input.

2.1.2.2.7 Emission factors

Once a New Source Performance Standard (NSPS) has been established, it is necessary for new sources constructed, modified, or reconstructed after a defined date to meet the standards. It is not possible to sample new sources to determine required collection or removal efficiencies. In these cases, knowledge of the emission factors for the specific regulated pollutant from these sources is used to estimate the approximate level of control required to meet the NSPS. The United States EPA has published a document, "Compilation of Air Pollutant Emission Factors," referred to as AP-42, since 1972. Supplements to AP-42 have been routinely published to add new emissions source categories and to update existing emission factors. This document is also provided on EPA's website on their CHIEF (Clearinghouse for Inventories and Emissions Factors) bulletin board [14].

EPA routinely updates AP-42 in order to respond to new emission factor needs of state and local air pollution control programs, industry, as well as the Agency itself. The current emission factors for facilities firing anthracite, bituminous and subbituminous coal, fuel oil, natural gas, liquefied petroleum gas, and lignite are provided in Appendix B. In addition, emission factors for stationary gas turbines are provided in Appendix B. The emission factors are from Chapter 1: External Combustion Sources, Sections 1.1 (Bituminous and Subbituminous Coal Combustion), 1.2 (Anthracite Coal

Table 2.7 Nitrogen oxide emission limits for new stationary combustion turbines

Combustion turbine type	Combustion turbine heat input at peak load (HHV)	NO_x emission standard
New turbine firing natural gas, electric generating	50 MM Btu/h	42 ppm @ 15% O_2 or 290 ng/J of useful output (2.3 lb/MWh)
New turbine firing natural gas, mechanical drive	50 MM Btu/h	100 ppm @ 15% O_2 or 690 ng/J of useful output (5.5 lb/MWh)
New turbine firing natural gas	50 MM Btu/h and 850 MM Btu/h	25 ppm @ 15% O_2 or 150 ng/J of useful output (1.2 lb/MWh)
New, modified, or reconstructed turbine firing natural gas	850 MM Btu/h	15 ppm @ 15% O_2 or 54 ng/J of useful output (0.43 lb/MWh)
New turbine firing fuels other than natural gas, electric generating	50 MM Btu/h	96 ppm @ 15% O_2 or 700 ng/J of useful output (5.5 lb/MWh)
New turbine firing fuels other than natural gas, mechanical drive	50 MM Btu/h	150 ppm @ 15% O_2 or 1,100 ng/J of useful output (8.7 lb/MWh)
New turbine firing fuels other than natural gas	50 MM Btu/h and 850 MM Btu/h	74 ppm @ 15% O_2 or 460 ng/J of useful output (3.6 lb/MWh)
New, modified, or reconstructed turbine firing fuels other than natural gas	850 MM Btu/h	42 ppm @ 15% O_2 or 160 ng/J of useful output (1.3 lb/MWh)
Modified or reconstructed turbine	50 MM Btu/h	150 ppm @ 15% O_2 or 1,100 ng/J of useful output (8.7 lb/MWh)
Modified or reconstructed turbine firing natural gas	50 MM Btu/h and 850 MM Btu/h	42 ppm @ 15% O_2 or 250 ng/J of useful output (2.0 lb/MWh)
Modified or reconstructed turbine firing fuels other than natural gas	50 MM Btu/h and 850 MM Btu/h	96 ppm @ 15% O_2 or 590 ng/J of useful output (4.7 lb/MWh)
Turbines located north of the Arctic Circle (latitude 66.5°N), turbines operating at less than 75% of peak load, modified and reconstructed offshore turbines, and turbine operating at <0°F	30 MW output	150 ppm @ 15% O_2 or 1,100 ng/J of useful output (8.7 lb/MWh)

(*Continued*)

Table 2.7 (Continued)

Combustion turbine type	Combustion turbine heat input at peak load (HHV)	NO_x emission standard
Turbines located north of the Arctic Circle (latitude 66.5°N), turbines operating at more than 75% of peak load, modified and reconstructed offshore turbines, and turbine operating at <0°F	30 MW output	96 ppm @ 15% O_2 or 590 ng/J of useful output (4.7 lb/MWh)
Heat recovery units operating independent of the combustion turbine	All sizes	54 ppm @ 15% O_2 or 110 ng/J of useful output (0.86 lb/MWh)

Combustion), 1.3 (Fuel Oil Combustion), 1.4 (Natural Gas Combustion), 1.5 (Liquefied Petroleum Gas), and 1.7 (Lignite Combustion) and from Chapter 3: Stationary Internal Combustion Sources 3.1 (Stationary Gas Turbines) of AP-42, Fifth Edition, Volume I, Supplements A through F and subsequent updates [15]. The emission factors have been developed for (not inclusive and not relevant for each fuel type):

- Various fuel types
- Various fuel firing configurations/systems
- Uncontrolled and controlled emissions
- Criteria gaseous pollutants (SO_x, NO_x, CO)
- Filterable particulate matter and condensable particulate matter
- Trace elements
- Various polynuclear organic matter (POM), polynuclear aromatic hydrocarbons (PAH), and organic compounds
- Acid gases (HCl and HF)
- Other gaseous pollutants such as CO_2, N_2O, and CH_4
- Cumulative ash particle size distribution and size-specific emissions

Emission factors and emissions inventories have long been fundamental tools for air quality management. Emission estimates are important for developing emission control strategies, determining applicability of permitting and control programs, ascertaining the effects of sources and appropriate mitigation strategies. Users include federal, state, and local agencies, consultants, and industry. Data from source-specific emission tests or continuous emission monitors are usually preferred for estimating a source's emissions because those data provide the best representation of the tested source's emissions. However, test data from individual sources are not always available and they may not reflect the variability of actual emissions over time. Consequently, emission factors are often the best or only method available for estimating emissions.

The passage of the Clean Air Act Amendments of 1990 and the Emergency Planning and Community Right-to-Know Act of 1986 has increased the need for both criteria and hazardous air pollutant emission factors and inventories. The Emission Factor and Inventory Group (EFIG), in EPA's Office of Air Quality Planning and Standards, develops and maintains emission-estimating tools. The AP-42 series is the principal means by which EFIG can document its emission factors.

Emission factors may be appropriate to use in a number of situations such as source-specific emission estimates for area-wide inventories. These inventories have many purposes including ambient dispersion modeling and analysis, control strategy development, in screening sources for compliance investigations, and in some permitting applications. Emission factors in AP-42 are neither EPA-recommended emission limits (e.g. Best Available Control Technology (BACT), or Lowest Achievable Emission Rate (LEAR)) nor standards (e.g. NSPS or NESHAP).

Figure 2.2 depicts various approaches to emission estimation, in a hierarchy of requirements and levels of sophistication that need to be considered when analyzing the trade-offs between cost of the estimates and the quality of the resulting estimates. More sophisticated and more costly emission determination methods may be necessary where risks of either adverse environmental effects or adverse regulatory outcomes are high. Less expensive estimation techniques such as emission factors and emission models may be appropriate and satisfactory where risks of using a poor estimate are low. Note that the reliability of the AP-42 emission factors are rated from A through E, which are a general indication of the robustness of that factor. This rating is assigned based on the estimated reliability of the tests used to develop the factor. In general, factors based on many observations, or on more widely accepted test procedures, are assigned higher rankings with A being the best. The emission factors rating is provided with the list of emission factors contained in Appendix B.

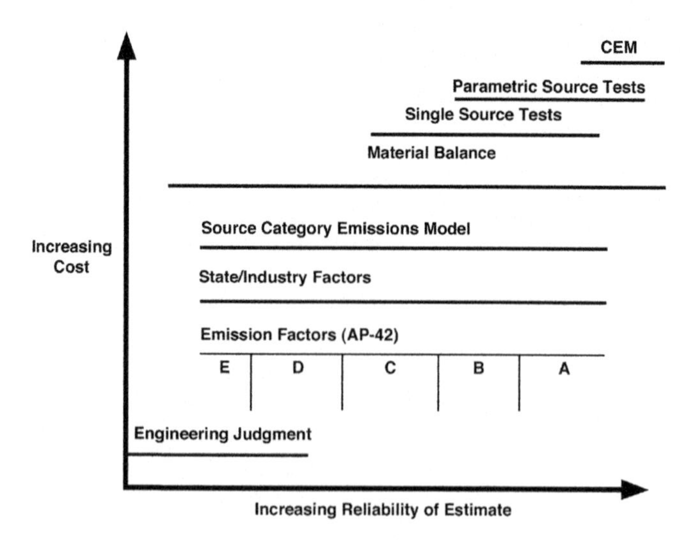

Figure 2.2 Approach to emission estimation [15].

2.1.2.3 National emission standards for hazardous air pollutants

The 1970 Clean Air Act Amendments also provided for national emission standards for hazardous air pollutants (NESHAPs). NESHAPs are stationary source standards for hazardous air pollutants (HAPs). HAPs are defined as those pollutants that are known or suspected to cause cancer or other serious health effects, such as reproductive effects or birth defects, or adverse environmental effects. Only seven standards were established between 1970 and 1990, as the NESHAPs were the subject of numerous suits and court decisions regarding how to address emission limits on carcinogenic pollutants [5]. None of the original seven NESHAPs, which are found in 40 CFR Part 61 and include asbestos, beryllium, mercury, vinyl chloride, benzene, radionuclides, and arsenic, specifically addressed fossil fuel-fired steam generating units. The Clean Air Act Amendments of 1990 made major changes in the approach taken to address hazardous air pollutants (HAPs) by significantly expanding EPA's authority to regulate HAPs. Section 112 of the Clean Air Act Amendment lists 187 HAPs to be regulated by source category. The NESHAPs regulated after the 1990 Clean Air Act Amendments are found in 40 CFR 63. NESHAPs have been established for fossil fuel-, with an emphasis on coal-, fired power plants, and stationary combustion turbines. These standards require application of technology-based emission standards referred to as Maximum Achievable Control Technology (MACT). Consequently, these post-1990 NESHAPs are also referred to as MACT standards. NESHAPs, HAPs, and MACT standards are discussed in more detail later in this chapter.

2.1.3 Clean air act amendments of 1977 and prevention of significant deterioration

By 1977 most areas of the country had still not attained the National Ambient Air Quality Standards (NAAQS) for at least one pollutant [5]. For those areas that had not attained a NAAQS (i.e. nonattainment areas), states were required to submit and have an approved state implementation plan (SIP) revision by July 1, 1979, that demonstrated how attainment would be achieved by December 31, 1982. This requirement was a precondition for the construction or modification of major emission sources in nonattainment areas after June 30, 1979. If a state could not attain primary standards for carbon monoxide or photochemical oxidants after implementation of all reasonably available measures, it was required to submit a second SIP revision by December 31, 1982 that would demonstrate how attainment would be achieved by December 31, 1987.

2.1.3.1 Prevention of significant deterioration

The concern over nonattainment areas and the controversy generated by the 1970 Clean Air Act Amendments' provision on standards preventing continuing deterioration of air quality, led to a set of guidelines issued in 1974 by EPA to prevent the significant deterioration of air quality in areas that were cleaner than that required by NAAQS – that is, that were in attainment of the NAAQS. This was necessary because some interpreted the 1970 act to mean that a region could not backslide in air quality even though the present air quality may be superior to the national standard. This interpretation would have stifled economic growth in a region (new industrial and commercial operations in the region could not contribute zero pollution), yet would have failed to force sources in the region to decrease their contaminant emissions. This concern led to the passage of the regulations on prevention of significant deterioration (PSD). The PSD regulations for attainment areas required that all of these areas be designated as Class I, II, or III, depending upon the degree of deterioration to be allowed, and that incremental limits were placed on the amount of increase in deterioration allowed. The classifications are [5]:

Class I Pristine areas including international parks, national parks, and national wilderness areas in which very little deterioration would be allowed
Class II Areas where moderate change will be allowed, but where stringent air quality constraints are desirable
Class III Areas where major growth and industrialization would be allowed

Congress specified which of the areas must be protected by the most stringent Class I designation, designated all other areas within the United States as Class II areas, and provided the option for redesignation of Class II areas to Class I or Class III areas by public referendum. Congress also specified the maximum allowable incremental increases in concentration of sulfur dioxide and particulate matter and charged EPA to determine comparable increments for hydrocarbons, carbon monoxide, photochemical oxidants, and nitrogen oxides. The PSD increments are listed in Table 2.8 [5].

A requirement was placed on major sources in the preconstruction PSD review process that specified that each major new plant must install best available control

Table 2.8 PSD increments

Pollutant	**Maximum allowable increase (μg per m³)**		
	Class I	**Class II**	**Class III**
Particulate Matter			
PM_{10}, annual geometric mean	5	19	37
PM_{10}, 24-hour maximum	10	37	75
Sulfur dioxide			
Annual arithmetic mean	2	20	40
24-hour maximum	5	91	182
3-hour maximum	25	512	700
Nitrogen dioxide			
Annual arithmetic mean	2.5	25	50

technology (BACT), which was defined to be at least as stringent as NSPS, to limit its emissions. Major sources subject to PSD review are those with the potential to emit 100 short tons or more per year of any regulated pollutant under the Clean Air Act Amendment of 1977 [5]. All sources emitting greater than 250 short tons per year are subject to PSD review. For non-NSPS sources, a BACT review document is prepared in the preconstruction review.

The 1974 Guidelines on PSD and the Clean Air Amendments of 1977 established the protocol for new sources and proposed modifications to major sources in attainment areas where they would be subjected to a new source review and would have to meet certain PSD requirements. This review requires that dispersion modeling be conducted of the proposed emissions from the sources to ensure that the emissions from the proposed facility would not exceed the increments listed in Table 2.8 or cause an exceedance of the NAAQS. In concept, sources are allowed to use some fraction of the increment, as determined by the state or local agency and based on dispersion modeling. The PSD increment can only be used to the extent that it does not cause the ambient concentration to exceed the NAAQS. Another important issue is the extent to which any single source would be allowed to use the available increment and the ramifications on long-term growth if a single source was allowed to use it all. Some state and local agencies address this on a case-by-case basis while others have taken the approach of allowing only a certain fraction of the increment, or the remaining increment, to be used in a single PSD application [5].

2.1.3.2 Nonattainment areas

It was feared that no industrial growth could occur in nonattainment areas, since these areas are in violation of one or more NAAQSs, and because PSD does not apply in these areas, there are no increments available. This was remedied on December 21, 1976, with an interpretive ruling, the Offsets Policy Interpretive Ruling, on the preconstruction review requirements for all new or modified stationary sources of air pollution in nonattainment areas [5]. This ruling, also known as the emission offset policy, requires that three conditions be met:

1. The source must meet the lowest achievable emission rate (LAER), defined as being more stringent than BACT.
2. All existing sources owned by the applicant in the same region must be in compliance or under an approved schedule to achieve compliance.
3. The source must provide an offset or reduction of emissions from other sources greater than the proposed emissions that the source contributes such that there is a net improvement in air quality.

The Clean Air Act Amendments of 1977 also provided for the banking of offsets. If the offsets achieved are considerably greater than the new source's emissions, a portion of this excess emission reduction (also known as emission reduction credits) can be banked by the source for use in future growth or traded to another source, depending on each state's offset/trading policy.

2.1.4 Clean air act amendments of 1990

In June 1989, President George H. W. Bush proposed major revisions to the Clean Air Act. Both the House of Representatives and the Senate passed Clean Air bills by large votes that contained the major components of the president's proposals. After a joint conference committee met to work out the differences in the bills, Congress voted out the package recommended by the conferees and President Bush signed the bill, the Clean Air Amendments of 1990 (1990 CAAA), on November 15, 1990. The 1990 CAAAs are the most substantive regulations adopted since the passage of the Clean Air Act Amendments of 1970. Specifically, the new law does the following:

- Encourages the use of market-based principles and other innovative approaches, like performance-based standards and emission banking and trading
- Provides a framework from which alternative clean fuels will be used by setting standards in the fleet and a California pilot program that can be met by the most cost-effective combination of fuels and technology
- Promotes the use of clean low sulfur coal and natural gas, as well as innovative technologies to clean high sulfur coal through the acid rain program
- Reduces enough energy waste and creates enough of a market for clean fuels derived from grain and natural gas to cut dependency on oil imports by one million barrels per day
- Promotes energy conservation through an acid rain program that gives utilities flexibility to obtain needed emission reductions through programs that encourage customers to conserve energy

The 1990 CAAA contains 11 major divisions, referred to as Titles I through XI, which either provided amendments to existing titles and sections of the Clean Air Act or provided new titles and sections. The titles for the 1990 CAAA are:

- Title I: Provisions for Attainment and Maintenance of National Ambient Air Quality Standards
- Title II: Provisions Relating to Mobile Sources
- Title III: Air Toxics
- Title IV: Acid Deposition Control
- Title V: Permits
- Title VI: Stratospheric Ozone and Global Climate Protection
- Title VII: Provisions Relating to Enforcement
- Title VIII: Miscellaneous Provisions
- Title IX: Clean Air Research
- Title X: Disadvantaged Business Concerns
- Title XI: Clean Air Deployment Transition Assistance

The titles that directly impact the use of fossil fuels – Titles I, III, IV, and V – are discussed in the following sections. The concepts of NAAQS, NSPS, and PSD, as defined in Title I, remained virtually unchanged; however, major changes have occurred in regulations and approaches used to address nonattainment areas in Title I, hazardous air pollutants in Title III, acid rain in Title IV, and permitting in Title V.

2.1.4.1 Title I: provisions for attainment and maintenance of national ambient air quality

Although the Clean Air Act Amendments of 1970 and 1977 brought about significant improvements in air quality, urban air pollution persisted and there were many cities out of attainment for ozone, carbon monoxide, and PM_{10}. Of these, the most widespread pollution problem is ozone (i.e. smog). One component of smog – hydrocarbons – comes from automobile emissions, petroleum refineries, chemical plants, dry cleaners, gasoline stations, house painting, and printing shops, while another key component – nitrogen oxides – comes from the combustion of fossil fuels for transportation, utilities, and industries [21].

The 1990 CAAA created a new, balanced strategy for the nation to address urban smog. The new law gave states more time to meet the air quality standard but it also required states to make constant progress in reducing emissions. It required the federal government to reduce emissions from cars, trucks, and buses, from consumer products such as hair spray and window washing compounds; and from ships and barges during loading and unloading of petroleum products. The federal government must also develop guidance that states need to control stationary sources.

The 1990 CAAA addresses the urban air pollution problems of ozone, carbon monoxide, and PM_{10}. Specifically, it clarifies how areas are designated and it redefined attainment. The 1990 CAAA also allows EPA to define the boundaries of nonattainment areas and establishes provisions defining when and how the federal government can impose sanctions on areas of the country that have not met certain conditions.

For the pollutant ozone, the 1990 CAAA establishes nonattainment area classifications ranked according to the severity of the area's air pollution problem. These classifications are marginal, moderate, serious, severe, and extreme and were designated based on the air quality of the nonattainment area during the period of 1987 to 1989. EPA assigns each nonattainment area one of these categories, thus triggering varying requirements the area must comply with in order to meet the ozone standard. Table 2.9 lists the ozone design value, which determines the classification, attainment deadlines, minimum size of a new or modified source that would be affected, and offset requirements [5].

Nonattainment areas have to implement different control measures depending upon their classification. Nonattainment areas with worse air quality problems must implement more control measures. For example, in areas classified as extreme, boilers with emission rates greater than 25 short tons per year are required to burn clean fuels or install advanced control technologies.

The 1990 CAAA also establishes similar programs for areas that do not meet federal health standards for carbon monoxide and PM_{10}. Areas exceeding the standards for these pollutants are divided into moderate and serious classifications. The classifications for nonattainment areas for carbon monoxide are shown in Table 2.9.

Areas that were nonattainment for PM_{10} at the time the 1990 CAAA was passed were designated as moderate areas with an attainment deadline of December 31,

Table 2.9 Classifications for nonattainment areas

Pollutant	Classification	Design value (ppm)	Attainment deadline	Major source (short tons VOCs/y for ozone; short tons CO/y for carbon monoxide)	Offset ratio for sources
Ozone	Marginal	0.121–0.138	11/15/1993	100	1.1 to 1
	Moderate	0.138–0.160	11/15/1996	100	1.15 to 1
	Serious	0.160–0.180	11/15/1999	50	1.2 to 1
	Severe	0.180–0.190	11/15/2005	25	1.3 to 1
	Severe	0.190–0.280	11/15/2007	25	1.3 to 1
	Extreme	>0.280	11/15/2007	10	1.5 to 1
Carbon monoxide	Moderate	9.1–16.4	12/31/1995	–	none
	Serious	>16.4	12/31/2000	50	none

1994 [5]. Areas classified as nonattainment subsequent to passage of the 1990 CAAA are designated moderate with six years to achieve compliance. Major sources in moderate areas are those that emit 100 short tons or more of particulate matter per year. Moderate areas require the adoption of reasonably available control measures (RACM).

Moderate areas which fail to reach attainment are redesignated as serious areas and have ten years from the date of designation as nonattainment to achieve attainment. For serious areas, major sources include those that emit 70 short tons or more of particulate matter per year. Serious areas must also adopt best available control measures (BACM).

2.1.4.2 Title III: air toxics

Hazardous air pollutants (HAPs), also known as toxic air pollutants or air toxics, are those pollutants that cause or may cause cancer or other serious health effects, such as reproductive effects or birth defects, or adverse environmental and ecological effects but are not specifically covered under another portion of the Clean Air Act. Most air toxics originate from human-made sources including mobile sources (e.g. cars, trucks, buses), stationary sources (e.g. factories, refineries, power plants), and indoor sources (e.g. building materials and activities such as cleaning) [21]. The Clean Air Act Amendments of 1977 failed to result in substantial reductions of the emissions of these very threatening substances. Over the history of the air toxics program, only seven pollutants had been regulated. Title III established a list of 189 (later modified to 187) HAPs associated with approximately 300 major source categories. The list of HAPs is provided in Appendix C [16]. Under Title III, a major source is defined as any new or existing source with the potential to emit, after controls, 10 short tons or more per year of any of the 187 HAPs or 25 short tons or more per year of any combination of those pollutants. These sources may release air toxics from equipment leaks, when materials are transferred from one location to another or during discharge through emissions stacks or vents.

EPA must then issue maximum achievable control technology (MACT) standards for each listed source category according to a prescribed schedule. These standards will be based on the best demonstrated control technology or practices within the regulated industry, and the prescribed schedule dictated that EPA must issue the standards for 40 source categories within two years, 25% of the source categories within five years, 50% of the source categories within seven years, and 100% of the source categories within ten years of passage of the new law. Eight years after MACT is installed on a source, EPA must examine the risk levels remaining at the regulated facilities and determine whether additional controls are necessary to reduce unacceptable residual risk [16].

The Bhopal, India tragedy, where an accidental release of methyl isocyanate at a pesticide-manufacturing plant in 1984 killed approximately 4,000 people and injured more than 200,000, inspired the 1990 CAAA requirement that factories and other businesses develop plans to prevent accidental releases of highly toxic chemicals. In addition, the Act established the Chemical Safety Board to investigate and report on accidental releases of HAPs from industrial plants.

Title III did not directly regulate air toxics from power plants but did state that regulation of air toxics from utility power plants will be based on scientific and engineering studies. Mercury is one pollutant that was identified for study and will be discussed in more detail later in this chapter. At power plants, compounds in the vapor phase (e.g. polycyclic organic matter) and those combined with or attached to particulate matter (e.g. arsenic) are subject to the Title III provisions [17].

2.1.4.3 Title IV: acid deposition control

The Acid Rain Program was established under Title IV of the 1990 CAAA. The program required major reductions of sulfur dioxide (SO_2) and nitrogen oxides (NO_x) emissions, the pollutants that cause acid rain. Using an innovative market-based or "cap and trade" approach to environmental protection, the program sets a permanent cap on the total amount of SO_2 that may be emitted by electric power plants nationwide. The cap is set at about one half of the amount of SO_2 emitted in 1980, and the trading component allows for flexibility for individual fossil fuel-fired combustion units to select their own methods of compliance. The program also set NO_x emission limitations for certain coal-fired electric utility boilers [18,19].

Under the Acid Rain Program, each unit must continuously measure and record its emissions of SO_2, NO_x, and CO_2, as well as volumetric flow and opacity [19]. In most cases, a continuous emissions monitoring (CEM) system must be used. Units report hourly emissions data to EPA on a quarterly basis. These data are then recorded in the Emissions Tracking System, which serves as a repository of emissions data for the utility industry. The emissions monitoring and reporting is critical to the program as they instill confidence in allowance transactions by certifying the existence and quantity of the commodity being traded and assure that NO_x averaging plans are working. Monitoring also ensures, through accurate accounting, that the SO_2 and NO_x emissions reduction goals are met.

2.1.4.3.1 The SO_2 program

Title IV of the 1990 CAAA called for a two-step program to reduce SO_2 emissions by 10 million short tons from all sources (8.4 million short tons from power plants) from 1980 levels and, when fully implemented in 2000, placed a cap of approximately 8.9 million short tons per year on SO_2 emissions (a level about one-half of the emissions from the power sector in 1980), forcing all generators that burn fossil fuels after 2000 to possess an emissions allowance for each ton of SO_2 they emit. By January 1, 1995, the deadline for Phase I, half of the total SO_2 reductions were to occur by requiring 110 of the largest SO_2-emitting power plants (with 263 boilers or units) located in 21 eastern and midwestern states to cut their emissions to an annual average rate of 2.5 lb SO_2/million Btu. These stations, specifically identified in the 1990 CAAA consisted of boilers with output greater than or equal to 100 MW and sulfur emissions of greater than 2.5 lb SO_2/million Btu. An additional 182 units joined Phase I of the program as substitution or compensating units, bringing the total of Phase I affected units to 445. Plants deciding to reduce SO_2 emissions by 90% were given to 1997 to meet the requirements. In Phase II, which began in 2000, EPA expanded the group of affected sources to include virtually all units greater than 25 MW in generating capacity and all new utility units. Phase II tightened the annual emissions limits imposed on large, higher emitting plants and also set restrictions on smaller, cleaner plants fired by coal, oil, and gas, encompassing over 2,000 units in all [19].

The Phase I reductions were accomplished by issuing the utilities that operated these units emission allowances equivalent to what their annual emission would have been at these plants in the years 1985–1987 based on burning a coal with emissions of 2.5 lb SO_2/million Btu. One allowance is equivalent to the emission of one short ton of SO_2 per year. Utilities were allowed the flexibility of determining which control strategies to be used on existing plants as long as the total emissions from all plants listed in Phase I and owned by the utility did not exceed the available allowances. The law was designed to let industry find the most cost-effective way to stay under the cap. This differs in approach from previous air quality regulations, such as NSPS and PSD, which are based on controlling emissions at their source and then monitoring to ensure compliance [20]. If a utility's emissions exceeded the available allowances, they were subject to fines assessed at $2,000/short ton of excess emissions with a requirement to offset the emissions in future years. Any emission reductions achieved that were in excess of those required could be banked by the utility for use at a later date or traded or sold to another utility.

The SO_2 component of the Acid Rain Program represents a dramatic departure from traditional regulatory approaches that establish source-specific emissions limitations; instead, the program uses an overall emissions cap for SO_2 that ensures emissions reductions are achieved and maintained and a trading system that facilitates lowest-cost emissions reductions. The program features tradable SO_2 emissions allowances, where one allowance is a limited authorization to emit one short ton of SO_2. A fixed number of allowances are issued by the government and they may be bought, sold, or banked for future use by utilities, brokers, or anyone else interested in holding them. Existing units are allocated allowances for each year; new units do not receive allowances and must buy them. New coal-fired boilers are subject to the NSPS, which remains in effect, i.e. 70 to 90+ % reduction, with the provision that they must acquire emission allowances to emit the residual SO_2 which is not controlled [4]. At the end of the year, all participants in the program are obliged to surrender to EPA the number of allowances that correspond to their annual SO_2 emissions [19].

2.1.4.3.2 The NO_x program

Title IV also required EPA to develop a NO_x reduction program and set a goal of reducing NO_x by 2 million short tons from 1980 levels. As with the SO_2 emission reduction requirements, the NO_x program was implemented in two phases, beginning in 1996 and 2000 [19]. The NO_x program embodies many of the same principles of the SO_2 trading program in its design: a results orientation, flexibility in the method to achieve emission reductions, and program integrity through measurement of the emissions. However, it does not cap NO_x emissions as the SO_2 program does nor does it utilize an allowance trading system, although NO_x trading programs have been implemented, which are discussed later in this chapter.

Emission limitations for the NO_x boilers provide flexibility for utilities by focusing on the emission rate to be achieved, expressed in pounds of NO_x per million Btu of heat input. Two options for compliance with the emission limitations are: (1) compliance with an individual emission rate for a boiler or (2) averaging of emission rates over two or more units, that have the same owner or operator, to meet an overall emission rate limitation. These options give utilities flexibility to meet the emission limitations in the most cost-effective way and allow for the further development of technologies to reduce the cost of compliance. If a utility properly installs and maintains the appropriate control equipment designed to meet the emission limitation established in the regulations but is still unable to meet the limitation, the NO_x program allows the utility to apply for an alternative emission limitation (AEL) that corresponds to the level that the utility demonstrates is achievable.

Phase I of the program, which was delayed a year due to litigation, began on January 1, 1996 and affected two types of boilers, which were among those already targeted for Phase I SO_2 reductions: dry bottom wall-fired boilers and tangentially fired boilers. The regulations to govern the Phase II portion of the program, which began in 2000, were promulgated December 19, 1996. These regulations set lower emission limits for Group 1 boilers and established NO_x limitations for Group 2 boilers, which include boilers applying cell-burner technology, cyclone boilers, wet bottom boilers, and other types of coal-fired boilers. The NO_x limitations and number of units affected are provided in Table 2.10 [19].

Table 2.10 Number of NO_x affected units by boiler type

Coal-fired boiler type (all coverage for boilers >25 MW unless otherwise noted)	Standard emission limit (lb/MM Btu)	Number of units
Group 1 boilers		
Tangentially fired	0.40	299
Dry bottom wall-fired	0.46	308
Group 2 boilers		
Cell burners	0.68	36
Cyclones >155 MW	0.86	55
Wet bottom >65 MW	0.84	26
Vertically fired	0.80	28
Total		752

2.1.4.4 Title V: permitting

The 1990 CAAA introduced a national permitting program to ensure compliance with all applicable requirements of the Clean Air Act and to enhance EPA's ability to enforce the Act [21]. Sources are required to submit applications for Title V operating permits through the state agencies to EPA. Air pollution sources subject to the program must obtain an operating permit, states must develop and implement the program, and EPA must issue permit program regulations, review each state's proposed program, and oversee the state's efforts to implement any approved program. EPA must also develop and implement a federal permit program when a state fails to adopt and implement its own program.

The following sources are required to submit Title V permits:

- Major sources as determined under Title I:
  100 short tons per year – and listed pollutant excluding CO
  10 to100 short tons year – or sources in nonattainment areas depending on the classification of Marginal to Extreme
- All NSPS, PSD review sources, and NESHAPS sources
- Major sources as determined under Title III:
  10 short tons per year any air toxic
  25 short tons per year multiple air toxics;
- Sources under Title IV
- Sources emitting ≥ 100 short tons per year of ozone depleting substances under Title VI
- Other sources required to have state or federal operating permits

2.1.5 Additional NO_x regulations and trading programs

2.1.5.1 Ozone transport commission NO_x budget program (1999–2002)

The Ozone Transport Commission (OTC) was established under the Clean Air Act Amendments of 1990 to help states in the Northeast and mid-Atlantic region meet the National Ambient Air Quality Standard (NAAQS) for ground-level ozone. Historically, ozone control strategies have focused on volatile organic emissions. In recognition of the important role of NO_x in ozone formation and transport, the OTC focused much of its efforts on regional summertime NO_x reduction strategies for the electric power industry and certain industrial sources [22,23].

Title I of the 1990 CAAA includes provisions designed to address both the continued nonattainment of the existing ozone NAAQS and the transport of air pollutants across state boundaries. These provisions allow downwind states to petition for tighter controls on upwind states that contribute to their NAAQS nonattainment status. In general, Title I nitrogen oxide provisions require areas with an ozone nonattainment region to:

- Require existing major stationary sources to apply reasonably available control technology (RACT), which is the lowest emission limitation that a particular source is capable of meeting by application of control technology that is reasonably available considering technological and economic feasibility
- Require new or modified stationary sources to offset their emissions and install controls representing the lowest achievable emission rate (LAER), which is the minimum emissions rate accepted by EPA for on major new or modified sources in nonattainment areas
- Require each state with an ozone nonattainment region to develop a State Implementation Plan that may, in some cases, include reductions in stationary source NO_x emissions beyond those required by the RACT provisions of Title I

Section 184 of the Clean Air Act delineates a multistate ozone transport region (OTR) in the northeast and requires specific additional nitrogen oxide and volatile organic compound controls for all areas in this region. It also established the Ozone Transport Commission (OTC) for the purpose of assessing the degree of ozone transport in the OTR and recommending strategies to mitigate the interstate transport of pollution. The OTR consists of the states of Connecticut, Delaware, Maine, Maryland, Massachusetts, New Hampshire, New Jersey, New York, Pennsylvania, Rhode Island, Vermont, the northern counties of Virginia, and the District of Columbia. The OTR states confirmed that they would implement RACT on major stationary sources of NO_x (Phase I), and agreed to a phased approach for additional controls, beyond RACT, for power plants (>25 MW) and other large fuel combustion sources (industrial boilers with a rated capacity >250 million Btu per hour input; Phases II and III (although Phase III was merged/replaced by the NO_x SIP Call Trading Program discussed in the following section)). This agreement, known as the OTC Memorandum of Understanding (MOU), was approved on September 27, 1994 with all OTR states, except for Virginia, signing the MOU.

The MOU established an emission trading system, based in part on the successful implementation of the national cap and trade Acid Rain Program for controlling SO_2, to reduce the costs of compliance with the control requirements under Phase II, which began on May 1, 1999, and Phase II, which began on May 1, 2003. The NO_x Budget Program applied to more than 1,000 large combustion facilities including 900 electric generating units and over 100 industrial units, such as steam boilers and process heaters [23]. The OTC program target caps for the summer season (May 1 through September 30) NO_x emissions were approximately 219,000 short tons in 1999 and 143,000 short tons in 2003, less than half of the 1990 baseline emission level of 490,000 short tons. The program is summarized in Table 2.11 [23].

In developing their NO_x Budget Program, the OTC states established three zones – Inner, Outer, and Northern – and required different emission rates for Phases II and III in the different zones [23]. The Inner Zone generally comprised the contiguous ozone nonattainment areas along the heavily populated corridor from the Washington, D.C. metropolitan area to sourthern New Hampshire. The OTC set the most stringent reduction targets for this zone. The Northern Zone contained all of Maine and Vermont, northeastern New York, and southern New Hampshire, and had the least stringent reduction target. The Outer Zone contained the rest of the OTC region.

The OTC NO_x Budget Program sources successfully reduced the regional ozone season emissions in 2002 by nearly 280,000 short tons or about 60%, from 1990 baseline levels [23]. The program demonstrated a viable NO_x trading system and successful reduction in NO_x despite increases in NO_x emissions from mobile sources and continued transport of pollutants into the region from states downwind of the OTC states. However, ambient ozone conditions remained a serious problem in the region and further reduction efforts both within and upwind of the OTC region were explored, as discussed in the following sections.

Table 2.11 OTC NO_x reduction strategy overview

Phase	Geographic distinctions	Emissions reduction target	Ozone season emission levels
1990 Baseline	Not applicable	Not applicable	473,000 short tons (approx. baseline level)
Phase I: RACT requirements beginning May 1, 1995	All areas	Approximate 40% reduction	290,000 short tons (approx., no budget applies)
Phase II: Trading program beginning May 1, 1995	Inner Zone	65% or 0.20 lb/MM Btu (based on 1990 heat input)	219,000 short tons (budget level)
	Outer Zone	55% or 0.20 lb/MM Btu (based on 1990 heat input)	
Phase III: Trading program beginning May 1, 2003[a]	Inner & Outer Zones	75% or 0.15 lb/MM Btu (based on 1990 heat input)	143,000 short tons (budget level)
	Northern Zone	55% or 0.20 lb/MM Btu (based on 1990 heat input)	

[a]Phase III was replaced by the NO_x SIP Call Trading Program on May 1, 2003.

2.1.5.2 NO_x budget trading program/ NOx SIP call (2003–2008)

The NO_x Budget Trading Program (NBP) was a market-based cap and trade program created to reduce the regional transport of NO_x emissions from power plants and other large combustion sources that contribute to ozone nonattainment in the eastern United States [24,25]. The NBP was established through the NO_x State Implementation Plan (SIP) call, promulgated on October 27, 1998.

The NBP was an ozone season (May 1 to September 30) cap and trade program for electric generating units and large industrial combustion sources, primarily boilers and turbines. Compliance with the NO_x SIP call was scheduled to begin on May 1, 2003 for 20 states and the District of Columbia; however, litigation delayed full implementation for 12 states not previously in the OTC NO_x Budget Program. The OTC states (and District of Columbia) listed previously, except for Maine, New Hampshire, and Vermont, which were excluded, adopted the original compliance date of May 1, 2003, to transition to the NO_x SIP call. Eleven states not previously in the OTC NO_x Budget Program (Alabama, Illinois, Indiana, Kentucky, Michigan, North Carolina, Ohio, South Carolina, Tennessee, Virginia, and West Virginia) began compliance on May 31, 2004. Missouri began compliance with the program on May 1, 2007. The states or areas of states affected are shown in Figure 2.3 [24].

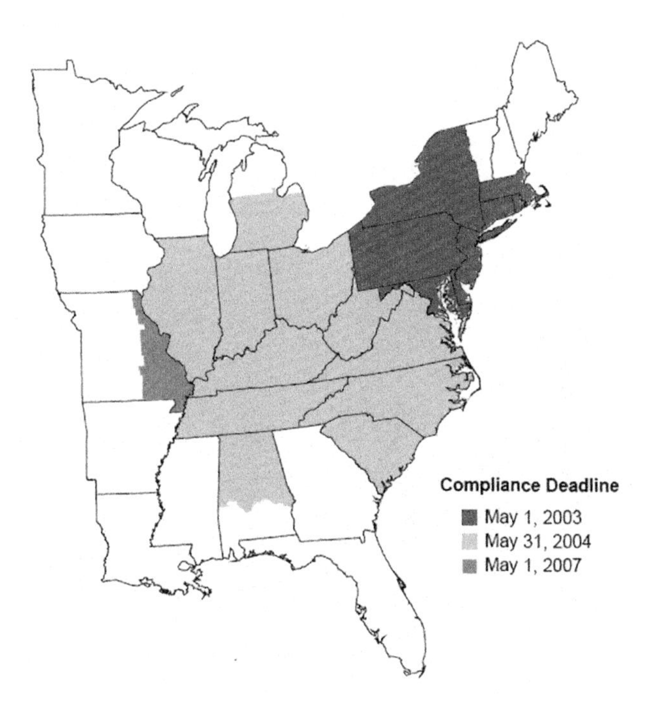

Figure 2.3 NO_x SIP Call Program implementation.

The NBP had several important features:

- Region-wide cap that was the sum of state emission budgets that EPA established under the NO_x SIP call
- Authorizations to emit, known as allowances, were allocated to affected sources based on state trading budgets
- Sources must surrender allowances to cover emissions while having the flexibility to lower allowance needs by adding emission controls, replacing existing controls with more advanced technologies, optimizing existing controls, or switching fuels
- Sources must utilize stringent, complete monitoring
- Automatic penalties were triggered if a source did not have sufficient allowances to cover its emissions
- If a source had excess allowances because it reduced emissions beyond required levels, it could sell the unused allowances or bank them for use in a future ozone season.

There were 2,568 affected units under the NBP in 2008, the last year of the program. Of these, 88% were electric generating units. A breakdown of the units is given in Table 2.12 [24].

In 2008, NBP sources emitted 481,420 short tons of NO_x during the summer ozone season [24,25]. Emissions in 2008 were 62% below 2000 levels, 75% below 1990 levels, and 9% below the 2008 cap. Figure 2.4 shows the total ozone season

Table 2.12 Number of units in the NBP by type in 2008

Type of units	Number of units (%)
Gas-fired electric generating units	1,089 (43%)
Coal-fired electric generating units	715 (28%)
Oil-fired electric generating units	433 (17%)
Unclassified electric generating units	3 (<1%)
Industrial	319 (12%)

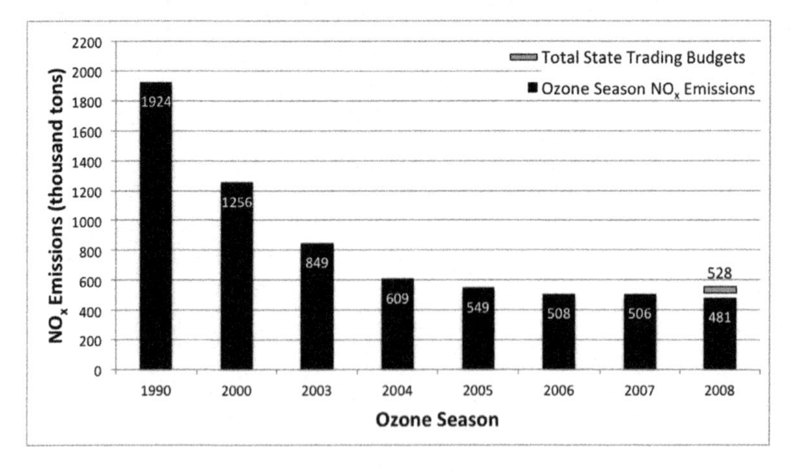

Figure 2.4 Ozone season NO_x emissions from all NBP sources.

Table 2.13 Comparison of ozone season NO_x emissions for all sources, 2003–2008

Unit by fuel type	Ozone season NO_x emissions (thousand short tons)					
	2003	2004	2005	2006	2007	2008
Coal	800	564	494	475	475	456
Oil	26	25	32	14	13	9
Gas	24	20	23	19	19	16
Total	849	609	549	508	506	481

NO_x emissions for all affected sources in the NBP region in 2008 compared to the pre-NBP baseline years of 1990 and 2000. Table 2.13 shows the NO_x emissions for the NBP compliance years by fuel type. The NO_x Budget Program was replaced in 2008 by the Clean Air Interstate Rule.

2.1.6 Clean air interstate rule (CAIR)/cross state air pollution rule (CSAPR)

On March 28, 2005, EPA issued the Clean Air Interstate Rule (CAIR) to permanently cap emissions of SO_2 and NO_x in the eastern United States and to address regional transport of ground-level ozone and fine particulate pollution [26,27]. It also requires certain states to limit ozone season NO_x emissions from May through September. CAIR includes three separate cap and trade programs to achieve the required reductions: the CAIR NO_x ozone season trading program, the CAIR NO_x annual trading program, and the CAIR SO_2 annual trading program. The CAIR NO_x ozone season and annual programs began in 2009, replacing the NO_x Budget Trading Program (NBP) in most states, while the CAIR SO_2 annual program began in 2010.

While the acid rain program (ARP) is a nationwide program affecting large fossil fuel-fired power plants across the country, CAIR covers 27 eastern states and the District of Columbia (D.C.) and requires reductions in annual emissions of SO_2 and NO_x from 24 states and D.C. and emission reductions of NO_x during the ozone season from 25 states and D.C. Figure 2.5 shows the former NBP affected 20 eastern states and D.C. (in outlined area), and states covered by the CAIR [27].

The CAIR SO_2 and NO_x annual programs generally apply to large electric generating units – boilers, turbines, and combined cycle units that primarily burn fossil fuels to generate electricity for sale. The CAIR NO_x ozone season program includes electric generating units as well as large industrial units that produce electricity or steam for internal use. The affected units in ARP and CAIR in 2012 are summarized in Table 2.14. In 2012, there were 3,336 affected electric generating units and industrial units at 952 facilities in the CAIR SO_2 and NO_x annual programs and 3,273 electric generating units and industrial units at 949 facilities in the CAIR NO_x ozone season program [27].

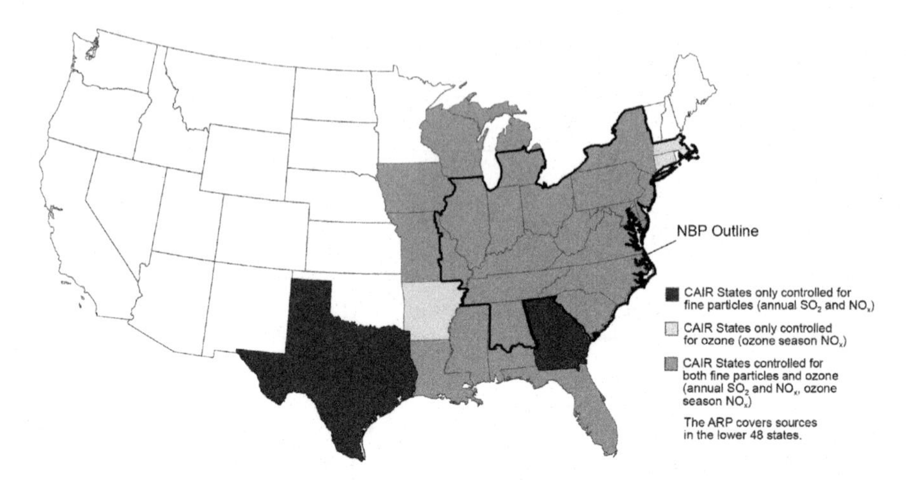

Figure 2.5 CAIR, ARP, and NBP states [27].

Table 2.14 **Affected units in CAIR and ARP (2012)**

Fuel	ARP SO_2 program	ARP NO_x program	CAIR NO_x ozone season program	CAIR NO_x & SO_2 annual programs
Coal EGUs[a]	1,003	885	812	869
Gas EGUs	2,430	11	1,719	2,009
Oil EGUs	187	0	514	427
Industrial units	4	0	198	0
Unclassified EGUs	10	0	1	4
Other Fuel EGUs	18	4	29	27
Total Units	3,652	900	3,273	3,336

[a]Electric generating units.

Table 2.15 shows a large reduction in annual SO_2 and NO_x emissions from CAIR and ARP sources between 2005 and 2012. Short tons of SO_2 emitted fell 68% from the 2005 level, and annual NO_x emissions dropped 53%. During the same period, ozone season NO_x emissions from CAIR sources decreased by approximately 37%. These reductions occurred while electricity demand remained relatively stable [27]. These reductions are further illustrated in Figures 2.6–2.8 [27].

On July 6, 2011, EPA finalized a rule, the Cross-State Air Pollution Rule (CSAPR), to further reduce air pollution and attain the 1997 ozone and fine particle and 2006 fine particle NAAQS [28]. This rule replaces the 2005 CAIR. A December 2008 court decision kept the requirements of CAIR in place temporarily but directed EPA to issue a new rule to implement Clean Air Act requirements concerning the transport of air pollution across state boundaries. The CSAPR was in response to the court action. The goal of the CSAPR is to improve air quality by reducing power plant emissions in 28 states that contribute to ozone and/or fine

Table 2.15 Comparison of emissions and emission rates for CAIR and ARP sources (2000–2012)

CAIR and ARP annual SO_2 trends										
Primary Fuel	SO_2 emissions (thousand short tons)					SO_2 rate (lb/MM Btu)				
	2000	2005	2009	2010	2012	2000	2005	2009	2010	2012
Coal	10,708	9,835	5,653	5,090	3,291	1.04	0.95	0.63	0.53	0.41
Gas	108	91	22	20	6	0.06	0.03	0.01	0.00	0.00
Oil	385	292	38	31	6	0.73	0.70	0.27	0.19	0.04
Other	3	4	8	26	16	0.22	0.27	0.27	0.53	0.29
Total	11,201	10,223	5,722	5,168	3,319	0.88	0.75	0.46	0.38	0.26
CAIR and ARP annual NO_x trends										
Primary Fuel	NO_x emissions (thousand short tons)					NO_x emissions (thousand short tons)				
	2000	2005	2009	2010	2012	2000	2005	2009	2010	2012
Coal	4,587	3,356	1,847	1,923	1,533	0.44	0.32	0.20	0.20	0.19
Gas	354	167	143	150	153	0.18	0.06	0.04	0.04	0.03
Oil	162	104	25	24	19	0.31	0.25	0.17	0.15	0.12
Other	2	6	5	7	7	0.25	0.42	0.12	0.13	0.13
Total	5,104	3,633	2,020	2,103	1,712	0.40	0.27	0.16	0.16	0.14
CAIR ozone season NO_x trends										
Primary Fuel	NO_x emissions (thousand short tons)					NO_x emissions (thousand short tons)				
	2000	2005	2009	2010	2012	2000	2005	2009	2010	2012
Coal	1,398	695	442	527	452	0.45	0.22	0.17	0.18	0.18
Gas	79	58	39	49	48	0.17	0.08	0.05	0.05	0.04
Oil	66	57	13	16	11	0.27	0.25	0.17	0.14	0.12
Other	1	2	2	2	3	0.15	0.17	0.14	0.12	0.13
Total	1,545	812	495	594	514	0.41	0.20	0.14	0.15	0.13

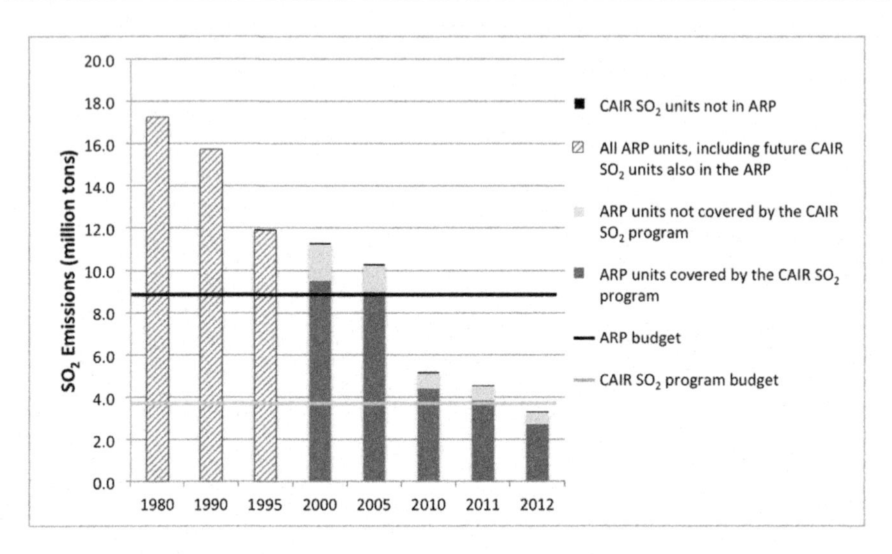

Figure 2.6 SO_2 emissions from CAIR SO_2 annual program and ARP sources (1980–2012).

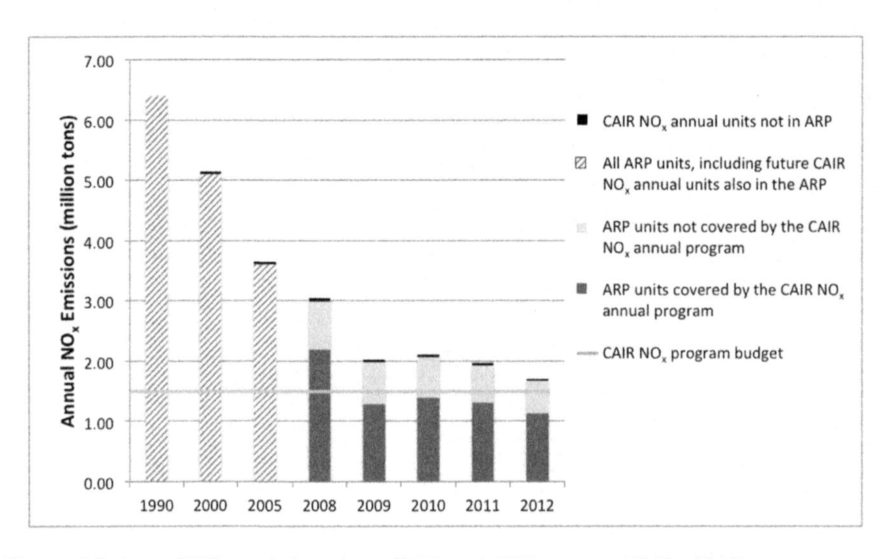

Figure 2.7 Annual NO_x emissions from CAIR and ARP sources (1990–2012).

particulate. The CSAPR requires 23 states to reduce SO_2 and NO_x emission to help downwind areas attain the 24-Hour and/or Annual $PM_{2.5}$ NAAQS. Twenty-five states are required to reduce ozone season NO_x emission to help downwind areas attain the 1997 8-Hour Ozone NAAQS. The rule addresses the upwind states' transport obligations under the 1997 annual $PM_{2.5}$ and 2006 24-Hour $PM_{2.5}$ standards. For 14 states, it also addresses upwind states obligations under the 1997 ozone NAAQS. Figure 2.9 is a map of the states affected by the transport rule [28].

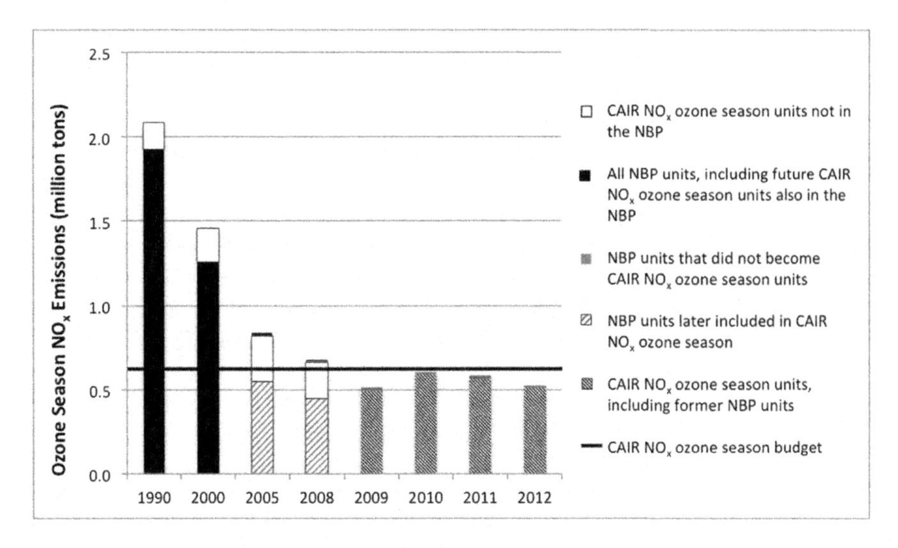

Figure 2.8 Ozone season NO_x emission from CAIR and ARP sources (1990–2012).

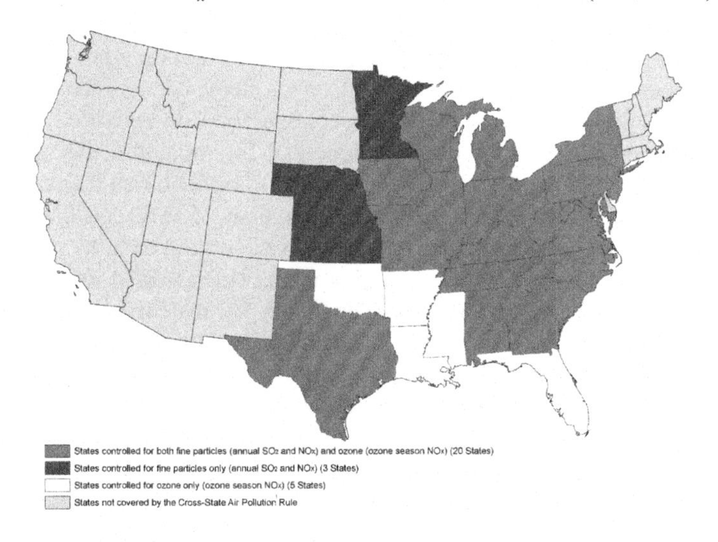

Figure 2.9 Map of states affected by CSAPR.

CSAPR includes three separate cap and trade programs to achieve these reductions: the CSAPR NO_x ozone seating trading program, the CSAPR NO_x annual trading program, and the CSAPR SO_2 annual trading program. The CSAPR trading programs were schedule to replace the CAIR trading programs starting January 1, 2012; however, on December 30, 2011, the U.S. Court of Appeals for the D.C. Circuit Court stayed CSAPR pending judicial review and on August 21, 2012 the court issued a decision vacating the rule [27]. On April 29, 2014, the U.S. Supreme Court reversed the D.C. Circuit opinion vacating CSAPR. EPA is currently reviewing (as of June 2014) the opinion and at this time, CAIR remains in place [29].

2.1.7 Clean air mercury rule/mercury air toxics standard/utility MACT rule

EPA prepared a Mercury Study Report, dated December 1997, which they submitted to Congress on February 24, 1998 as a requirement of the Section 112(n)(1)(B) of the 1990 CAAA [30]. The report provided an assessment of the magnitude of United States mercury emissions by source, the health and environmental implications of those emissions, and the availability and cost of control technologies. The report identified electric utilities as the largest remaining source of mercury emissions in the air as EPA has regulated mercury emissions from municipal waste combustors, medical waste incinerators, and hazardous waste combustion.

EPA also submitted the Utility Hazardous Air Pollutant Report to Congress on February 24, 1998 in which EPA examined 67 air toxics emitted from 52 fossil fuel-fired power plants and concluded that mercury is the air toxic of greatest concern [31]. Although not conclusive, the report finds evidence suggesting a link between utility emissions and the methylmercury found in soil, water, air, and fish from contaminated waters. The report identified the need for additional information on the amount of mercury in United States coals and mercury emissions from coal-fired power plants. Specifically, data identified by EPA included obtaining additional data on the quantity of mercury emitted from various types of generating units, the amount of mercury that is divalent versus elemental, and the effect of pollution control devices, fuel type, and plant configuration on emissions and speciation. To obtain these data, EPA issued a three-part Information Collection Request (ICR) for calendar year 1999 [32]. Part I collected information on the size and configuration of all coal-fired utility boilers greater than 25 MW and their pollution control devices. Part II obtained data quarterly on the origin, quantity, and analyses of coal shipments delivered to the generating units, which totaled more than 1,100, including a minimum of three analyses per month for mercury and chlorine contents, together with any other available analyses such as ash and sulfur contents and heating value. Part III required emission tests on 84 generating units selected at random from 36 categories representing different plant configurations and coal rank to measure total and speciated mercury concentrations in the flue gas before and after the final air pollution control device upstream of the stack.

On December 14, 2000, EPA announced that it would begin regulating mercury emissions from power plants [33]. On December 15, 2003, EPA proposed a rule to permanently cap and reduce mercury emissions from power plants. The schedule required a final rule by December 2004, and implementation of controls by the end of 2007.

On March 15, 2005, EPA issued the first-ever federal rule to permanently cap and reduce mercury emissions from coal-fired power plants, making the United States the first country in the world to regulate mercury emissions from coal-fired power plants [34]. The rule, called the Clean Air Mercury Rule (CAMR), created a market-based cap-and-trade program that would permanently cap utility mercury emissions in two phases. The first cap was 38 tons and emissions were to be reduced by taking advantage of co-benefit reductions, i.e. mercury reductions achieved by reducing SO_2 and NO_x emissions under CAIR (see previous section). In the second phase, due in 2018, coal-fired power plants were to be subjected to a second cap, which would reduce emissions to 15 tons upon full implementation.

The CAMR, however, was challenged in the courts because EPA determined that the December 2000 Regulatory Finding on the Emissions of Hazardous Air Pollutants from Electric Utility Steam Generating Units lacked foundation and that recent information demonstrated that it was not appropriate or necessary to regulate coal- and oil-fired utility units under Section 112 of the Clean Air Act, and subsequently removed those utility units from the Section 112(c) list of source categories [34]. On February 8, 2008, the U.S. Court of Appeals for the D.C. Circuit Court vacated EPA's rule removing power plants from the Clean Air Act list of sources of hazardous pollutants and vacated the CAMR. Consequently, EPA decided to develop emissions standards for power plants under the Clean Air Act (Section 112), consistent with the D.C. Circuit's opinion on the CAMR.

Section 112 of the Clean Air Act requires EPA to regulate emissions of hazardous air pollutants, including mercury, nickel, arsenic, acid gases, and other toxic pollutants, through the establishment of maximum achievable control technology (MACT) standards. On December 16, 2011, EPA finalized federal standards to reduce mercury and other toxic air pollution from coal and oil-fired power plants and they were published in the Federal Register on February 16, 2012 [35,36]. The rule also revises New Source Performance Standards, under Section 111 of the CAA for certain fossil fuel fired electric generating units, including particulate matter, SO_2, and NO_x limits.

The Utility MACT Rule, known as the Mercury and Air Toxics Standards (MATS), applies to electric generating units larger than 25 MW that burn coal or oil for the purposes of generating electricity for sale and distribution through the national electric grid to the public. There are about 1,400 coal- and oil-fired electric generating units at 600 power plants covered by these standards. Power plants are currently the dominant emitters of mercury (53% or 53 short tons/year), acid gases (75%), and many toxic metals (20–60%) [36]. MATS requires overall reductions in mercury emissions of 90%, as well as reductions in acid gases (88%), particulate matter, and sulfur dioxide (41% beyond the reductions expected from CSAPR) [36]. Affected facilities are generally required to comply with the standards for HAPs by 2015; however, the rule allows for compliance extensions until 2016 on a case-by-case basis.

2.1.8 Industrial boiler major source and area source rules

On March 21, 2011, EPA published NESHAP rules in the Federal Register for major source boilers and area source boilers to cut emissions of HAPs such as mercury, dioxin, and lead. At the same time, EPA also announced its intent to reconsider those standards under a Clean Air Act process that allows additional public review and comment, and ensures full transparency in its rulemaking [37]. On December 2, 2011, EPA released proposed amendments to the released rules and on December 20, 2012, EPA finalized a reconsideration process for its Clean Air Act pollution standards for major source boilers (40 CFR 63, Subpart DDDDD) and area source boilers (40 CFR 63, Subpart JJJJJJ) [37–41]. A major source is defined as a facility that emits or has the potential to emit 10 or more short tons per year (tpy) of any single air toxic or 25 tpy or more of any combination of air toxics. An area source is defined as a facility that emits less than 10 tpy of any single air toxic or less than 25 tpy of any combination of air toxics. EPA estimates that of the approximately 1.5 million boilers in the U.S., less than 1% (≈2,300) will need to meet numerical emission limits under these rules. Another 13% (about 197,000 boilers) will need to follow work practice standards such as annual tune-ups, to minimize air toxics. About 86%, or approximately 1.3 million boilers, are not covered by the rules.

There are approximately 14,000 major source boilers and process heaters in the U.S. firing coal, fuel oil, natural gas, biomass, refinery gas, or other gas to produce steam. All of those are required to conduct periodic tune-ups, and about 12% (≈1,680) will be required to meet numerical emission limits. Major source boilers are located at large sources of air pollutants, including refineries, chemical plants, and other industrial facilities, but are also used to provide heat for commercial facilities and institutions. Emission limits were established for carbon monoxide (CO) as a surrogate for organic HAPs, hydrogen chloride (HCl) as a surrogate for acid gas HAP, mercury, and total selected metals (sum of arsenic, beryllium, cadmium, chromium, lead, manganese, nickel, and selenium) or filterable particulate matter as a surrogate for nonmercury metallic HAPs. Dioxins/furans emissions are addressed through periodic tune-up work practices. Compliance dates are January 31, 2016 for existing sources and January 31, 2013, or upon start-up, whichever is later, for new sources (new sources are defined as sources that began operation on or before June 4, 2010). Table 2.16 lists the emission limits for boilers and process heaters for major sources [40].

There are approximately 1.3 million boilers located at area source facilities that operate on natural gas. They are not covered by these rules. The area source standards cover approximately 183,000 boilers located at 92,000 area source facilities firing coal, fuel oil, biomass or nonwaste materials. The majority of area source boilers are located at commercial and institutional facilities, including commercial establishments, medical centers, educational facilities, and municipal buildings. Some industrial boilers used in manufacturing, processing, mining, refining, or other industries are also included. Of the 183,000 covered units, approximately 182,400 (over 99%) need only to conduct periodic tune-ups, and some of these need to perform a one-time energy assessment. Approximately 600 coal-burning units (less than 1%), that represent the largest of these sources, are required to meet emission limits. Compliance dates are March 21, 2014 for existing sources, and May 20, 2014 for new sources that began operations on or before May 20, 2011. Table 2.17 lists the emission limits for area source boilers that must meet numerical emission limits [41]. Table 2.18 lists the work practice standards for those units that are not subject to numeric emission limits [41].

Table 2.16 Emission limits for major source boilers and process heaters

Subcategory	Filterable PM (or TSM) (lb/MM Btu)[a]	HCl (lb/MM Btu)[a]	Mercury (lb/MM Btu)[a]	CO (ppm @3% O_2)[a]
Existing units				
Coal Stoker	0.040 (0.000053)	0.022	0.0000057	160
Coal Fluidized Bed (FB)	0.040 (0.000053)	0.022	0.0000057	130
Coal FB with FB Heat Exchanger	0.040 (0.000053)	0.022	0.0000057	140
Coal-Burning Pulverized Coal	0.040 (0.000053)	0.022	0.0000057	130
Biomass Wet Stoker/ Sloped Grate/ Other	0.037 (0.00024)	0.022	0.0000057	1,500
Biomass Kiln-Dried Stoker/Sloped Grate/Other	0.32 (0.0004)	0.022	0.0000057	460
Biomass Fluidized Bed	0.11 (0.0012)	0.022	0.0000057	470
Biomass Suspension Burner	0.051 (0.0065)	0.022	0.0000057	2,400
Biomass Dutch Ovens/Pile Burners	0.28 (0.002)	0.022	0.0000057	770
Biomass Fuel Cells	0.02 (0.0058)	0.022	0.0000057	1,100
Biomass Hybrid Suspension Grate	0.44 (0.00045)	0.022	0.0000057	2,800
Heavy Liquid	0.062 (0.0002)	0.0011	0.000002	130
Light Liquid	0.0079 (0.000062)	0.0011	0.000002	130
Non-Continental Liquid	0.27 (0.00086)	0.0011	0.000002	130
Gas 2 (other process gases)[b]	0.0067 (0.00021)	0.0017	0.0000079	130
New units				
Coal Stoker	0.0011 (0.000023)	0.022	0.0000008	130
Coal Fluidized Bed (FB)	0.0011 (0.000023)	0.022	0.0000008	130
Coal FB with FB Heat Exchanger	0.0011 (0.000023)	0.022	0.0000008	140
Coal-Burning Pulverized Coal	0.0011 (0.000023)	0.022	0.0000008	130
Biomass Wet Stoker/ Sloped Grate/Other	0.03 (0.000026)	0.022	0.0000008	620

(Continued)

Table 2.16 (Continued)

Subcategory	Filterable PM (or TSM) (lb/MM Btu)[a]	HCl (lb/MM Btu)[a]	Mercury (lb/MM Btu)[a]	CO (ppm @3% O_2)[a]
Biomass Kiln-Dried Stoker/Sloped Grate/Other	0.03 (0.0004)	0.022	0.0000008	460
Biomass Fluidized Bed	0.0098 (0.000083)	0.022	0.0000008	230
Biomass Suspension Burner	0.03 (0.0065)	0.022	0.0000008	2,400
Biomass Dutch Ovens/Pile Burners	0.0032 (0.000039)	0.022	0.0000008	330
Biomass Fuel Cells	0.02 (0.000029)	0.022	0.0000008	910
Biomass Hybrid Suspension Grate	0.026 (0.00044)	0.022	0.0000008	1,100
Heavy Liquid	0.013 (0.000075)	0.00044	0.00000048	130
Light Liquid	0.0011 (0.000029)	0.00044	0.00000048	130
Noncontinental Liquid	0.023 (0.00086)	0.00044	0.00000048	130
Gas 2 (other process gases)[b]	0.0067 (0.00021)	0.0017	0.0000079	130

[a]lb/MM Btu heat input; 3-run average.
[b]Gaseous fuels other than natural gas and refinery gas.

Table 2.17 Emission limits for area source boilers

Subcategory	Filterable PM[a] (lb/MM Btu input)	Mercury (lb/MM Btu input)	Carbon Monoxide (ppm @ 3% O_2)
New coal-fired boilers ≥30 MM Btu/h[b]	0.03	0.000022	420
New coal-fired boilers, 10–30 MM Btu/h	0.42	0.000022	420
New biomass-fired boilers, ≥30 MM Btu/h	0.03	NA[c]	NA
New biomass-fired boilers, 10–30 MM Btu/h	0.07	NA	NA
New oil-fired boilers, ≥10 MM Btu/h	0.03	NA	NA
Existing coal-fired boilers, ≥10 MM Btu/h	NA	0.000022	420

[a]Particulate matter.
[b]Heat input capacity.
[c]Not available.

Table 2.18 Work practice standards, emission reduction measures, and management practices for area source boilers

Subcategory	Work practices
Existing or new coal-fired, new biomass-fired, or new oil fired boilers with ≥10 MM Btu/h heat input	Minimize startup and shutdown periods, follow manufacturer's recommended procedures
Existing coal-fired boilers with <10 MM Btu/h heat input that do not meet the definition of limited-use boiler, or use an O_2 trim system	Conduct an initial boiler tune-up, and conduct a boiler tune-up biennially
New coal-fired boilers with <10 MM Btu/h heat input that do not meet the definition of limited-use boiler, or use an O_2 trim system	Conduct a tune-up of the boiler biennially
Existing oil-fired boilers with >5 MM Btu/h heat input that do not meet the definition of limited-use boiler, or use an O_2 trim system	Conduct an initial boiler tune-up, and conduct a boiler tune-up biennially
New oil-fired boilers with >5 MM Btu/h heat input that do not meet the definition of limited-use boiler, or use an O_2 trim system	Conduct a tune-up of the boiler biennially
Existing biomass-fired boilers that do not meet the definition of limited-use boiler, or use an O_2 trim system	Conduct an initial boiler tune-up, and conduct a boiler tune-up biennially
New biomass-fired boilers that do not meet the definition of limited-use boiler, or use an O_2 trim system	Conduct a tune-up of the boiler biennially
Existing seasonal boilers	Conduct an initial boiler tune-up, and conduct a boiler tune-up every 5 years
New seasonal boilers	Conduct a boiler tune-up every 5 years
Existing limited-use boilers	Conduct a boiler tune-up every 5 years
New limited-use boilers	Conduct an initial boiler tune-up, and conduct a boiler tune-up every 5 years
Existing oil-fired boilers with heat input ≤5 MM Btu/h	Conduct an initial boiler tune-up, and conduct a boiler tune-up every 5 years
New oil-fired boilers with heat input ≤5 MM Btu/h	Conduct a boiler tune-up every 5 years
Existing coal-fired, biomass-fired, or oil-fired boilers with an O_2 trim system that would otherwise be subject to a biennial tune-up	Conduct an initial boiler tune-up, and conduct a boiler tune-up every 5 years
New coal-fired, biomass-fired, or oil-fired boilers with an O_2 trim system that would otherwise be subject to a biennial tune-up	Conduct a boiler tune-up every 5 years
Existing coal-fired, biomass-fired, or oil-fired boilers with heat input of ≥ 10 MM Btu/h, not including limited-use boilers	Must have a one-time energy assessment performed by a qualified energy assessor

2.1.9 New source review

The New Source Review (NSR) program is one of many programs created by the Clean Air Act to reduce emissions of air pollutants, particularly "criteria" pollutants" that are emitted from a wide variety of sources and have an adverse impact on human health and the environment. The NSR program was established in parts C (Prevention of Significant Deterioration of Air Quality) and D (Plan Requirements for Nonattainment Areas) of Title I (Air Pollution Prevention and Control) of the 1977 Clean Air Act Amendments and modified in the 1990 Amendments to protect public health and welfare, as well as national parks and wilderness areas, as new sources of air pollution are built and when existing sources are modified in a way that significantly increases air pollutant emissions. Specifically, NSR's purpose is to ensure that when new sources are built or existing sources undergo major modifications the air quality improves if the change occurs where the air currently does not meet Federal air quality standards and air quality is not significantly degraded where the air currently meets federal standards [42]. NSR is also referred to as construction permitting or preconstruction permitting and the three types of NSR requirements are [43]:

- Prevention of Significant Deterioration (PSD) permits which are required for new major sources or a major source making a major modification in an attainment area
- Nonattainment NSR permits which are required for new major sources or major sources making a major modification in a nonattainment area
- Minor source permits which are for pollutants from stationary sources that do not require PSD or nonattainment NSR permits and are used to prevent the construction of sources that would interfere with attainment or maintenance of a NAAQS or violate the control strategy in nonattainment areas

The original intent of the NSR program was to ensure that major new facilities that are sources of emissions, or existing facilities that are modified and result in increased emissions, would install state-of-the-art controls. Subsequently, EPA provided interpretive guidance that complicated the review program and expanded it to include maintenance or improvement. While the determination of whether an activity is subject to the major NSR program is fairly straightforward for a newly constructed source, the determination of what should be classified as a modification subject to major NSR presents a more difficult issue. Consequently, installation of new technology, greater energy efficiency, and improved environmental performance at facilities was being inhibited. In addition, there was much controversy between industry and EPA over what "triggers" applicability of the major NSR program, which led to litigation between Wisconsin Electric Power Company (WEPCO) and EPA. In 1992, EPA promulgated revisions to the applicability regulations creating special rules for physical and operational changes at electric utility steam generating units.

On July 23, 1996, a number of changes were proposed to the existing major NSR requirements as part of a larger regulatory package [44]. This was followed by a Notice of Availability published by EPA in the Federal Register on July 24, 1998 requesting comment on three of the proposed changes. After public comment, on

December 31, 2002, EPA issued an NSR final rule to improve the NSR program and a proposed rule to provide a regulatory definition of routine maintenance, repair, and replacement. In summary, the final rule that became effective March 3, 2003:

- Reforms the emissions accounting system for determining when a change is triggered (now the trigger is based on actual emissions);
- Allows already controlled "clean units" to make changes without triggering NSR;
- Broadens the exclusion from NSR for projects intended for pollution control; and
- Sets new rules for establishing plant-wide applicability limits of specific pollutants under the program.

EPA took action to improve the NSR program after performing a comprehensive review of the program and, in June 2002, issued a Report to the President on NSR [42]. The report concluded that the program, as was administered and as related to the energy sector, impeded or resulted in the cancellation of projects that would maintain or improve the reliability, efficiency, or safety of existing power plants. EPA issued the final rule improvements after the culmination of a 10-year process that included pilot studies and the engagement of state and local governments, environmental groups, private sector representatives, academia, and concerned citizens in an open and far-reaching public rulemaking process. In addition, the nation's governors and environmental commissioners, on a bipartisan basis, called for NSR reform.

The final rule implements the following major improvements to the NSR program [42]:

- Plantwide Applicability Limits (PALs): To provide facilities with greater flexibility to modernize their operations without increasing air pollution, facilities that agree to operate within strict site-wide emissions caps called PALs will be given flexibility to modify their operations without undergoing NSR, so long as the modifications do not cause emissions to violate their plantwide cap;
- Pollution Control and Prevention Projects: To maximize investments in pollution prevention, companies that undertake certain specified environmentally beneficial activities will be free to do so upon submission to their permitting authority of a notice, rather than having to wait for adjudication of a permit application. EPA is also creating a simplified process for approving other environmentally beneficial projects;
- Clean Unit Provision: To encourage the installation of state-of-the-art air pollution controls, EPA will give plants that attain clean unit status flexibility in the future if they continue to operate within permitted limits. This flexibility is an incentive for plants to voluntarily install the best available pollution controls. Clean units must have an NSR permit or other regulatory limit that requires the use of the best air pollution technologies; and
- Emissions Calculation Test Methodology: To provide facilities with a more accurate procedure for evaluating the effect of a project on future emissions, the final regulations improve how a facility calculates whether a particular change will result in a significant emissions increase and thereby trigger NSR permitting requirements.

EPA's proposed rule would make improvements to the routine maintenance, repair, and replacement exclusion currently contained in EPA's regulations. These proposed improvements will be subject to a full and open public rulemaking process. Since 1980, EPA regulations have excluded from NSR review all repairs and maintenance activities that are routine but a complex analysis must be made to determine what activities meet the standard. This has deterred companies from conducting repairs and replacements that are necessary for the safe, efficient, and reliable operation of facilities.

After issuing the new NSR final and proposed rules, certain environmental groups and state and local governments petitioned EPA to reconsider specific aspects of the final NSR reform rule. EPA announced on July 25, 2003 that it would reconsider parts of the NSR final rule [82]. EPA's notice responds in part to the petitions by requesting comment on six limited areas. EPA then solicited comments on the following six areas [45]:

- EPA's report titled: "Supplemental Analysis of the Environmental Impact of the 2002 Final NSR Improvement Rules," which concluded that the NSR improvement rule will likely result in greater environmental benefits than the prior program;
- The decision to allow certain sources of air emissions to maintain Clean Unit status after an area is redesignated from attainment to nonattainment for one of the six criteria air pollutants;
- EPA's inclusion of the reasonable possibility standard as it pertains to the need to maintain records and file certain reports when project actual emissions following a physical or operational change;
- The method for assessing air emissions from process units built after the 24-month baseline period used to establish PAL limits;
- The decision to allow a PAL to supersede existing emissions limits established for NSR applicability purposes. Compliance with the PAL is then used to determine if NSR requirements apply in the future; and
- The method of measuring emission increases when existing emission units are replaced.

On October 27, 2003, EPA published its final NSR Equipment Replacement Rule [83]. This final version of the rule applied only to equipment replacement. It was to become effective December 26, 2003, and states would have up to three years to revise their state implementation plans to reflect these requirements. The regulation specified that replacement components must be functionally equivalent to existing components in that there are no changes in the basic unit design or to pollutant emitting capacity. It also set a 20% limit on replacement cost for equipment. If these restrictions are exceeded, the replacement work is subject to the NSR process. However, on December 24, 2003, the Court of Appeals for the District of Columbia Circuit stayed the rule and it did not become effective December 26, 2003. On June 24, 2004, EPA granted reconsideration of certain aspects of the equipment replacement provision (ERP). On June 6, 2005, EPA issued its final response on the reconsideration of the ERP and stated that it has decided not to change any aspect of the ERP as it was originally issued in 2003. EPA clarified the legal basis and support for selecting 20% for the threshold criterion [46,47].

2.1.10 Fine particulate matter

Epidemiological research has revealed a consistent statistical correlation between levels of airborne fine particulate matter ($PM_{2.5}$) and adverse respiratory and cardiopulmonary effects in humans [48]. This has resulted in EPA's promulgation of NAAQS that limit the allowable mass concentrations of $PM_{2.5}$. Attainment of the $PM_{2.5}$ NAAQS requires an annual average mass concentration of less than 12 $\mu g/m^3$ and a daily maximum concentration of less than 35 $\mu g/m^3$. The daily maximum was reduced from 65 to 35 $\mu g/m^3$ on September 21, 2006 when EPA issued the strongest national air quality standards for particle pollution in the U.S.'s history [49]. This was implemented because EPA identified 39 areas as not meeting the standards for $PM_{2.5}$ [49] and it is known that ambient $PM_{2.5}$ has also been found to contribute significantly to the impairment of long-range visibility (regional haze) in many areas of the United States [50]. EPA issued a Regional Haze Rule in 1999 that established goals for reducing regional haze in areas of the United States where long-range visibility has been determined to have exceptional value (Class I areas) and has outlined methods for achieving these goals [50] and followed it up with a fine particulate pollution standard [49]. On December 14, 2012, EPA revised the annual NAAQS for $PM_{2.5}$ and decreased it from 15 $\mu g/m^3$ to 12 $\mu g/m^3$ [49].

It is generally recognized that coal-fired power plants can be important contributors to ambient $PM_{2.5}$ mass concentrations and regional haze. Consequently, EPA required additional restrictions of coal power plant emissions in the 2006 time frame as they develop SIPs for achieving and/or maintaining compliance with the $PM_{2.5}$ NAAQS. This was followed on April 25, 2014 when EPA finalized a rule to classify areas designated nonattainment for the 1997 and/or 2006 fine particle pollution standard as "Moderate" and set a deadline of December 31, 2014, for states to submit any remaining nonattainment-related state implementation plans (SIPs) and nonattainment NSR SIPs required under the Clean Air Act [51].

2.1.11 Climate change/greenhouse gas emissions

Climate change has been on the agenda of Congress, beginning with the 108th Congress (2003–2005), and the Bush (George W. Bush) administration [52–55]. At that time, Congress considered several legislative initiatives with key issues being considered including better coordination of government research, setting caps on greenhouse gas (GHG) emissions, and whether reporting systems should be voluntary or mandatory. Climate change debates occurred in connection with standalone climate change legislation, as well as during discussions of energy policy and climate change legislation. There have been several bills proposed addressing climate change research and data management, managing risks of climate change, reporting of GHG emissions, and stabilization and caps on GHG emissions [56].

The second Bush administration also took steps in addressing GHG emissions, although for the most part climate change legislation stalled under the second Bush administration. On June 25, 2003, United States Department of Energy Secretary Spencer Abraham and energy ministers from around the world signed the first international framework for research and development on the capture and storage of CO_2 emissions. This initiative is called the Carbon Sequestration Leadership Forum and includes the following participants – Australia, Brazil, Canada, China, Columbia, India, Italy, Japan, Mexico, Norway, Russian Federation, United

Kingdom, United States, and European Commission [53]. This was soon followed with the announcement on July 23, 2003 of the Bush Administration's unveiling of a long-term strategic plan to study global change [54,55]. Presented by Commerce Secretary Don Evans and DOE Secretary Spencer Abraham, the 10-year plan sets five goals, including:

- Identifying the natural variability in the Earth's climate;
- Understanding the forces that cause global warming;
- Reducing the uncertainties in climate forecasting;
- Improving the understanding of sensitivity and adaptability of ecosystems to climate change; and
- Developing more exact methods for calculating the risks of global warming.

In response to the Consolidated Appropriations Act (H.R. 2764; Public Law 110–161), EPA has issued the Final Mandatory Reporting of Greenhouse Gases Rule, which was signed by the Administrator on September 22, 2009 [57]. The rule requires reporting of greenhouse gas (GHG) emissions from large sources and suppliers in the United States, and is intended to collect accurate and timely emissions data to inform future policy decisions. Under the rule, sources that emit 25,000 metric tons or more of GHG emissions are required to submit annual reports to EPA. The gases covered include carbon dioxide (CO_2), methane (CH_4), nitrous oxide (N_2O), hydrofluorocarbons (HFC), perfluorocarbons (PFC), sulfur hexafluoride (SF_6), and other fluorinated gases including nitrogen fluoride (NF_3), and hydrofluorinated ethers (HFE). It is the intent of the reporting system to identify where GHGs are coming from and guide development of policies and programs to reduce the emissions.

On May 13, 2010, EPA issued its final "Tailoring Rule" setting air permitting requirements for large stationary sources of greenhouse gas emissions under the Prevention of Significant Deterioration (PSD) and Title V permitting requirements of the Clean Air Act [58]. The rule, published in the Code of Federal Regulations on June 3, 2010, set thresholds for GHG emissions that define when permits under the New Source Review Prevention of PSD and Title V Operating Permit programs are required for new and existing industrial facilities [59]. The rule addressed emission of six GHGs: carbon dioxide, methane, nitrous oxide (N_2O), hydrofluorocarbons, perfluorocarbons, and sulfur hexafluoride (SF_6). The Tailoring Rule consists of two steps. Step 1 (January 2, 2011 to June 30, 2011) required that best available control technology apply to projects that increased net GHG emissions by at least 75,000 short tons per year carbon dioxide equivalent (CO_2 eq, which is discussed in the next section), but only if the project also significantly increases emissions of at least one non-GHG pollutant. The second step of the Tailoring Rule (July 1, 2011 to June 30, 2013) phased in additional large sources of GHG emissions. New sources as well as existing sources not already subject to Title V that emit, or have the potential to emit, at least 100,000 short tons per year CO_2 eq became subject to the PSD and Title V requirements.

On April 13, 2012, EPA proposed a new source performance standard for emissions of carbon dioxide for new affected fossil fuel-fired electric generating units by establishing a single standard applicable to all new fossil fuel-fired electric generating units serving intermediate and base load power demand [60]. After consideration of the information provided in more than 2.5 million comments on the proposal, as well as consideration of continuing changes in the electricity sector, EPA issued a new proposal to establish separate standards for fossil fuel-fired electric steam generating units (utility boilers and IGCC units) and for natural gas-fired stationary combustion turbines. On September 20, 2013, EPA withdrew the April 2012 proposal and proposed a New Source Performance Standard, which was published in the Federal Register on January 8, 2014 [60]. The proposed standards reflect separate determinations of the best system of emission reduction (BSER) adequately demonstrated for utility boilers and IGCC units and for natural gas-fired stationary combustion turbines. The proposed standards, which have not been finalized yet, consist of three subcategories and the emission limits are:

- 1,000 lb CO_2/MWh for new combustion turbines with a heat input rating greater than 850 MM Btu/h
- 1,100 lb CO_2/MWh for new combustion turbines with a heat input rating less than or equal to 850 MM Btu/h
- 1,100 lb CO_2/MWh for new fossil fuel-fired boilers and IGCC units

On June 2, 2014, EPA released two new carbon pollution rules, which were published in the Federal Register on June 18, 2014 [61,62]. One addressed modified or reconstructed electric utility generating units [61] and the other addressed existing stationary sources [62]. Comments on the proposed rules were due on or before October 16, 2014 at which time EPA was to make a decision on finalizing the rules. Table 2.19 summarizes the proposed standards for modified and reconstructed fossil fuel-fired electric steam generating units and natural gas-fired stationary combustion turbines [61]. EPA also proposed emission guidelines for states to use in developing plans to limit CO_2 emission from existing fossil fuel-fired electric generating units. States must then submit plans to the EPA [62]. Nationwide, by 2030, existing electric generating units are required to reduce CO_2 emissions by 30% from 2005 CO_2 emission levels. Proposed state plan dates are:

- June 30, 2016 – Initial plan or complete plan due
- June 30, 2017 – Complete individual plan due if state is eligible for one-year extension
- June 30, 2018 – Complete multistate plan due if state is eligible for two-year extension

2.2 Air quality and emissions trends

This section summarizes the air quality and emissions trends in the United States for the criteria pollutants – NO_2, ozone, SO_2, particulate matter, carbon monoxide, and lead, along with acid rain, air toxics, specifically mercury, and CO_2. Air quality is based on actual measurements of pollutant concentrations in the ambient air at monitoring sites. Trends are derived by averaging direct measurements from these

Table 2.19 Summary of BSER and proposed standards for affected sources

Affected units	BSER	Standard
Modified utility boilers and IGCC units	Most efficient generation at the affected source achievable through a combination of best operating practices and equipment upgrades	**Co-Processed Alternative #1** 1. Source would be required to meet a unit-specific emission limit determined by the unit's best historical annual CO_2 emission rate (from 2002 to the date of the modification) plus an additional 2% emission reduction: the emission limit will be no lower than: a. 1,900 lb CO_2/MWh-net for sources with heat input >2,000 MM Btu/h OR b. 2,100 lb CO_2/MWh-net for sources with heat input ≤2,000 MM Btu/h
Modified utility boilers and IGCC units	Most efficient generation at the affected source achievable through a combination of best operating practices and equipment upgrades	**Co-Processed Alternative #2** Source would be required to meet a unit-specific emission limit dependent upon when the modification occurs 1. Sources that modify prior to becoming subject to a CAA 111(d) plan would be required to meet a unit-specific emission limit determined by the unit's best historical annual CO_2 emission rate (from 2002 to the date of the modification) plus an additional 2% emission reduction: the emission limit will be no lower than: a. 1,900 lb CO_2/MWh-net for sources with heat input >2,000 MM Btu/h OR b. 2,100 lb CO_2/MWh-net for sources with heat input ≤2,000 MM Btu/h

(*Continued*)

Table 2.19 (Continued)

Affected units	BSER	Standard
		2. Sources that modify after becoming subject to a CAA 111(d) plan would be required to meet a unit-specific emission limit determined by the 111(b) implementing authority from the results of an energy efficiency improvement audit.
Modified natural gas-fired stationary combustion turbines	Efficient NGCC[a] technology	1. Sources with heat input >850 MM Btu/h would be required to meet an emission limit of 1,000 lb CO_2/MWh-gross 2. Sources with heat input ≤ 850 MM Btu/h would be required to meet an emission limit of 1,100 lb CO_2/MWh-gross
Reconstructed utility boilers and IGCC units	Most efficient generating technology at the affected source	1. Sources with heat input >2,000 MM Btu/h would be required to meet an emission limit of 1,900 lb CO_2/MWh-net 2. Sources with heat input ≤ 2,000 MM Btu/h would be required to meet an emission limit of 2,100 lb CO_2/MWh-net
Reconstructed natural gas-fired stationary combustion turbines	Efficient NGCC[a] technology	1. Sources with heat input >850 MM Btu/h would be required to meet an emission limit of 1,000 lb CO_2/MWh-gross 2. Sources with heat input ≤ 850 MM Btu/h would be required to meet an emission limit of 1,100 lb CO_2/MWh-gross

[a]Natural gas combined cycle.

monitoring stations on a yearly basis. Emissions of ambient pollutants and their precursors are estimated based on actual monitored readings or engineering calculations of the amounts and types of pollutants emitted by vehicles, factories, stationary combustion, and other sources.

The U.S. EPA evaluates the status and trends in the nation's air quality on a yearly basis. EPA tracks air pollution by evaluating the air quality measured from over 5,200 ambient air monitors located at over 3,000 sites across the nation that are operated primarily by state, local, and tribal agencies [63]. In addition, EPA has been tracking emissions from all sources for more than 40 years and has data back to 1970. From 1970 to 2012, aggregate national emissions of the six common pollutants – particles, ozone, lead, carbon monoxide, nitrogen dioxide, and sulfur dioxide – alone have decreased an average of 72% while gross domestic product increased by 219% [64]. In the most recent summary (for the year 2012) of trends in air quality in the United States, EPA states aggregate emissions of the six principal (i.e. criteria) air pollutants tracked nationally have been reduced by 67% since 1980 [65]. During this same period, United States gross domestic product increased 133%, energy consumption increased 19%, the population increased 38%, and vehicle miles traveled increased 92%. From 1990 to 2008, air toxics decreased by 62%. Carbon dioxide emissions however, increased by 19% between 1980 and 2012. The reduction in emissions in criteria air pollutants between 1970 and 2013 is illustrated in Figure 2.10 [from 64,66], which shows the emissions for 1970, 1980, 2000, and 2013. This is further illustrated in Figure 2.11, which shows the emissions for each year from 2002 to 2012 [67]. Air quality has improved across the U.S. since the Clean Air Act was amended in 1990 and Figure 2.12 shows the national trends in the criteria pollutants between 1990 and 2010, relative to their respective national air quality standards (as of 2010) [68]. A detailed breakdown of the sources (i.e. stationary combustion, industrial processes, transportation, and miscellaneous) of the 2013 criteria pollutants, and ammonia (NH_3) emissions is provided in Table 2.20. From Table 2.20 it is observed that, of the total emissions of each pollutant, stationary combustion (all fuel sources including nonfossil fuel feedstocks) contributes approximately 6% of the CO, 28% of the NO_x, 5% of the PM_{10}, 13% of the $PM_{2.5}$, 82% of the SO_2, 4% of the VOC, and 2% of the NH_3 emissions. The emissions trends for stationary fuel combustion, including electrical utilities, industrial/commercial boilers, and residential units are illustrated in Figure 2.13 for the years 2002 through 2012 [67].

Despite this progress, about 36 million short tons of pollution are emitted into the air each year in the United States, and approximately 142 million people live in counties where air monitored in 2012 was unhealthy because of high levels of at least one of the six criteria air pollutants [65]. Most of the areas that experienced the unhealthy air did so because of particulate matter and/or ground-level ozone.

2.2.1 Six principal pollutants

Under the Clean Air Act, as discussed previously, EPA established air quality standards to protect human health and public welfare. EPA has set national air quality standards for six principal or criteria air pollutants, which include nitrogen dioxide, ozone, sulfur

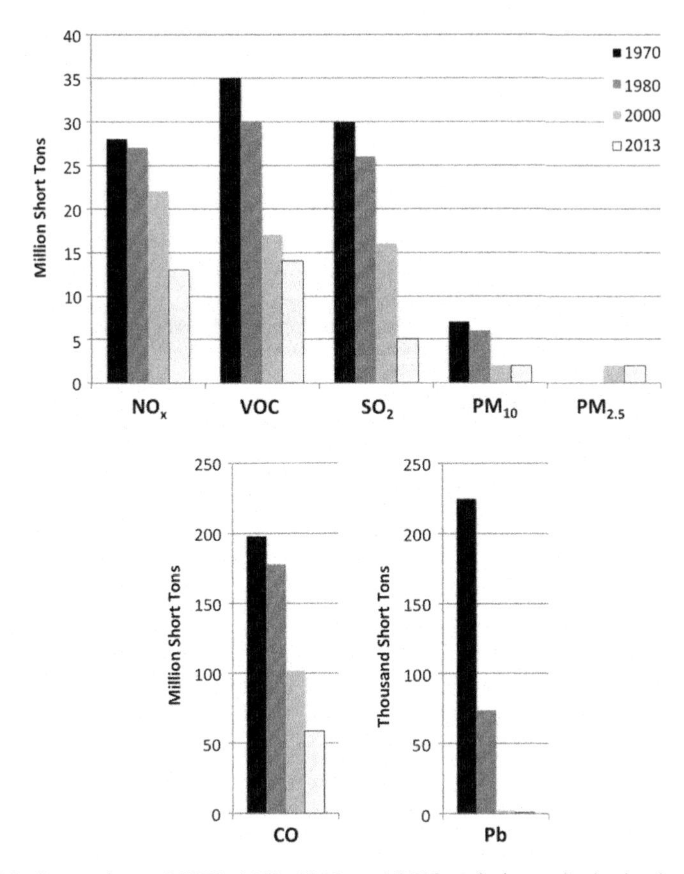

Figure 2.10 Comparison of 1970, 1980, 2000, and 2013 emissions of criteria air pollutants.

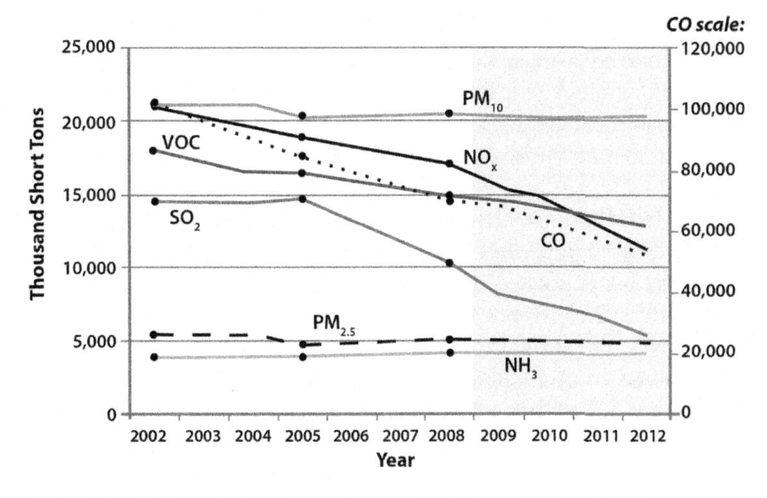

Figure 2.11 National air emissions, 2002–2012 (excludes wildfires) [67].

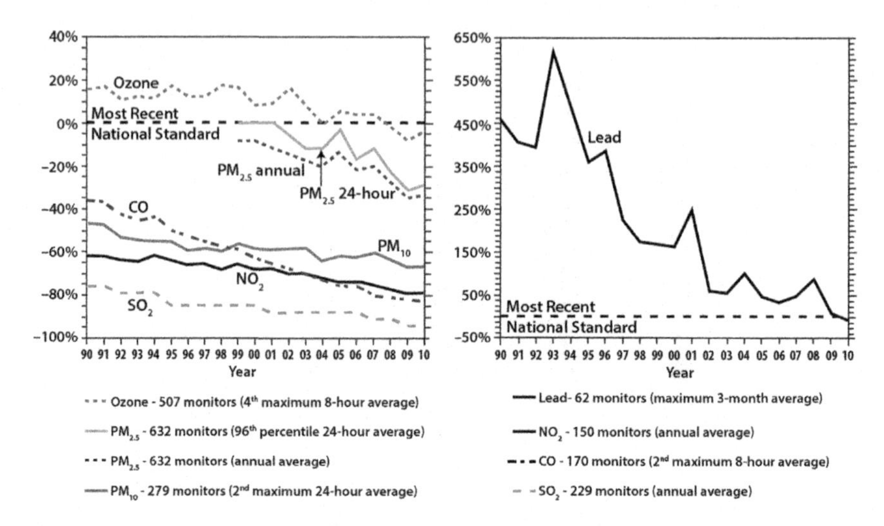

Figure 2.12 Comparison of national levels of the six common pollutants to the most recent (as of 2010) national ambient air quality standards, 1990–2010.

dioxide, particulate matter, carbon monoxide, and lead. Four of these pollutants, NO_2, SO_2, CO, and lead, result primarily from direct emissions from a variety of sources. Particulate matter results from direct emissions but is also commonly formed when emissions of nitrogen oxides, sulfur oxides, ammonia, organic compounds, and other gases react in the atmosphere. Ozone is not directly emitted but is formed when nitrogen oxides and volatile organic compounds react in the presence of sunlight.

2.2.1.1 Nitrogen dioxide

Nitrogen oxides (NO_x), the term used to describe the sum of NO, NO_2, and other oxides of nitrogen, contribute to the formation of ozone, particulate matter, haze, and acid rain. While EPA traces national emissions of NO_x, the national monitoring network measures ambient concentrations of NO_2 (nitrogen dioxide) for comparison to national air quality standards. The major sources of anthropogenic NO_x emissions are high-temperature combustion processes, such as those that occur in vehicles and power plants.

Over the period 1980–2012, monitored levels of NO_2 decreased 60%, from 1990 to 2012 decreased 46%, and from 2000 to 2012 decreased 29% [65]. All areas of the country that once violated the NAAQS for NO_2 now meet the standard.

Figures 2.14 [65], 2.15 [65], 2.16 [65], and 2.17 [66] illustrate the trends of NO_2 air quality and NO_x emissions by sources. Of the approximately 13 million short tons of NO_x emitted from all sources in 2013, stationary fuel combustion contributed approximately 3.7 million short tons of which 3.1 million tons were emitted from power plants and industrial sources (see Table 2.20) [66]. All recorded concentrations in 2012 were well below the level of the daily one-hour standard of 0.100 ppm (100 ppb).

Table 2.20 National emissions totals, 2013 (thousand short tons)[a]

Source category/Subcategory	CO	NO_x	PM_{10}	$PM_{2.5}$[b]	SO_2	VOC	NH_3
Stationary fuel combustion	4,632	3,686	976	830	4,245	630	105
Electric utilities	784	1,825	276	202	3,257	41	25
Industrial	927	1,301	271	207	763	109	13
Other	2,921	561	429	421	224	480	67
Industrial and other Processes	3,098	1,262	1,142	566	599	7,100	153
Chemical & allied product manufacturing	167	50	21	16	126	79	23
Metals processing	766	71	63	48	145	34	1
Petroleum & related industries	686	690	41	32	116	2,490	2
Other industrial processes	336	349	762	274	186	328	53
Solvent utilization	1	1	4	4	0	2,815	1
Storage and transport	26	19	51	20	9	1,222	6
Waste disposal & recycling	1,114	83	200	172	17	132	68
Transportation	39,441	7,736	464	369	107	4,147	115
Highway vehicles	24,796	5,010	268	185	29	2,161	112
Off-highway vehicles	14,645	2,725	196	184	78	1,986	3
Miscellaneous	26,262	435	18,277	4,492	219	5,867	3,935
Total (major source categories; wildfires Included in categories)	73,433	13,119	20,861	6,259	5,170	17,744	4,308
Wildfires	14,949	240	1,493	1,267	105	3,320	233
Total (major source categories; wildfires excluded)	58,939	12,924	19,123	4,992	5,065	14,423	4,075

[a]Categories and subcategories may not total due to rounding.
[b]With condensibles.

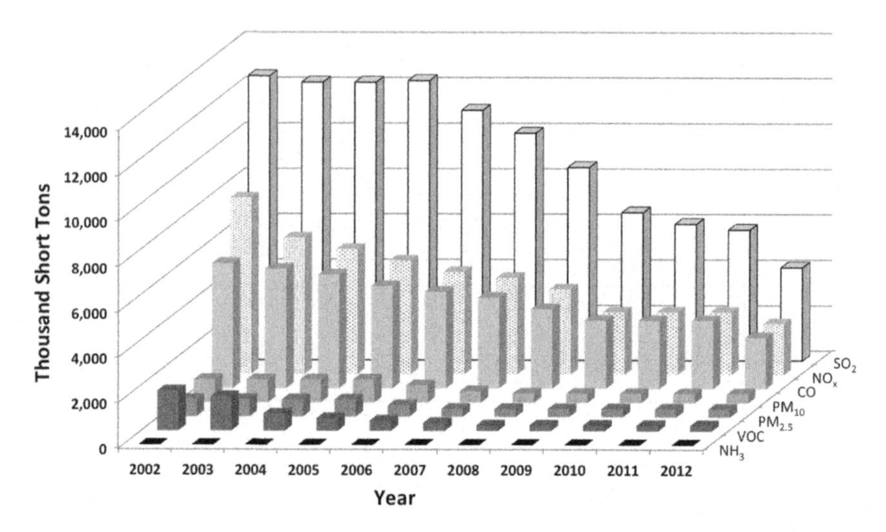

Figure 2.13 National air emissions, fuel combustion sector, 2002–2012.

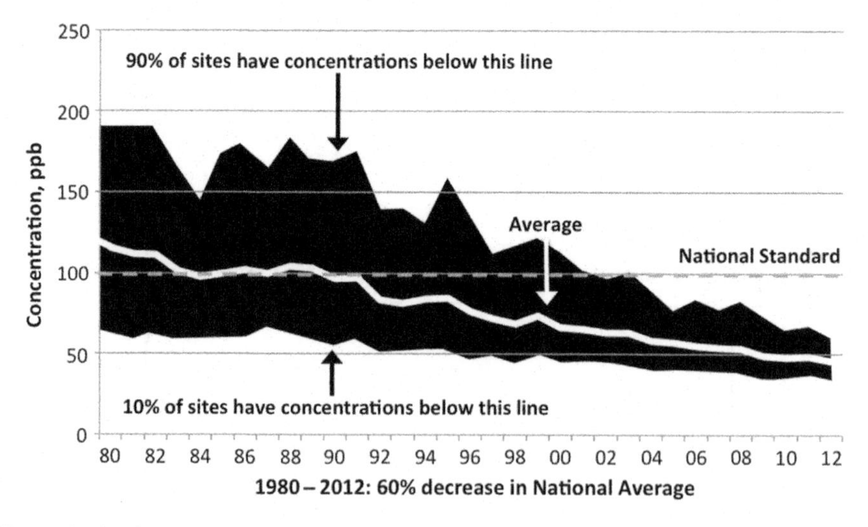

Figure 2.14 NO_2 air quality from 1980 to 2012 (Daily maximum 1-hour average).

2.2.1.2 Ozone

Ground-level ozone, which is the primary constituent of smog, continues to be a pollution problem throughout many areas of the United States. Ozone is not emitted directly into the air but is formed by the reaction of volatile organic compounds (VOCs) and NO_x in the presence of heat and sunlight. The trends of VOC emissions and their sources, for the period 1970–2013, are shown in Figure 2.18 [65]. Fuel combustion contributes less than 5% of the VOC emissions with power stations comprising less than one-fourth of the 5% [66].

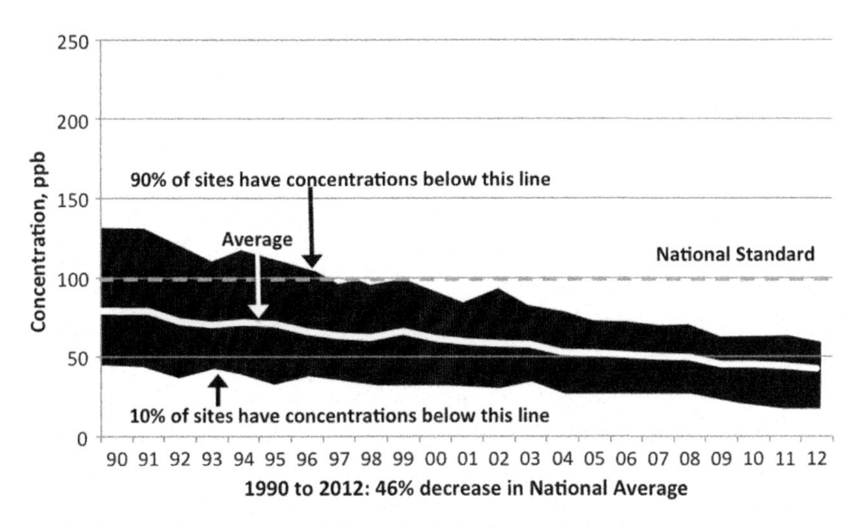

Figure 2.15 NO_2 air quality from 1990 to 2012 (Daily maximum 1-hour average).

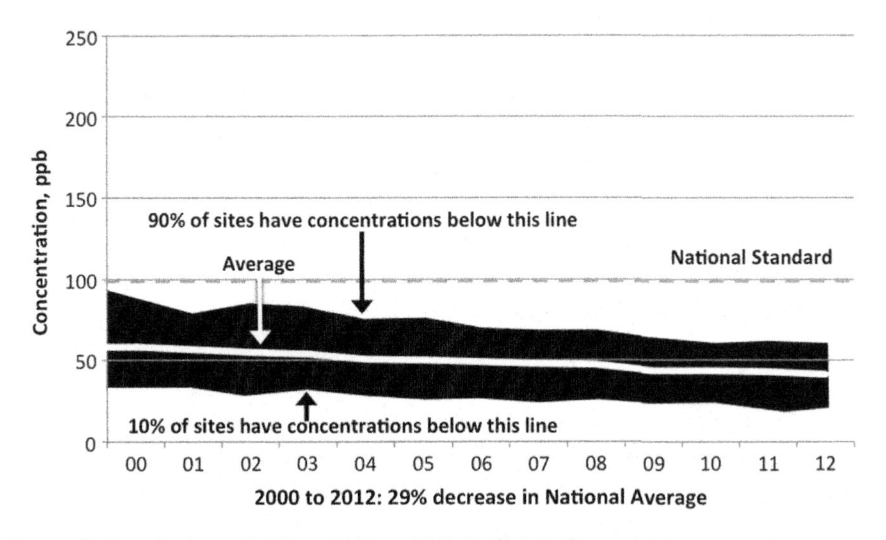

Figure 2.16 NO_2 air quality from 2000 to 2012 (Daily maximum 1-hour average).

Figures 2.19–2.21 show the trends of the 8-hour ozone standards. Over the period 1980–2012, monitored levels of ozone decreased 25%, from 1990 to 2012 decreased 14%, and from 2000 to 2012 decreased 9% [65]. Many areas measured concentrations above the national air quality standard for ozone (0.075 ppm).

2.2.1.3 Sulfur dioxide

Nationally, average sulfur dioxide (SO_2) ambient concentrations, based on the daily maximum 1-hour average, decreased 78% from 1980 to 2012 and 72% from 1993

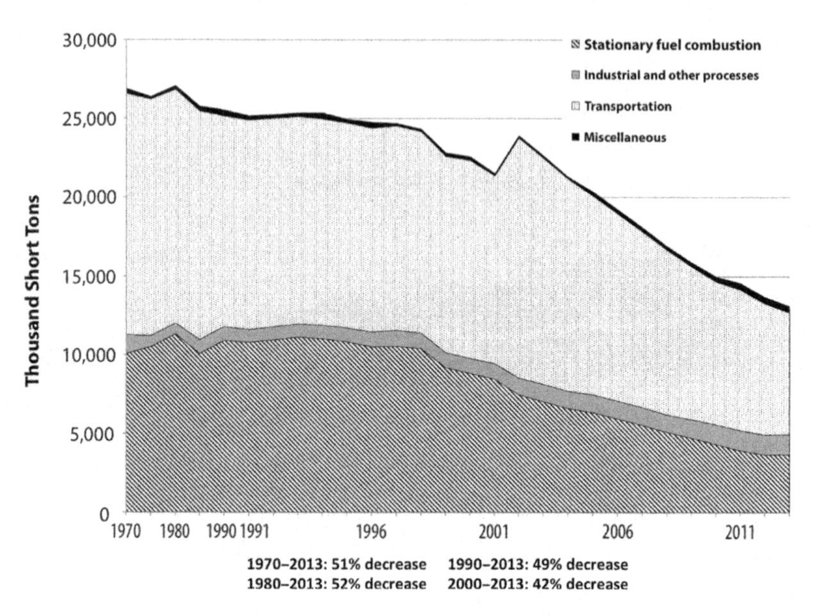

Figure 2.17 NO_x emissions from 1970 to 2013.

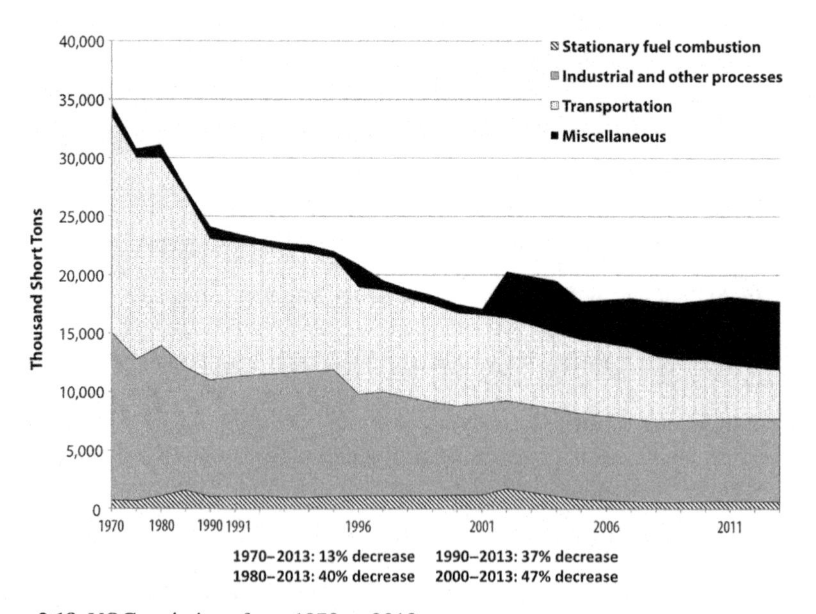

Figure 2.18 VOC emissions from 1970 to 2013.

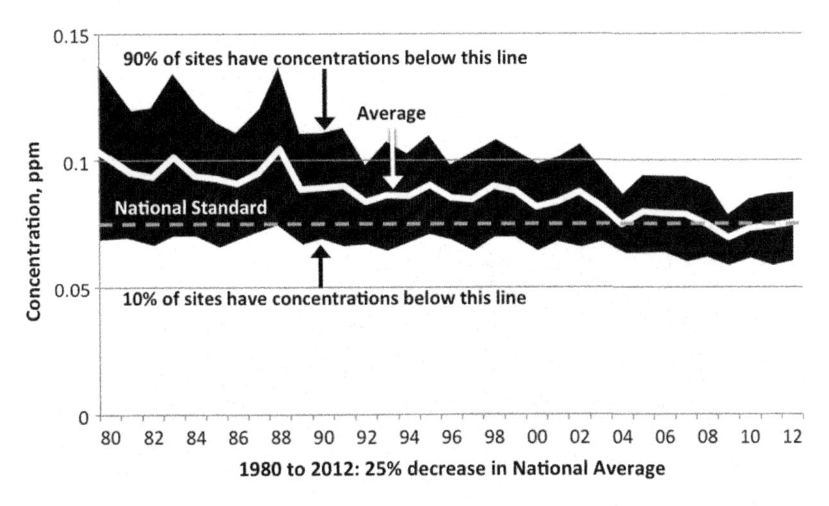

Figure 2.19 Ozone air quality from 1980 to 2012 (Daily maximum 8-hour average).

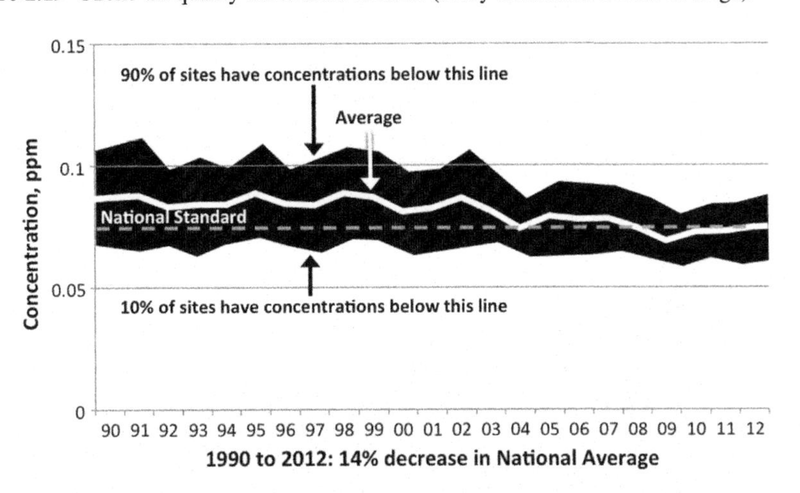

Figure 2.20 Ozone air quality from 1990 to 2012 (Daily maximum 8-hour average).

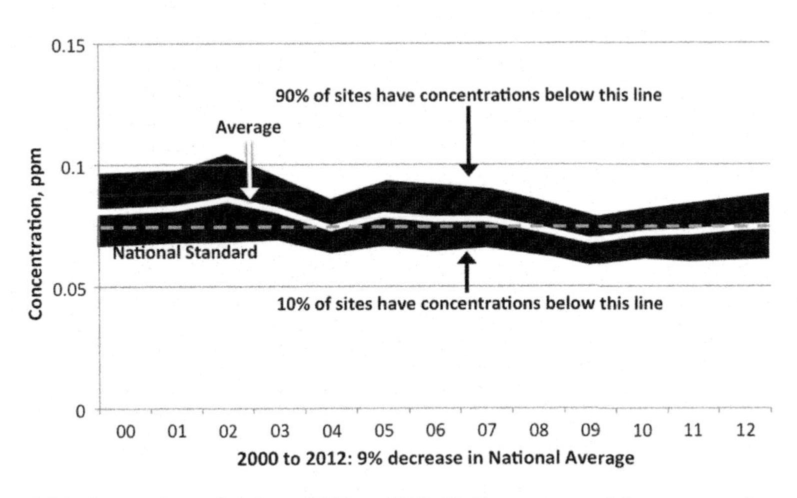

Figure 2.21 Ozone air quality from 2000 to 2012 (Daily maximum 8-hour average).

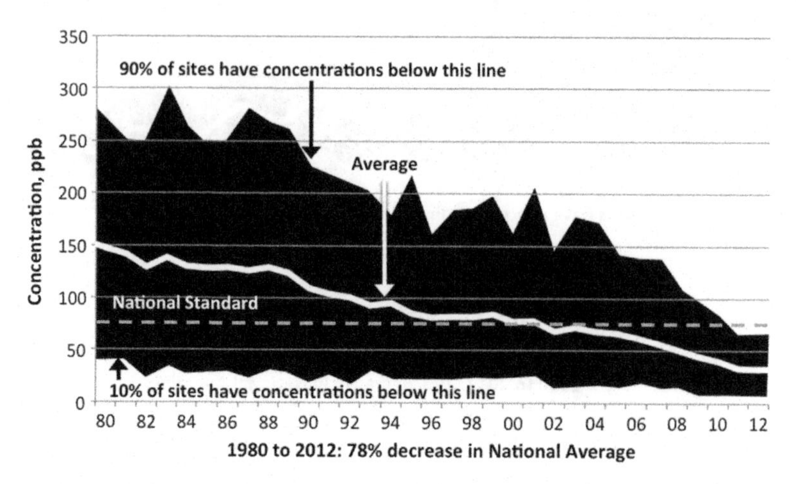

Figure 2.22 SO_2 air quality from 1980 to 2012 (Daily maximum 1-hour average).

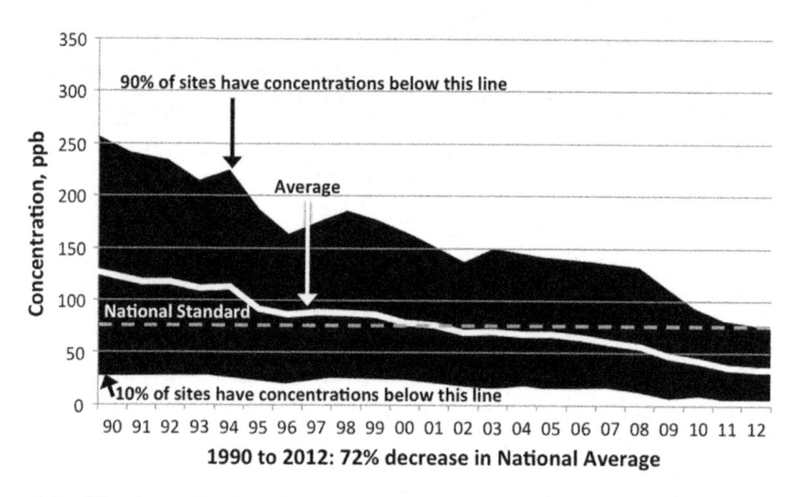

Figure 2.23 SO_2 air quality from 1990 to 2012 (Daily maximum 1-hour average).

to 2002, as shown in Figures 2.22 and 2.23 [65]. SO_2 emissions decreased 80% from 1980 to 2013 and 79% from 1990 to 2013. Reductions in SO_2 concentrations and emissions since 1990 are due to controls implemented under EPA's Acid Rain Program beginning in 1995. As shown in Figure 2.24 [66], stationary fuel combustion, mainly coal and oil, accounts for most of the total SO_2 emissions (see Table 2.20). Nationally, concentrations of SO_2 declined 54% between 2000 and 2012, as shown in Figure 2.25 [65]. The average concentrations in 2012 were below the daily maximum 1-hour average standard of 0.100 ppm (100 ppb).

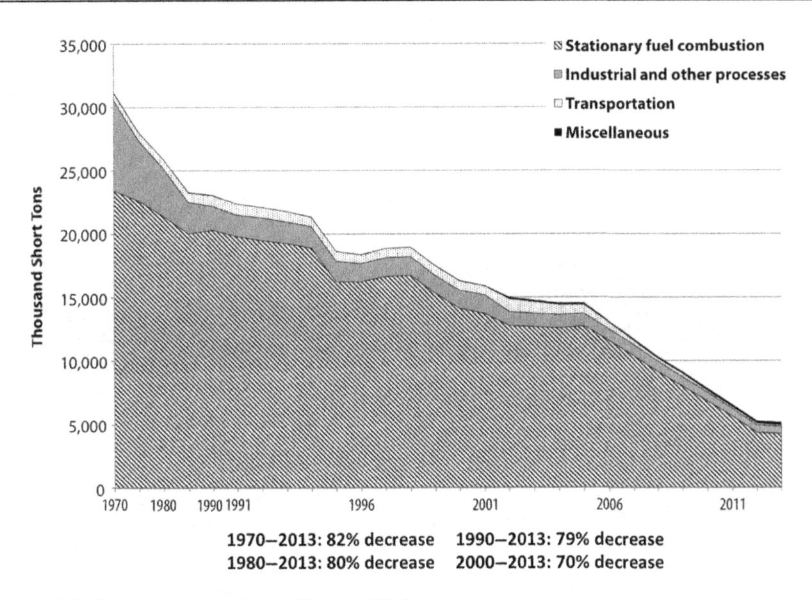

Figure 2.24 SO_2 emissions from 1970 to 2013.

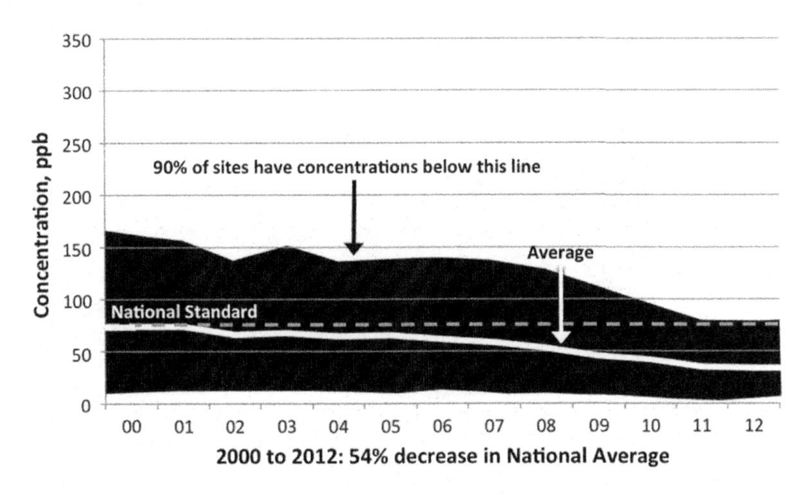

Figure 2.25 SO_2 air quality from 2000 to 2012 (Daily maximum 1-hour average).

2.2.1.4 Particulate matter

Between 1990 and 2012, PM_{10} concentrations decreased 39%, while PM_{10} emissions decreased 25% [65,66]. This can be seen in Figures 2.26 [65] and 2.27 [66], respectively. Nationally, PM_{10} concentrations declined by 27% between 2000 and 2012, as shown in Figure 2.28 [65]. Fuel combustion accounts for about half of total particulate emissions (see Figure 2.27), while electric utilities account for approximately 1% of the total particulate matter emitted [66].

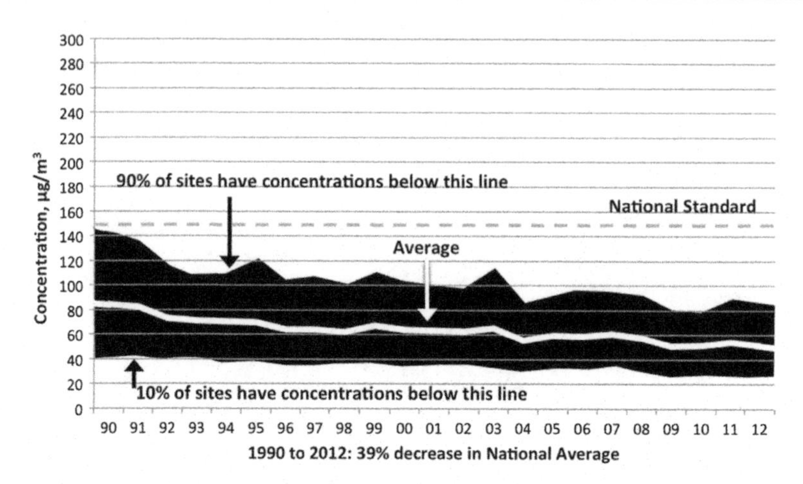

Figure 2.26 PM_{10} air quality from 1990 to 2012 (Maximum 24-hour standard).

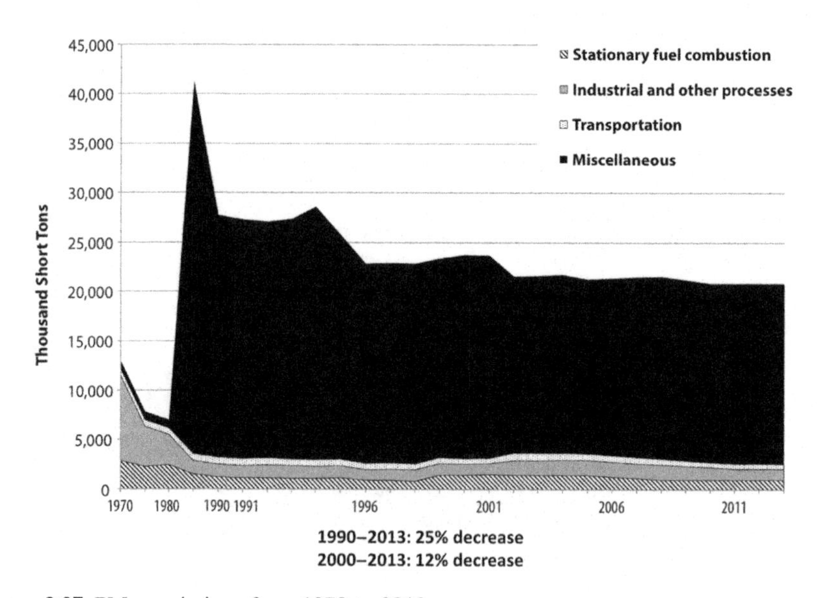

Figure 2.27 PM_{10} emissions from 1970 to 2013.

Between 2000 and 2012, $PM_{2.5}$ concentrations decreased 27% as shown in Figure 2.29 [65]. Figure 2.30 shows that direct $PM_{2.5}$ emissions from anthropogenic sources decreased 17% nationally between 2000 and 2012 [65]. Figure 2.30 tracks only directly emitted particles and does not account for secondary particles, which are primarily sulfates and nitrates, formed when emissions of NO_x, SO_2, ammonia, and other gases react in the atmosphere [66].

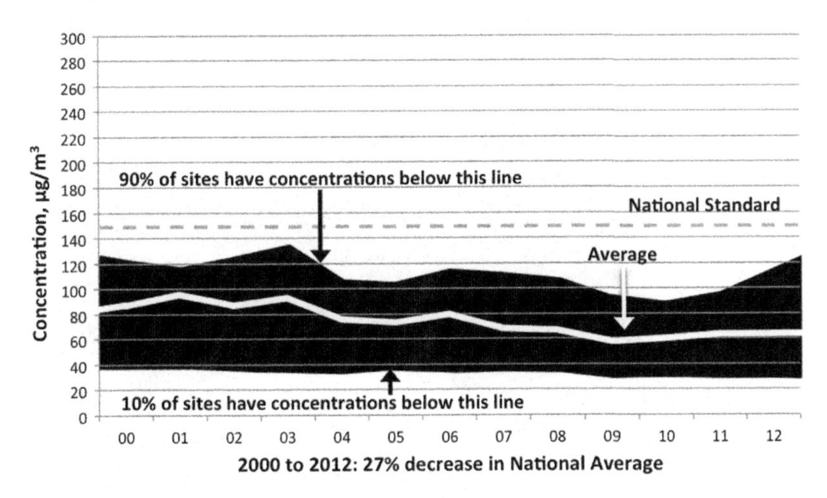

Figure 2.28 PM_{10} air quality trend from 2000 to 2012 (Maximum 24-hour standard).

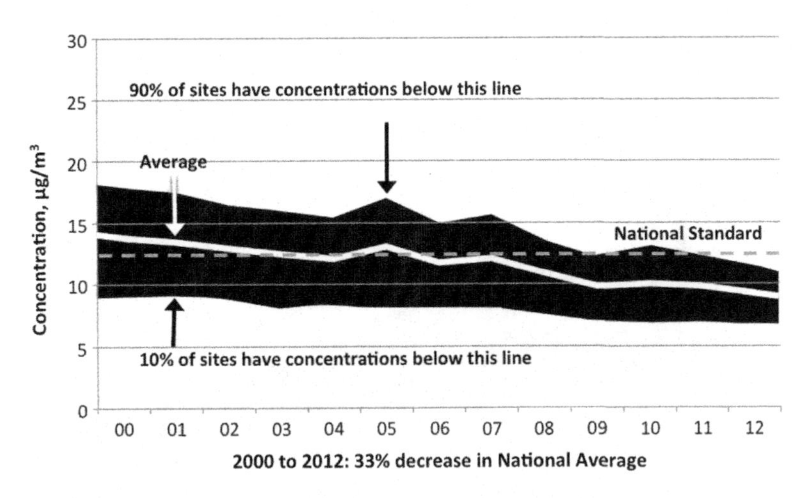

Figure 2.29 $PM_{2.5}$ annual air quality trend from 2000 to 2012 (annual average).

2.2.1.5 Carbon monoxide

Carbon monoxide (CO) is a component of motor vehicle exhaust, which contributes about 54% of all CO emissions nationwide. Other sources of CO emissions include industrial processes, nontransportation fuel combustion, and natural sources such as wildfires.

Nationally, the 2012 ambient average CO concentration was nearly 83% lower than that for 1980, which is illustrated in Figure 2.31 [65]. CO emissions decreased

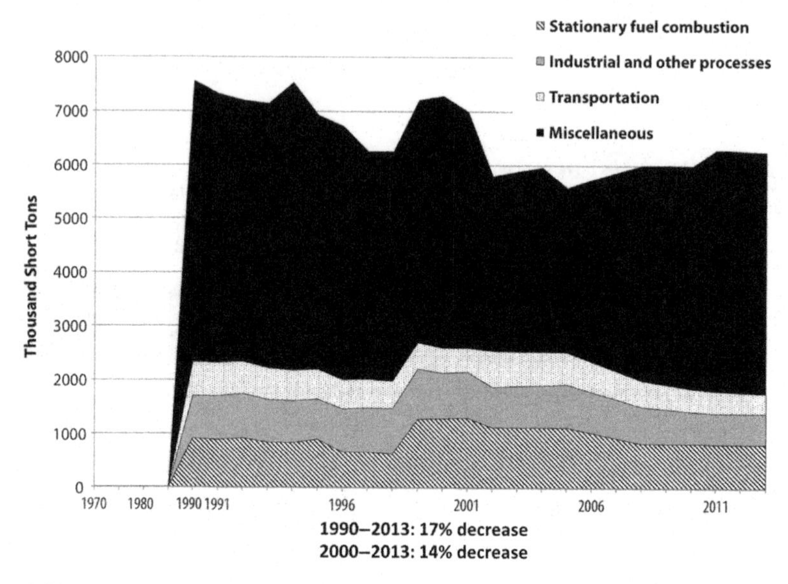

Figure 2.30 Direct $PM_{2.5}$ emissions from 1990 to 2013.

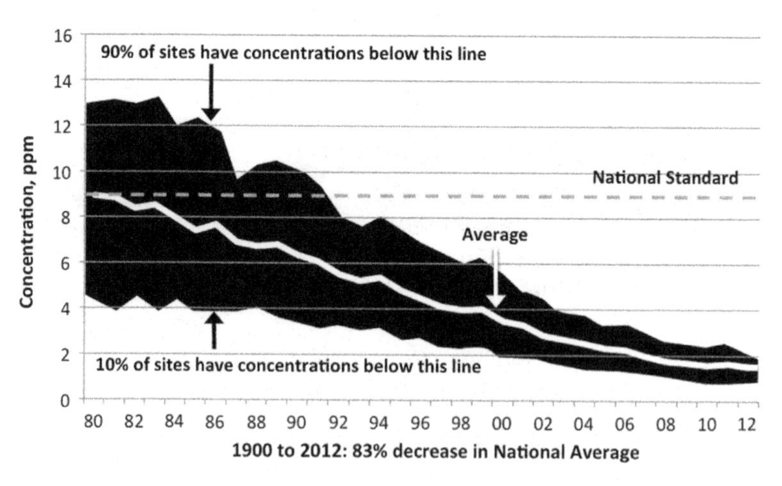

Figure 2.31 CO air quality from 1980 to 2012 (Maximum 8-hour average).

about 75% over the period 1990 to 2012, despite an approximately 92% increase in vehicle miles traveled, as shown in Figure 2.32 [65]. CO concentrations declined 57% between 2000 and 2012 as shown in Figure 2.33 [65].

Transportation sources are the largest contributors to CO emissions with fuel combustion accounting for about 8% of the CO emissions. Electric utilities account for less than 1% of the total CO emissions (see Table 2.20) [66]. The trend in CO emissions is shown in Figure 2.34 [66].

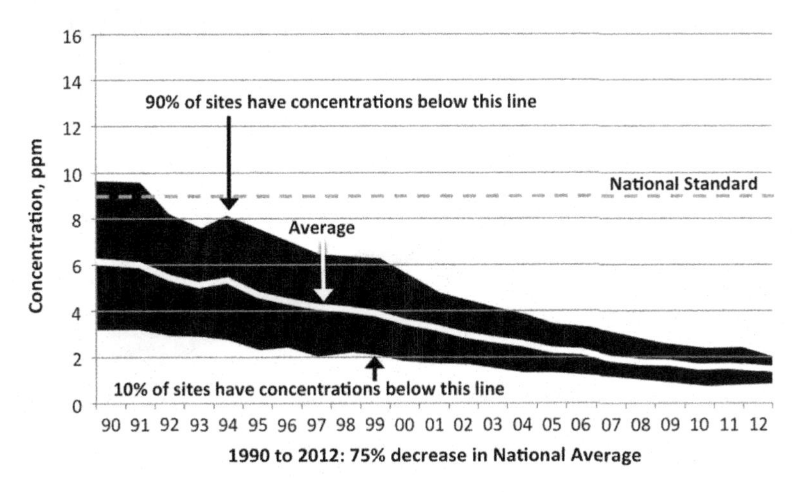

Figure 2.32 National CO air quality trend from 1990 to 2012 (Maximum 8-hour average).

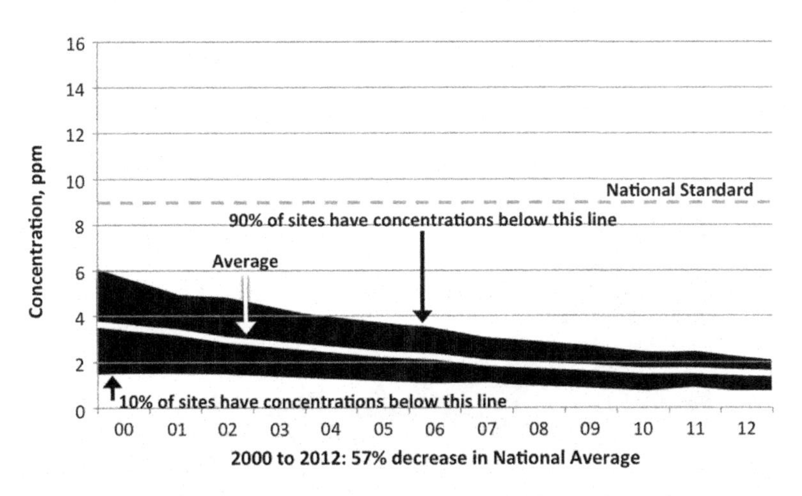

Figure 2.33 National CO air quality trend from 2000 to 2012 (Maximum 8-hour average).

2.2.1.6 Lead

In the past, automotive sources were the major contributor of lead (Pb) emissions to the atmosphere. The emissions of lead from the transportation sector have greatly declined over the last 20 years as leaded gasoline was phased out. Today, industrial processes, primarily metals processing, are the major sources of lead emissions to the atmosphere. As a result of the phase-out of leaded gasoline, lead concentrations and emissions have decreased significantly as shown in Figures 2.35 [65], 2.36 [65], 2.37 [65], and 2.38 [66]. Nationally, the 2012 ambient average lead concentration was 91%

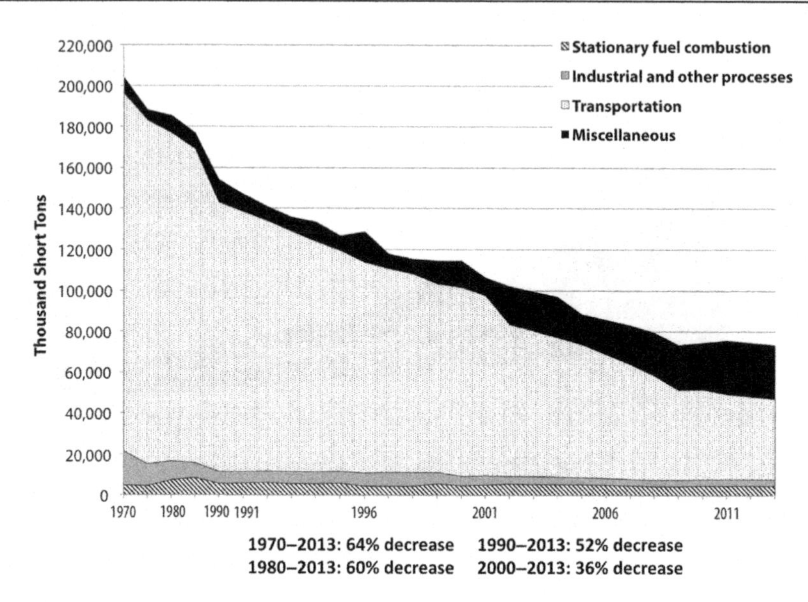

Figure 2.34 CO emissions from 1970 to 2013.

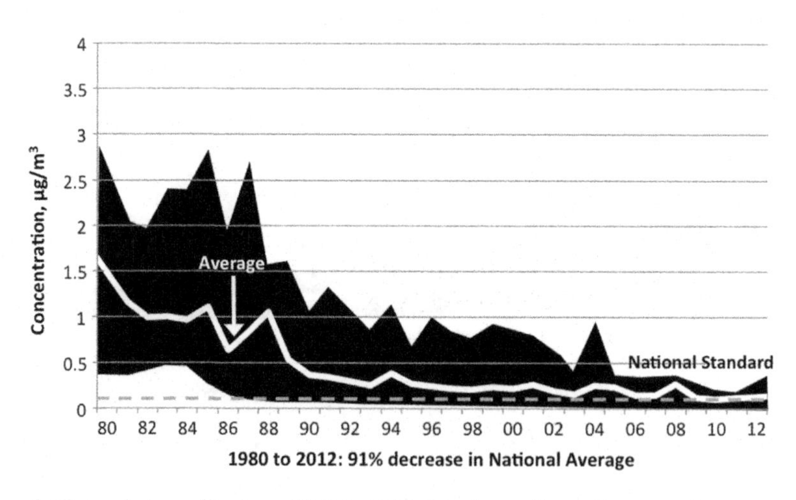

Figure 2.35 Lead air quality from 1980 to 2012 (Maximum 3-month average).

lower than that for 1980, which is illustrated in Figure 2.35. Lead emissions decreased about 87% over the period 1990 to 2012 as shown in Figure 2.36. Lead concentrations declined 52% between 2000 and 2012 as shown in Figure 2.37. Lead emissions from electric utilities are less than 10% of the total, i.e. less than 500 short tons [69], and the only violations of the lead NAAQS that occur today are near large industrial sources such as lead smelters and battery manufacturers [63].

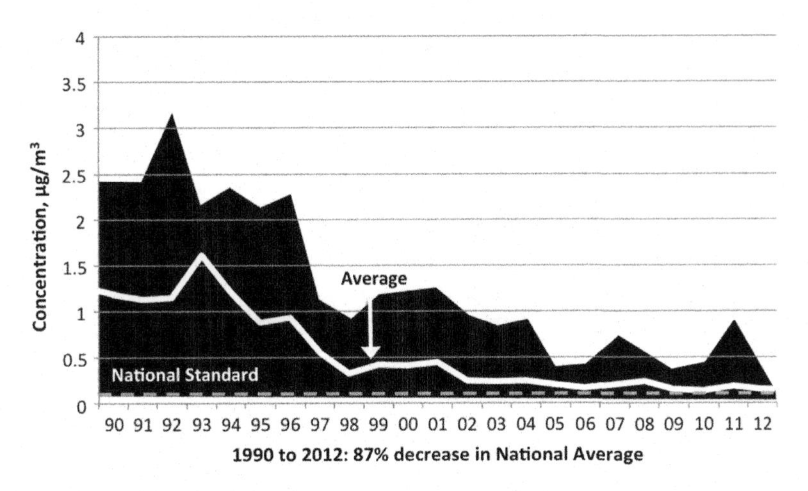

Figure 2.36 Lead air quality from 1990 to 2012 (Maximum 3-month average).

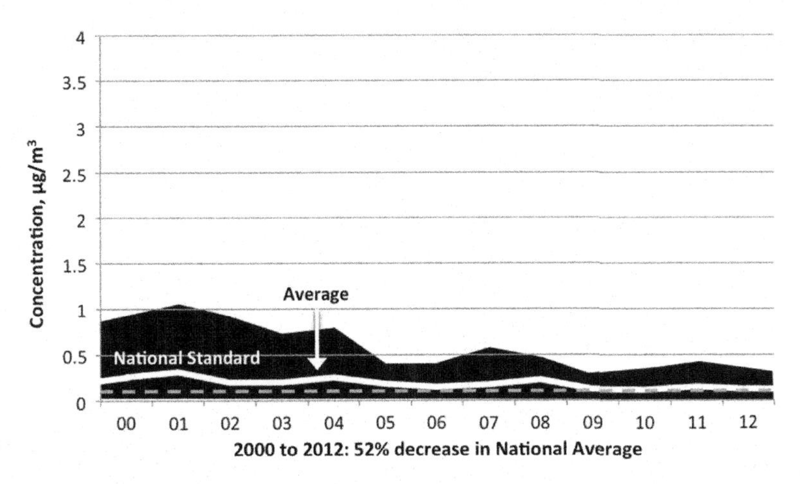

Figure 2.37 National lead air quality trend from 2000 to 2012.

2.2.2 *Acid rain*

Acid rain or acidic deposition occurs when emissions of sulfur dioxide and nitrogen oxides in the atmosphere react with water, oxygen, and oxidants to form acidic compounds. These compounds then fall to earth in either dry form (gas and particles) or wet form (rain, snow, and fog). In the United States, about 79% of annual SO_2 emissions and 24% of NO_x emissions are produced by electric utility plants that burn fossil fuels [66].

The EPA's acid rain program (addressing SO_2 and NO_x emissions), NO_x Budget Trading Program (addressing NO_x and ozone), and the Clean Air Interstate Rule (CAIR)/Cross State Air Pollution Rule (addressing SO_2, NO_x, fine particulate

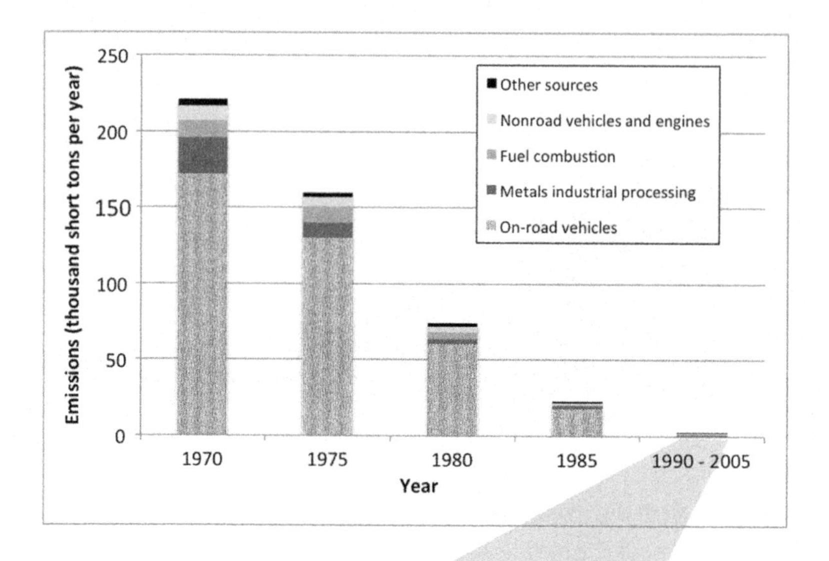

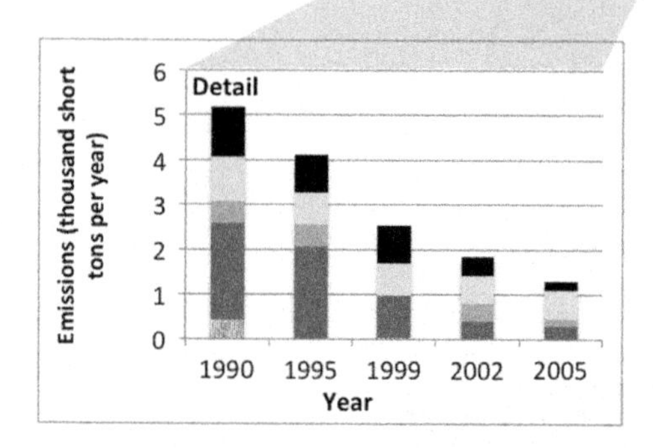

Figure 2.38 Lead emissions by source category.

matter, and ozone), all of which are discussed in detail earlier and generally apply to large electric generating units, have had significant impact in reducing acid rain. The results of these programs are summarized in Figures 2.6–2.8, and Table 2.15. Between 2005 and 2012, annual SO_2 and NO_x emissions from CAIR and acid rain program sources decreased 68 and 53%, respectively, from the 2005 level. These reductions occurred while electricity demand remained relatively stable [27]. In the case of the SO_2 emissions, this is in addition to major reductions in SO_2 emissions from the start of the acid rain program.

National Atmospheric Deposition Program/National Deposition Trends Network (NADP/NTN) monitoring data show significant improvement in the primary acid deposition indicators [70,71]. The sulfur indicator is assessed using the atmospheric deposition levels of sulfate (wet deposition) and sulfur (dry and total deposition). The nitrogen indicator is measured using levels of wet inorganic nitrogen deposition (combined deposition of inorganic nitrate and ammonium ions in wet deposition) and total

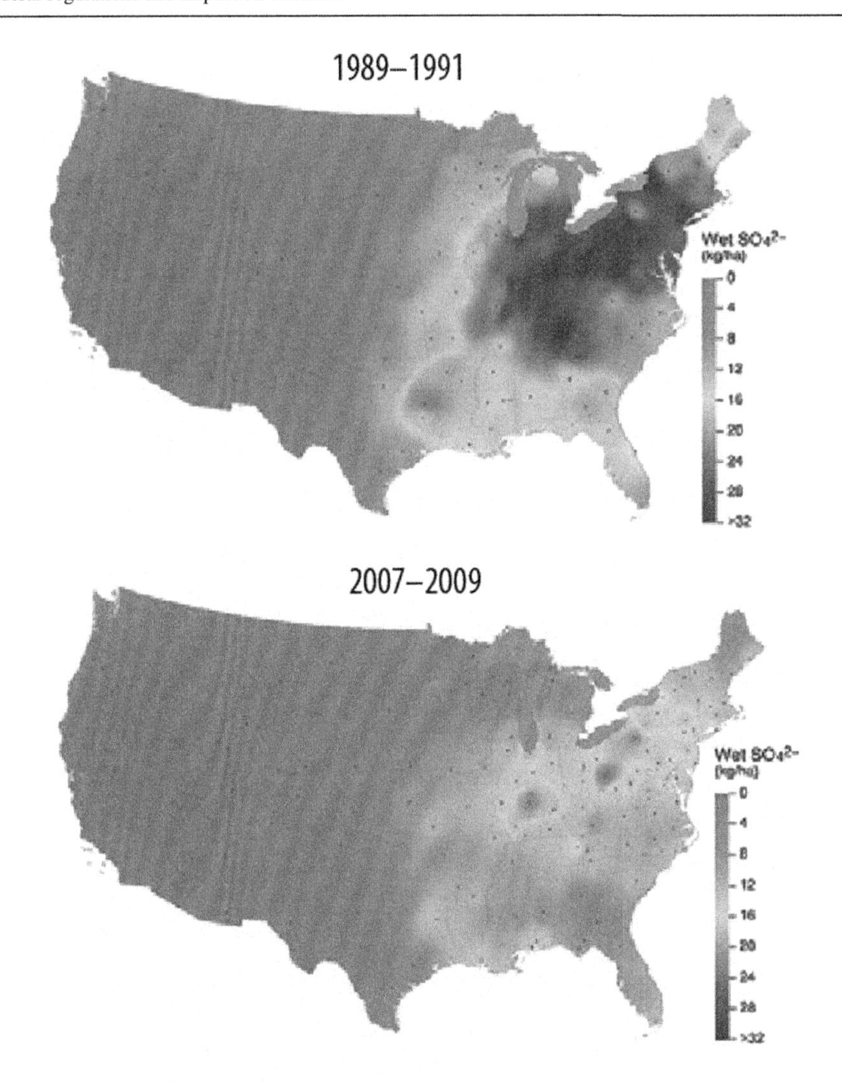

Figure 2.39 Annual mean wet sulfate deposition.

inorganic nitrogen deposition (combined deposition of inorganic nitrate and ammonium deposition in precipitation, dry particulate deposition, and gaseous nitric acid).

Between the 1989 to 1991 and 2007 to 2009 observation periods, decreases in wet sulfate deposition (sulfate that falls to the earth through rain, snow, and fog) has decreased since the implementation of the acid rain program in much of the Ohio River Valley and northeastern U.S. Some of the greatest reductions have occurred in the mid-Appalachian region, including Maryland, New York, most of Pennsylvania, West Virginia, and Virginia. Less dramatic reductions have been observed across much of New England, portions of the southern Appalachian Mountains, and some areas of the Midwest. Average decreases in wet deposition of sulfate averaged more than 43% for the eastern U.S. (see Figure 2.39) [70].

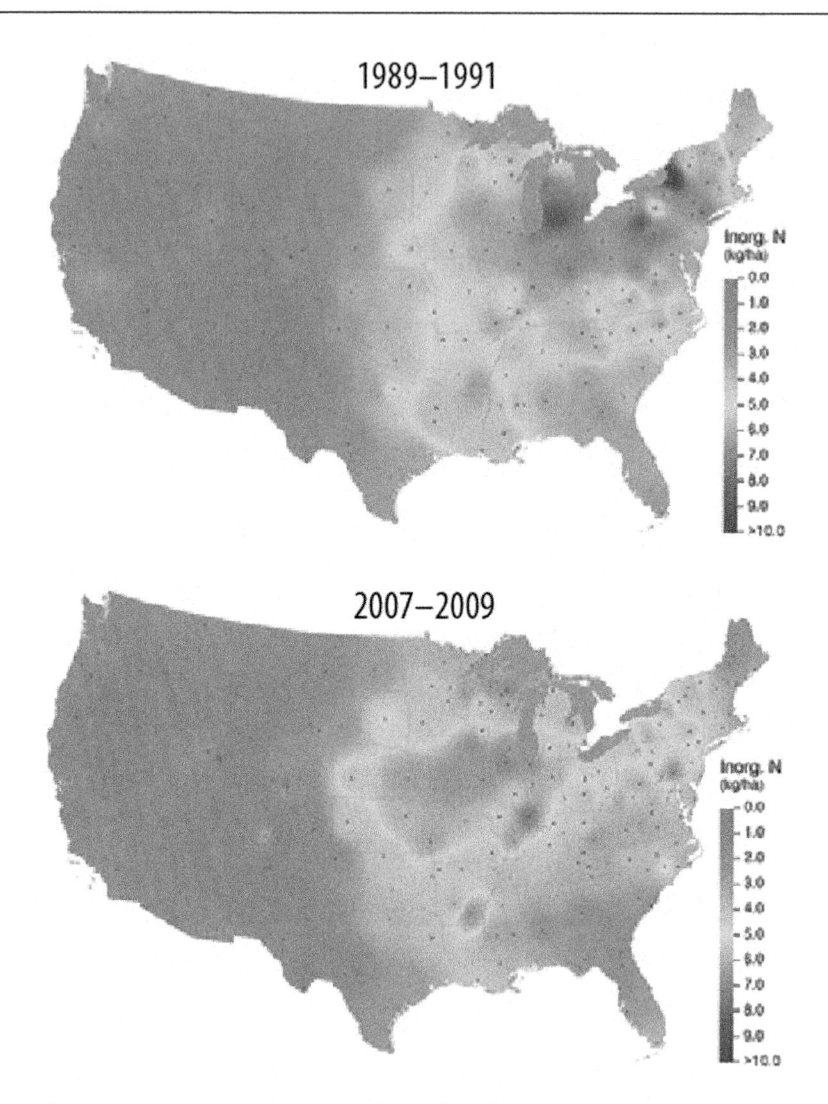

Figure 2.40 Annual mean wet inorganic nitrate deposition.

Along with wet sulfate deposition, wet sulfate concentrations have also decreased by similar percentages. A strong correlation between large-scale SO_2 emission reductions and large reductions in sulfate concentrations in precipitation has been noted in the Northeast, one of the areas most affected by acid deposition. The reduction in total sulfur deposition (wet plus dry) has been even more dramatic than that of wet deposition in the mid-Atlantic and Midwest, with reductions of 50 and 53%, respectively [71].

Nitrogen deposition has decreased since the 1990s (see Figure 2.40) but this has been less pronounced than that for sulfur. Nitrogen deposition shows greater spatial variation in temporal trends and is less certain than that of sulfur deposition for the following reasons [70]:

- Decline of NO_x emissions has been less than the decline in SO_2 emissions
- The proportion of nationwide NO_x emission that originates from electric generating units is much lower (about 20%) than the proportion of SO_2 emissions that originates from these units (about 70%)
- NO_x emissions originate from many nonacid-rain sources, including motor vehicles
- The contribution of ammonia emissions to overall nitrogen deposition originates largely from agricultural sources and is not regulated by the Clean Air Act
- A large number of chemical species of nitrogen contribute to total nitrogen deposition, but are not well measured by existing monitoring networks

Between the 1989 to 1991 and 2007 to 2009 observation periods, decreases in wet inorganic nitrogen (nitrate and ammonium) deposition levels in the eastern U.S. ranged from 16% in the Midwest to 27% in the mid-Atlantic region and Northeast. Decreases in dry and total inorganic nitrogen deposition generally have been greater than those of wet deposition, with a 23% and 31% decrease in total nitrogen deposition for the Midwest and the mid-Atlantic region, respectively.

2.2.3 Hazardous air pollutants/air toxics

Under the 1990 Clean Air Act Amendments, Section 112(n)(1)(A), EPA was mandated by Congress to perform a study of hazardous air pollutants (HAPs) to determine control strategies for utilities if appropriate and necessary [31]. EPA began work in 1991 to develop and collect the information and data to prepare this study. At that time, only a small amount of reliable data on HAP emissions from utilities was available. Emissions tests were not available for each utility in the United States so EPA used average annual emissions estimates in assessing long-term exposure to individual HAPs [31]. The study, Study of Hazardous Air Pollutant Emissions from Electric Utility Steam Generating Units, was published in 1988 and was used to report the estimated quantity of hazardous air pollutants being emitted from fossil fuel-fired power plants and the impact of pollution controls [31]. In this study, HAP emissions test data were gathered from 52 utility units (i.e. boilers), including a range of coal-, oil-, and natural gas-fired utility boilers. The emissions tests data, along with facility specific information (e.g. boiler type, control devices, fuel usage) were used to estimate HAP emissions from all 684 utility plants in the United States. These utilities are fueled primarily by coal (59%), oil (12%), or natural gas (29%). Many plants have two or more units and several plants burn more than one type of fuel (e.g. contain both coal- and oil-fired units). In 1990, there were 426 plants that burned coal as one of their fuels, 137 plants that burned oil, and 267 plants that burned natural gas. The overall summary of the study is presented in Table 2.21, which lists nationwide utility emissions estimates for 13 priority HAPs [31]. Table 2.22 contains estimated emissions for nine priority HAPs from characteristic utility units.

Table 2.21 **Nationwide utility emissions for 13 priority HAPs**[a]

HAP	Nationwide HAP emission estimates (short tons per year)[b]								
	Coal			Oil			Natural gas		
	1990	1994	2010	1990	1994	2010	1990	1994	2010
Arsenic	61	56	71	5	4	3	0.15	0.18	0.25
Beryllium	7.1	7.9	8.2	0.46	0.4	0.23	NM[c]	NM	NM
Cadmium	3.3	3.2	3.8	1.7	1.1	0.9	–	–	–
Chromium	73	62	87	4.7	3.9	2.4	–	–	–
Lead	75	62	87	11	8.9	5.4	0.43	0.47	0.68
Manganese	164	168	219	9.3	7.3	4.7	–	–	–
Mercury	46	51	60	0.25	0.2	0.13	0.0015	0.0017	0.024
Nickel	58	52	69	390	320	200	2.2	2.4	3.5
HCl	143,000	134,000	155,000	2,900	2,100	1,500	NM	NM	NM
HF	20,000	23,000	26,000	140	280	73	NM	NM	NM
Acrolein	25	27	34	NM	NM	NM	NM	NM	NM
Dioxins[d]	0.000097	0.00012	0.00020	0.00001	0.000009	0.000003	NM	NM	NM
Formaldehyde	35	29	155,000	19	9.5	9.5	36	39	57

[a]Radionuclides are the one priority HAP not included on this table because radionuclide emissions are measured in different units (i.e. curies per year) and, therefore, would not provide a relevant comparison to the other HAPs shown.
[b]The emissions estimates in this table are derived from model projections based on a limited sample of specific boiler types and control scenarios. Therefore, there are uncertainties in these numbers.
[c]NM = Not measured.
[d]These emissions estimates were calculated using the toxic equivalency (TEQ) approach, which is based on the summation of the emissions of each congener after adjusting for toxicity relative to 2,3,7,8-tetrachlorodibenzo-p-dioxin (i.e. 2,3,7,8-TCDD).

Table 2.22 Estimated emissions for 9 priority HAPs from characteristic utility units (1994; short tons per year)[a]

HAP	Fuel (unit size; MWe)		
	Coal (325)	Oil (160)	Natural gas (240)
Arsenic	0.0050	0.0062	0.0003
Cadmium	0.0023	0.0014	NC[b]
Chromium	0.11	0.0062	NC
Lead	0.021	0.014	NC
Mercury	0.05	0.0012	NC
Hydrogen chloride	190	9.4	NC
Hydrogen fluoride	14	NC	NC
Dioxins[c]	0.00000013	0.00000023	NC
Nickel	NC	1.7	0.004

[a]There are uncertainties in these numbers. Based on an uncertainty analysis, the EPA predicts that the emissions estimates are generally within a factor of roughly three of actual emissions.
[b]Not calculated.
[c]These emissions estimates were calculated using the toxic equivalency (TEQ) approach, which is based on the summation of the emissions of each congener after adjusting for toxicity relative to 2,3,7,8-tetrachlorodibenzo-p-dioxin (i.e., 2,3,7,8-TCDD).

In summary, the Utility Hazardous Air Pollutant Report to Congress analyzed 66 other air pollutants (other than mercury, which is discussed in the next section) from 684 power plants that are 25 MW or larger and burning coal, oil, or natural gas [31]. The report noted potential health concerns about utility emissions of dioxin, arsenic, hydrogen chloride, hydrogen fluoride, and nickel, although uncertainties exist about the health data and emissions for these pollutants.

Since 1998, EPA has been collecting more data from sources (i.e. through the National Emissions Inventory) and, with assistance from state agencies, ambient monitors. These data collection efforts have resulted in more reliable data for use in legislating emissions levels and controls.

EPA relies on modeling studies to supplement air toxic monitoring data and to better define trends in toxic air pollutants [68]. One such study, the National-Scale Air Toxic Assessment (NATA), is a nationwide study of ambient levels, inhalation exposures, and health risks associated with emissions of 177 toxic air pollutants plus diesel particulate. Figure 2.41 shows the trends from 2003 to 2010 in ambient air monitoring levels for some of the important air toxic pollutants [68]. When the median % change per year (marked by an x for each pollutant shown) is below zero, the majority of sites in the U.S. show a decrease in concentrations.

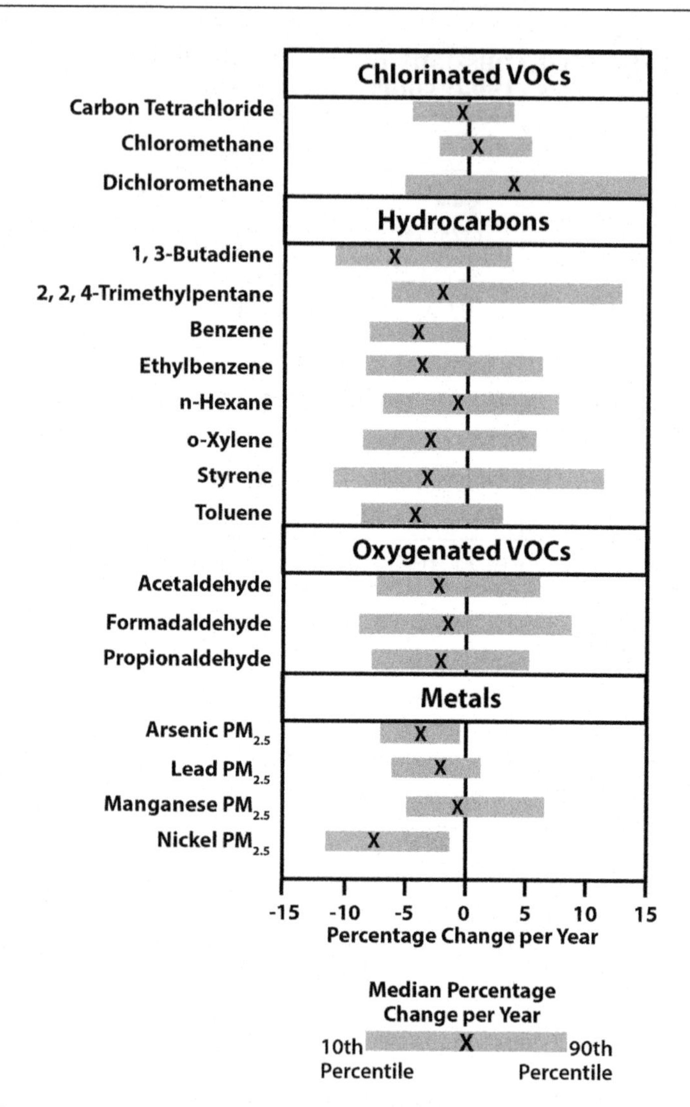

Figure 2.41 Distribution of changes in ambient concentrations at U.S. toxic air pollutant monitoring sites, 2003–2010 (% change in annual average concentrations).

EPA has provided a summary of selected HAPs from stationary and mobile sources from the most recent review and analysis of the National Emissions Inventory (NEI), which was published in 2013 for 2008 NEI data [67]. Table 2.23 lists the various stationary emissions sources, including several fuel combustion sources, and compares the stationary emissions totals with the total mobile sources

Table 2.23 **Selected HAP emissions totals (in short tons) for various stationary sources and total mobile sources**[a]

Pollutant	Agr.	Dust-road const.	Fire-Ag. field burning	FC-comm./ inst.	FC-elec. gen.	FC-ind. boilers	FC-resid.	Ind. proc.	Misc.	Solvent	Total stationary	Total mobile
Ethylbenzene	–[b]	0	10	9	112	81	0	1,228	3,195	6,016	10,650	84,208
Acetaldehyde	–	0	612	46	412	2,472	9,744	4,890	3,336	1,025	22,537	49,697
Acrolein	–	–	3,957	47	308	2,114	1,065	361	920	110	8,881	4,998
Formaldehyde	–	0	414	600	1,565	12,471	17,987	4,856	1,550	374	39,818	89,191
Tetrachlorethylene	–	–	–	2	24	18	0	255	246	5,317	5,861	
1,4-Dichloroben.[c]	–	–	–	0	1	1	0	37	17	1,115	1,170	
1,3-Butadiene	–	–	170	2	4	198	2,778	673	58	1	3,883	21,333
Chromium compounds	15	35	0	4	209	39	3	201	1	12	520	33
Lead	–	0	1	7	59	48	5	248	12	5	386	578
Arsenic	0	0	0	2	65	18	2	32	0	1	121	28
Total Selected HAPs	15	35	5,164	719	2,759	17,460	31,584	12,781	9,335	13,976	93,827	250,066

[a]Excluding wildfires and prescribed fires; Categories are:
Agriculture.
Fire-agriculture field burning.
Fuel combustion – commercial/institutional.
Fuel combustion – electric generation.
Fuel combustion – industrial boilers.
Fuel combustion – residential.
Industrial processes.
Miscellaneous.
[b]Not determined.
[c]1,4-Dichlorobenzene.

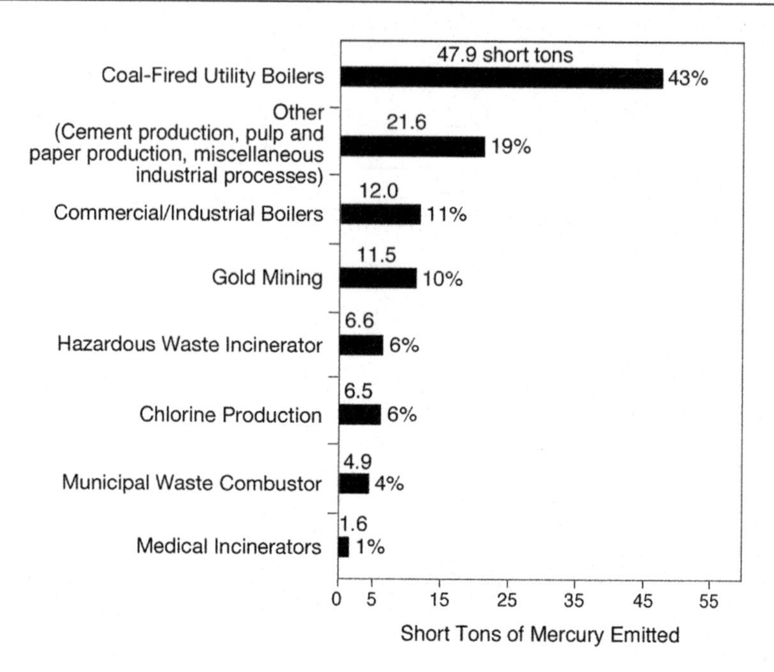

Figure 2.42 Estimated annual mercury emissions in the U.S., 1999.

emissions [modified from 67]. Specifically, the fuel combustion categories have high proportions of total HAP emissions from formaldehyde and acetaldehyde. Interestingly, total mobile sources exhibited significantly higher emissions levels than total stationary sources for ethylbenzene, acetaldehyde, formaldehyde, 1,3-butadiene, and lead. For the emissions reported, total mobile sources had lower acrolein, chromium compounds, and arsenic levels that total stationary sources; however, the acrolein emissions level for total fuel combustion categories was lower than the total mobile sources emissions levels.

2.2.4 Mercury

Prior to 1999, anthropogenic mercury emissions were unknown. EPA conducted an Information Collection Request (ICR) in 1999 to assess mercury emissions from coal-fired power plants (as discussed previously). From the ICR, EPA's best estimate of annual anthropogenic United States emissions of mercury in 1999 was about 113 short tons [72]. Approximately two-thirds of these emissions are from combustion sources, including waste and fossil fuel combustion. This is illustrated in Figure 2.42 [72]. Contemporary anthropogenic emissions are only one part of the

Table 2.24 Summary of 1999, 2005, and 2008 mercury emissions (short tons)[a]

Sector	Year		
	1999	2005	2008
Utility coal boilers	47.9	52.2	29.5
Electric arc furnaces	–[b]	7	4.7
Portland cement nonhazardous waste	–	7.5	4.2
Industrial-commercial-institutional boilers and process heaters	12.0	6.4	4.5
Chlor-alkali plants	6.5	3.1	1.3
Municipal waste combustors	4.9	2.3	1.3
Gold mining	11.5	2.5	1.7
Mobile sources	–	1.2	1.7
Other categories	21.6[c]	18	10.3
Hazardous waste	6.6	3.2	1.3
Commercial/industrial solid waste	–	1.1	0.02
Sewage sludge	–	0.3	0.45
Hospital/medical/infectious waste	1.6	0.2	0.1
Total (all categories)	112.6	105	61

[a]Data for 2005 and 2008 are from 2008 NEI data review; data for 1999 was previous EPA estimate.
[b]1999 categories were not identical to 2005 and 2008 categories.
[c]In 1999 data, other categories included cement production, pulp and paper production, and miscellaneous industrial processes.

mercury cycle. Releases from human activities today are adding to the mercury reservoirs that already exist in land, water, and air, both naturally and as a result of previous human activities. One estimate of the total annual global input to the atmosphere from all sources including natural, anthropogenic, and oceanic emissions is 1230 to 2890 metric tons [73,74]. United States sources are estimated to have contributed about 3% of the 1999 total. Mercury emissions from U.S. coal-fired boilers are estimated to be 48 short tons per year.

In more recent estimates (2010), annual global mercury emissions emitted worldwide are in the range of 5,000 to 8,000 metric tons per year [75]. This estimate includes mercury that is re-emitted. Total anthropogenic emissions of mercury to the atmosphere in 2010 are estimated at 1,960 metric tons with coal burning emitting approximately 475 metric tons [76]. Estimates of U.S. mercury emissions, in 2008, are 61 short tons, of which coal-fired utility boilers and industrial-commercial-institutional boilers and process heaters are estimated to emit 29.5 and 4.5 short tons per year, respectively, as shown in Table 2.24 [68,72]. Table 2.24 contains the 1999 mercury emissions estimate along with more recent data from EPA's 2008 NEI analysis, which compares 2005 and 2008 data. From Table 2.24, reductions in coal-fired boiler mercury emissions have not been as significant as reductions in several other categories, which is one reason mercury legislation (i.e., Utility Boiler MACT and Industrial Boiler Major Source and Area Source Rules) has been recently passed (as discussed earlier in the chapter). The 29.5 short tons of mercury emitted from utility coal boilers in 2008 were from units greater than 25 MW in size.

2.2.5 Greenhouse gases/carbon dioxide

The Earth naturally absorbs and reflects incoming solar radiation and emits longer wavelength terrestrial (thermal) radiation back into space [77]. On average, the absorbed solar radiation is balanced by the outgoing terrestrial radiation emitted to space. A portion of this terrestrial radiation though, is absorbed by gases in the atmosphere. These gases, known as greenhouse gases (GHG), have molecules that have the right size and shape to absorb and retain heat. These gases include water vapor (H_2O), carbon dioxide (CO_2), methane (CH_4), nitrous oxide (N_2O), and to a lesser extent halocarbons consisting of hydrofluorocarbons (HFCs), hydrochlorofluorocarbons (HCFCs), perfluorocarbons (PFCs), and sulfur hexafluoride (SF_6). The energy from this absorbed terrestrial radiation warms the Earth's surface and atmosphere, creating what is known as the natural greenhouse effect, which makes the Earth habitable.

Although the Earth's atmosphere consists of mainly oxygen and nitrogen, i.e. over 99% of the dry atmosphere, neither plays a significant role in enhancing the greenhouse effect because both are essentially transparent to terrestrial radiation [77]. The greenhouse gases comprise the remaining approximately 1% of the atmosphere, of which over 97% is water vapor. Methane, carbon dioxide, nitrous oxide, and other gases comprise the balance of the greenhouse gases. The composition of global greenhouse gas emissions in 2012 includes [78]:

- Carbon dioxide (CO_2) – 82.5%
- Methane (CH_4) – 8.7%
- Nitrous oxide (N_2O) – 6.3%
- High global warming potential gases (HFCs, PFCs, HCFCs, and SF6) – 2.5%.

Carbon dioxide, methane, and nitrous oxide are continuously emitted to and removed from the atmosphere by natural processes on Earth. Anthropogenic activities, however, can cause additional quantities of these and other greenhouse gases to be emitted or sequestered, thereby changing their global average atmospheric concentrations. Natural activities such as respiration by plants or animals and seasonal cycles of plant growth and decay generally do not alter average atmospheric greenhouse gas concentrations over decadal timeframes [77]. Climatic changes, which are long-term fluctuations in temperature, precipitation, wind, and other elements of the Earth's climate system, resulting from anthropogenic activities can have positive or negative feedback effects on these natural systems.

Overall, the most abundant and dominant greenhouse gas in the atmosphere is water vapor. Water vapor varies in concentration in the atmosphere both spatially and temporally, and can range from 0 to 4% of the total atmosphere. The highest concentrations of water vapor are found near the equator over the oceans and tropical rain forests. Water vapor concentrations can approach 0% near cold polar regions and subtropical continental deserts. Human activities, however, are not believed to directly affect the average global concentration of water vapor, although this currently is being debated [77].

The fifth most abundant gas in the atmosphere is carbon dioxide at 0.038% (note that argon is the fourth at 0.93% but argon is not a greenhouse gas). In nature, carbon dioxide is cycled between various atmospheric, oceanic, land biotic, marine biotic, and mineral reservoirs. Of all the greenhouse gases, human activity has the largest influence on carbon dioxide, which is a product from the combustion of fossil fuels, the cement industry, and land use change and forestry.

Methane is primarily produced through anaerobic decomposition of organic matter in biological systems. Agricultural processes such as wetland rice cultivation, enteric fermentation in animals, and the decomposition of animal wastes emit methane, as does the decomposition of municipal solid wastes [77]. Methane is also emitted during the production and distribution of natural gas and petroleum. Methane is released as a byproduct of coal mining and, to a lesser extent, incomplete fossil fuel combustion.

Anthropogenic sources of nitrous oxide include agricultural soils, especially through the use of fertilizers, fossil fuel combustion, especially from mobile sources, nylon and nitric acid production, wastewater treatment, waste combustion, and biomass burning [64]. Halocarbons that contain chlorine (e.g. chlorofluorocarbons, hydrofluorocarbons, methyl chloroform, and carbon tetrachloride) and bromine (e.g. halons, methyl bromide, and hydrobromofluorocarbons), perfluorocarbons, and sulfur hexafluoride (SF_6) are manmade chemicals and not products of combustion. They are, however, powerful greenhouse gases.

A concept, global warming potentials (GWPs), has been devised to evaluate the relative effects of emissions over a given time period in the future [79]. GWPs take account of the differing times that gases remain in the atmosphere, their greenhouse effect while in the atmosphere, and the time period over which climatic changes are of concern. GWPs are intended as a quantified measure of the globally averaged relative radiative forcing impacts of a particular greenhouse gas [77,80,81]. It is defined as the cumulative radiative forcing – both direct and indirect effects – integrated over a period of time from the emission of a unit mass of gas relative to some reference gas. Table 2.25 summarizes selected greenhouse gases, their major anthropogenic sources, and their GWPs [77]. EPA uses the GWPs from the 1995 IPCC (Intergovernmental Panel on Climate Change) Second Assessment Report (SAR) [77]. Table 2.25 contains the 1995 SAR GWPs, the 2007 IPCC 4th Assessment Report [81], and the recent 2013 IPCC 5th Assessment Report GWPs [80]. Carbon dioxide has been chosen as the reference

Table 2.25 **Selected greenhouse gases, major anthropogenic sources, and global warming potentials (100-year)**

Industrial designation, common name or chemical name	Chemical formula	Major anthropogenic source	Global warming potential		
			1995 IPCC second assessment report	2007 IPCC 4th assessment report	2013 IPCC 5th assessment report
Carbon dioxide	CO_2	Fossil fuel combustion, iron and steel production, cement manufacture	1	1	1
Methane	CH_4	Landfills, enteric fermentation, natural gas systems, coal mining, manure management	21	25	28
Nitrous oxide	N_2O	Agriculture soil management, mobile sources, nitric and adipic acid production, manure management, stationary sources, human sewage	310	298	265
HFC-23	CHF_3	Substitution of ozone depleting substances, semiconductor manufacture, mobile air conditioners, HCFC (hydrochlorofluorocarbons)-22 production	11,700	14,800	12,400
HFC-125	CHF_2CF_3		2,800	3,500	3,170
HFC-134a	CH_2FCF_3		1,300	1,430	1,300
HFC-143a	CH_3CF_3		3,800	4,470	4,800
HFC-152a	CH_3CHF_2		140	124	138
HFC-227EA	CF_3CHFCF_3		2,900	3,220	3,350
HFC-236fa	$CF_3CH_2CF_3$		6,300	9,810	8,060
HFC-43-10mee	$CF_3CHFCHFCF_2CF_3$		1,300	1,640	1,650
PFC-14	CF_4	Substitution of ozone depleting substances, semiconductor manufacture, aluminum production	6,500	7,390	6,630
PFC-116	C_2F_6		9,200	12,200	11,100
PFC-31-10	C_4F_{10}		7,000	8,860	9,200
PFC-51-14	C_6F_{14}		7,400	9,300	7,910
Sulfur hexafluoride	SF_6	Electrical transmission and distribution systems, magnesium casting	23,900	22,800	23,500

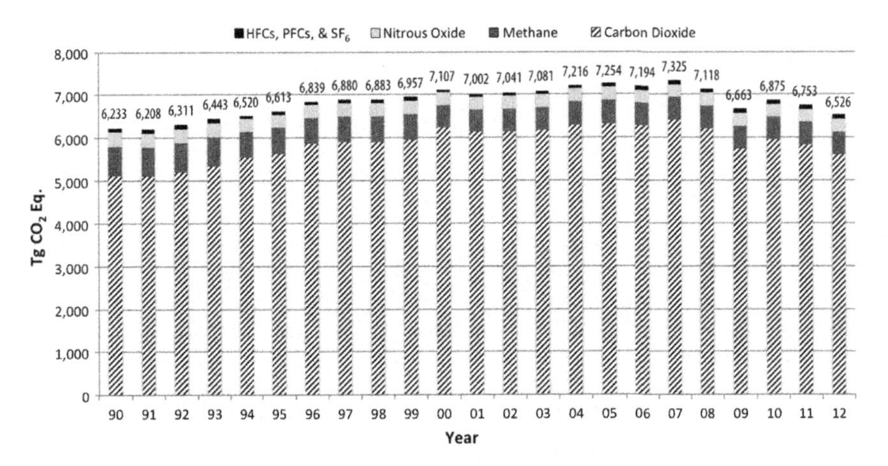

Figure 2.43 U.S. greenhouse gas emissions from 1990 to 2012.

gas and, as an example, methane's GWP (based on the IPCC 5th Assessment Report) is 28, which means that methane is 28 times better at trapping heat in the atmosphere than carbon dioxide. GWPs are typically reported on a 100-year time horizon. The United States EPA uses a time period of 100 years for policy making and reporting purposes.

The greenhouse gas emissions of direct greenhouse gases in the United States inventory are reported in terms of equivalent emissions of carbon dioxide, using teragrams (one teragram is equal to 10^{12} grams or one million metric tons) of carbon dioxide equivalents (Tg CO_2 eq). The relationship between gigagrams (Gg) of a gas and Tg CO_2 eq can be expressed as follows:

$$\text{Tg CO}_2\text{ eq} = (\text{Gg of gas}) \times (\text{GWP}) \times \left(\frac{\text{Tg}}{1,000\text{Gg}}\right) \tag{2.15}$$

In 2012, total U.S. greenhouse gas emissions were 6,525.6 Tg or one million metric tons, Tg CO_2 eq. Trends in U.S. greenhouse gas emissions are shown in Figure 2.43 [78]. The decrease over the last couple of years was due to a decrease in carbon intensity of fuels consumed by the power generation producers to generate electricity due to a decrease in the price of natural gas, a decrease in transportation sector emissions, and warmer winter conditions. Trends in U.S. greenhouse gas emissions for the various greenhouse gases and sources are listed in Table 2.26 with emissions from fossil fuel combustion specifically noted.

As the largest source of U.S. greenhouse gas emissions, CO_2 from fossil fuel combustion has accounted for approximately 78% of GWP-weighted emissions since 1990, and is approximately 78% of total GWP-weighted emissions in 2012 [78]. The five major fuel-consuming sectors contributing to CO_2 emissions from fossil fuel combustion are electricity generation, transportation, industrial, residential, and commercial. Figures 2.44 and 2.45 and Table 2.27 summarize CO_2 emissions from fossil fuel combustion by end-use sector [78].

Table 2.26 **Recent trends in U.S. greenhouse gas emissions (emissions from fossil fuel combustion are specifically listed where relevant), in Tg or million metric tons CO_2 eq**

Gas/source	1990	2005	2008	2009	2010	2011	2012
CO_2	5,108.7	6,112.2	5,936.9	5,506.1	5,722.3	5,592.2	5,383.2
Fossil Fuel Combustion	4,745.1	5,752.9	5,593.4	5,225.7	5,404.9	5,271.1	5,072.3
Electricity generation	1,820.8	2,402.1	2,360.9	2,146.4	2,259.2	2,158.5	2,022.7
Transportation	1,494.0	1,891.7	1,816.5	1,747.7	1,765.0	1,747.9	1,739.5
Industrial	845.1	827.6	804.1	727.5	775.6	768.7	774.2
Residential	338.3	357.9	346.2	336.4	334.8	324.9	288.9
Commercial	219.0	223.5	224.7	223.9	220.7	221.5	197.4
U.S. Territories	27.9	50.0	41.0	43.8	49.6	49.6	49.6
CH_4	635.7	585.7	606.0	596.5	585.5	578.3	567.3
Stationary Combustion	7.5	6.6	6.6	6.6	6.4	6.3	5.7
N_2O	398.6	415.8	423.3	412.2	409.3	417.2	410.1
Stationary Combustion	12.3	20.6	21.1	20.8	22.5	21.6	22.0
HFCs	36.9	119.8	136.0	135.1	144.0	148.6	151.2
PFCs	20.6	5.6	5.1	3.3	3.8	6.0	5.4
SF_6	32.6	14.7	10.7	9.6	9.8	10.8	8.4
Total	6,233.2	7,253.8	7,118.1	6,662.9	6,874.7	6,753.0	6,525.6
Net Emissions (sources and sinks)	5,402.1	6,223.1	6,137.1	5,701.2	5,906.7	5,772.7	5,546.3

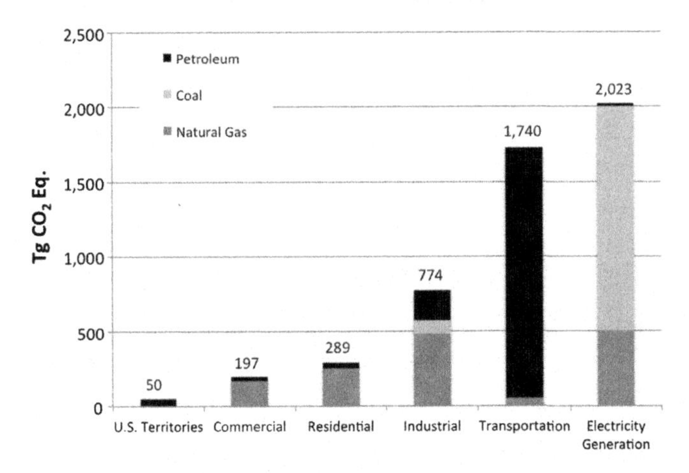

Figure 2.44 2012 CO_2 emissions from fossil fuel combustion by sector and fuel type.

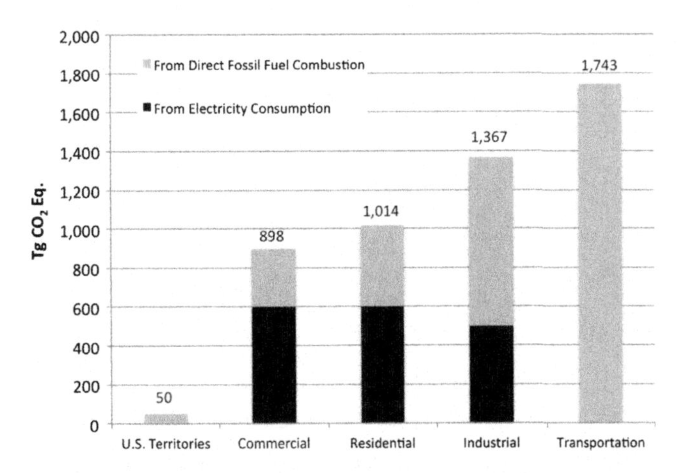

Figure 2.45 2012 end use sector emission of CO_2 from fossil fuel combustion.

Table 2.27 **CO_2 Emissions from fossil fuel combustion by fuel consuming end-use sector (Tg or million metric tons CO_2 eq.)**

Gas Source	1990	2005	2008	2009	2010	2011	2012
Transportation	1,497.0	1,896.5	1,821.2	1,752.2	1,769.5	1,752.1	1,743.4
Combustion	1,494.0	1,891.7	1,816.5	1,747.7	1,765.0	1,747.9	1,739.5
Electricity	3.0	4.7	4.7	4.5	4.5	4.3	3.9
Industrial	1,531.8	1,564.6	1,501.4	1,329.5	1,416.6	1,393.6	1,367.1
Combustion	845.1	827.6	804.1	727.5	775.6	768.7	774.2
Electricity	686.7	737.0	697.3	602.0	641.1	624.9	592.9
Residential	931.4	1,214.7	1,189.2	1,122.9	1,175.2	1,115.9	1,014.3
Combustion	338.3	357.9	346.2	336.4	334.8	324.9	288.9
Electricity	593.0	856.7	842.9	786.5	840.4	791.0	725.5
Commercial	757.0	1,027.2	1,040.8	977.4	993.9	959.8	897.9
Combustion	219.0	223.5	224.7	223.9	220.7	221.5	197.4
Electricity	538.0	803.7	816.0	753.5	773.7	738.3	700.4
U.S. Territories	27.9	50.0	41.0	43.8	49.6	49.6	49.6
Total	4,745.1	5,752.9	5,593.4	5,225.7	5,404.9	5,271.1	5,072.3
Electricity Generation	1,820.8	2,402.1	2,360.9	2,146.4	2,259.2	2,158.5	2,022.7

References

[1] Reitze Jr. AW. The Legislative History of U.S. Air Pollution Control. Houston Law Review; 1999.

[2] C&EN (Chemical & Engineering News). Recognizing pollution's hazard's, C&E News 2003;81(15):54–55.

[3] Dunn S. King coal's weakening grip on power. World watch 1999, p. 10–19.

[4] Hobson C. Complying with EPA Regulations: How Big a Challenge Is It? Elcon: Electricity Consumers Resource Council. Southern Company; 2012.

[5] Wark K, Warner CF, Davis WT. Air Pollution: Its Origin and Control. 3rd ed. Menlo Park, California: Addison Wesley Longman, Inc.; 1998.

[6] EPA (United States Environmental Protection Agency). National Ambient Air Quality Standards (NAAQS), <www.epa.gov/air/criteria.html>; 2012.

[7] EPA (United States Environmental Protection Agency). Subpart D – standards of performance for fossil-fuel-fired steam generators for which construction is commenced after August 17, 1971, Federal Register, December 23, 1971.

[8] EPA (United States Environmental Protection Agency). Title 40: Part 60 – standards of performance for new stationary sources, subpart D – standards of performance for fossil-fuel-fired steam generators, <www.e-CFR>, <www.ecfr.gov>, as of May 27, 2014.

[9] EPA (United States Environmental Protection Agency). Title 40: Part 60 – standards of performance for new stationary sources, subpart Da – standards of performance for fossil-fuel-fired steam generators, <www.e-CFR>, <www.ecfr.gov>, as of May 27, 2014.

[10] EPA (United States Environmental Protection Agency). Title 40: Part 60 – standards of performance for new stationary sources, subpart Db – standards of performance for fossil-fuel-fired steam generators, <www.e-CFR>, <www.ecfr.gov>, as of May 27, 2014.

[11] EPA (United States Environmental Protection Agency). Title 40: Part 60 – standards of performance for new stationary sources, subpart Dc – standards of performance for fossil-fuel-fired steam generators, <www.e-CFR>, <www.ecfr.gov>, as of May 27, 2014.

[12] EPA (United States Environmental Protection Agency). Title 40: Part 60 – standards of performance for new stationary sources, subpart GG – standards of performance for stationary gas turbines, <www.e-CFR>, <www.ecfr.gov>, as of May 27, 2014.

[13] EPA (United States Environmental Protection Agency). Title 40: Part 60 – standards of performance for new stationary sources, subpart KKKK – standards of performance for stationary combustion turbines, <www.e-CFR>, <www.ecfr.gov>, as of May 27, 2014.

[14] EPA (United States Environmental Protection Agency). Technology transfer network clearinghouse for inventories and emissions factors (CHIEF), emissions factors & AP 42, compilation of air pollutant emission factors. Available from: <www.epa.gov/ttn/chief/ap42>.

[15] AP-42, Emission Factors, Fifth Edition, Chapter 1: external combustion sources and chapter 3: stationary internal combustion, EPA office of air quality planning and standards and office of air and radiation, 1995 with latest revisions in 2010.

[16] EPA (United States Environmental Protection Agency). CAA: original list of hazardous air pollutants, <www.epa.gov/ttn/atw/188polls.html>; 2013.

[17] Makanski J. Clean air act amendments: the engineering response. Power 1991;135 (6):11–66.

[18] EPA (United States Environmental Protection Agency). 2012 Progress report: SO_2 and NO_x emissions, compliance, and market analysis; 2014.

[19] EPA (United States Environmental Protection Agency). Acid rain program, <www.epa.gov/AIRMARKET/progsregs/arp/>; 2012.

[20] Leone M. Cleaning the air the market-based way. Power 1990;129(10):9–10.

[21] EPA (United States Environmental Protection Agency). Overview: the clean air act amendments of 1990, <www.epa.gov/oar/caa/index.html>; 2014.

[22] Smith DN, McIlvried HG, Mann AN. Understanding NO_x and how it impacts coal.. Coal Age 2000;105(11):35.

[23] EPA (United States Environmental Protection Agency). Ozone transport commission NO_x budget program, 1999–2002 progress report, office of air and radiation, U.S. Government Printing Office, Washington, D.C.; 2003.

[24] EPA (United States Environmental Protection Agency). The NO_x budget training program: 2008 highlights, office of air and radiation, U.S. Government Printing Office, Washington, D.C.; 2009.

[25] EPA (United States Environmental Protection Agency). The NO_x budget training program: 2008 emission, compliance, and market data, office of air and radiation, U.S. Government Printing Office, Washington, D.C.; 2009.

[26] EPA (United States Environmental Protection Agency). Clean air interstate rule, acid rain program, and former NO_x budget trading program 2011 progress report, <www.epa.gov/airmarkets/progress>; 2013.

[27] EPA (United States Environmental Protection Agency). Clean air interstate rule, acid rain program, and former NO_x budget trading program 2012 progress report; 2013.

[28] EPA (United States Environmental Protection Agency), Cross-state air pollution rule (CSAPR) basic information, <www.epa.gov/crossstaterule/basic>; 2014.

[29] EPA (United States Environmental Protection Agency). Cross-state air pollution rule, <www.epa.gov/airtransport/CSAPR>; 2014.

[30] EPA (United States Environmental Protection Agency). Mercury study report to congress. (Office of Air Quality Planning & Standards and Office of research and Development, U.S. Government Printing Office, Washington, D.C.; 1997).

[31] EPA (United States Environmental Protection Agency). Study of hazardous air pollutant emissions from electric utility steam generating units – final report to congress. (Office of Air Quality Planning & Standards, U.S. Government Printing Office, Washington, D.C.; 1998).

[32] EPA (United States Environmental Protection Agency). EPA ICR No. 1858: information collection request for electric utility steam generating unit mercury emissions information collection effort; 1999.

[33] EPA (United States Environmental Protection Agency). Regulatory finding on the emissions of hazardous air pollutants from electric utility steam generating units, federal register, Vol. 65, No. 245, 2000, p. 79825–79831.

[34] EPA (United States Environmental Protection Agency). Clean air mercury rule, <www.epa.gov/air/mercuryrule/>; 2009.

[35] EPA (United States Environmental Protection Agency). Mercury and air toxics standards, <www.epa.gov/mats/>; 2012.

[36] EPA (United States Environmental Protection Agency). Mercury and air toxics standards (MATS) cleaner power plants, <www.epa.gov/airquality/powerplanttoxics/powerplants.html>; 2014.

[37] EPA (United States Environmental Protection Agency). Boiler MACT technical assistance; 2012.

[38] EPA (United States Environmental Protection Agency). Boiler compliance at area sources, <www.epa.gov/boilercomplaince/>; 2014.

[39] Power Engineering, Industrial Boiler MACT, 2013; 117(10).

[40] EPA (United States Environmental Protection Agency). Regulatory finding on the national emission standards for hazardous air pollutants for major sources: industrial, commercial, and institutional boilers and process heaters, Final Rule, 40 CFR Part 63, Federal Register, 2014;78(21):7138–7213.
[41] EPA (United States Environmental Protection Agency). Regulatory finding on the national emission standards for hazardous air pollutants for area sources: industrial, commercial, and institutional boilers and process heaters, Final Rule, 40 CFR Part 63, Federal Register, 2014;78(22):7488–7521.
[42] EPA (United States Environmental Protection Agency). New source review: report to the president; 2002.
[43] EPA (United States Environmental Protection Agency). New source review basic information, <www.epa.gov/nsr/info.html>; 2013.
[44] EPA (United States Environmental Protection Agency). Prevention of significant deterioration (PSD) and nonattainment new source review (NSR): final rule and proposed rule, Federal Register, 2002;67(251):80186–80289.
[45] EPA (United States Environmental Protection Agency). New source review; 2003.
[46] EPA (United States Environmental Protection Agency). EPA completes reconsideration of new source review equipment replacement provision fact sheet; 2005.
[47] EPA (United States Environmental Protection Agency). Prevention of significant deterioration (PSD) and nonattainment new source review (NSR): equipment replacement provision of the routine maintenance, repair and replacement exclusion: reconsideration, federal register, 2005;70(111):33838–33850.
[48] DOE (United States Department of Energy). Atmospheric aerosol source-receptor relationships: the role of coal-fired power plants – project facts, (Office of Fossil Energy, National Energy Technology Laboratory; 2003).
[49] EPA (United States Environmental Protection Agency) Particulate matter regulatory actions, <http://epa.gov/air/particlepollution/standards.html>; 2014.
[50] EPA (United States Environmental Protection Agency). Fact sheet – final regional Haze regulations for protection of visibility in National Parks and Wilderness Areas; 1999.
[51] EPA (United States Environmental Protection Agency). Identification of nanattainment classification and deadlines for submission of state implementation plan provisions for the 1997 and 2006 fine particle national ambient air quality standards fact sheet; 2014.
[52] CEP (Chemical Engineering Progress). Congress considers climate change 2003;99 (6):23.
[53] C&EN (Chemical & Engineering News). Global initiative on CO_2 storage 2003:81 (26):19.
[54] C&EN (Chemical & Engineering News). Climate-change plan released 2003;81(30):39.
[55] CDT (Centre Daily Times). White house seeks more data on global climate change 2003:A10.
[56] Miller BG. Clean coal engineering technology. Oxford, United Kingdom: Butterworth-Heinemann; 2011.
[57] Power Engineering. Climate Bill Likely Won't Pass in 2009, <http://pepei.pennet.com>; 2009.
[58] EPA (United States Environmental Protection Agency). Prevention of significant deterioration and title V greenhouse gas tailoring rule, federal register 2010;75 (106):31514–31608.
[59] EPA (United States Environmental Protection Agency). Final rule: prevention of significant deterioration and title V greenhouse gas tailoring fact sheet; 2010.

[60] EPA (United States Environmental Protection Agency). Standards of performance for greenhouse gas emissions from new stationary sources: electric utility generating units: proposed rule, federal register 2014;79(5):1430–1519.

[61] EPA (United States Environmental Protection Agency). Carbon pollution standards for modified and reconstructed stationary sources: electric utility generating units: proposed rule, federal register 2014;79(117):34960–34994.

[62] EPA (United States Environmental Protection Agency). Carbon pollution emission guidelines for existing stationary sources: electric utility generating units: proposed rule, federal register 2014;79(117):34830–34958.

[63] EPA (United States Environmental Protection Agency). Latest findings on national air QUALITY 2002 status and trends (Office of Air Quality Planning and Standards, U.S. Government Printing Office, Washington, D.C.; 2003).

[64] EPA (United States Environmental Protection Agency). Progress cleaning air and improving people's health, <www.epa.gov/air/caa/progress.html>; 2014.

[65] EPA (United States Environmental Protection Agency). Air trends, <www.epa.gov/airtrends>; 2014.

[66] EPA (United States Environmental Protection Agency). National emissions inventory (NEI), 1970–2013 average annual emissions, all criteria pollutants, <www.epa.gov/ttnchie1/trends/>; 2014.

[67] EPA (United States Environmental Protection Agency). 2008 National emissions inventory: review, analysis, and highlights; 2013.

[68] EPA (United States Environmental Protection Agency). Our nation's air status and trends through 2010; 2012.

[69] EPA (United States Environmental Protection Agency). Report on the environment, <http://www.epa.gov/ncea/roe/>; 2014.

[70] EPA (United States Environmental Protection Agency). National acid precipitation assessment program report to congress 2011: an integrated assessment; 2011.

[71] EPA (United States Environmental Protection Agency). Acid rain and related programs, 2009 environmental results, <www.epa.gov/airmarkets/progress/>; 2010.

[72] EPA (United States Environmental Protection Agency). Controlling power plant emissions: emissions progress, <http://www.epa.gov/mercury/control_emissions/emisisons.htm>; 2009.

[73] EPA (United States Environmental Protection Agency). Mercury emissions: the global context, <http://www.epa.gov/mercury/control_emissions/global.htm>; 2009.

[74] United Nations Environmental Programme (UNEP). Global atmospheric mercury assessment: sources, emissions and transport; 2008.

[75] EPA (United States Environmental Protection Agency). Mercury emissions: the global context, <http://www2.epa.gov/international-cooperation/mercury-emissions-global-context#learn>; 2014.

[76] UNEP (United Nations Environment Programme). Global mercury assessment 2013: sources, emissions, releases and environmental transport; 2013.

[77] EPA (United States Environmental Protection Agency). Greenhouse gases and global warming potential values. (U.S. Greenhouse Gas Inventory Program Office of Atmospheric Programs, U.S. Government Printing Office, Washington, D.C.; 2002).

[78] EPA (United States Environmental Protection Agency). Inventory of U.S. greenhouse gas emissions and sinks: 1990–2012; 2014.

[79] Smith IM, Nilsson C, Adams DMB. Greenhouse gases – perspectives on coal. London: IEA Coal Research; 1994. p. 10

[80] Climate Change 2013: the physical science basis. Contribution of working group I to the fifth assessment report of the intergovernmental panel on climate change, Cambridge University Press, Cambridge, United Kingdom and New York, NY, USA, 1535 p. 2013.
[81] Climate change 2007: the physical science basis. Contribution of working group I to the fourth assessment report of the intergovernmental panel on climate change, Cambridge University Press, Cambridge, United Kingdom and New York, NY, USA, 996 p. 2007.
[82] EPA (United States Environmental Protection Agency). EPA announces improvements to new source review program; 2003.
[83] CEP (Chemical Engineering Progress). EPA finalizes new source review rule, chemical engineering progress, 2003;99(12):24.

Particulate formation and control technologies

3

3.1 Introduction

Particulate matter (PM) is the general term used for a mixture of solid particles and liquid droplets found in the air. Some particles are large or dark enough to be seen as soot or smoke while others are so small they cannot be seen with the naked eye. These small particles, which come in a wide range of sizes, originate from many different stationary and mobile sources as well as natural sources [1]. Fine particles, those less than 2.5 μm (i.e. $PM_{2.5}$), result from fuel combustion from motor vehicles, power generation, industrial facilities, and residential fireplaces and wood stoves. Coarse particles, those larger than 2.5 μm but classified as less than 10 μm (i.e. PM_{10}), are generally emitted from sources such as vehicles traveling on unpaved roads, materials handling, crushing and grinding operations, as well as windblown dust [1]. Some particles are emitted directly from their sources, such as smokestacks and cars. In other cases, gases such as SO_2, NO_x, and volatile organic compounds (VOCs) react with other compounds in the air to form fine particles. Several terms are used to classify airborne particles. The definitions of these terms are presented in Table 3.1 [2].

In general, airborne particles range in size from 0.001 to 500 μm, with the bulk of the particulate mass in the atmosphere ranging from 0.1 to 10 μm. Particles below 0.1 μm in size display a behavior similar to that of molecules and are characterized by large random motions caused by collisions with gas molecules. Particles larger than 1 μm but smaller than 20 μm tend to follow the motion of the gas in which they are entrained. Particles larger than 20 μm have significant settling velocities and are airborne for relatively short periods of time [2]. Figure 3.1 shows the range of particle sizes for various materials and a range of particle sizes for which various types of collection equipment are appropriate [modified from 3]. Equipment relevant to particulate matter from fossil fuel-fired stationary systems is discussed in this chapter.

Particulate matter is responsible for reduction in visibility. Visibility is principally affected by fine particles that are formed in the atmosphere from gas-phase reactions. Although these particles are not directly visible, carbon dioxide, water vapor, and ozone in increased concentrations changes the absorption and transmission characteristics of the atmosphere [2].

Particulate matter can cause damage to materials, depending upon its chemical composition and physical state [2]. Particles will soil painted surfaces, clothing, and curtains merely by settling on them. Particulate matter can cause corrosive damage to metals either by intrinsic corrosiveness or by the action of corrosive chemicals absorbed or adsorbed by inert particles.

Fossil Fuel Emissions Control Technologies. DOI: http://dx.doi.org/10.1016/B978-0-12-801566-7.00003-8

Table 3.1 Definitions of terms that describe airborne particulate matter

Term	Definition
Particulate matter	Any material except uncombined water that exists in the solid or liquid state in the atmosphere or gas stream at standard condition.
Aerosol	A dispersion of microscopic solid or liquid particles in gaseous media.
Dust	Solid particles larger than colloidal size capable of temporary suspension in air.
Fly ash	Finely divided particles of ash entrained in flue gas. Particles may contain unburned fuel.
Fog	Visible aerosol.
Fume	Particles formed by condensation, sublimation, or chemical reaction, predominantly smaller than 1 μm.
Mist	Dispersion of small liquid droplets of sufficient size to fall from the air.
Particle	Discrete mass of solid or liquid matter.
Smoke	Small gasborne particles resulting from combustion.
Soot	An agglomeration of carbon particles.
PM_{10}	Refers to particles with an aerodynamic diameter less than or equal to a nominal 10 μm.
$PM_{2.5}$	Refers to particles with an aerodynamic diameter less than or equal to a nominal 2.5 μm. Also referred to as the fine fraction of PM_{10}.
$PM_{10-2.5}$	Refers to particles with an aerodynamic diameter less than or equal to a nominal 10 μm but greater than 2.5 μm. Also referred to as the coarse fraction of PM_{10}.

Little is known of the effects of particulate matter in general on vegetation [2]. The combination of particulate matter and other pollutants such as sulfur dioxide may affect plant growth. Coarse particles, such as dust, may be deposited directly onto leaf surfaces and reduce gas exchange, increase leaf surface temperature, and decrease photosynthesis. Toxic particles that consist of elements such as arsenic or fluorine fall onto agricultural soils or plants and are ingested by animals, which can be harmful to the animal's health.

Particulate matter alone or in combination with other pollutants constitutes a very serious health hazard. The pollutants enter the human body mainly via the respiratory system. Inhalable particulate matter includes both fine and coarse particles. These particles can accumulate in the respiratory system and are associated with numerous health effects [1]. Exposure to coarse particles is primarily associated with the aggravation of respiratory conditions such as asthma. Fine particles are most closely associated with such health effects as increased hospital admissions and emergency room visits for heart and lung disease, increased respiratory symptoms and disease, decreased lung function and even premature death. Sensitive groups that appear to be at greatest risk include the elderly, individuals with cardiopulmonary disease, such as asthma, and children [1].

Emissions standards for particulate matter were first introduced in Japan, the United States, and Western European nations in the early to mid-1900s.

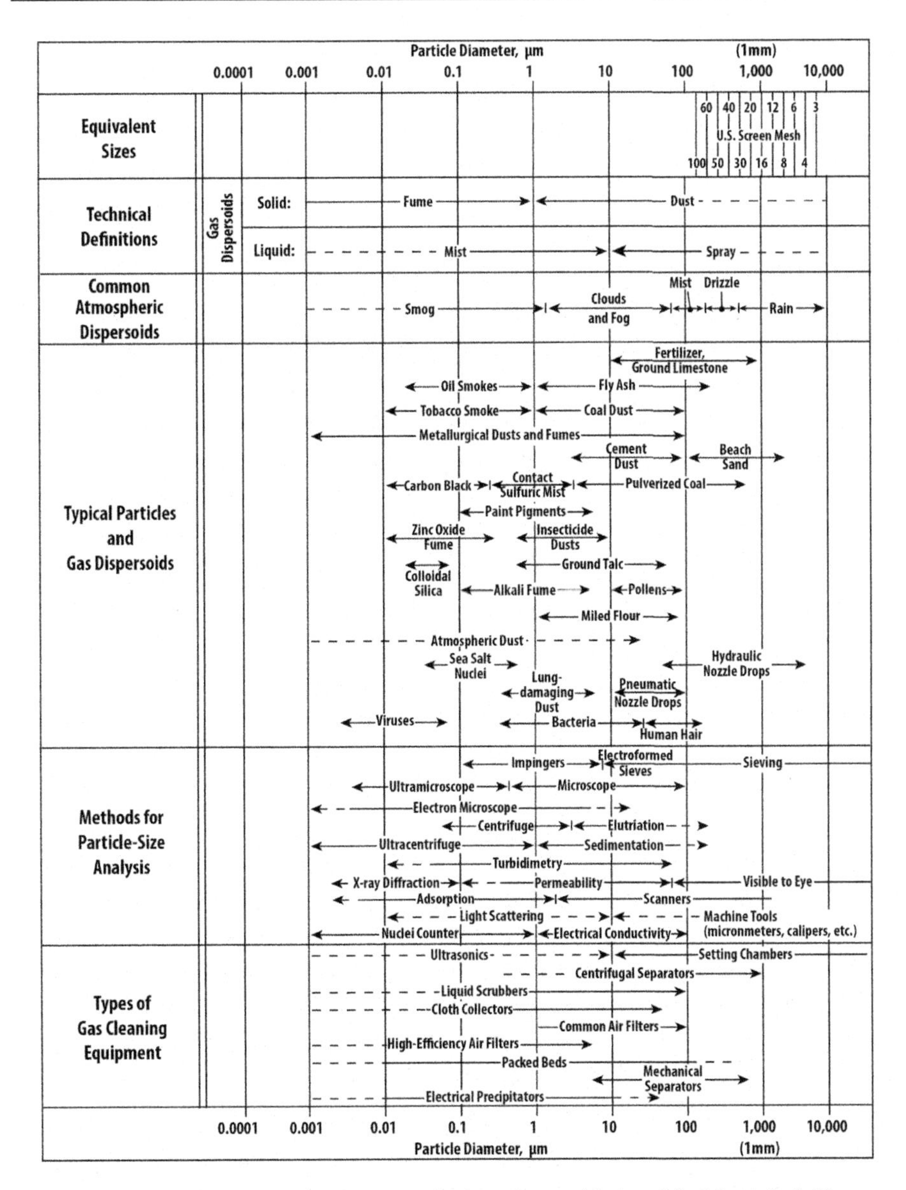

Figure 3.1 Characteristics of particles and particle dispersoids [modified from Ref. 3].

The following decades found many countries also setting standards for particulate emissions including those in Asia, Eastern Europe, Australia, and India. The importance of utilizing coal in an environmentally friendly manner for power generation has led to the introduction or proposal of particulate emissions standards in more than 40 nations [4]. In recent years, there has been increasing concern for control of fine particulate matter and existing particulate emissions standards have progressively become more stringent over the years and the most stringent measures are associated with wealthy countries such as Japan and those in North America and Western Europe [4].

Particulate matter emissions from stationary fuel combustion (i.e. electric utility boilers, industrial applications, and others) in the United States have decreased significantly since the implementation of the 1970 Clean Air Act Amendments (see Figure 2.10 and Table 2.20 [5,6]). In 2013, approximately 21 million short tons of dry particulate matter, reported as PM_{10} (i.e. particles with an aerodynamic diameter less than or equal to 10 microns), were emitted from inventoried point and area sources of which 976,000 short tons (or $\approx$4.6% of the total) were emitted by solid fuel combustion sources [5,6]. Of the total PM emissions from solid fuel combustion sources, 276,000 short tons (or $\approx$1.3% of the total PM_{10} emissions) were emitted from electric utilities with coal being the major source of PM emissions from electricity generation activities. This is a substantial decrease from a total of approximately 1.7 million short tons of PM_{10} being emitted from coal-fired power plants in 1970, especially since coal consumption for electricity generation has increased more than 150% over this period, and the reduction is due to the application of particulate control technologies. Similarly, annual emissions of dry particulates smaller than 2.5 μm (i.e. $PM_{2.5}$) in 2013, which is a subset of PM_{10}, were 202,000 short tons or approximately 3.2% of the total primary $PM_{2.5}$ emitted from all sources. As of 2012 (the most recent data as of January 2011), there were more than 1,200 coal-fired electric generators equipped with particulate collectors with a total of nearly 356,000 MW generating capacity [7,8].

There are several particulate control technologies for fossil fuel-fired power plants, which include electrostatic precipitators (ESPs), fabric filters (baghouses), wet particulate scrubbers, mechanical collectors (cyclones), and hot-gas particulate filtration [9]. Of these, ESPs and fabric filters are currently the technologies of choice as they can meet current and pending legislation particulate matter levels while cleaning large volumes of flue gas, achieve very high collection efficiencies, and remove fine particles. When operating properly, ESPs and baghouses can achieve overall collection efficiencies of 99.9% of primary particulates (over 99% control of PM_{10} and 95% control of $PM_{2.5}$), thereby achieving the 1978 New Source Performance Standards required limit of 0.03 lb PM/million Btu [10].

Particle formation and characterization is discussed in this chapter along with particulate control mechanisms and technologies. The primary particulate matter collection devices used in the power generation industry – ESPs and fabric filters (baghouses) – are discussed in this chapter. In addition, cyclone separators and wet collectors/scrubbers are discussed as are hybrid systems that are under development, where ESPs and fabric filters are combined in a single overall system. Operating principles, design basics, factors that affect performance, and advantages/disadvantages of the collection devices will be presented.

3.2 Particle formation and characterization

This section presents a brief overview of the formation of fly ash particles from fossil fuel (specifically coal and fuel oil)-fired stationary power systems. Particle formation and the need for particulate control devices are especially important for coal and fuel oil-fired systems and to a lesser extent gaseous fuels. Examples of particle compositions are also provided.

3.2.1 Coal fly ash formation and characterization

Coal generally contains from 5 to 20 weight percent inorganic mineral matter and inorganic elements; the inorganic elements can be bound organically in the coal or present in the form of simple salts. Mineral matter in coal commonly includes alumino-silicate clays, silicates, carbonates, and disulfides as major components. The majority of the inorganic matter in high-rank coals (e.g. anthracite and bituminous coal) occurs in the form of minerals of various types and sizes that are closely associated with the organic matter (included minerals) or they occur excluded from the organic matter (excluded minerals). Organically-bound inorganic elements such as Na, K, Ca, and Mg, which are distributed within the coal macerals and are generally the more volatile species, are commonly found in lower rank coals such as lignites and subbituminous coals. The modes of occurrence of minerals and other inorganic constituents in coal are illustrated in Figure 3.2 [modified from 11].

During the combustion process, these constituents undergo chemical and physical transformations to form ash. The mechanisms of ash formation have been studied extensively and include fusion, agglomeration, shedding, fragmentation, vaporization, and condensation [12–14]. These stages are not sequential and are influenced by the combustion conditions and the coal characteristics. Examples of two simplified mechanisms for producing fly ash (which must then be captured by control technologies discussed in this chapter) are provided in Figures 3.3 [modified from 14] and 3.4 [modified from 15], which, respectively, depict mechanisms for pulverized coal and fluidized-bed combustion, the two primary combustion technologies for power production.

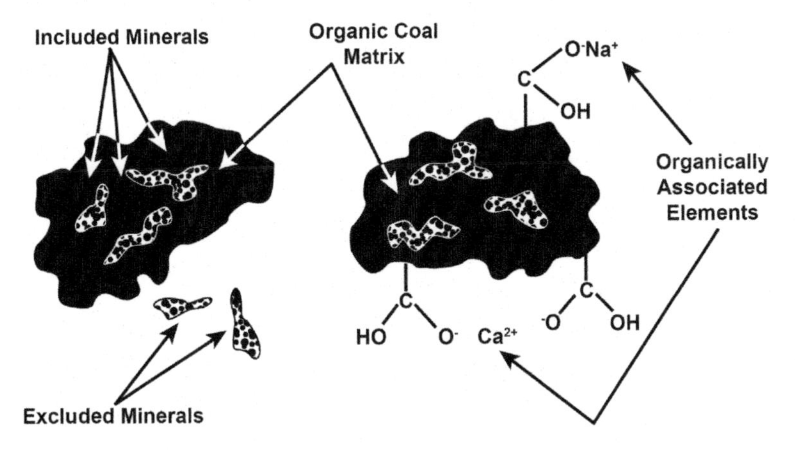

Figure 3.2 Modes of occurrence for minerals and other inorganic constituents in coal.

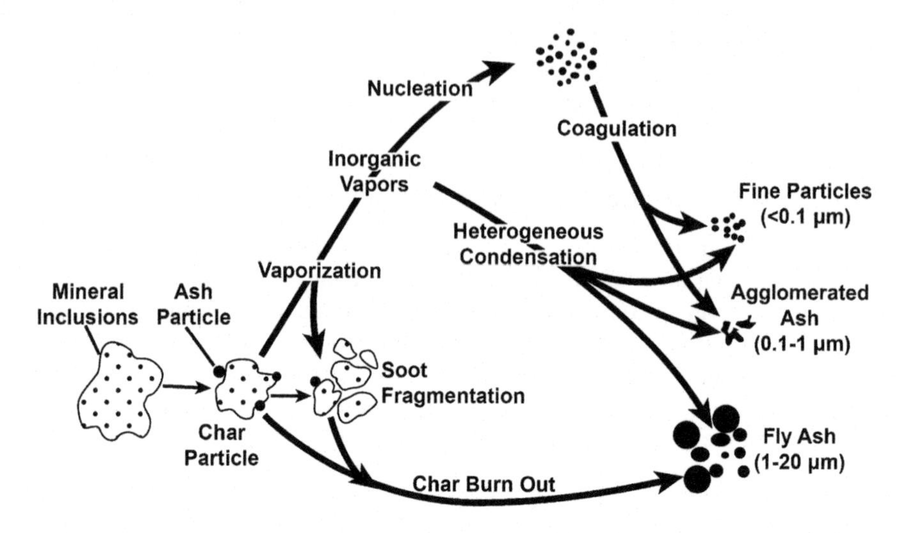

Figure 3.3 Formation of particles in pulverized coal combustion.

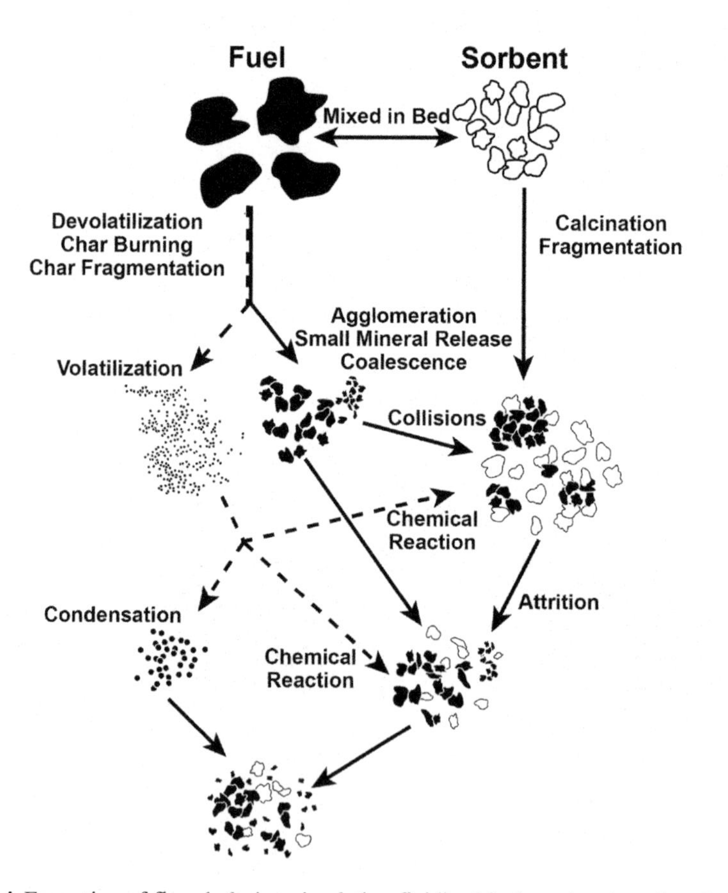

Figure 3.4 Formation of fly ash during circulating fluidized-bed combustion of coal.

All coals contain small concentrations of trace elements. Although these elements are present in small concentrations in the coal, i.e. parts per million (ppm) by weight, the large amount of coal burned annually mobilizes tons of these pollutants as particles or gases.

As previously discussed, Title III of the United States Clean Air Act Amendments (CAAA) of 1990 designates 187 hazardous air pollutants (HAPs). Included in the list are 11 trace elements: antimony (Sb), arsenic (As), beryllium (Be), cadmium (Cd), chromium (Cr), cobalt (Co), lead (Pb), manganese (Mn), mercury (Hg), nickel (Ni), and selenium (Se). In addition, barium (Ba) is regulated by the Resources Conservation and Recovery Act, and boron (B) and molybdenum (Mo) are regulated by Irrigation Water Standards [16]. Vanadium (V) is regulated based on its oxidation state and vanadium pentoxide (V_2O_5) is a highly toxic regulated compound. Other elements, such as fluorine (F) and chlorine (Cl), which produce acid gases (i.e. HF and HCl) upon combustion, and radionuclides such as radon (Rn), thorium (Th) and uranium (U) are also of interest. A detailed discussion of the health and environmental effects of trace elements can be found elsewhere [17].

Partitioning of the trace elements in the bottom ash, ash collected in the air pollution control device, and fly ash and gaseous constituents emitted into the atmosphere depends on many factors including the volatility of the elements, temperature profiles across the system, pollution control devices, and operating conditions [18,19]. Numerous studies have shown that trace elements can be classified into three broad categories based on their partitioning during coal combustion. A summary of these studies is presented by Clarke and Sloss [19]. Figure 3.5 illustrates the classification scheme for selected elements (modified after [16] and [19]).

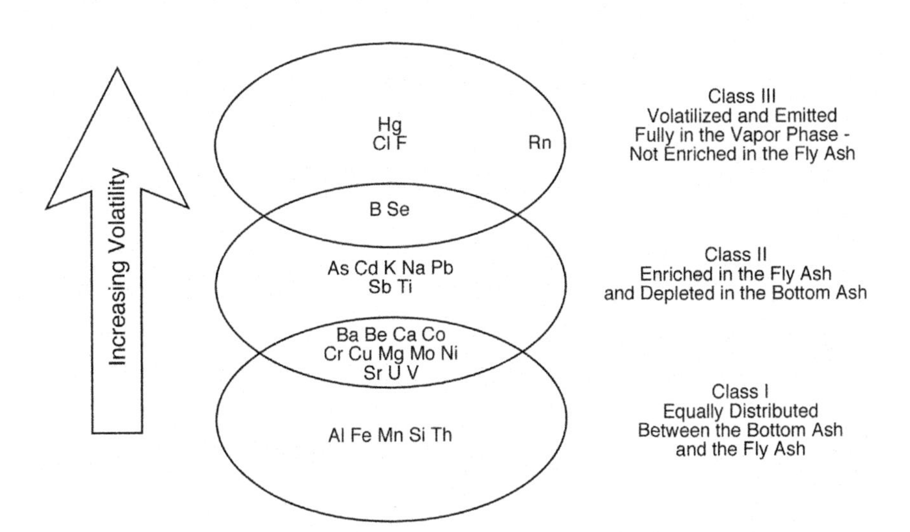

Figure 3.5 Classification scheme for selected trace elements relative to their volatility and partitioning in power plants.

Class I elements are the least volatile and are concentrated in the coarse residues (i.e. bottom ash) or equally partitioned between coarse residues and finer particles (i.e. fly ash). Class II elements will volatilize in the boiler but condense downstream and are concentrated in the finer-sized particles. Class III elements are the most volatile and exist entirely in the vapor phase. Overlap between classifications exists, which can be a function of fuel, combustion system design, and operating conditions, especially temperature [19].

Control of Class I elements is directly related to control of total particulate matter emissions, while control of Class II elements depends on the collection of fine particulate matter. Because of the volatility of Class III elements, particulate controls have only a limited impact on emissions of these elements. Capture of Class III elements is discussed in the chapters on mercury control and acid gas control.

A range of bulk fly ash chemical composition for coals is provided in Table 3.2. Because of the variability in coal mineral matter composition, which is a function of coal rank and varies between mines and within seams in a mine, the data are presented as a range. The data have been compiled from several sources [17,20–22]. The composition of fly ashes from heavy fuel oil combustion is also included in Table 3.2 for comparison. Note that fly ash also contains unburned carbon and this can vary from tenths of a percent (by weight) of the fly ash from efficient combustion systems to very high values (in excess of 50 weight percent) where combustion is inefficient or the initial fuel contains very low concentrations of inorganic species such as in fuel oils and the particulate matter consists primarily of soot/coke (i.e. unburned carbon residue).

3.2.2 Fuel oil fly ash formation and characterization

In general, particulate matter emissions depend predominately on the grade of the fuel oil fired. Combustion of lighter distillate fuel oils results in lower particulate matter formation than does combustion of heavier residual oils because the heavier fuels oils contain more asphaltenes, inorganic species, and sulfur than do the distillate oils [23]. Distillate fuel oils (No. 1 and 2) contain trace amounts of ash (as determined by a proximate analysis) while heavier fuel oils (Nos. 4 to 6) contain 0.1 to 0.5 weight percent ash with No 6 fuel oil containing the highest levels of ash (see Table 1.11).

The fundamental particle formation processes during the combustion of fuel oils, specifically heavy fuel oils, are the same as those for pulverized coal, although there are distinct differences between the two fuels. Oils do not typically contain significant extraneous or included mineral matter and heteroatoms are the only source of ash. The metals in heavy fuel oils are generally inherently bound with the organic molecule. A simplified mechanism for formation of particles observed in the flue gas for heavy fuel oil combustion is shown in Figure 3.6 [modified from 24].

Typical fuel oils contain iron (Fe), nickel (Ni), vanadium (V), and zinc (Zn), in addition to aluminum (Al), calcium (Ca), magnesium (Mg), silicon (Si), and sodium (Na). This is reflected in the composition of fly ash produced from fuel oil combustion (see Table 3.2).

Table 3.2 **Compositional ranges of inorganic species of coal fly ash and heavy fuel oil fly ash from combustion systems (ppm)**

Element	Coal fly ash	Heavy fuel oil fly ash[a]
Major/Minor elements		
Aluminum (Al)	70,000–280,000	
Calcium (Ca)	9,500–360,000	
Iron (Fe)	34,000–235,000	4,300
Silicon (Si)	150,000–605,000	
Magnesium (Mg)	3,900–66,000	
Potassium (K)	6,200–21,000	
Sodium (Na)	1,100–26,000	
Sulfur (S)	1,300–390,000	585–126,000
Titanium (Ti)	3,500–27,000	
Trace elements		
Antimony (Sb)	BDL[b]–16	
Arsenic (As)	22–260	26
Barium (Ba)	380–5,100	
Beryllium (Be)	2–26	
Boron (B)	120–1,000	
Cadmium (Cd)	BDL–4	0.8
Chromium (Cr)	27–300	7
Copper (Cu)	62–220	87–287
Lead (Pb)	11–230	57–342
Manganese (Mn)	91–700	125
Mercury (Hg)	0.01–0.51	0.42
Molybdenum (Mo)	4.8–60	85–720
Nickel (Ni)	47–230	90–13,800
Selenium (Se)	1.8–18	3.8
Strontium (Sr)	270–3,100	
Thallium (Tl)	BDL–45	
Uranium (U)	BDL–19	
Vanadium (V)	BDL–360	6,700–61,000
Zinc (Zn)	63–680	72.5

[a]In some cases, no values available or single values available.
[b]Below detection level.

3.2.3 Natural gas and liquefied petroleum gas fly ash formation

Particulate matter emissions from natural gas and liquefied petroleum gas (LPG) are typically low. Particulate matter in natural gas combustion is usually larger molecular weight hydrocarbons that are not fully combusted. Particulate matter emissions from LPG result from soot, aerosols formed by condensable emitted species, or boiler scale dislodged during combustion.

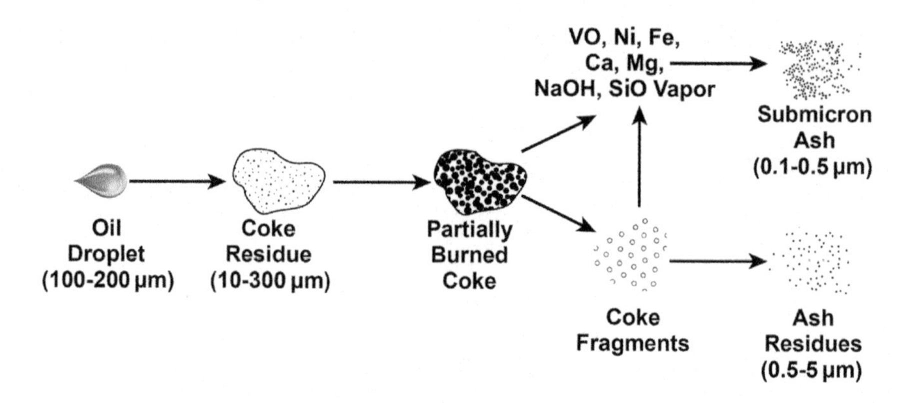

Figure 3.6 Simplified mechanism for formation of particles during heavy fuel oil combustion.

3.3 Mechanisms of particulate collection

The forces or mechanisms utilized for dust collection may be classified as 1) gravity settling, 2) inertial deposition or inertial impaction, 3) flow-line interception or direct interception, 4) diffusional deposition, 5) electrostatic deposition, 6) thermal precipitation, and 7) agglomeration [2,25]. Gravity settling employs gravitational forces to remove particles in a settling chamber. Figure 3.7 illustrates examples of inertial deposition/impaction, flow-line interception, and diffusional deposition [modified from 2,26]. Inertial impaction is associated with relatively larger particles colliding with the interceptor due to inertia while smaller particles remain in the gas stream. In direct interception, some of the smaller particles may contact the interceptor at the point of closest approach. In diffusional deposition, very small particles (usually less than 0.3 μm) impinge upon the collector as a result of random molecular (Brownian) motion or diffusion. Electrostatic precipitation is based on the mutual attraction between particles of one electrical charge and a collection device of opposite polarity. Thermal deposition is only a minor factor in particulate collection equipment because the thermophoretic force is small. Agglomeration is a mechanism where the average particle size is increased, often by electrostatic or ultrasonic devices, to improve collection efficiency.

Most forms of particulate collection equipment use more than one of these collection mechanisms, and in some instances the controlling mechanism may change when, to classify, the collector is operated over a wide range of conditions [26]. Therefore, it is more convenient to classify the particulate collection equipment according to type rather than according to collecting mechanism.

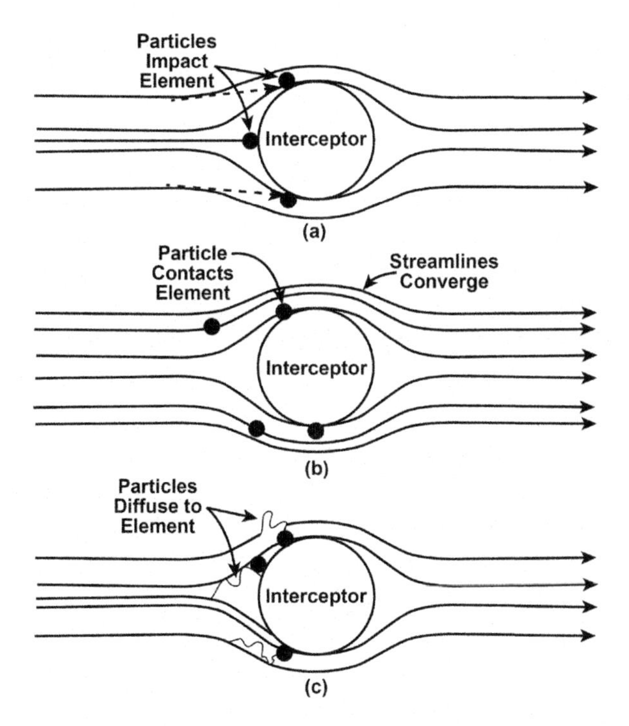

Figure 3.7 Mechanisms for particle deposition on collector bodies: (a) inertial impaction; (b) direct interception; (c) diffusional deposition.

3.4 Particulate control technologies

Particulate collection equipment basically consists of the following:

- Gravity settling chambers
- Impingement separators
- Cyclone (centrifugal) separators
- Electrostatic precipitators
- Fabric filters
- Wet collectors/scrubbers

Of these classes, gravity-settling chambers are typically limited to particles larger than 325 mesh (44 μm) or the required chamber size becomes excessive to collect smaller particles. Consequently, this is not used as a particulate control device in fossil fuel-fired stationary heat and power applications. Similarly, with the increased requirement for very high performance dust collectors, there is little application for dry impingement collectors. Therefore, this section focuses on cyclone separators, electrostatic separators (ESPs), fabric filters, wet collectors/scrubbers, and hybrid systems, as these are the primary types of equipment used in fossil-fuel fired heat and power systems.

Table 3.3 Examples of particulate removal efficiencies for several particulate control devices

Particulate	Particle size (μm)			
Control device	**<1**	**1–3**	**3–10**	**>10**
High efficiency ESP	96.5	98.25	99.1	99.5
Fabric filter	100	99.75	>99.95	>99.95
Venturi scrubber	>70	99.5	>99.8	>99.8
Multicyclones	11	54	85	95

Natural gas systems do not have particulate control devices because natural gas does not contain mineral matter. Similarly, distillate fuel oils contain only trace quantities of mineral matter. Heavier grades of fuel oil and all ranks of coal use particulate control devices. The type of collector used is also a function of the size/firing rate of the heat and power system. Large power plants firing coal or oil typically use one of two major types of particle collection devices, which are ESPs or fabric filters. Smaller-sized units may use cyclones for their particulate control. Wet scrubbers, typically venturi scrubbers, are also used to capture particulate matter. Some of the situations under which the four basic types of particulate control equipment are used include [2]:

- ESPs are typically used when very high efficiencies are required for removing fine particulate matter and very large volumes of gas are to be handled. Table 3.3 lists some examples of ranges of collection efficiency for some different particulate removal devices; more information on removal efficiencies is presented in later sections when discussing various particulate removal devices.
- Fabric filters are typically used when very high efficiencies are required, the gas is always above its dewpoint, volumes are reasonably low, and temperatures are relatively low.
- Cyclone separators are used when the dust is coarse, concentrations are fairly high, classification is desired, and very high efficiency is not required.
- Wet scrubbers are typically used when fine particles need to be removed at a relatively high efficiency, cooling of the flue gas and moisture content in the flue gas are not objectionable, and gaseous as well as particulate matter need to be removed.

Table 3.4 presents a summary of particulate emissions control techniques pertinent to fossil fuel-fired stationary heat and power systems [modified from 26]. The table includes minimum particle sizes, ranges of efficiency, and advantages and disadvantages of each type of collector. The advantages and disadvantages listed in Table 3.4 give an indication of the considerations that enter into a determination of the cost of a particular device. The various control devices are discussed in more detail in the following sections.

3.4.1 Cyclone separators

Cyclone separators are gas-cleaning devices that utilize a centrifugal force created by a spinning gas stream to separate particulate matter from the gas. The separator

Table 3.4 Summary of particulate emissions control techniques for fossil fuel-fired stationary heat and power systems

Control device	Minimum particle size (μm)	Efficiency (%, mass basis)	Advantages	Disadvantages
Cyclone	5–25	50–90	Simplicity of design and maintenance Little floor space required Dry, continuous disposal of collected particles Low-to-moderate pressure loss Handles large particles	Much head room required Low collection efficiency of small particles Sensitive to variable dust loadings and flow rates
Wet Collectors Spray towers Cyclonic Venturi	 >10 >2.5 >0.5	 <80 <80 <80	Simultaneous gas absorption and particle removal Ability to cool and clean high-temperature, moisture-laden gases Corrosive gases and mists can be recovered and neutralized Reduced dust explosion risk Efficiency can be varied	Corrosion and erosion problems Added cost of wastewater treatment and reclamation Low efficiency on submicron particles Contamination of effluent stream by liquid entrainment Freezing problems in cold weather Reduction in buoyancy and plume rise Water vapor contributes to visible plume
Electrostatic Precipitator	<1	95–99	99+% efficiency obtainable Very small particles can be collected	Relatively high initial cost Precipitators are sensitive to variable dust

(*Continued*)

Table 3.4 (Continued)

Control device	Minimum particle size (μm)	Efficiency (%, mass basis)	Advantages	Disadvantages
			Particles may be collected wet or dry Pressure drops and power requirements are small compared to other high-efficiency collectors Maintenance is nominal unless corrosive materials are handled Few moving parts Can be operated at high temperatures	loadings, flow rates or resistivity Precautions are required to safeguard personnel from high voltage Collection efficiencies can deteriorate gradually and imperceptibly
Fabric Filtration	<1	>99	Dry collection possible Decrease of performance is noticeable Collection of small particles possible High efficiency possible	Sensitivity to filtering velocity High-temperature gases must be cooled Affected by relative humidity (condensation) Susceptibility of fabric to chemical attack

unit may be a single chamber, and a number of small chambers in parallel or series, or a dynamic unit similar to a blower [2,18,26]. They are used for the collection of medium-sized and coarse particles. Their relatively simple construction, along with no moving parts, result in lower capital and maintenance costs than for bagfilters and ESPs. On their own, they do not meet stringent emissions standards but are used as precleaners for other, more expensive control devices.

Cyclone separators are usually used for removing particles 10 μm in size and larger. However, conventional cyclones seldom remove particles with an efficiency

Table 3.5 Collection efficiency of particle size ranges for different types of cyclones

Particle size range	Cyclone type		
	Conventional	High efficiency	High volume
<5 μm	<50%	30–70%	<25%
5–20 μm	50–80%	80–95%	25–65%
20–50 μm	80–95%	95–99%	65–95%
>50 μm	95–99%	95–99%	95–99%

greater than 90% unless the particle size is 25 μm or larger. High-efficiency cyclones can remove particles down to 5 μm. Regardless of the design, the fractional removal efficiency of any cyclone drops significantly beyond a certain particle size. This is shown in Table 3.5 [modified from 2].

3.4.1.1 Operating principles

Conventional cyclones can be categorized as 1) reverse-flow cyclones (tangential inlet and axial inlet), 2) straight-through-flow cyclones, and 3) impeller collectors [27]. A standard tangential inlet vertical reverse flow cyclone separator is shown in Figure 3.8. The gas flow is forced to follow the curved geometry of the cyclone while the inertia of particles in the flow causes them to move toward the outer wall, where they collide and then drop due to gravity. The particles slide down the walls and into the storage hopper. At the bottom of the cyclone the clean gas flow reverses to form a smaller inner core that leaves at the top of the unit. A vortex finder tube extending downward into the cylinder aids in directing the inner vortex out of the device.

In a straight-through-flow cyclone, the inner vortex of gas leaves at the bottom with initial centrifugal motion being imparted by vanes at the top. This type of cyclone is often used as a precleaner to remove large particles [27]. This unit operates with low-pressure drop and can handle large volumetric flow rates.

In the impeller collector, gases enter normal to a many-bladed impeller and exit by the impeller around its circumference while the particles are thrown into an annular slot around the periphery of the device [27]. This unit is compact in design but has a tendency for particles plugging.

Cyclone collection efficiency increases with increasing particle size, particle density, inlet gas velocity, cyclone body length, number of gas revolutions, and smoothness of the cyclone wall [27]. Cyclone efficiency decreases with increasing cyclone diameter, gas outlet diameter, and gas inlet area. The design of a cyclone represents a compromise among collection efficiency, pressure drop, and size. Higher efficiencies require higher pressure drops and larger unit sizes. The major variables that affect the fractional collection efficiency of a cyclone are shown in Eq. (3.1). As the particle-laden gas enters the cyclone, it spins through N_e revolutions in the main outer vortex before entering the inner vortex and passing upward toward the exit of the cyclone. As an approximation, N_e is given by

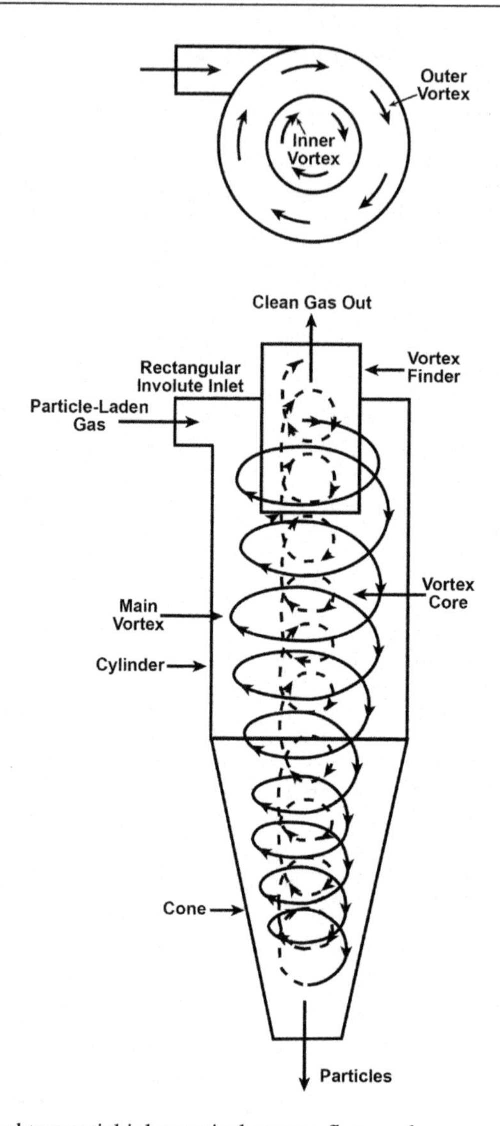

Figure 3.8 A standard tangential inlet vertical reverse flow cyclone separator.

$$N_e = \frac{1}{H}\left(L_c + \frac{Z_c}{2}\right) \tag{3.1}$$

where L_c is the height of the main upper cylinder, Z_c is the height of the lower cone, and H is the height of the rectangular inlet. See Figure 3.9 for cyclone separator proportions, which are those of the classic design of Shepherd and Lapple [modified from Refs. 18 and 26].

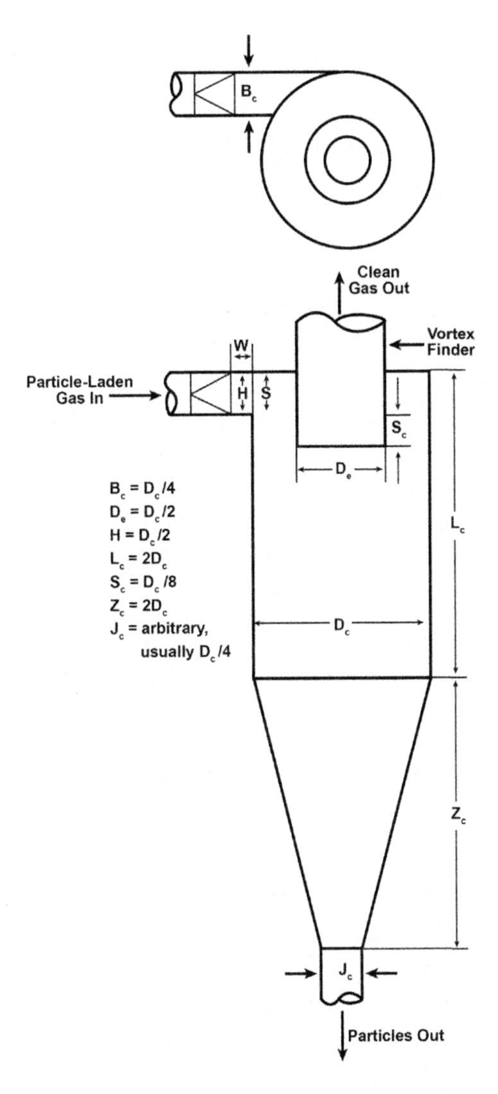

Figure 3.9 Cyclone separator proportions.

Cyclones are usually designed with geometric similarity based on the classic work by Shepherd and Lapple that still serve as the basis for cyclone design. Cyclones are designed such that the ratios of the dimensions remain constant at different diameters, and these dimensions can be expressed in terms of the body diameter D_c. The dimensions shown in Figure 3.9 are standard cyclone proportions. Table 3.6 shows cyclone proportions for high-efficiency and high-throughput types [modified from Ref. 18].

Table 3.6 Characteristics of cyclones

	Cyclone type		
	High efficiency	**Conventional**	**High throughput**
Height of inlet			
H/D_c	0.44–0.5	0.5	0.78–0.8
Width of inlet			
W/D_c	0.2–0.21	0.25	0.35–0.375
Diameter of gas exit			
D_e/D_c	0.4–0.5	0.5	0.75
Length of vortex finder			
S/D_c	0.5	0.6–0.625	0.85–0.875
Length of body			
L_c/D_c	1.4–1.5	1.75–2.0	1.5–1.7
Length of cone			
Z_c/D_c	2.5	2.0	2.0–2.5
Diameter of dust outlet			
J_c/D_c	0.375–0.4	0.25–0.4	0.375–0.4

3.4.1.2 Determining cyclone efficiency

Cyclone collection efficiency is a strong function of particle size and increases with increasing particle size. The collection efficiency of a single particle size can be determined either by a semiemperical approach, developed by Lapple in 1951 [28], in which the gas flow is considered laminar, or by one of several theoretical equations that have been developed since, specifically the theory developed by Leith and Licht [29]. The semiemperical approach has limitations in that the gas flow in a cyclone is not simply laminar and theoretical equations have been developed to account for this flow, which is also not fully turbulent because the boundary layer has significant depth [18,27].

3.4.1.2.1 Laminar flow approach

Assuming laminar flow, the fractional collection efficiency is defined as

$$\eta_d = \frac{\pi N_e \rho_p {d_p}^2 V_g}{9\mu W} = \frac{\pi N_e \rho_p d_p^2 Q}{9\mu H W^2} \tag{3.2}$$

where η_d is the fractional efficiency, N_e the effective number of turns, ρ_p the particle density, d_p the particle diameter, V_g the gas velocity, μ the gas viscosity, Q the volumetric flow rate of gas, and H and W defined as in Figure 3.9. This equation has one main deficiency in that it predicts that there is a finite value of d_p beyond which the collection efficiency is always 100% whereas experimental evidence shows that efficiency approaches 100% asymptotically with increasing particle diameter.

3.4.1.2.2 Cut diameter approach

To address the finite value problem in Eq. (3.2), equations have been developed realizing the size of particle collected with an efficiency of 50% to other parameters of interest. Lapple developed an empirical expression for the collection efficiency based on Eq. (3.2) and the particle cut size, d_{50}. If η_d in Eq. (3.2) is set to 0.5, solving for d_p gives

$$d_{p,50} = \left[\frac{9\mu W}{2\pi N_e V_g \rho_p}\right]^{1/2} = \left[\frac{9\mu W^2 H}{2\pi N_e \rho_p Q}\right]^{1/2} \quad (3.3)$$

where $d_{p,50}$ is the particle size collected with 50% efficiency. Lapple correlated data from cyclones of similar proportions and generated the curve in Figure 3.10 [28]. This graph has been fitted to an algebraic equation by others, which makes it more convenient for computer modeling [18]:

$$\eta_j = \frac{1}{\left[1 + (d_{p50}/d_{pj})^2\right]} \quad (3.4)$$

where d_{p50} is the particle cut diameter, d_{pj} the particle diameter in range j, and η_j the fractional efficiency in range j.

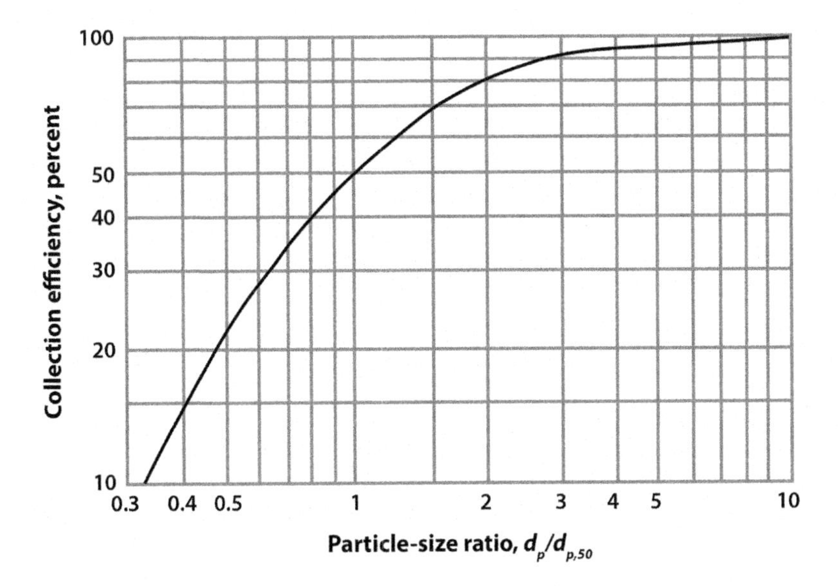

Figure 3.10 Cyclone efficiency versus particle size ratio.

3.4.1.2.3 More advanced approach

As previously mentioned, the laminar flow model has limitations in that the gas flow in the cyclone is neither laminar nor fully turbulent; therefore, other models have been developed. A theory developed by Leith and Licht [29] has proved useful in practical cyclone design. In their model, cyclone collection efficiency is given by

$$\eta(D_p) = 1 - \exp(-MD_p^N) \tag{3.5}$$

where

$$N = 1/(n+1) \tag{3.6}$$

$$n = 1 - (0.67D_c^{0.14})\left(\frac{T}{283}\right)^{0.3} \tag{3.7}$$

and

$$M = 2\left[\frac{KQ}{D_c^3}\frac{\rho_p(n+1)}{18\mu}\right]^{N/2} \tag{3.8}$$

where D_c is the cyclone diameter in meters, T is the gas temperature in Kelvin, D_p is the particle diameter in centimeters, ρ_p is the particle density in g/cm^3, Q is the volumetric flow rate in m^3/s, μ is the gas viscosity in g/cm-s and K is a geometric configuration parameter that depends only on the relative dimensions of the unit. For the conventional cyclone design configuration shown in Figure 3.9, $K = 402.9$ [27].

3.4.1.3 Pressure drop

Pressure drop is an important operating parameter because increased pressure drop results in higher operating costs. Higher cyclone efficiencies typically come with higher pressure drop; therefore, a compromise must be made between collection efficiency and pressure drop. Often, a cyclone must be designed around a given system pressure drop. There are several pressure drop equations in the literature but most have not been found to be more accurate than the equation developed by Lapple:

$$\Delta P = \frac{1}{2}\rho_g V_g^2 H_v \tag{3.9}$$

where ΔP is the pressure drop (Pa), ρ the gas density (kg/m^3), V_g the gas velocity (m/s) and H_v the number of velocity heads defined by

$$H_v = K\frac{HW}{D_e^2} \tag{3.10}$$

where H, W, and D_e are defined in Figure 3.9 and K is an empirical constant with a value of 16 for a tangential inlet cyclone and 7.5 for one with an inlet vane [18].

More complex equations can be found elsewhere and take into account variables such as inlet contraction, particle acceleration, barrel friction, gas flow reversal, and exit contraction [26]. These pressure drops are added together to determine a total pressure drop; however, the actual pressure drop observed is a function of the solids loading. The pressure drop is high when the gas is free of solids and then decreases as the solids loading increases. Correlations have been developed where the ratio of true ΔP/calculated ΔP is plotted as a function of solids loading (lb of solids/second times square foot of cyclone inlet area) to generate true ΔP values.

3.4.1.4 Factors that affect cyclone performance

There are many factors that can affect cyclone performance. A cyclone must be airtight for proper performance. Any leakage in a cyclone can cause a loss in collection efficiency. Erosion can affect cyclone performance by altering the flow patterns. Erosion can also lead to cyclone failure by eroding welds or areas of concentrated solids impingement. Fouling of a cyclone can occur by plugging the dust outlet or by the build-up of dust on the walls, thereby affecting flow patterns and causing ash reentrainment. Cyclone roughness (e.g. large weld beads) reduces cyclone efficiency. Cyclone efficiency is a strong function of particle size and if the particle size decreases from the design specifications then cyclone efficiency will also decrease. Similarly, cyclone efficiency is a function of particle density and if the particle density decreases from that for which the cyclone was designed, then collection efficiency will decrease. Dust collection increases with increasing dust loading. If the dust loading should decrease from the design parameter, then cyclone efficiency will decrease. Physical properties of the gas can have some effect on the cyclone performance. Increasing the gas temperature decreases its density and increases its viscosity, thereby reducing collection efficiency (see Eq. (3.2)).

3.4.2 Electrostatic precipitators

Particulate and aerosol collection by electrostatic precipitation is based on the mutual attraction between particles of one electrical charge and a collection electrode of opposite polarity. This concept was pioneered by F. G. Cottrell in 1910 [2]. The advantages of this technology are the ability to handle large gas volumes (ESPs have been built for volumetric flow rates up to 4,000,000 ft^3/min), achieve high collection efficiencies (which vary from 99 to 99.9%), maintain low pressure drops (0.1–0.5 inches of water column), collect fine particles (0.5–200 μm), and operate at high gas temperatures (gas temperatures up to 1,200°F can be accommodated). In addition, the energy expended in separating particles from the gas stream acts solely on the particles and not on the gas stream.

ESPs have been used in the control of particulate emissions from coal-fired boilers used for steam generation for about 60 years [2]. Initially, all ESPs were installed downstream of the air preheaters at temperatures of 270–350°F, and are referred to as cold-side ESPs. ESPs have been installed upstream of air preheaters where the temperature is in the range of 600–750°F (i.e. hot-side ESPs) as a result of using low-sulfur fuels with lower fly ash resistivity.

In the early 1970s, ESPs were the preferred choice for a high-efficiency particulate control device [18]. Nearly 90% of United States coal-based electric utilities use ESPs to collect fine particles [30].

3.4.2.1 Operating principles

There are several basic geometries used in the design of ESPs but the common design used in the power generation industry is the plate and wire configuration. In this design, shown schematically in Figure 3.11, the ESP consists of a large hopper-bottomed box containing rows of plates forming passages through which the flue gas flows. Centrally located in each passage are electrodes energized with high-voltage (45–70 kV), negative-polarity, direct current (dc) provided by a transformer-rectifier set [31]. Examples of various designs of rigid discharge electrodes are shown schematically in Figure 3.12 [32]. The most commonly used discharge electrode in the United States is the weighted wire electrode while the rigid frame electrode is commonly used in Europe [32]. The flow is usually horizontal and the passageways are typically 8–10 inches wide. The height of a plate varies from 18 to 40 feet with a length of 25 to 30 feet. The ESP is designed to reduce the flue gas flow from 50–60 feet/s to less than 10 feet/s as it enters the ESP so the particles can be effectively collected.

Electrostatic precipitation consists of three steps: 1) charging the particles to be collected via a high-voltage electric discharge; 2) collecting the particles on the

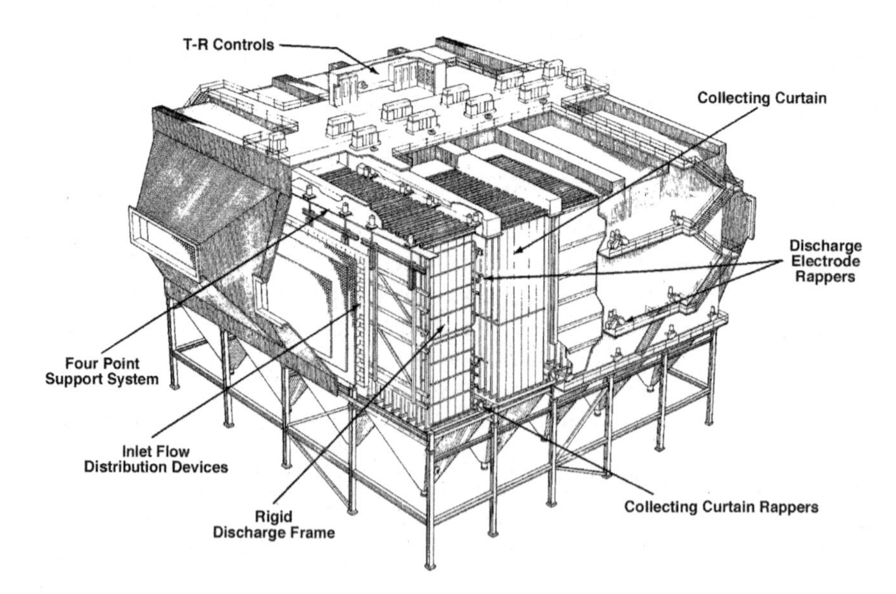

Figure 3.11 Electrostatic precipitator.

surface of an oppositely charged collection surface; and 3) cleaning the collection surface. These are illustrated in Figure 3.13 (modified from [9]).

The electrodes discharge electrons into the flue gas stream, ionizing the gas molecules. These gas molecules, with electrons attached, form negative ions. The gas is heavily ionized in the vicinity of the electrodes, resulting in a visible blue corona effect. The fine particles are then charged through collisions with the

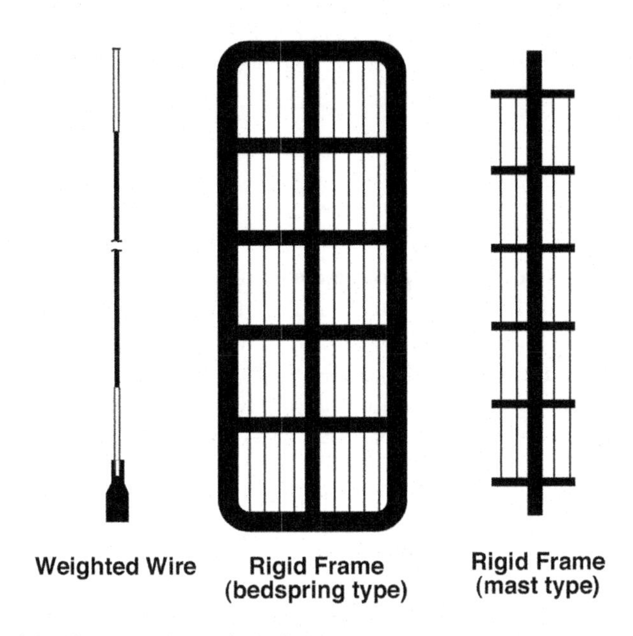

Figure 3.12 Rigid discharge electrode designs.

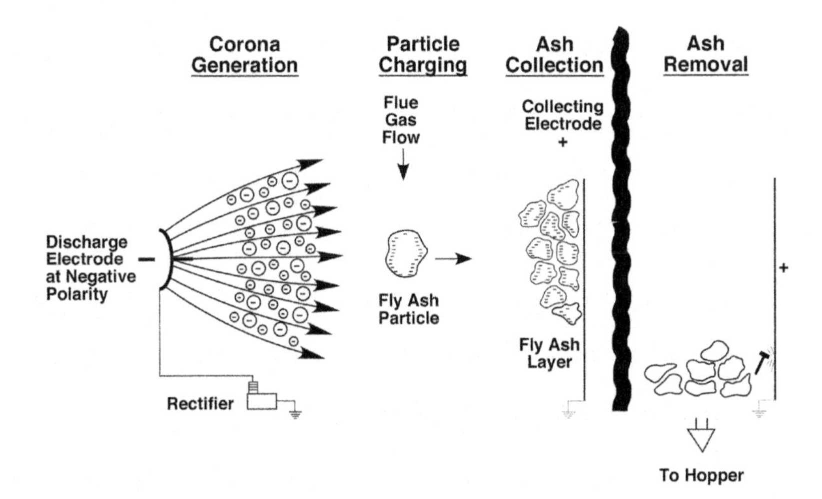

Figure 3.13 Basic concept of charging and collecting particles in an ESP.

negatively charged gas ions, resulting in the particles becoming negatively charged. Under the large electrostatic force, the negatively charged ash particles migrate out of the gas stream toward the grounded plates where they collect, forming an ash layer. These plates are periodically cleaned by a rapping system to release the layer into the ash hoppers as an agglomerated mass.

The speed at which the migration of the ash particles takes place is known as the migration or drift velocity. It depends upon the electrical force on the charged particle as well as the drag force developed as the particle attempts to move perpendicular to the main gas flow toward the collecting electrode [2]. The drift velocity, w, is defined as

$$w = \frac{2.95 \times 10^{-12} p E_c E_p d_p}{\mu_g} K_c \tag{3.11}$$

where w is in meters per second, p is the dielectric constant for the particles (which typically lies between 1.50 and 2.40), E_c is the strength of the charging field (v/m), E_p is the collecting field strength (v/m), d_p is the particle diameter (μm), K_C is Cunningham correction factor for particles with a diameter less than roughly 5 μm (dimensionless), and μ_g is the gas viscosity (kg/m s).

The Cunningham correction factor in Eq. (3.11) is defined as [2]

$$K_C = 1 + \frac{2\lambda}{d_\mathrm{p}}\left[1.257 + 0.400 e^{\left(\frac{-0.55 d_p}{\lambda}\right)}\right] \tag{3.12}$$

where λ is the mean free path of the molecules in the gas phase. This quantity is given by

$$\lambda = \frac{\mu_g}{0.499 \rho_g u_m} \tag{3.13}$$

where u_m is the mean molecular speed (m/s) and ρ_g is the gas density (kg/m^3). From the kinetic theory of gases, u_m is given by

$$u_m = \left[\frac{8 R_u T}{\pi M}\right]^{\frac{1}{2}} \tag{3.14}$$

where M is the molecular weight of the gas, T is temperature (°K), and R_u is the universal gas constant (8.31×10^3 m^2/s^2 mole °K).

The drift velocity is used to determine collection efficiency using the Deutsch-Anderson equation:

$$\eta = 1 - e^{\left(-\frac{wA}{Q}\right)} \tag{3.15}$$

where w is the drift velocity, A is the area of collection electrodes, and Q is the volumetric flow rate. The units of w, A, and Q must be consistent since the factor wA/Q is dimensionless.

The ratio, A/Q, is often referred to as the specific collection area (SCA) and is the most fundamental ESP size descriptor [32]. Collection efficiency increases as SCA and w increase. The value of w increases rapidly as the voltage applied to the emitting voltage is increased; however, the voltage cannot be increased above that level at which an electric short circuit, or arc, is formed between the electrode and ground.

The collecting plates are periodically cleaned to release the layer into the ash hoppers as an agglomerated mass by a mechanical (rapping) system in a dry ESP or by water washing in the case of a WESP. The hopper system must be adequately designed to minimize ash reentrainment into the gas stream until the hopper is emptied. The strength of the electric field and ash bonding on the plates, mass gas flow, and the striking energy must be matched to ensure that ash is not reentrained into the gas stream [33]. The ideal situation is where the electric field holding the ash layer that is directly adjacent to the plate is of such strength that the strike energy just breaks this bond and gravity dislodges the particulate matter into the ash hopper.

3.4.2.2 *Factors that affect esp performance*

There are several factors that affect ESP performance. Of these, fly ash resistivity is the most important.

3.4.2.2.1 Fly ash resistivity

Fly ash resistivity plays a key role in dust-layer breakdown and the ESP performance. Resistivity is dependent on the flue gas temperature and chemistry, and the chemical composition of the ash itself. Electrostatic precipitation is most effective in collecting dust in the resistivity range of 10^4 to 10^{10} ohm-cm [2]. In general, resistivities above 10^{11} ohm-cm are considered to be a problem because the maximum operating field strength is limited by the fly ash resistivity. Back corona, the migration of positive ions generated in the fly ash layer towards the emitting electrodes, which neutralize the negatively charged particles, will result if the ash resistivity is greater than 10^{12} ohm-cm. If the fly ash resistivity is below 2×10^{10} ohm-cm, it is not considered to be a problem because the maximum operating field strength is limited by other factors other than resistivity.

Examples of low and high resistivity fly ashes are shown in Figure 3.14 where resistivity is plotted as a function of temperature for four United States lignite (two from North Dakota) and subbituminous samples (two from the Powder River Basin) [34]. The differences in fly ash resistivity are due to variations in ash composition. The low resistivity fly ashes were produced from coals that contained higher levels of sodium in the coal ash. Higher sodium levels result in lower resistivity. Similarly, higher concentrations of iron result in lower resistivity. Higher levels of calcium and magnesium have the opposite effect on resistivity. This is illustrated in Figure 3.15 where the fly ash resistivities of two Texas lignites are shown along with

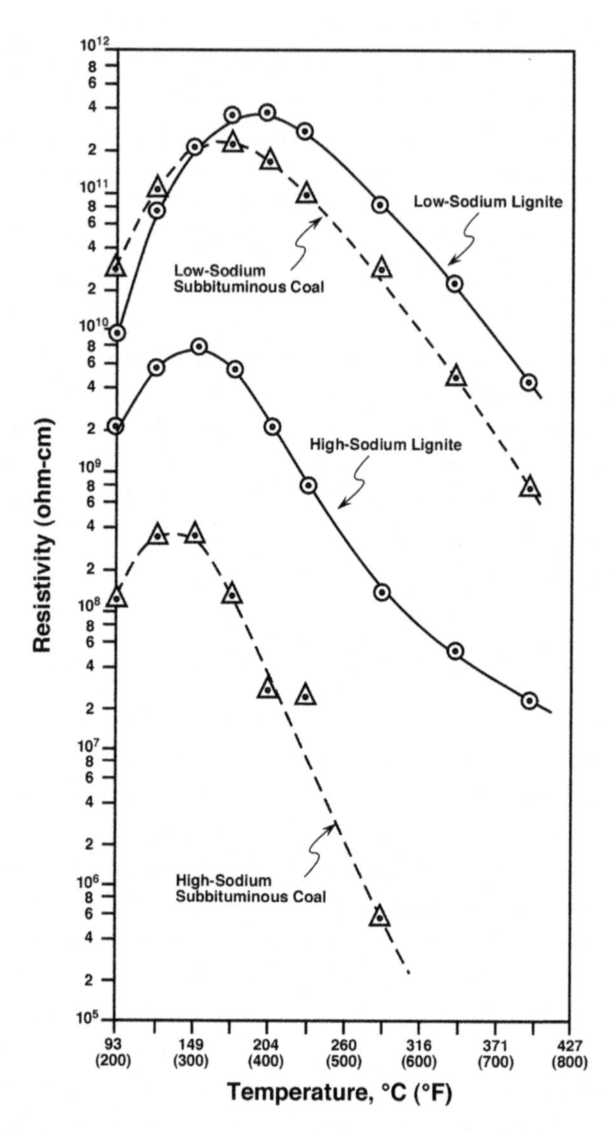

Figure 3.14 Illustration of effect of ash composition on fly ash resistivity for coals from the same geographical location.

the fly ash resistivities from the same two coals when injecting limestone for SO_2 control [35]. The addition of calcium through sorbent injection resulted in increasing the fly ash resistivities.

Flue gas properties also affect fly ash resistivity. The two properties that have the most influence on ash resistivity are temperature and humidity. The effect of temperature can be observed in Figures 3.14 and 3.15. Similarly, as moisture content in the flue gas is increased, the fly ash resistivity decreases.

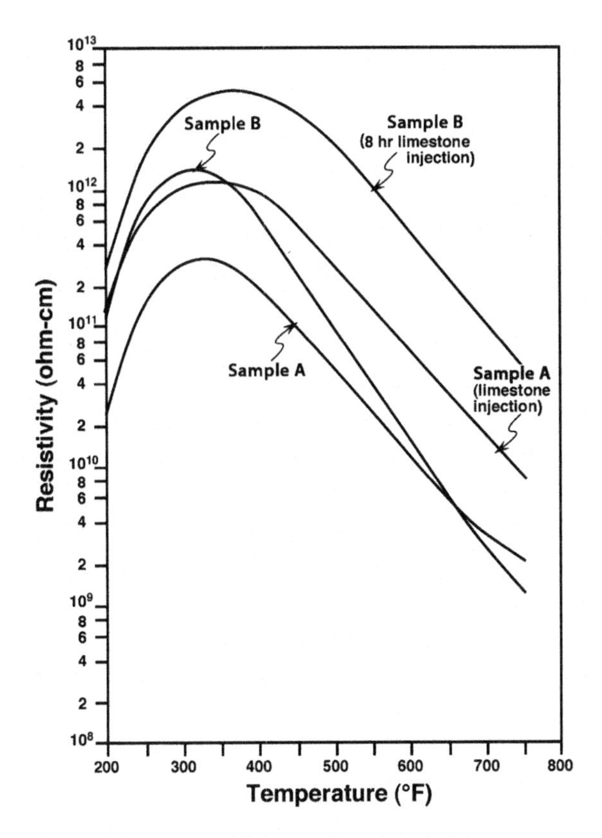

Figure 3.15 The effect of limestone addition on fly ash resistivity.

The dome-shaped curves shown in Figures 3.14 and 3.15 are typical of fly ashes. The shape of the curves is due to a change in the mechanism of conduction through the bulk layer of particles as the temperature is varied [2]. The predominant mechanism below 300°F is surface conduction where the electric charges are carried in a surface film adsorbed on the particle. As the temperature is increased above 300°F, the phenomenon of adsorption becomes less effective and the predominant mechanism is volume or intrinsic conduction. Volume conduction involves passage of electric charge through the particles.

3.4.2.2.2 Other factors

The three primary mechanical deficiencies in operating units are gas sneakage, fly ash reentrainment, and flue gas distribution [32]. Flue gas sneakage, i.e. flue gas that is bypassing the effective region of the ESP, increases the outlet dust loading. Reentrainment occurs when individual dust particles are not collected in the hoppers but are caught up in the gas stream, increasing the dust loading to the ESP and resulting in higher outlet dust loadings. Nonuniform flue gas distribution throughout the entire cross-section of the ESP decreases the collection ability of the unit.

There are many additional factors that can affect the performance of an ESP including the quality and type of fuel. Changes in coal and ash composition, grindability, and the burner/boiler system are important. Fly ash resistivity increases with decreasing sulfur content, an issue that must be considered when switching to lower sulfur coals. Moisture content and ash composition affect resistivity as discussed earlier. Changes in coal grindability can affect pulverizer performance by altering particle size distribution, which in turn can impact combustion performance and ESP performance. Modifications to the boiler system can affect temperatures or combustion performance and thereby impact ESP performance.

3.4.2.3 Methods to enhance esp performance

Difficulties in collecting high-resistivity fly ash and fine particulates has led to very large units being specified, unacceptable increases in ESP power consumption, and, in extreme cases, the use of fabric filters in lieu of ESPs [32]. As a result, concepts have been developed to overcome technical limitations and maintain competitiveness with fabric filters. This includes [32]: pulse energization where a high-voltage pulse is superimposed on the base voltage to enhance ESP performance during operation under high-resistivity conditions; intermittent energization where the voltage to the ESP is turned off during selected periods allowing for a longer period between each energization cycle and limiting the potential for back corona; and wide plate spacing to reduce capital and maintenance costs and allow for thicker discharge electrodes and increased current density.

Another approach to achieving electrical resistivities in the desired range is the addition of conditioning agents to the flue gas stream. This technique is applied commercially to both hot-side and cold-side ESPs. Conditioning modifies the electrical resistivity of the fly ash and/or its physical characteristic by changing the surface electrical conductivity of the dust layer deposited on the collecting plates, increasing the space charge on the gas between the electrodes, and/or increasing dust cohesiveness to enlarge particles and reduce rapping reentrainment losses [32]. Over 200 utility boilers are equipped with some form of conditioning in the United States [32].

The most common conditioning agents are sulfur trioxide (SO_3), ammonia (NH_3), compounds related to them, and sodium compounds. Sulfur trioxide is most widely applied for cold-side ESPs while sodium compounds are used for hot-side ESPs [32]. While results vary between coal and system, the injection of 10–20 ppm of SO_3 can reduce the resistivity to a value that will permit good collection efficiencies. In select cases, SO_3 injection of 30–40 ppm has resulted in reductions of fly ash resistivity of 2–3 orders of magnitude (e.g. from 10^{11} to $\approx 10^8$ ohm-cm) [2]. Disadvantages of SO_3 injection systems include the possibility of plume color degradation. Disadvantages for sodium compounds are the potential problems with increased deposition and interference from certain fuel constituents, which affects the economics of the injection [2]. Combined SO_3-NH_3 conditioning is used with the SO_3 adjusting the resistivity downward while the NH_3 modifies the space-charge effect, improves agglomeration, and reduces rapping reentrainment losses [2].

3.4.2.4 Wet ESPs

Dry ESPs, which have been discussed up to this point, have been successfully used for many years in utility applications for coarse and fine particulate removal. Dry ESPs can achieve 99 + percent collection efficiency for particles 1–10 μm in size. However, dry ESPs cannot remove toxic gases and vapors that are in a vapor state at 400°F, cannot efficiently collect very small fly ash particles, cannot handle moist or sticky particulate that would stick to the collection surface, require much space for multiple fields due to reentrainment of particles, and rely on mechanical collection methods to clean the plates, which require maintenance and periodic shutdowns [36].

Wet electrostatic precipitators (WESPs) address these issues and are a viable technology to collect finer particulate than existing technology while also collecting aerosols. WESPs have been commercially available since their first introduction by F. G. Cottrell in 1907 [37]. An example of a WESP is shown in Figure 3.16. However, most of their use has been in small, industrial-type settings as opposed to

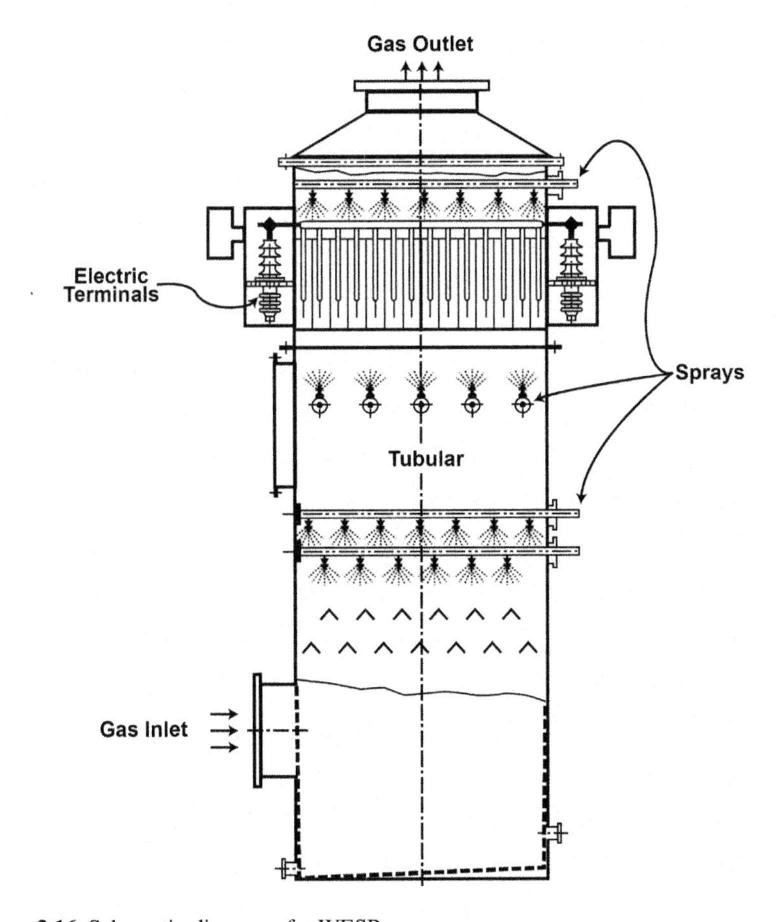

Figure 3.16 Schematic diagram of a WESP.

utility power plants. WESPs have been in service for ≈ 100 years in the metallurgical industry and in many other applications. They are used to control acid mists, submicron particulate (as small as 0.01 μm with 99.9% removal), mercury, metals, and dioxins/furans as the final polishing device within a multi-pollutant control system [36]. When integrated with upstream air pollution control equipment, such as an SCR, dry ESP and wet scrubber, multiple pollutants can be removed with the WESP acting as the final polishing device.

WESPs operate in the same three-step process as dry ESPs – charging, collecting, and cleaning of the particles from the collecting electrode [38]. However, cleaning of the collecting electrode is performed by washing the collection surface with liquid, rather than by mechanically rapping the collection plates.

WESPs operate in a wet environment in order to wash the collection surface; therefore, they can handle a wider variety of pollutants and gas conditions than dry ESPs [38]. WESPs find their greatest use where:

- The gas in question has a high moisture content;
- The gas stream includes sticky particulate;
- The collection of submicron particulate is required;
- The gas stream has acid droplets of mist; and
- The temperature of the gas stream is below the moisture dew point.

WESPs continually wet the collection surface and create a dilute slurry that flows down the collecting wall to a recycle tank, never allowing a layer of particulate cake to build up [38]. As a result, captured particulate is never reentrained. Also, when firing low-sulfur coal, which produces a high resistivity dust, the electrical field does not deteriorate and power levels within a WESP can be dramatically higher than in a dry ESP – 2,000 watts/1,000 scfm versus 100–500 watts/1,000 scfm, respectively.

Similar to a dry ESP, WESPs can be configured either as tubular precipitators (i.e. the charging electrode is located down the center of a tube) with vertical gas flow or as plate precipitators with horizontal gas flow [39]. For a utility application, tubular WESPs are appropriate as a mist eliminator above a flue gas desulfurization scrubber, while the plate type can be employed at the back end of dry ESP train for final polishing of the gas.

3.4.3 Fabric filters

Historically, ESPs have been the principal control technology for fly ash emissions in the electric power industry. Small, relatively inexpensive ESPs could be installed to meet early federal and state regulations. However, as particulate control regulations have become more stringent, ESPs have become larger and more expensive. Also, increased use of low-sulfur coal has resulted in the formation of fly ash with higher electric resistivity that is more difficult to collect. Consequently, ESP size and cost have increased to maintain high collection efficiency [40]. As a result, interest in baghouses has increased. Baghouses are a technology with extremely high collection efficiency (i.e. 99.9 to 99.99+ percent), are capable of filtering large volumes of flue gas, and whose size and efficiency are relatively independent of the type of coal burned [40]. Baghouses are essentially huge vacuum cleaners consisting of a large number of long, tubular filter bags arranged in parallel flow paths. As the ash-laden flue gas passes through these filters, the particulate is removed. Advantages of fabric filters include: high collection efficiency over a broad range of particle sizes; flexibility in design provided by the availability of various cleaning methods and filter media; wide range of volumetric capacities in a single installation, which may range from 100 to 5 million ft^3/min; reasonable operating pressure drops and power requirements; and ability to handle a variety of solid materials [2]. Disadvantages of baghouses include large footprints so space factors may prohibit consideration of baghouses, possibility of an explosion or fires if sparks are present in the vicinity of a baghouse, and hydroscopic materials usually cannot be handled, owing to cloth cleaning problems.

The first utility baghouse in the United States was installed on a coal-fired boiler in 1973 by the Pennsylvania Power and Light Company at its Sunbury Station [40]. This baghouse, as well as the next several baghouses installed, were small and it was not until 1978 that the first large baghouse was installed on a utility boiler. This baghouse serviced a 350 MW pulverized coal-fired boiler at the Harrington Station of Southwestern Public Service Company. Starting in 1978, there has been a steady increase in the installation of utility commitments to baghouse technology and there are more than 110 baghouses in operation on utility boilers in the United States servicing more than $\approx$26,000 MW of generating capacity [40,41].

3.4.3.1 Filtration mechanisms

Filtration occurs when the particulate-laden flue gas is forced through a porous, solid medium, which captures the particles. In a baghouse, this solid medium is the filter bag and/or the residual dust cake on the bag. The important filtering mechanisms are three aerodynamic capture mechanisms – direct interception, inertial impaction, and diffusion. Electrostatic attraction may also play a role with certain types of dusts/fiber combinations [2].

Direct interception occurs if the gas streamlines carrying the particles are close to the filter elements for contact. Inertial impaction occurs when the particles have sufficient momentum and cannot follow the gas stream when the stream is diverted by the filter element and the particles strike the filter. Diffusion results when the particle mass is very low and Brownian diffusion superimposes random motion on the streamline trajectory, thereby increasing the particle's probability of contacting and being captured by the filter [40]. Particles may be attracted to or repulsed by filters due to a variety of Coulombic and polarization forces. Particles larger than 1 μm are removed by impaction and direct interception, whereas particles from 0.001 to 1 μm are removed mainly by diffusion and electrostatic separation [2].

The effectiveness of a filter in capturing particles is reported in terms of collection efficiency or particle penetration. Particle penetration, P, is defined as the ratio of the particle concentration (mass or number of particles per unit volume of gas), also referred to as dust loading, on the outlet of the filter (i.e. cleaned flue gas stream) to that on the inlet side of the filter (i.e. dirty flue gas stream). Collection efficiency, η, is defined as:

$$\eta = 1 - P \tag{3.16}$$

Typically, both penetration and collection efficiency are multiplied by 100 and reported as a percent.

The filtration process can be divided into three distinct time regimes: 1) filtration by a clean fabric, which occurs only once in the life of a bag; 2) establishment of a residual dust cake, which occurs after many filtering and cleaning cycles; and 3) steady-state operation in which the quantity of particulate matter removed during the cleaning cycles equals the amount collected during each filter cycle [40]. In general, the initial collection efficiency of new filters is quite low (<99% and can be on the order of 75–90%), whereas a conditioned bag (i.e. a bag that has retained residual particles in the fibers of the filter that cannot be removed by cleaning) may have a collection efficiency of greater than 99.99%. A dust cake will form on the filters where the adhesive and cohesive forces acting between the particles and filter elements, and among the particles, respectively, are sufficiently strong to allow particulate agglomerates to bridge the filter pores. The accumulated dust cake forms a secondary filter of much higher efficiency than the clean fabric. On a seasoned bag, residual dust cakes generally weigh 10–20 times as much as the ash deposited during an average cleaning cycle [40].

3.4.3.2 Operating principles

Baghouses remove particles from the flue gas within compartments arranged in parallel flow paths with each compartment containing several hundred large, tube-shaped filter bags. Figure 3.17 is a cutaway view of a typical 10-compartment baghouse [40]. A baghouse on a 500 MW coal-fired unit may be required to handle in excess of 2 million ft^3/min of flue gas at temperatures of 250–350°F. From an inlet manifold, the dirty flue gas, with typical dust loadings from 0.1 to 10 grains/ft^3 of gas (0.23 to 23 grams/m^3), enters hopper inlet ducts that route it into individual compartment hoppers. From each hopper, the gas flows upward through the bags where the fly ash is deposited. The clean gas is drawn into an outlet manifold, which carries it out of the baghouse to an outlet duct. Periodic operation requires shutdown of portions of the baghouse at regular intervals for cleaning. Cleaning is accomplished in a variety of ways, including mechanical vibration or shaking, pulse jets of air, and reverse air flow.

The two fundamental parameters in sizing and operating baghouses are the air-to-cloth (A/C) ratio and pressure drop across the filters. Other important factors that affect the performance of the fabric filter include the flue gas temperature, dew point, and moisture content, and particle size distribution and composition of the fly ash [41].

The A/C ratio, which is a fundamental fabric filter descriptor denoting the ratio of the volumetric flue gas flow (ft^3/min) to the amount of filtering surface area (ft^2), is reported in units of ft/min [30]. For fabric filters, it has been generally

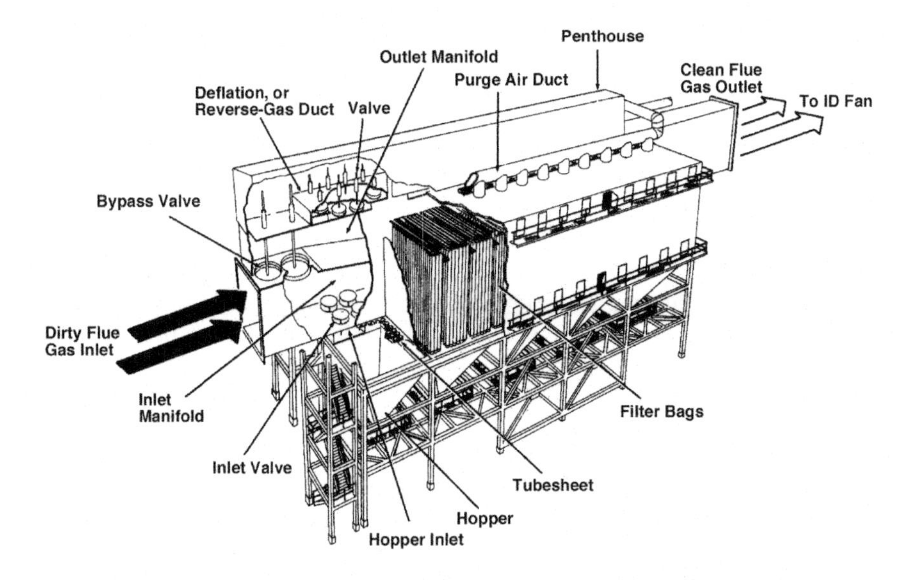

Figure 3.17 Cutaway view of a typical 10-compartment baghouse.

observed that the overall collection efficiency is enhanced as the A/C ratio, i.e. superficial filtration velocity decreases. Factors to be considered with the A/C ratio include type of filter fabric, type of coal and firing method, fly ash properties, duty cycle of the boiler, inlet fly ash loading, and cleaning method [42]. The A/C ratio determines the size of the baghouse and hence the capital cost.

Pressure drop is a measure of the energy required to move the flue gas through the baghouse. Factors affecting pressure drop are boiler type (which influence the fly ash particle size), filtration media, fly ash properties, and flue gas composition [42]. The pressure drop is an important parameter as it determines the capital cost and energy requirements of the fans.

As the filter cake accumulates on the supporting fabric, the removal efficiency typically increases; however, the resistance to flow also increases. For a clean filter cloth, the pressure drop is about 0.5 inches water column (W.C.) and the removal efficiency is low. After sufficient filter cake buildup, the pressure drop can increase to 2–3 inches W.C. with the removal efficiency greater than 99% [2]. When the pressure drop reaches 5–6 inches W.C., it is usually necessary to clean the filters.

The pressure drop for both the cleaned filter and the dust cake, ΔP_T, may be represented by Darcy's equation [2]:

$$\Delta P_T = \Delta P_R + \Delta P_C = \frac{\mu_g x_R V}{K_R} + \frac{\mu_g x_C V}{K_C} \tag{3.17}$$

where ΔP_R is the conditioned residual pressure drop, ΔP_C is the dust cake pressure drop, K_R and K_C are the filter and dust cake permeabilities, respectively, V is the superficial velocity, μ_g is the gas viscosity, and x_R and x_C are the filter and dust cake thicknesses, respectively. The permeabilities K_R and K_C are difficult quantities to predict with direct measurements since they are functions of the properties of the filter and dust such as porosity, pore size distribution, and particle size distribution. Therefore, in practice ΔP_R is usually measured after the bags are cleaned and ΔP_C is determined using the equation

$$\Delta P_C = K_2 C_i V^2 t \tag{3.18}$$

where C_i is the dust loading and, along with V, is assumed constant during the filtration cycle, t is the filtration time, and K_2, the dust resistance coefficient is estimated from

$$K_2 = \frac{0.00304}{(d_{g,mass})^{1.1}} \left(\frac{\mu_g}{\mu_{g,70^oF}}\right) \left(\frac{2600}{\rho_p}\right) \left(\frac{V}{0.0152}\right)^{0.6} \tag{3.19}$$

where d_g is the geometric mass median diameter (m), μ_g is the gas viscosity (kg/m-s), ρ_p is the particle density (kg/m^3), and V is the superficial velocity (m/s).

3.4.3.3 Basic types of fabric filters

There are three basic types of baghouses: reverse-gas, shake-deflate, and pulse-jet. They are distinguished by the cleaning mechanisms and by their A/C ratio. Air-to-cloth ratios for fabric filters range from a low of 1.0 to 12.0 ft/min depending on the type of cleaning mechanism used and characteristics of the fly ash [2]. The two most common baghouse designs are the reverse-gas and pulse-jet types.

Ash that accumulates on the bags in excess of the desired residual dust cake must be removed by periodic bag cleaning to reduce the gas flow resistance, and hence induced draft fan power requirements, and to reduce bag weight. In United States utility baghouses, cleaning is done off-line by isolating individual compartments for cleaning.

3.4.3.3.1 Reverse-gas

Reverse-gas fabric filters are generally the most conservative design of the fabric filter types. These typically operate at low A/C ratio ranging from 1.5 to 3.5 ft/min [2,41]. Fly ash collection is on the inside of the bags as the flue gas flow is from the inside of the bags to the outside as illustrated in Figure 3.18 [40]. Reverse-gas baghouses use off-line cleaning where compartments are isolated and cleaning air is passed from the outside of the bags into the inside causing the bags to partially collapse to release the collected ash. The dislodged ash falls into the hopper. A simplified schematic showing the cleaning cycle is illustrated in Figure 3.19 (modified from [9]). A variation of the reverse-gas cleaning method is the use of sonic energy for bag cleaning. With this method, low-frequency (<250–300 Hz), high sound pressure (0.3–0.6 inches W.C.) pneumatic horns are sounded simultaneously with the normal reverse-gas flow to add energy to the cleaning process. Reverse-gas fabric filters are widely used in the United States with approximately 90% of the utility baghouse employing this reverse-gas cleaning [40].

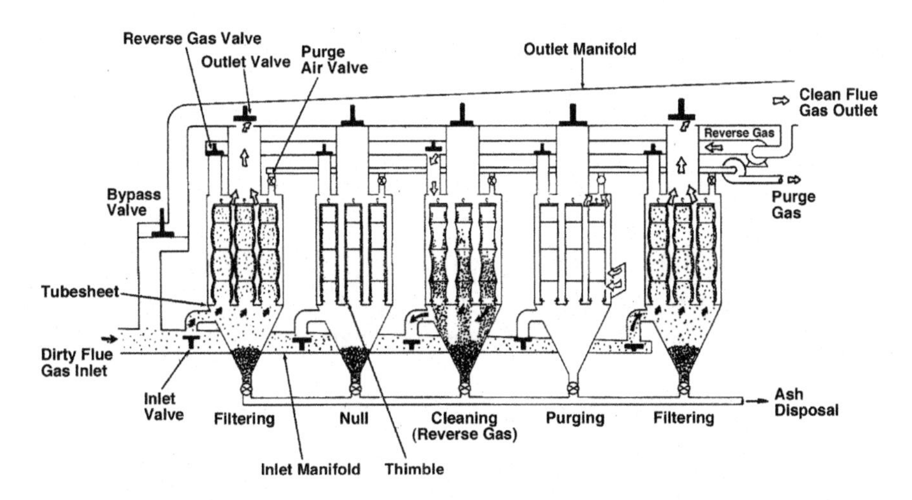

Figure 3.18 Schematic diagram of the compartments in a reverse-gas baghouse illustrating the flue gas and cleaning air flows during the various cycles of operation.

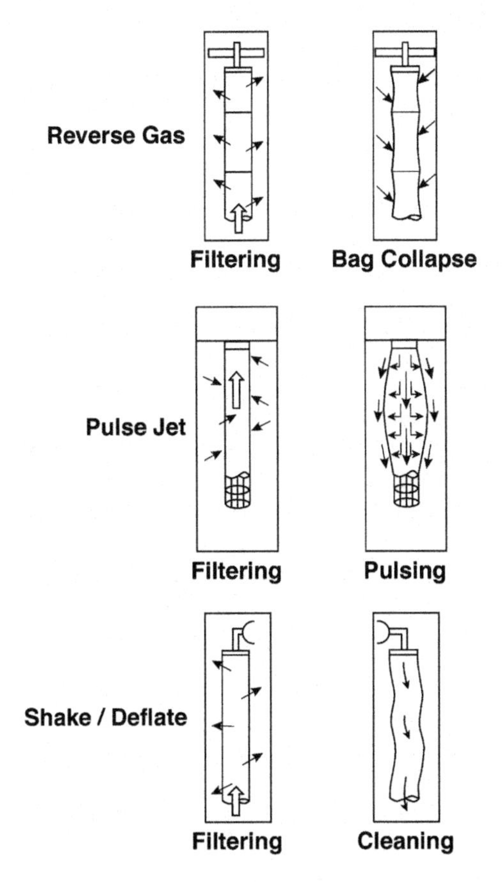

Figure 3.19 Simplified schematic diagrams of baghouse cleaning mechanisms.

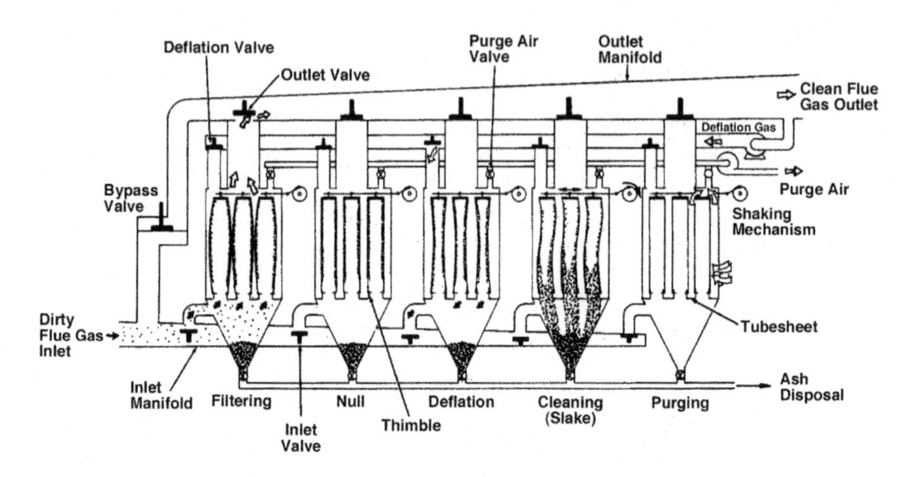

Figure 3.20 Schematic diagram of the compartments in a shake-deflate baghouse illustrating the flue gas and cleaning air flows during the various cycles of operation.

3.4.3.3.2 Shake-deflate

Shake-deflate baghouses are another low A/C type system (2 to 4 ft/min) and they collect dust on the inside of the bags similar to the reverse-gas systems [2]. With shake-deflate cleaning, a small quantity of filtered gas is forced backward through the compartment being cleaned, which is done off-line. The reversed filtered gas relaxes the bags but does not completely collapse them. As the gas is flowing or immediately after it is shut off, the tops of the bags are mechanically shaken for 5–20 s at frequencies ranging from 1 to 4 Hz and at amplitudes of 0.75 to 2 inches [40]. The operating cycles of a shake-deflate baghouse are illustrated in Figure 3.20 [40] with a simplified cleaning cycle illustrated in Figure 3.19. Operating experience with shake-deflate baghouses in utility service has been good [32].

3.4.3.3.3 Pulse-jet

In pulse-jet fabric filters, the flue gas flow is from the outside of the bag inward. This is illustrated in Figure 3.21 [40]. The A/C ratio is higher than reverse-air units and is typically 3 to 4 ft/min allowing for a more compact installation, but the ratio can vary from 2 to 5 ft/min [2]. Cleaning is performed with a high-pressure burst of air into the open end of the bag as shown in the simplified schematic of Figure 3.19. Pulse-jet systems required metal cages on the inside of the bags to prevent bag collapse. Bag cleaning can be performed on-line by pulsing selected bags while the remaining bags continue to filter the flue gas. Three cleaning methods have evolved for the pulse-jet systems [42]:

1) high pressure (40–100 psig), low volume pulse;
2) intermediate pressure (15–30 psig) and volume pulse; and
3) low pressure (7.5–10 psig), high volume pulse.

The first method is used mainly in the United States while the latter two methods are used mainly in larger boilers in Australia, Canada, and Western Europe [42].

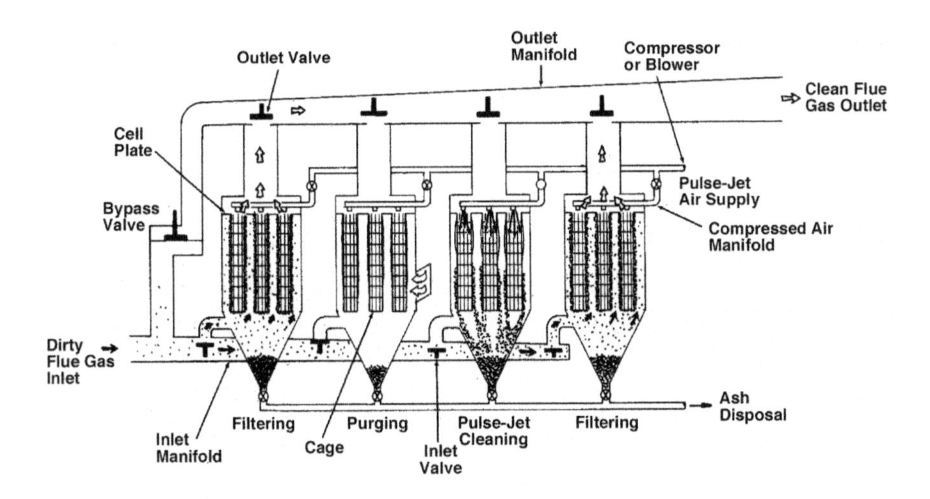

Figure 3.21 Schematic diagram of the compartments in a pulse-jet baghouse illustrating the flue gas and cleaning air flows during the various cycles of operation [40].

Pulse-jet cleaning results in lower resistance to gas flow than the other two baghouse types, thereby allowing smaller baghouses to filter the same volume of flue gas. Despite this, pulse-jet cleaning is not the preferred choice in United States for utility boilers because of concerns that the more rigorous cleaning method results in lower particulate collection efficiency and shorter bag life. Pulse-jet baghouses are used in the United States, as well as Japan, for industrial boilers [42]. In Canada and Europe, pulse-jet systems are used in industrial plants and some large-sized utility plants. Much work has been done on improving fabrics for the filters and the pulse-jet technology is becoming more attractive to utilities.

3.4.3.4 Fabric filter characteristics

Fabric filters are made from woven, felted, and knitted materials with filter weights that generally range from as low as 5 ounces/yd^2 to as high as 25 ounces/yd^2 [2]. Filtration media are selected depending on the type of baghouse, their efficiency in capturing particles, system operating temperature, physical and chemical nature of the fly ash and flue gas, durability for a long bag life, and the cost of the fabric. There is a tendency towards using needle felts or polytetrafluorethylene (PTFE) membranes on woven glass, due to their ability to withstand higher temperatures (during system upsets which result in temperature excursions) and improve bag performance [9]. To protect bags against chemical attack, the fabrics are usually coated with other materials such as Teflon, silicone, graphite, and GORE-TEX® [40].

The most common bag material in coal-fired utility units with pulse-jet fabric filters is polyphenylene sulfide (PPS) needle felt. In addition to PPS, fiberglass, acrylic, polyester, polypropylene, Nomex®, P84®, special high temperature fiberglass media, membrane covered media, and ceramic are used in various applications. A summary of selected filter media characteristics is provided in Box 3.1 [modified from 43]. Typical bag size is 127 or 152 mm (5 or 6 inches) diameter round or oval with a length of 3 to 8 m (10 to 26 ft).

3.4.3.5 Factors that affect baghouse performance

Key factors in proper baghouse design and operation are flue gas flow and properties, fly ash characteristics, and coal composition [32]. The baghouse must minimize pressure drop, maintain appropriate temperature and velocity profiles, and distribute the ash-laden flue gas evenly to the individual compartments and bags.

Particle size distribution of the fly ash and loading of the flue gas varies with type of combustion system [42]. Stoker-fired units produce ash with high carbon content, moderate loading, and large particle size distribution (compared to other combustion systems). Pulverized coal-fired systems produce ash with low carbon content, high loading, and fine particle size distribution. Cyclone-fired units produce ash with low carbon content, moderate loadings, and very fine particle size distribution. Fluidized-bed systems generally produce ash with high carbon content, high loading, and fine particle size distribution [32].

Sulfur content of the coal has been correlated to fabric filter operation. The cohesiveness of ash produced from high-sulfur coals is greater than from Western low-sulfur coals [32]. Also, maintaining the baghouse above the acid dew point is critical in high-sulfur coal applications.

The fly ash properties are important since they affect the adhesion and cohesion characteristics of the dust cake. This in turn affects the properties of the residual dust cake, collection efficiency, and cleanability of the bags.

3.4.3.6 Methods to enhance filter performance

The most recognized method to enhance fabric filter performance is the application of sonic energy, which was discussed previously. Virtually all reverse-gas baghouses have included sonic horns [32].

Gas conditioning has been explored for improving filter performance although this is not done commercially [32]. Low concentrations of ammonia and/or sulfur trioxide have been added in test programs to control fine particulate emissions and reduce pressure drop when firing low-rank fuels.

3.4.4 Wet collectors/scrubbers

Wet collectors, or scrubbers, form a class of devices in which a liquid, usually water, is used to assist or accomplish the collection of particles or aerosols. Due to excessive pressure drop and stringent particulate regulations, wet particulate scrubbers are now infrequently used in large fossil fuel-fired power plants as a primary collection device [43]. However, on most coal-fired units where wet flue gas

Box 3.1 Fabric Filter Media Characteristics

Glass

Glass fabrics offer outstanding performance in high heat applications. In general, by using a proprietary finish they become resistant to acids, except by hydrofluoric and hot phosphoric in their most concentrated forms. They are attached by strong alkalis at room temperature and weak alkalis at higher temperatures. Glass is vulnerable to damage caused by abrasion and flex. However, the proprietary finishes can lubricate the fibers and reduce the internal abrasion caused by flexing. Maximum operating temperature is 260°C (500°F).

Polyphenylene sulfide (PPS)

Polyphenylene sulfide fibers offer excellent resistance to acids, good to excellent resistance to alkalis, have excellent stability and flexibility, and provide excellent filtration efficiency. Maximum operating temperature is 190°C (375°F).

Acrylic

The resistance of homopolymer acrylic fibers is excellent in organic solvents, good in oxidizing agents and mineral and organic acids, and fair in alkalis. They dissolve in sulfuric acid concentrations. Maximum operating temperature is 127°C (260°F).

Polyester

Polyester fabrics offer good resistance to most acids, oxidizing agents, and organic solvents. Concentrated sulfuric and nitric acids are the exception. Polyesters are dissolved by alkalis at high concentrations. Maximum operating temperature is 132°C (270°F).

Polypropylene

Polypropylene fabrics offer good tensile strength and abrasion resistance. They perform well in organic and mineral acids, solvents, and alkalis. Polypropylene is attacked by nitric and chlorosulfonic acids, and sodium and potassium hydroxide at high temperatures and concentrations. Maximum operating temperature is 93°C (200°F).

Nomex®

Nomex fabrics resist attack by mild acids, mild alkalis, and most hydrocarbons. Resistance to sulfur oxides above the acid dew point at temperatures above 150°F is better than polyester. Flex resistance of Nomex is excellent. Maximum continuous operating temperature is 204°C (400°F).

P-84®

P-84 fabrics resist common organic solvents and avoid high pH levels. They provide good acid resistance. P-84 offers superior collection efficiency due to irregular fiber structure. Maximum operating temperature is 260°C (500°F).

Table 3.7 **Summary of particle scrubbers**

Scrubber type	Description
Spray/Tower	Particles are collected by liquid drops that have been atomized by spray nozzles. Horizontal and vertical gas flows are used, as well as spray introduced concurrent, countercurrent, or cross-flow to the gas. Collection efficiency depends on droplet size, gas velocity, liquid/gas ratio, and droplet trajectories. For droplets falling at their terminal velocity, the optimum droplet diameter for fine particle collection lies in the range 100 to 500 μm. Gravitational settling scrubbers can achieve cut diameters of about 2.0 μm. The liquid/gas ratio is in the range of 0.0001 to 0.01 m^3/m^3 of gas treated.
Venturi	A moving gas stream is used to atomize liquids into droplets. High gas velocities (60 to 120 m/s) lead to high relative velocities between gas and particles and promote collection.
Cyclonic	Drops can be introduced into the gas stream of a cyclone to collect particles. The spray can be directed outward from a central manifold or inward from the collector wall.

desulfurization (FGD) scrubbers are required in series with a high efficiency collector for control of acid gas emissions, the extra particulate removal of the FGD scrubber is an added benefit (FGD scrubbers are discussed in a later chapter). On smaller coal and heavy fuel oil-fired boilers, wet collectors are sometimes employed. This section discusses spray chambers, venturi scrubbers, and cyclonic scrubbers. Table 3.7 summarizes the particle scrubbing devices discussed in this section [modified from Ref. 27].

Scrubbing is a very effective means of removing small particles from a gas. Removal of particles results from collisions between dry particles and water drops thereby increasing the size of the particles to be removed. Fine particulates, ranging from 0.1 to 20 μm, can be effectively removed from a gas stream by wet collectors. The mechanism used in the removal equipment can be inertial impaction, gravitational settling, Brownian diffusion, diffusiophoresis, electrostatics, and thermophoresis. For particles greater than about 0.5 μm diameter, inertial impaction is usually the primary collection mechanism, and for particles smaller than about 0.05 μm diameter, Brownian diffusion is the primary collection mechanism [18].

The collection efficiency of wet collectors is related to the total energy loss in the equipment; the higher the scrubber power per unit of volume of gas treated, the better is the collection efficiency [27]. Almost all the energy is introduced in the gas; therefore, the energy loss can be measured by the pressure drop of gas through the unit.

The major advantage of wet collectors is the wide variety of types, allowing the selection of a unit suitable to the specific removal problem. As disadvantages, high pressure drops, and therefore energy requirements, must be maintained, and large volumes of scrubbing liquid must be handled and disposed.

3.4.4.1 Spray chambers/towers

A spray tower uses liquid droplets formed by flowing the liquid through spray nozzles, which is then contacted with the flue gas containing the particles to be removed. Figure 3.22 is a schematic of a simplified spray tower configuration. Horizontal and vertical gas flow paths have been used with the spray introduced such that the liquid drops travel in a concurrent, countercurrent, or cross-flow direction with respect to the gas direction. In the case of an upward flow chamber such as that shown in Figure 3.22, the flue gas flows upward and the particles collide with liquid droplets and are captured on the drops. In the vertical countercurrent gravitational spray tower, the large drops fall due to gravity while the smaller drops with settling velocities less than the upward gas velocity will travel upward to the mist eliminator.

The overall spray chamber efficiency for a specific size particle and droplet diameter is

$$\eta_{\text{overall}} = 1 - \exp\left[-\frac{3}{2}\eta_I \frac{Q_L}{Q_G}\frac{z}{d_D}\frac{V_T}{V_T - V_g}\right] \tag{3.20}$$

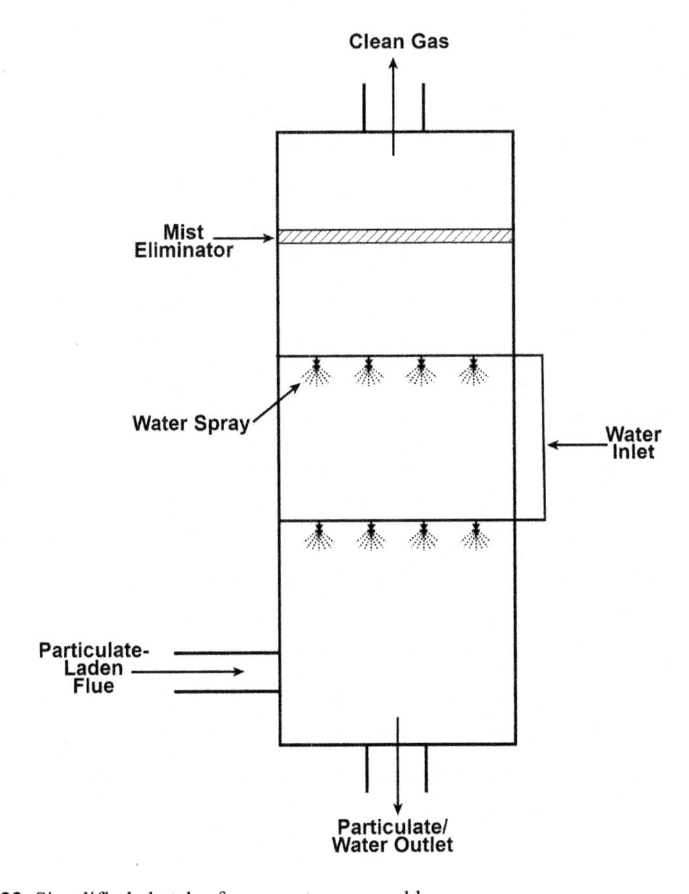

Figure 3.22 Simplified sketch of a spray tower scrubber.

where Q_L/Q_G is the ratio of the liquid/gas flowrates (in consistent units), z is the tower height, V_T is the droplet's theoretical terminal settling velocity, d_D is the diameter of the falling drops, V_g is the upward gas velocity, and η_I is the efficiency due to impaction [2,27]. The droplet's terminal velocity, V_T, is defined as

$$V_T = \frac{g d_p^2 \rho_p}{18 \mu_g} \tag{3.21}$$

where g is the gravitational constant, d_p is the particle diameter, ρ_p is the particle density, and μ_g is the gas viscosity.

The efficiency due to impaction, η_I, is defined as [2,27]:

$$\eta_I = \frac{\eta_{vis} + \eta_{pot}\left(\frac{Re}{60}\right)}{1 + \frac{Re}{60}} \tag{3.22}$$

where Re is the Reynolds number, η_{vis} and η_{pot} are the viscous and potential flow efficiencies, respectively, which are read from Figure 3.23 [modified from Ref. 2] for the specific impaction number N_I. The Reynolds number is defined by

$$Re = \frac{D_p V_T \rho_g}{\mu_g} \tag{3.23}$$

where D_p is the particle diameter, V_T is the particle terminal velocity, ρ_g is the gas density, and μ_g is the gas viscosity. The specific impaction number, N_I, is approximated by [2]

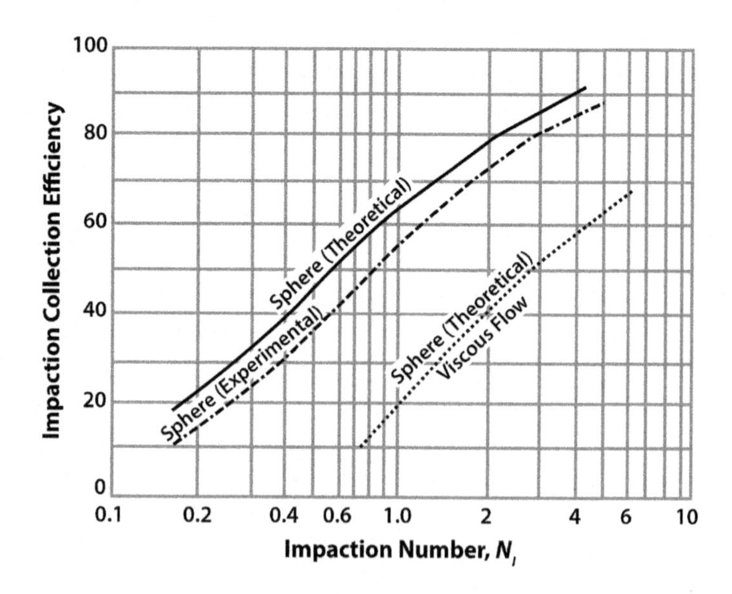

Figure 3.23 Theoretical and experimental impaction collection efficiencies for spheres.

$$N_I = \frac{d_p^2 \rho_p K_C (V_p - V_D)}{18 \mu_g d_D} \tag{3.24}$$

where V_p is the particle velocity, V_D is the droplet velocity, and K_C is the Cunningham correction factor, which is defined in Eq. (3.12).

3.4.4.2 Venturi scrubbers

Venturi scrubbers are used when high collection efficiencies are required and when most of the particles are smaller than 2 μm in diameter [27]. Venturi scrubbers use a constricted gas flow section or throat, which causes the gas and particle stream to increase in velocity followed by a diverging section where the gases decrease in velocity. A generalized venturi configuration is shown in Figure 3.24 [modified from 18 and 27]. The gases and particles reach velocities ranging from

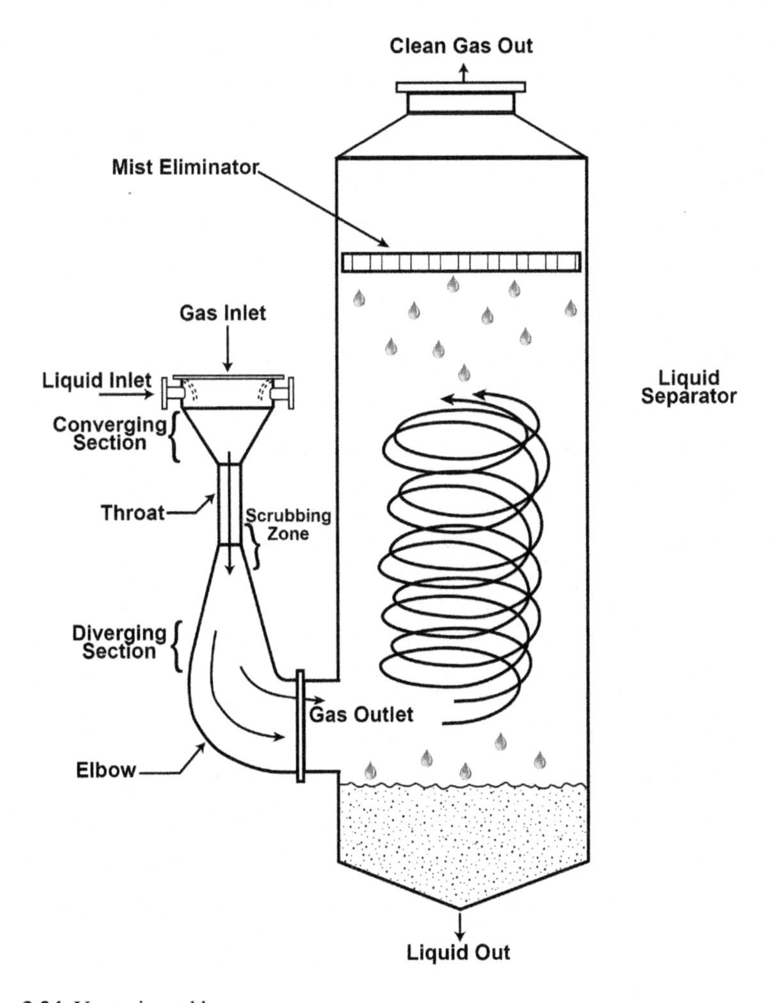

Figure 3.24 Venturi scrubber.

100 to 600 ft/s and impact upon the slower moving water droplets and the inertial impaction particle collection mechanism predominates [2,18]. The collection process is essentially complete by the end of the throat. Venturi scrubbers operate at much higher velocities than ESPs or baghouses and hence are physically much smaller and can be made of corrosion-resistant materials.

The overall efficiency of a venturi scrubber is defined as [27]

$$\eta_t = 1 - exp\left[\frac{1}{55}\frac{Q_L}{Q_G}\frac{V_g \rho_L D_d}{\mu_g}F(K_p f)\right] \tag{3.25}$$

where

$$F(K_p f) = \frac{1}{K_p}\left[-0.7 - K_p f + 1.4\ln\left(\frac{K_p f + 0.7}{0.7}\right) + \frac{0.49}{0.7 + K_{pf}}\right] \tag{3.26}$$

and $K_p = 2$ St. St, Stokes number, is V_T from Eq. (3.21), f is an empirical parameter that has been found to be 0.5 in gas-atomized spray scrubbers, Q_L and Q_G are the volumetric flow rates of the liquid and gas streams, respectively, and D_d is the droplet diameter.

3.4.4.3 Cyclonic scrubbers

Cyclonic scrubbers are wet cyclones with the inlet gas flow through a tangential entry that is similar to the cyclones discussed earlier. Water is introduced through either banks of nozzles that are installed in a ring fashion inside a conventional dry cyclone or through an axially located multiple nozzle, as shown in Figure 3.25 [2]. The spray acts on the particles in the outer vortex, and particulate-laden liquid particles are thrown outward against the wet inner wall of the cyclone [2,18]. The particulate-laden liquid flows down the walls of the cyclone to the bottom, where it is removed. In general, wet cyclones have a collection efficiency of 100% for droplets of 100 μm or larger, around 99% for droplets from 50 to 100 μm, and from 90 to 98% for droplets between 5 and 50 μm [2].

3.4.5 Hybrid systems

Hybrid systems have been under development for more than 15 years, especially since coal users such as utilities are required to meet increasingly tighter emissions regulations for particulate matter as well as sulfur dioxide. Fly ash resistivity and dust loadings are affected by switching to low sulfur coals or injecting sorbents for sulfur dioxide control, which in turn can reduce ESP efficiency. The desire to reduce fine particulate emissions is also leading to innovative technologies to reduce particulate emissions. There are several hybrid systems that have been under development – namely the Advanced Hybrid Collector (AHPC), Compact Hybrid

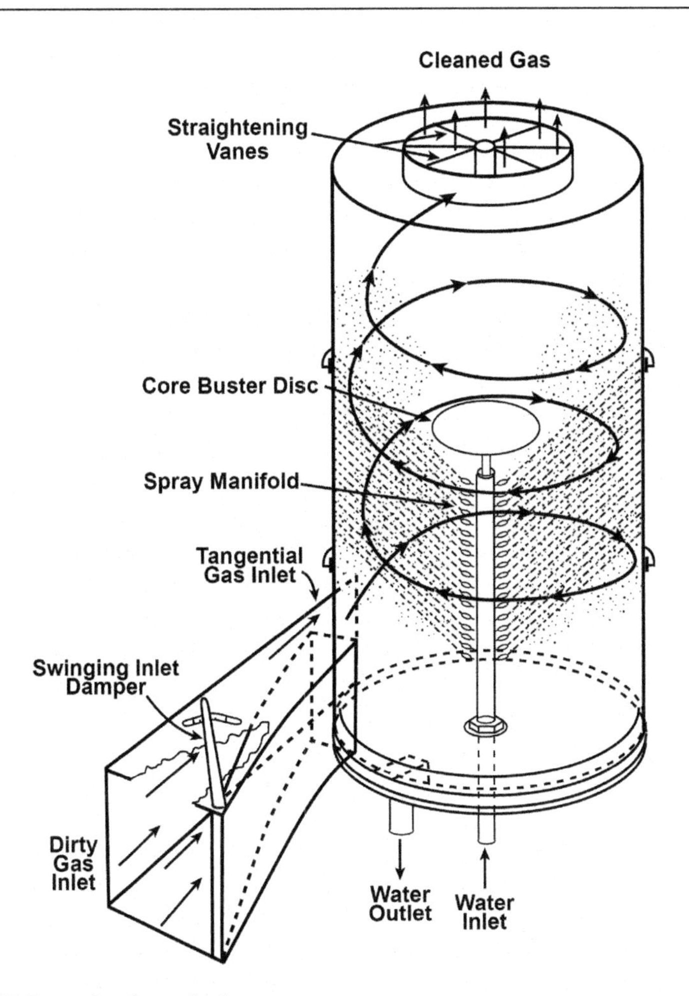

Figure 3.25 Example of a cyclonic spray tower.

Particulate Collector (COHPAC), multistage collector (MSC), and a PM screen [18,44–49]. Of these, only the COHPAC, discussed below, is commercially available. The others are at various stages of development including full-scale trials but have not progressed yet to the point of commercialization and are not discussed here.

3.4.5.1 Compact hybrid particulate collector

The Compact Hybrid Particulate Collector (COHPAC), developed by the Electric Power Research Institute (EPRI), involves the installation of a pulse-jet fabric filter (PJFF) downstream of the ESP or retrofitted into the last field of an ESP [50,51]. Since the pulse-jet collector is operating as a polisher for achieving lower particulate emissions, the low dust loading to the baghouse allows the filter to be operated at high A/C ratios (5–8 ft/min), allowing a small footprint on-site, longer bag life, lower pressure drops, and lower parasitic load. This system allows for the ability to retrofit existing units and achieve high efficiencies at relatively low cost. There are two design variations, as shown in Figure 3.26. COHPAC I has a separate fabric filter system to the ESP. COHPAC II replaces the last plate electrodes in an ESP with a fabric filter system.

The COHPAC technology has been demonstrated at the utility scale, including full-scale operation at Alabama Power's E.C. Gaston Station (272 MW) and TXU Energy's (formerly TU Electric) Big Brown Plants (2 units, each 575 MW) [45,50]. Results from COHPAC operation have been positive, achieving high collection

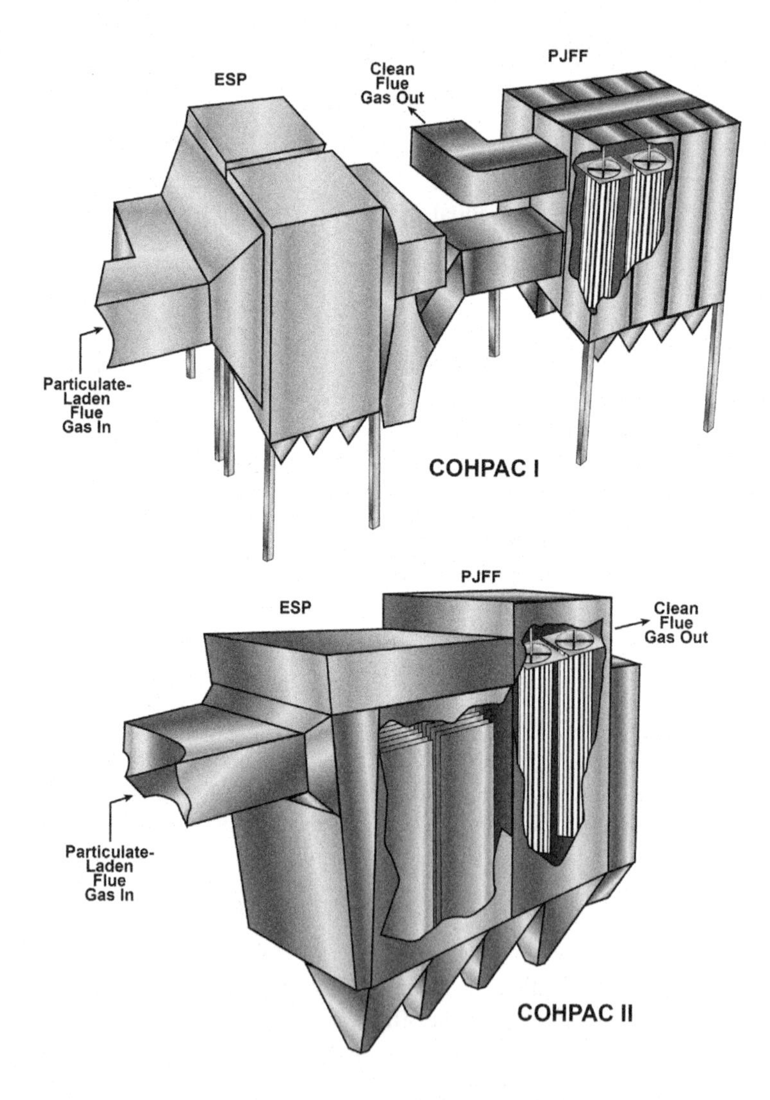

Figure 3.26 Schematics of COHPAC I and COHPAC II Systems.

efficiencies (as high as 99.9%), allowing for fuel flexibility, reducing opacity, increased bag filter life, and decreasing operating costs. This has led to units being installed commercially since 2000.

3.5 Economics of particulate matter control

As with other pollution control technology costs, the costs for particulate control systems are site specific and are influenced by the required emission limit, type of fuel and its characteristics, and volume of flue gas being treated. This section summarizes the costs for cyclones, ESPs, fabric filters, venturi scrubbers, and hybrid systems using published data.

3.5.1 Cyclones

Table 3.8 lists the capital costs and operating maintenance (O&M) costs for a variety of particulate matter removal devices including cyclones. These data were compiled from costs reported by EPA [52–60]. The capital costs for cyclones vary between $2.20 to $3.50 per scfm while the O&M costs vary from $0.70 to $8.50 per scfm [52]. The costs do not include costs for disposal or transport of collected fly ash.

3.5.2 ESPs

EPA reported capital costs and O&M costs for dry and wet ESPs are listed in Table 3.8 [53–56]. Generally, the capital and O&M costs for wet ESPs are higher than for dry ESPs.

IEA reports that the capital cost for a new ESP is between $40–60 kW with the higher cost associated with higher collection efficiencies [61]. Since most coal-fired

Table 3.8 Capital costs and operating and maintenance costs for particulate matter removal devices

Removal device	Capital cost ($/scfm)	O&M cost ($/scfm)
Cyclone	2.20–3.50	0.70–8.50
Dry ESP, Wire-Pipe	20–125	4–9
Dry ESP, Wire-Plate	10–33	3–35
Wet ESP, Wire-Pipe	40–200	6–10
Wet ESP, Wire-Plate	20–40	5–40
Fabric Filter-Pulse Jet	6–26	5–24
Fabric Filter-Reverse Air	9–85	6–27
Sonic horns	0.51–0.61	
Fabric Filter-Mechanical Shaker	8–72	4–24
Sonic horns	0.51–0.61	
Venturi Scrubber	2.50–21	4.40–120

power plants are already fitted with ESPs, much of the published data relate to costs for upgrading existing ESPs. ESP rebuilds are less costly today due to greater market competition, the emergence of new construction techniques, and the use of wide plate spacing requiring fewer collecting plates. Wide plate spacing is one of the most economic and effective approaches to replacing internals. The cost benefits result from the need for fewer internal elements, materials and erection savings due to reuse of part of the original casing, and the weight savings effects on the existing support structures and foundations. Costs for upgrading ESPs has been estimated at about $12/kW per field for a 500 MW unit with the increased operating costs estimated to be $100,000 per year [61].

Flue gas conditioning has proved to be more cost effective than adding new fields. With difficult-to-collect fly ashes, conditioning allows operation without adding new fields. The reduction in ESP size with conditioning also lowers the operating cost because fewer fields and hoppers are used, decreasing the number of heaters and consequently the power consumption [17]. A native SO_3 conditioning system for a 500 MW power plant requires a capital cost of $4.50/kW. Adding an anhydrous ammonia conditioning system to an existing SO_3 system would cost about $1/kW for a 50 MW unit with the operating cost increasing by $50,000/year [61].

Staehle et al. [37] performed an economic analysis for using WESPs for SO_3 control at three different levels of control – 50, 80, and 95%. The capital costs for the three levels of control were $10, $15, and $20/kW, respectively. The total operating costs, based on 8,000 hours of operation per year, were $120,000, $160,000, and $200,000 per year, respectively.

3.5.3 Fabric filters

EPA reported capital costs and O&M costs for pulse-jet, reverse air, and mechanical shaker fabric filters are listed in Table 3.8 [57–59]. Pulse-jet fabric filters tend to have lower capital and O&M costs of the three types.

Fabric filters are reported to cost between $50 and $70/kW [61]. Reverse-gas baghouses have higher capital and operating costs than pulse-jet baghouses because reverse-gas baghouses operate at a lower A/C ratio. Fabric filters are generally more expensive than ESPs for collection efficiencies up to 99.5%; however, baghouses become more cost effective for higher collection efficiencies. In addition, high resistivity fly ashes need to be upgraded to achieve high collection efficiencies and baghouses have economic advantages over ESPs for fly ash resistivity greater than 10^{13}–10^{14} ohm-cm. Operating costs for baghouses are also higher than ESPs due to bag replacement and auxiliary power requirements.

3.5.4 Venturi scrubbers

EPA reported capital costs and O&M costs for venturi scrubbers are listed in Table 3.8 [60]. The capital costs for venturi scrubbers vary between $2.50 to $21 per scfm, while the O&M costs vary from $4.40 to $120 per scfm [52].

3.5.5 COHPAC

A cost analysis performed for a COHPAC that uses a pulse-jet bag filter following an ESP estimated that the capital cost varied from $57–70/kW and operating cost varied from $320,000–$570,000 per year, both depending on the unit size [61]. The analysis was performed for upgrading ESPs at a few coal-fired units ranging in size from 150–300 MW.

References

[1] EPA (United States Environmental Protection Agency). Latest findings on national air quality: 1997 status and trends. Washington, DC: Office of Air Quality Planning and Standards; 1998.

[2] Wark K, Warner CF, Davis WT. Air pollution: its origin and control. 3rd ed. Menlo Park, California: Addison Wesley Longman, Inc; 1998.

[3] Lapple CE. Stanford Research Institute Journal, 5, 1961.

[4] Zhu Q. Developments in particulate control. London: IEA Clean Coal Centre; 2003.

[5] EPA (United States Environmental Protection Agency). Air Trends, www.epa.gov/airtrends; April 21, 2014.

[6] EPA (United States Environmental Protection Agency). National Emissions Inventory (NEI), 1970–2013 Average Annual Emissions, All Criteria Pollutants, www.epa.gov/ttnchie1/trends/; February 27, 2014.

[7] EIA (United States Energy Information Agency). Electric power annual 2012 (U.S. Department of Energy, Office of Coal, Nuclear, Electric and Alternate Fuels, U.S. Government Printing Office, Washington, D.C., December 2013).

[8] EIA (United States Energy Information Agency). Annual energy review 2011, (U.S. Department of Energy, Office of Energy Statistics, U.S. Government Printing Office, Washington, D.C., September 2012).

[9] Soud HN, Mitchell SC. Particulate Control Handbook for Coal-Fired Plants. London: IEA Coal Research; 1997.

[10] DOE (United States Department of Energy). Description – PM emissions control, www.netl.doe.gov/coalpower/environment/pm/description.html; last updated December 2, 2003.

[11] Benson SA, Jones ML, Harb JN. Ash formation and deposition Chapter 4 In: Smoot LD, editor. Fundamentals of coal combustion – for clean and efficient use, coal science and technology 20. Amsterdam: Elsevier Science Publishers; 1993. p. 299–373

[12] Sarofim AF, Howard JB, Padia AS. The physical transformation of the mineral matter in pulverized coal under simulated combustion conditions. Combustion Science and Technology 1977;16:187–209.

[13] Bryers RW. Fireside behavior of minerals in coal, symposium on slagging and fouling in steam generators. Brigham Young University, 1987.

[14] Beer JM. Stationary combustion. Proceedings of the 22nd Symposium (International) on Combustion, Pittsburgh, Pennsylvania, 1988.

[15] Lind T. Ash formation in circulating fluidised bed combustion of coal and solid biomass. Ph.D. Thesis. Technical Research Centre of Finland, 1999.

[16] Miller SJ, Ness SR, Weber GF, Erickson TA, Hassett DJ, Hawthorne SB, et al. A comprehensive assessment of toxic emissions from coal-fired power plants: phase I results from the U. S. Department of Energy Study Final Report, 1996.
[17] Miller BG. Clean coal engineering technology. Oxford, United Kingdom: Butterworth-Heinemann; 2011696
[18] Davis WT, editor. Air pollution engineering manual. 2nd ed. New York: John Wiley & Sons, Inc; 2000.
[19] Clarke LE, Sloss LL. Trace elements – emissions from coal combustion and gasification. London: IEA Coal Research; 1992.
[20] Electric Power Research Institute. Coal ash: characteristics, management and environmental issues, technical update – coal combustion products – environmental issues. September 2009.
[21] Al-Degs US, Ghrir A, Khoury H, Walker GM, Sunuk M, Al-Gouti MA. Characterization and utilization of fly ash from heavy fuel oil generated in power stations. Fuel Processing Technology 2014;123:41–6.
[22] Al-Malack MH, Bukhari AA, Al-Amoudi OS, Al-Muhanna HH, Zaidi TH. Characteristics of fly ash produced at power and water desalination plants firing fuel oil. Int J Environ Res 2013;7(2):455–66.
[23] Miller CA, and Linak WP. Primary particles generated by the combustion of heavy fuel oil and coal, review of research results from epa's national risk management research laboratory, Prepared for U.S. Environmental Protection Agency, Office of Research and Development, November 2002.
[24] Walsh PM, Mormile DJ, and Piper BF. Dependence of particulate matter emissions from boilers on quality of residual fuel oil. American Chemical Society National Meeting, 1992.
[25] Perry RH, Chilton CH. Chemical engineers' handbook. 5th ed. The McGraw-Hill Inc; 1973.
[26] Green DW, Perry RH. Perry's chemical engineers' handbook. 8th ed. The McGraw-Hill Companies Inc; 2007.
[27] Flagan RC, Seinfeld HH. Fundamentals of air pollution engineering. Englewood Cliffs, New Jersey: Prentice-Hall Inc; 1988.
[28] Lapple CE. Processes use many collection types. Chemical Engineering 1951;58 (8):144–51.
[29] Leith D, and Licht W. The collection efficiency of cyclone type particle collectors – a new theoretical approach, in AIChE Symposium Series, No. 126, Vol. 68, 196–206, 1972.
[30] DOE (United States Department of Energy). Controlling air toxics with electrostatic precipitators fact sheet. (Office of Fossil Energy, Washington, D.C., 1997).
[31] B&W. Electrostatic precipitator product sheet, PS151 2M A 12/82. (The Babcock & Wilcox Company, Barberton, Ohio, December 1982).
[32] Elliot TC, editor. Standard handbook of powerplant engineering. New York: McGraw-Hill Publishing Company; 1989.
[33] Miller BG, Tillman DA, editors. Combustion engineering issues for solid fuel systems. Burlington, Massachusetts: Academic Press; 2008.
[34] Miller BG. unpublished data, 1986.
[35] Miller BG, Miller SJ, Lamb GP, and Luppens JA. Sulfur capture by limestone injection during combustion of pulverized panola county texas lignite. Proceedings of Gulf Coast Lignite Conference, 1984.

[36] Buckley W, and Ray I. Application of wet electrostatic precipitation technology in the utility industry for PM2.5 control. Proceedings of the EPRI-DOE-EPA Combined Power Plant Air Pollution Control MEGA Symposium, 2003.
[37] Staehle RC, Triscori RJ, Ross G, Kumar KS, and Pasternak E. The past, present and future of wet electrostatic precipitators in power plant applications. Proceedings of the EPRI-DOE-EPA Combined Power Plant Air Pollution Control MEGA Symposium, 2003.
[38] Altman R, Offen G, Buckley W, Ray I. Wet electrostatic precipitation demonstrating promise for fine particulate control – part I. Power Engineering 2001;105(1):37–9.
[39] Altman R, Buckley W, Ray I. Wet electrostatic precipitation demonstrating promise for fine particulate control – part II. Power Engineering 2001;105(2):42–4.
[40] Bustard CJ, Cushing KM, Pontius DH, Smith WB, Carr RC. Fabric Filters for the Electric Utility Industry, Volume 1 General Concepts. Palo Alto, California: Electric Power Research Institute; 1988.
[41] DOE (United States Department of Energy). National energy technology laboratory, NETL Coal Power Database 2000, 2002.
[42] Soud HN. Developments in particulate control for coal combustion. London: IEA Coal Research; 1995.
[43] Kitto J, Stultz S, editors. Steam, its generation and use. 41st ed. The Babcock & Wilcox Company; 2005.
[44] Nicol K. Recent developments in particulate control. London: IEA Clean Coal Centre; 2013.
[45] Cushing KM, Harrison WA, Chang RL. Performance response of COHPAC I baghouse during operation with normal and artificial changes in inlet fly ash concentration and during injection of sorbents for control of air toxics. Proceedings of the EPRI-DOE-EPA Combined Utility Air Pollution Control Symposium: The MEGA Symposium: Volume III: Particulates and Air Toxics, 1997.
[46] Gebert R, Rinschler C, Davis D, Leibacher U, Studer P, Eckert W, et al. Commercialization of the advanced hybrid filter technology, conference on air quality III: mercury, trace elements, and particulate matter. (University of North Dakota, Grand Forks, North Dakota, 2002).
[47] DOE (United States Department of Energy). Advanced hybrid particulate collector fact sheet. (Office of Fossil Energy, Washington, D.C., June 2001).
[48] Blankinship S. Hybrid filter technology weds ESPs with bag filters. Power Engineering 2002;106(2):9.
[49] DOE (United States Department of Energy). Demonstration of a Full-Scale Retrofit of the Advanced Hybrid Particulate Collector (AHPC) Collector Fact Sheet. (Office of Fossil Energy, Washington, D.C., February 2003).
[50] Miller RL, Harrison WA, Prater DB, and Chang R. Alabama Power Company E.C. Gaston 272 MW Electric Steam Plant – Unit No. 3 Enhanced COHPAC I Installation. Proceedings of the EPRI-DOE-EPA Combined Utility Air Pollution Control Symposium: The MEGA Symposium: Volume III: Particulates and Air Toxics, 1997.
[51] Bustard CJ, Sjostrom SM, and Chang R. Predicting COHPAC performance. Proceedings of the EPRI-DOE-EPA Combined Utility Air Pollution Control Symposium: The MEGA Symposium: Volume III: Particulates and Air Toxics, 1997.
[52] EPA (United States Environmental Protection Agency). Air pollution control technology fact sheet – cyclones. technology transfer network. http://www.epa.gov/ttn/catc/cica/index.html; 2007.

[53] EPA (United States Environmental Protection Agency). Air pollution control technology fact sheet – dry electrostatic precipitator (ESP) – wire-pipe type. Technology Transfer Network, http://www.epa.gov/ttn/catc/cica/index.html; 2007.
[54] EPA (United States Environmental Protection Agency). Air pollution control technology fact sheet – dry electrostatic precipitator (ESP) – wire-plate type. Technology Transfer Network, http://www.epa.gov/ttn/catc/cica/index.html; 2007.
[55] EPA (United States Environmental Protection Agency). Air pollution control technology fact sheet – wet electrostatic precipitator (ESP) – wire-pipe type. Technology Transfer Network, http://www.epa.gov/ttn/catc/cica/index.html; 2007.
[56] EPA (United States Environmental Protection Agency). Air pollution control technology fact sheet – wet electrostatic precipitator (ESP) – wire-plate type. Technology Transfer Network, http://www.epa.gov/ttn/catc/cica/index.html; 2007.
[57] EPA (United States Environmental Protection Agency). Air pollution control technology fact sheet – fabric filter – pule-jet cleaned type. Technology Transfer Network, http://www.epa.gov/ttn/catc/cica/index.html; 2007.
[58] EPA (United States Environmental Protection Agency). Air pollution control technology fact sheet – fabric filter – reverse air cleaned type. Technology Transfer Network, http://www.epa.gov/ttn/catc/cica/index.html; 2007.
[59] EPA (United States Environmental Protection Agency). Air pollution control technology fact sheet – fabric filter – mechanical shaker cleaned type. Technology Transfer Network, http://www.epa.gov/ttn/catc/cica/index.html; 2007.
[60] EPA (United States Environmental Protection Agency). Air pollution control technology fact sheet – venturi scrubber. Technology Transfer Network, http://www.epa.gov/ttn/catc/cica/index.html; 2007.
[61] Wu Z. Air pollution control costs for coal-fired power stations. London: IEA Coal Research; 2001.

Sulfur oxides formation and control

4

4.1 Introduction

Gaseous emissions of sulfur oxides from coal combustion are mainly sulfur dioxide (SO_2) and, to a much lesser extent sulfur trioxide (SO_3) and gaseous sulfates. Sulfur dioxide is a nonflammable, nonexplosive, colorless gas that causes a taste sensation at concentrations from 0.1 to 1.0 ppmv (part per million by volume) in air [1]. At concentrations greater than 3.0 ppm, the gas has a pungent, irritating odor. Sulfur dioxide is partly converted to sulfur trioxide or to sulfuric acid (H_2SO_4) and its salts by photochemical or catalytic processes in the atmosphere. Sulfur trioxide and water form sulfuric acid.

Environmental effects of sulfur compounds include impaired visibility, damage to materials, damage to vegetation, and deposition as acid rain. Fine particles in the atmosphere reduce the visual range by scattering and absorbing light [2]. Aerosols of sulfuric acid and other sulfates comprise from 5 to 20% of the total suspended particulate matter in urban air and, hence, contribute to the reduction in visibility.

Sulfur compounds are responsible for major damage to materials. Sulfur oxides generally accelerate metal corrosion by first forming sulfuric acid either in the atmosphere or on the metal surface. Sulfur dioxide is the most detrimental pollutant that contributes to metal corrosion [2]. Temperature and relative humidity also significantly influence the rate of corrosion. Sulfurous and sulfuric acids are capable of damaging a wide variety of building materials including limestone, marble, roofing slate, and mortar. Textiles made of nylon are also susceptible to pollutants in the atmosphere.

In general, the damage to plants from air pollution usually occurs in the leaf structure since the leaf contains the building blocks for the entire plant [2]. Sulfur dioxide enters the leaf where the plant cells convert it to sulfite and then into sulfate. Apparently when excessive sulfur dioxide is present, the cells are unable to oxidize sulfite to sulfate fast enough and disruption of the cell structure begins. Spinach, lettuce, and other leafy vegetables are most sensitive as are cotton and alfalfa. Pine needles are also affected, with either the needle tip or the whole needle becoming brown and brittle.

Acidic deposition or acid rain occurs when emissions of sulfur dioxide and oxides of nitrogen in the atmosphere react with water, oxygen, and oxidants to form acidic compounds. These compounds fall to the earth in either dry form (i.e. gas and particles) or wet form (i.e. rain, snow, and fog).

The acidity of water is reported in terms of the pH, where pH is the logarithm (base 10) of the molar concentration of hydrogen ions:

$$pH = -\log_{10}[H^+] \quad (4.1)$$

Fossil Fuel Emissions Control Technologies. DOI: http://dx.doi.org/10.1016/B978-0-12-801566-7.00004-X

Pure water contains a hydrogen ion concentration that is approximately 10^{-7} molar or pH = 7, which is referred to as neutral pH. Water droplets formed in the atmosphere, however, normally have a pH of ≈ 5.6 because atmospheric carbon dioxide is dissolved in the rain and forms carbonic acid (H_2CO_3). When sulfur dioxide or nitrogen oxides are also dissolved in the water, the pH becomes lower and average yearly pH values range from 4.0 to greater than 5.0 in the United States. Sulfur dioxide can be absorbed from the gas phase into an aqueous droplet creating acidic conditions as follows [2]:

$$SO_2(g) \Leftrightarrow SO_2(aq) \tag{4.2}$$

$$SO_2(aq) + H_2O \Leftrightarrow HSO_3^- + H^+ \tag{4.3}$$

$$HSO_3^- \Leftrightarrow SO_3^{2-} + H^+ \tag{4.4}$$

$$SO_3^{2+} + H_2O \Leftrightarrow SO_3^{2-} + 2H^+ \tag{4.5}$$

There are several effects of acid rain that are of concern. There is an acidification of natural water sources, which can have a devastating effect on fish. Trout and salmon are particularly sensitive to a low pH [2]. Reproduction in many fish fails to occur at a pH less than 5.5. A decrease in plankton and bottom fauna is also observed as the pH is lowered, which reduces the food supply for the fish. Leaching of nutrients occurs from the soil. This demineralization can lead to loss in productivity of crops and forests, or a change in natural vegetation. Vegetation itself can be directly damaged. An increase in corrosion to materials is also observed.

Sulfur dioxide and other oxides of sulfur have been studied extensively; however, many questions concerning the effects of sulfur dioxide upon health remain unanswered [2]. Few epidemiological studies have been able to differentiate adequately the effects of individual pollutants because sulfur oxides tend to occur in the same kinds of atmospheres as particulate matter and high humidity.

High concentrations of sulfur dioxide can result in temporary breathing impairment for asthmatic children and adults who are active outdoors [3]. Short-term exposures of asthmatic individuals to elevated sulfur dioxide levels, while at moderate exertion, may result in reduced lung function that may be accompanied by such symptoms as wheezing, chest tightness, or shortness of breath. Other effects that have been associated with longer-term exposures to high concentrations of sulfur dioxide, in conjunction with high levels of particulate matter, include respiratory illness, alterations in the lung's defenses, and aggravation of existing cardiovascular disease. Those that may be affected under these conditions include individuals with cardiovascular disease or chronic lung disease, as well as children and the elderly.

Sulfur dioxide is one of the most abundant air pollutants emitted in the United States totaling about 5 million short tons in 2013 (see Table 2.20) with stationary fuel combustion accounting for approximately 84% of the total anthropogenic emissions (i.e. 4.2 million short tons). Of the 4.2 million tons, electric generating utilities are the largest source of SO_2 emissions compared to industry with electric utilities accounting for 83%, or 3.3 million tons of all SO_2 emissions. Sulfur dioxide emissions have, however, decreased 80% and 79% for the periods from 1980 to 2013 and from 1990 to 2013, respectively [4]. Reductions in SO_2 emissions and concentrations since 1990 are primarily due to controls implemented under the United States Environmental Protection Agency's (EPA) Acid Rain Program beginning in 1995. As of 2012, there were 468 coal-fired electric generators equipped with scrubbers with a total of more than 185,000 MW generating capacity or 66% of MWh generated [5,6].

The chemistry of sulfur dioxide formation is reviewed in this chapter, followed by technologies used to control SO_2 emissions. Control technologies will focus on commercially available and commercially used systems. Emerging technologies are not discussed. Industry deployment of SO_2 removal processes in the U.S. is discussed, as are the economics of flue gas desulfurization.

4.2 Chemistry of sulfur oxide (SO_2/SO_3) formation

Sulfur in coal occurs in three forms: as pyrite, organically-bound to the coal, or as sulfates. The sulfates represent a very small fraction of the total sulfur while pyritic and organically-bound sulfur comprise the majority. The distribution between pyritic and organic sulfur is variable with up to approximately 40% of the sulfur being pyritic. During combustion, the pyritic and organically-bound sulfur are oxidized to sulfur dioxide with a small amount of sulfur trioxide (SO_3) being formed. The SO_2/SO_3 ratio is typically 40:1 to 80:1 [2].

The overall reaction for the formation of sulfur dioxide is

$$S + O_2 \rightarrow SO_2 \quad \Delta H_f = -128,560\ \mathrm{Btu/lb\ mole} \tag{4.6}$$

and the overall reaction for the formation of sulfur trioxide is

$$SO_2 + \tfrac{1}{2}O_2 \leftrightarrow SO_3 \quad \Delta H_f = -170,440\ \mathrm{Btu/lb\ mole} \tag{4.7}$$

It is proposed that sulfur monoxide, SO, is formed early in the reaction zone from sulfur containing molecules and is an important intermediate product [7]. The major SO_2 formation reactions are believed to be

$$SO + O_2 \rightarrow SO_2 + O \tag{4.8}$$

and

$$SO + OH \rightarrow SO_2 + H \tag{4.9}$$

with the highly reactive O and H atoms possibly entering the reaction scheme later.

The reactions involving SO_3 are reversible. The major formation reaction for SO_3 is the three-body process

$$SO_2 + O + M \rightarrow SO_3 + M \quad (4.10)$$

where M is a third body which is an energy absorber [7]. The major steps for removal of SO_3 are thought to be the following:

$$SO_3 + O \rightarrow SO_2 + O_2 \quad (4.11)$$

$$SO_3 + H \rightarrow SO_2 + OH \quad (4.12)$$

$$SO_3 + M \rightarrow SO_2 + O + M \quad (4.13)$$

4.3 Sulfur dioxide control

Methods to control sulfur dioxide emissions from fossil fuel-fired heat and power plants include switching to a lower sulfur fuel, cleaning coal to remove the sulfur-bearing components such as pyrite, or installing flue gas desulfurization systems. In the past, building tall stacks to disperse the pollutants was a control method; however, this practice is no longer an alternative as tall stacks do not remove the pollutants, they only dilute the concentrations to reduce the ground-level emissions to acceptable levels.

When neither fuel switching nor coal cleaning is an option, flue gas desulfurization (FGD) is selected to control sulfur dioxide emissions from coal-fired power plants. FGD has been in commercial practice since the early 1970s. It became the most widely used technique to control sulfur dioxide emissions next to the firing of low-sulfur fuels. In the case of coal firing, initially operators were switching to low-sulfur coals. With the increased production of shale gas, natural gas is becoming the fuel of choice when switching from coal to a fuel source with lower emissions. Many existing FGD systems are currently in use, and many others are under development. This section summarizes the application of FGD systems in the United States. FGD processes are generally classified as wet scrubbers or dry scrubbers but can also be categorized as follows [8]:

- wet scrubbers;
- spray dryers;
- dry (sorbent) injection processes;
- regenerable processes;
- circulating fluid-bed and moving-bed scrubbers; and
- combined SO_2/NO_x removal systems.

Based on the nature of the waste/byproduct generated, a commercially available throwaway FGD technology may be categorized as wet or dry. A wet FGD process produces a slurry waste or a saleable slurry byproduct. A dry FGD process application results in a solid waste, the transport and disposal of which is easier compared to the waste/byproduct from wet FGD applications. Regenerable FGD processes produce a concentrated SO_2 byproduct, usually sulfuric acid or elemental sulfur. With the implementation of mercury emissions legislation, there has been a focus on mercury removal in FGD systems. This is discussed in a later chapter.

Postcombustion control of sulfur dioxide emissions from pulverized coal combustion began in the early 1970s in the United States and Japan. Western Europe followed in the 1980s. In the 1990s, the application of FGD became more widespread and countries in Central and Eastern Europe, Asia, and others have installed FGD systems. According to IEA, by the end of 2008, more than 1400 coal-fired power generating units worldwide with a total capacity of over 502.5 GWe were fitted with FGD systems and more than 230 units with a capacity of 102 GWe were planning to have FGD systems installed [9]. The sorbent/reagent of choice is predominately limestone or lime for these units. Wet FGD is the predominate technology used worldwide for the control of SO_2 from utility power plants firing all ranks of coal [10]. In 2012, IEA reported that 80% of the FGD systems worldwide were wet based [11]. Wet FGD is also installed on systems that use heavy fuel oil.

Of the United States SO_2 control technology population, 83% are wet FGD processes, which are installed on 180 GW of electric capacity [12]. Dry FGD processes comprise 17% (i.e., 38 GWe) of the total 218 GW electric power base that is scrubbed.

A variety of FGD processes exist and the selection of a system is dependent upon site-specific consideration, economics, and other criteria. Elliot provides a ranking of various FGD processes used in the United States in Tables 4.1 and 4.2 where cost, performance, and flexibility of application are assessed [13].

With respect to coal firing, wet limestone systems have been installed across the United States at plants of all sizes, firing all ranks of coal with sulfur contents varying from low to high. Wet lime systems have been installed at power plants of all sizes firing both low-and high-sulfur coals with plants predominately in the Ohio River valley [13]. Some plants in the West use wet lime systems where the cost of lime delivered to the plant is less than limestone. Some plants firing high-sulfur coal in the Midwest have selected wet sodium-based dual-alkali systems. Dry scrubbing systems have typically been selected at power plants firing low-sulfur coals. Generally dry scrubbing systems are considered more economical for power plants firing low-sulfur coal, while wet-based systems are selected for high-sulfur coal applications.

4.4 Techniques to reduce sulfur dioxide emissions

The primary methods used to control sulfur dioxide emissions from coal-fired power plants are to switch to a lower sulfur fuel or install flue gas desulfurization systems and, to a lesser extent, clean coal to remove the sulfur-bearing components. These techniques are discussed in this section with an emphasis on flue gas desulfurization technologies.

Table 4.1 Assessment relative to cost and performance

FGD process	Criterion[a]				
	Operating cost	Capital cost	SO_2 removal	Reliability	Commercial use
Limestone					
Natural oxidation	M	M	M	M	H
Forced oxidation	M	M	M	M	H
MgO-lime	M	M	H	H	H
High-calcium lime	M	M	M	M	M
Dual-alkali					
Lime	M	M	H	H	M
Limestone	L	M	H	–	–
Dry scrubbing	H	M	L	H	H
Dry Injection	M	L	L	–	–
Wellman-Lord	H	H	H	M	M
Regenerable MgO	M	H	H	M	L

[a]H = high; M = medium; L = low.

Table 4.2 Assessment with respect to flexibility of application

FGD process	Criterion[a]				
	High sulfur	Low sulfur	Retrofit ease	Waste management	SO_2/NO_x removal
Limestone					
Natural oxidation	H	H	M	L	L
Forced oxidation	H	M	M	L	L
MgO-lime	H	L	M	L	M
High-calcium lime	H	M	M	L	L
Dual-alkali					
Lime	H	L	M	L	M
Limestone	H	L	M	L	L
Dry scrubbing	M	H	M	M	M
Dry Injection	L	H	H	L	L
Wellman-Lord	H	L	L	M	M
Regenerable MgO	H	M	M	H	M

[a]H = high; M = medium; L = low.

4.4.1 Using low-sulfur fuels

One option for reducing sulfur dioxide emissions is to switch to fuels containing less sulfur. Fuel switching includes using natural gas, liquefied natural gas, low-sulfur fuel oils, or low-sulfur coals in place of high-sulfur coals.

Fuel switching to lower sulfur coals has been an option used by many power generators to achieve emissions compliance. In the United States, the replacement of high-sulfur Eastern or Midwestern bituminous coals with lower sulfur Appalachian region bituminous coals or Powder River Basin coals is a control option that is widely exercised. The option of using lower sulfur coal has resulted in a large increase in Western coal production and use. Table 4.3 illustrates the distribution of sulfur in United States coals by region as of January 1, 1997, the last year these data were reported in this manner [14]. Although Table 4.3 is dated, it provides a good relative distribution of the sulfur content in coal in the United States by region. Low-sulfur coals are also imported to the United States, specifically to coastal areas such as Florida or the Eastern seaboard.

The relationship between sulfur content in the coal and pounds of sulfur per million Btu is provided in Table 4.4 for comparison [14]. This listing, developed by the United States Department of Energy's Energy Information Agency (EIA), is used for approximate correlations with New Source Performance Standards (NSPS) and 1990 Clean Air Act Amendments criteria. With the exception of the low-sulfur coal, which meets NSPS requirements, the medium- and high-sulfur coals require control strategies. This includes emission reduction technologies or offsets through sulfur dioxide allowances.

The use of natural gas, as a lower emissions fuel, for electricity production is increasing. Natural gas-fired generation is projected to grow 3.1% a year through 2038, increased natural gas fired capacity by 348,000 MW according to a study by Black & Veatch [15]. Natural gas-fired combined cycle plants are expected to account for 53% of U.S. power production by 2038 compared to about 22% in 2014. Coal-fired generation is expected to decrease from about 43% in 2014 to 21% in 2038. This change is being driven by low gas prices due to the increased production of shale gas in the U.S., stricter regulations for coal-fired plants, the integration of growing amounts of renewable energy, and coal-fired power unit retirements [15,16]. Nearly 23,000 MW of coal-fired generating capacity was retired in the United States from 2009 to early 2014. U.S. power producers plan to shutter 27,000 MW of coal capacity between 2014 and 2022 [16]. In addition to these retirements, approximately 11,200 MW of coal capacity is being targeted for conversion to other fuels, primarily natural gas.

4.4.2 Coal cleaning

Coal preparation, or beneficiation, is a series of operations that remove mineral matter (i.e. ash) from coal. Preparation relies on different mechanical operations, which will not be discussed in detail, to perform the separation, such as size reduction, size classification, cleaning, dewatering and drying, waste disposal, and

Table 4.3 **Estimated recoverable coal reserves in the United States by sulfur range and major coal-producing region (remaining as of January 1, 1997)**

Coal-producing region	Sulfur content categories (pounds of sulfur per million Btu)							
	Low sulfur ($\leq$0.60)		Medium sulfur (0.61–1.67)		High sulfur ($\geq$1.68)		Total	
	Million short tons	Percent of total	Million short tons	Percent of total	Million short tons	Percent of total	Million short tons	Percent of total
Appalachia	11,675	11.6	20,337	24.0	23,283	25.9	55,295	20.1
Interior	769	0.8	10,041	11.8	57,966	64.4	68,776	25.0
Western	87,775	87.6	54,529	64.2	8,768	9.7	151,072	54.9
Total	100,219		84,907		90,017		275,143	

Table 4.4 **Comparison of sulfur content in coal with pounds of sulfur per Million Btu**

Qualitative rating	Pounds of sulfur per million Btu	Approximate range of coal sulfur content (%)	
		High-grade bituminous coal	High-grade lignite
Low Sulfur	≤0.4 to 0.6	≤0.5 to 0.8	≤0.3 to 0.5
Medium Sulfur	0.61 to 1.67	0.8 to 2.2	0.5 to 1.3
High Sulfur	1.68 to >2.50	2.2 to >3.3	1.3 to >1.9

pollution control. Coal preparation processes, which are physical processes, are designed mainly to provide ash removal, energy enhancement, and product standardization [13]. Sulfur reduction is achieved because the ash material removed contains pyritic sulfur. Coal cleaning is used for moderate sulfur dioxide emissions control as physical coal cleaning is not effective in removing organically bound sulfur. Chemical coal cleaning processes are being developed to remove the organic sulfur; however, these are not used on a commercial scale. An added benefit of coal cleaning is that several trace elements including antimony, arsenic, cobalt, mercury, and selenium, are generally associated with pyritic sulfur in raw coal and they too are reduced through the cleaning process. As the inert material is removed, the volatile matter content, fixed carbon content, and heating value increase, thereby producing a higher-quality coal. The moisture content, from residual water from the cleaning process, can also increase, which lowers the heating value but this is usually minimal so as to have little impact on coal quality. Coal cleaning does add additional cost to the coal price; however, there are several benefits to reducing the ash content, which includes lower sulfur content, less ash to be disposed, lower transportation costs since more carbon and less ash is transported (since coal cleaning is usually done at the mine and not the power plant), and increases in power plant peaking capacity, rated capacity, and availability [14]. Developing circumstances are making coal cleaning more economical and a potential sulfur control technology, and include [13]:

- higher coal prices and transportation costs;
- diminishing coal quality because of less selective mining techniques;
- the need to increase availability and capacity factors at existing boilers;
- more stringent air quality standards; and
- lower costs for improving fuel quality versus investing in extra pollution control equipment.

4.4.3 *Wet flue gas desulfurization*

Wet scrubbers (wet FGD) are the most common FGD method currently in use (or under development) and include a variety of processes, the use of many sorbents, and are manufactured by a large number of companies. The sorbents used by wet scrubbers include calcium-, magnesium-, or sodium-based sorbents, ammonia, or seawater. The calcium-based scrubbers are by far the most popular and this technology is discussed in this section along with the use of sodium- and magnesium-based sorbents and seawater and ammonia scrubbing processes.

4.4.3.1 Absorption with chemical reaction

There are a variety of approaches for removing gaseous pollutants from effluent streams, one of which is absorption with chemical reaction. Absorption is an operation involving mass transfer of a soluble vapor component to a solvent liquid in equipment that promotes contact between the gas and the liquid. The driving force for the absorption is the difference between the partial pressure of the soluble gas in the gas stream and its vapor pressure above the surface of the liquid. Gases of limited solubility, such as SO_2, can be absorbed readily in an alkaline solution. The liquid stream is broken into small droplets or thin films to provide a large surface area for mass transfer. The most commonly used devices are columns containing packing or regularly spaced plates, open spray chambers and towers, and a combination of sprayed and packed chambers. Countercurrent contact of liquid and gas is used to maximize the driving forces. Many commercial gas absorption processes involve systems in which chemical reactions occur in the liquid phase. These reactions generally enhance the rate of absorption and increase the capacity of the liquid solution to dissolve the solute.

4.4.3.2 Limestone- and lime-based scrubbers

Wet scrubbing with limestone and lime are the most popular commercial FGD systems. The inherent simplicity, the availability of an inexpensive sorbent (limestone), production of a usable byproduct (gypsum), reliability, availability, and the high removal efficiencies obtained (which can be as high as 99%) are the main reasons for this popularity. Capital costs are typically higher than other technologies, such as sorbent injection systems; however, the technology is known for its low operating costs as the sorbent is widely available and the system is cost effective.

In a limestone/lime wet scrubber, the flue gas is scrubbed with a 5–15% (by weight) slurry of calcium sulfite/sulfate salts along with calcium hydroxide ($Ca(OH_2)$) or limestone ($CaCO_3$). Calcium hydroxide is formed by slaking lime (CaO) in water according to the reaction:

$$CaO(s) + H_2O(l) \rightarrow Ca(OH)_2(s) + \text{heat} \qquad (4.14)$$

In the limestone and lime wet scrubbers, the slurry containing the sulfite/sulfate salts and the newly added limestone or calcium hydroxide is pumped to a spray tower absorber and sprayed into it. The sulfur dioxide is absorbed into the droplets of slurry and a series of reactions occur in the slurry. The reactions between the calcium and the absorbed sulfur dioxide create the compounds calcium sulfite hemihydrate ($CaSO_4 \cdot \frac{1}{2}H_2O$) and calcium sulfate dihydrate ($CaSO_4 \cdot 2H_2O$). Both of these compounds have low solubility in water and precipitate from the solution. This enhances the absorption of sulfur dioxide and further dissolution of the limestone or hydrated lime.

The reactions occurring in the scrubbers are complex. Simplified overall reactions for limestone- and lime-based scrubbers are:

$$\mathrm{SO_2(g) + CaCO_3(s) + \tfrac{1}{2}\,H_2O(l) \rightarrow CaSO_3 \cdot \tfrac{1}{2}H_2O(s) + CO_2(g)} \tag{4.15}$$

for a limestone scrubber and

$$\mathrm{SO_2(g) + Ca(OH)_2(s) + H_2O(l) \rightarrow CaSO_3 \cdot \tfrac{1}{2}\,H_2O(s) + \tfrac{3}{2}H_2O(l)} \tag{4.16}$$

for a lime scrubber.

The calcium sulfite hemihydrate can be converted to the calcium sulfate dihydrate with the addition of oxygen by the reaction:

$$\mathrm{CaSO_3 \cdot \tfrac{1}{2}\,H_2O(s) + \tfrac{3}{2}\,H_2O(l) + \tfrac{1}{2}O_2(g) \leftrightarrow CaSO_4 \cdot 2H_2O(s)} \tag{4.17}$$

The actual reactions that occur, however, are much more complex and include a combination of gas-liquid, solid-liquid, and liquid-liquid ionic reactions. In the limestone scrubber, the following reactions describe the process [18]. In the gas-liquid contact zone of the absorber (see Figures 4.1 and 4.2 for a wet FGD absorber module cutaway and a typical schematic diagram of a limestone scrubber system, respectively [19]), sulfur dioxide dissolves into the aqueous state

$$\mathrm{SO_2(g) \leftrightarrow SO_2(l)} \tag{4.2}$$

and is hydrolyzed to form ions of hydrogen and bisulfate

$$\mathrm{SO_2(l) + H_2O(l) \leftrightarrow HSO_3^- + H^+} \tag{4.3}$$

The limestone dissolves in the absorber liquid forming ions of calcium and bicarbonate

$$\mathrm{CaCO_3(s) + H^+ \leftrightarrow Ca^{++} + HCO_3^-} \tag{4.18}$$

which is followed by acid-base neutralization

$$\mathrm{HCO_3^- + H^+ \leftrightarrow CO_2(l) + H_2O(l)} \tag{4.19}$$

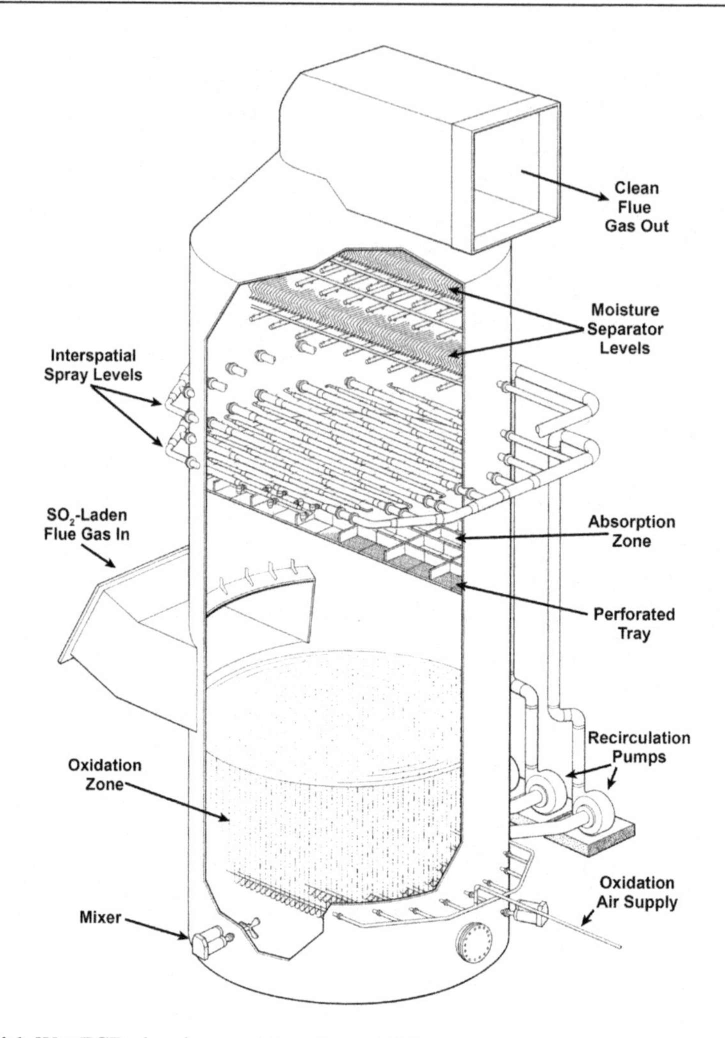

Figure 4.1 Wet FGD absorber module cutaway [19].

stripping of the CO_2 from the slurry

$$CO_2(l) \leftrightarrow CO_2(g) \tag{4.20}$$

and dissolution of the calcium sulfite hemihydrate

$$CaSO_3 \cdot \tfrac{1}{2}H_2O(s) \leftrightarrow Ca^{++} + HSO_3^- + \tfrac{1}{2}H_2O(l) \tag{4.21}$$

In the reaction tank of a scrubber system, the solid limestone is dissolved into the aqueous state (Reaction (4.18)), acid-base neutralization occurs (Reaction

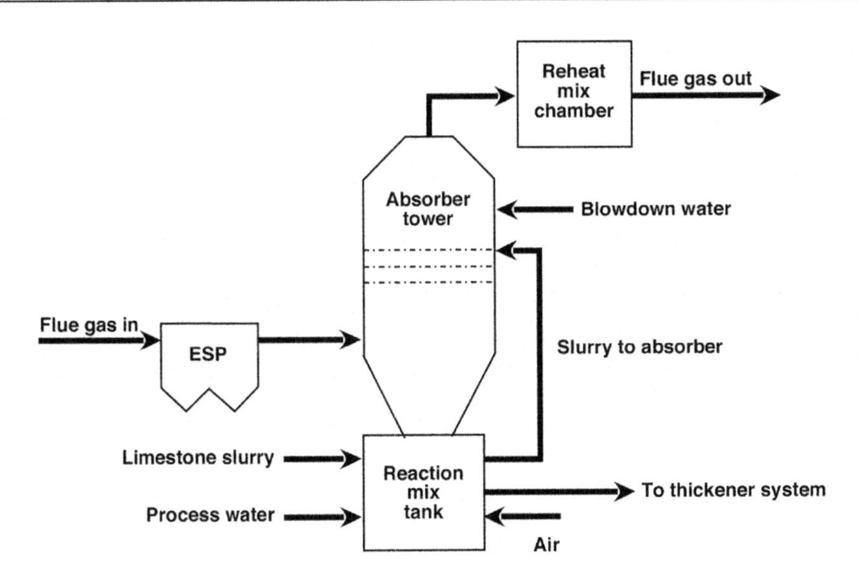

Figure 4.2 Limestone scrubber system with forced oxidation.

(4.19)), the CO_2 is stripped out (Reaction (4.20)), and the calcium sulfite hemihydrate is precipitated by the reaction

$$Ca^{++} + HSO_3^- + \tfrac{1}{2}H_2O(l) \leftrightarrow CaSO_3 \cdot \tfrac{1}{2}H_2O(s) + H^+ \tag{4.22}$$

Equation (4.23) expresses the mass transfer rate of SO_2 from the gas phase to the liquid or aqueous phase and its mass transfer rate is expressed by

$$\frac{d(Gy)}{dV} = k_g \alpha (y - y^*) \tag{4.23}$$

where G is the molar gas flow rate (moles/s), y is the mole fraction of SO_2 in flue gas, k_g is the gas film mass transfer coefficient (moles/m^2s), α is the interfacial surface area (m^2/m^2), y^* is the equilibrium SO_2 concentration at the gas/liquid interface, and V is the volume of the gas/liquid regime (m^3) [19].

In a limestone-based wet scrubber, the rate-limiting reaction in the gas/liquid contact zone is the dissolution of the limestone (Reaction (4.18)). The reaction rate for limestone dissolution is

$$\frac{d[CaCO_3]}{dt} = k_c\left([H^+] - [H^+]_{eq}\right) Sp_c [CaCO_3] \tag{4.24}$$

where $[CaCO_3]$ is the calcium carbonate concentration in the slurry (moles/l), k_c is the reaction rate constant, $[H^+]$ is the hydrogen ion concentration (moles/l), $[H^+]_{eq}$ is the equilibrium hydrogen ion concentration at the limestone surface (moles/l), and Sp_c is the specific surface area of the limestone in the slurry.

The dissolution of the calcium sulfite in the gas-liquid contact zone in the absorber is necessary in order to minimize scaling of the calcium sulfite hemihydrate in the absorber [5]. The equilibrium pH for calcium sulfite is ≈ 6.3 at a CO_2 partial pressure of 0.12 atmospheres, which is the typical concentration of CO_2 in flue gas. Typically the pH is maintained below this level to keep the calcium sulfite hemihydrate from dissolving (i.e. keep Reaction (4.21) from proceeding to the right).

The slurry returning from the absorber to the reaction tank can have a pH as low as 3.5, which is increased to 5.2 to 6.2 by the addition of freshly prepared limestone slurry into the tank [2]. The pH in the reaction tank must be maintained at a pH that is less than the equilibrium pH of calcium carbonate in water, which is 7.8 at 77°F.

The reaction equations for the lime scrubber are similar to those for the limestone scrubber, with the exception that the following reactions are substituted for Reactions (4.18) and (4.19), respectively [18]:

$$Ca(OH)_2(s) + H^+ \leftrightarrow CaOH^+ + H_2O(l) \tag{4.25}$$

$$CaOH^+ + H^+ \leftrightarrow Ca^{++} + H_2O(l) \tag{4.26}$$

4.4.3.2.1 Limestone with forced oxidation

Limestone scrubbing with forced oxidation (LSFO) is one of the most popular systems in the commercial market. A limestone slurry is used in an open spray tower with in-situ oxidation to remove SO_2 and form a gypsum sludge. The major advantages of this process, relative to a conventional limestone FGD system (where the product is calcium sulfite rather than calcium sulfate (gypsum)), are easier dewatering of the sludge, more economical disposal of the scrubber product solids, and decreased scaling on the tower walls. LSFO is capable of greater than 90% SO_2 removal [20].

In the LSFO system, the hot flue gas exits the particulate control device, usually an ESP, and enters a spray tower where it comes into contact with a sprayed dilute limestone slurry. The SO_2 in the flue gas reacts with the limestone in the slurry via the reactions listed earlier to form the calcium sulfite hemihydrate. Compressed air is bubbled through the slurry, which causes this sulfite to be naturally oxidized and hydrated to form calcium sulfate dihydrate. In a limestone system with forced oxidation the overall reaction in the reaction tank is [19]

$$CaCO_3(s) + H^+ + HSO_3^- + 1/2O_2 + H_2O \rightarrow CaSO_4 \cdot 2H_2O(s) + CO_2(g) \tag{4.27}$$

The rate of gypsum crystallization in the reaction tank can be expressed by

$$\frac{d[CaSO_4 \cdot 2H_2O]}{dt} = k(R-1)Sp_g[CaSO_4 \cdot 2H_2O] \tag{4.28}$$

where k is the crystallization rate constant, R equals $(A_{Ca^{2+}} \cdot A_{SO_4^{2-}}/K_{sp})$, $A_{Ca^{2+}}$ is the activity of the Ca^{2+} ion, $A_{SO_4^{2-}}$ is the activity of the SO_4^{2-} ion, K_{sp} is the solubility product of gypsum, and Sp_g is the specific surface area of gypsum [19]. R is a measure of the level of supersaturation: if R is greater than 1, the solution is supersaturated with gypsum; if R is less than 1, the solution is subsaturated in gypsum.

The calcium sulfate can be first dewatered using a thickener or hydrocyclones then further dewatered using a rotary drum filter. The gypsum is then transported to a landfill for disposal. The formation of the calcium sulfate crystals in a recirculation tank slurry also helps to reduce the chance of scaling.

The absorbing reagent, limestone, is normally fed to the open spray tower in an aqueous slurry at a molar feed rate of 1.1 moles of $CaCO_3$/mole of SO_2 removed. This process is capable of removing more than 90% of the SO_2 present in the inlet flue gas. The advantages of LSFO systems are [20]:

- Lower scaling potential on tower internal surfaces due to the presence of gypsum seed crystals and reduced calcium sulfate saturation levels. This in turn allows a greater reliability of the system;
- The gypsum product is filtered easier than the calcium sulfite ($CaSO_3$) produced with conventional limestone systems;
- A lower chemical oxygen demand in the final disposed product;
- The final product can be safely and easily disposed in a landfill;
- The forced oxidation allows the limestone utilization to be greater than conventional systems;
- Low cost of the raw material (limestone) used as an absorbent; and
- LSFO is an easier retrofit than natural oxidation systems since the process uses smaller dewatering equipment.

A disadvantage of this system is the high energy demand due to the relatively higher liquid-to-gas ratio necessary to achieve the required SO_2 removal efficiencies.

4.4.3.2.2 Limestone with forced oxidation producing a wallboard gypsum byproduct

In the limestone/wallboard (LS/WB) gypsum FGD process, a limestone slurry is used in an open spray tower to remove SO_2 from the flue gas. The flue gas enters the spray tower where the SO_2 reacts with the $CaCO_3$ in the slurry to form calcium sulfite. The calcium sulfite is then oxidized to calcium sulfate in the absorber recirculation tank. The calcium sulfate produced with this process is of a higher quality so that it may be used in wallboard manufacture.

There are a few differences with this process in order to achieve a higher quality gypsum. The LS/WB system uses horizontal belt filters to produce a drier product and provides sufficient cake washing to remove residual chlorides. Since the byproduct is a higher quality, the use of the product handling system is replaced with a byproduct conveying and temporary storage equipment. Sulfuric acid addition is used in systems with an external oxidation tank. The acid is used to control the pH of the slurry and neutralizes unreacted $CaCO_3$.

The limestone feed rate in this process is 1.05 moles $CaCO_3$/mole of SO_2 removed, which is slightly lower than the feed rate for the LSFO system [20]. Other advantages of this process are that the disposal area is kept to a minimum since most of the byproduct is reusable. The gypsum can be sold to cement plants and agricultural users. Also, SO_2 removal is slightly enhanced because of the high sulfite to sulfate conversion.

There are some disadvantages to this process. There are few full-scale operating systems that actually produce quality gypsum in the United States. To produce quality gypsum, specific process control and tight operator attention are constantly needed to ensure that chemical impurities do not lead to off-specification gypsum. Another disadvantage is the inability to use cooling tower blowdown as system make-up water due to chloride limits in the gypsum byproduct.

4.4.3.2.3 Limestone with inhibited oxidation

In the limestone with inhibited oxidation process, the hot flue gas exits the particulate control device and enters an open spray tower where it comes into contact with a dilute $CaCO_3$ slurry. This slurry contains thiosulfate ($Na_2S_2O_3$), which inhibits natural oxidation of the calcium sulfite. The calcium sulfite is formed from the reaction with SO_2 in the flue gas and the $CaCO_3$ slurry. The slurry absorbs the SO_2, and then drains down to a recirculation tank below the tower. By inhibiting natural oxidation of the sulfite, gypsum scaling on process equipment is reduced along with gypsum relative saturation. The gypsum relative saturation is reduced below 1.0. Thiosulfate is either added directly as $Na_2S_2O_3$ to the feed tank or is generated in situ by the addition of emulsified sulfur. In some cases, thiosulfate has the ability to increase the dissolution of the calcium carbonate and enlarge the size of the sulfite crystals to improve solids dewatering [20].

This process is capable of removing more than 90% of the SO_2 in the flue gas. The calcium sulfite slurry product is thickened, stabilized with fly ash and lime, and then sent to a landfill. The calcium carbonate feed rate is 1.10 moles Ca/mole of SO_2 removed.

The effectiveness of thiosulfate is site specific since the amount of thiosulfate required to inhibit oxidation strongly depends on the chemistry and operating conditions of each FGD system. Variables such as saturation temperature, dissolved magnesium, chlorides, flue gas inlet SO_2 and O_2 concentrations, and slurry pH affect the thiosulfate effectiveness [20].

Thiosulfate has been shown to increase limestone utilization when added to the system. This occurs because the thiosulfate reduces the gypsum relative saturation level, which in turn reduces the level of calcium dissolved in the liquor. The dissolution rate is increased by lowering the calcium concentration in the slurry. Thiosulfate also improves the dewatering characteristics of the sulfite product. By preventing the high concentrations of sulfate, the thiosulfate allows the calcium sulfite to form larger, single crystals. This increases the crystal's settling velocity and improves the filtering characteristics, which results in a higher solid content of dewatered product.

There are a few disadvantages of the process. The thiosulfate/sulfur reagent requires additional process equipment and storage facilities. Also, the reagent can cause corrosion of many stainless steels under scrubber conditions. Another disadvantage is that the thiosulfate is fairly temperature dependent, thereby requiring the system to operate within a particular temperature range.

4.4.3.2.4 Magnesium enhanced lime

In the magnesium enhanced lime (MagLime) process, the hot flue gas exits the particulate control device and enters a spray tower where it comes into contact with a magnesium sulfite/lime slurry. Magnesium lime, such as thiosorbic lime (which contains 4–8% MgO), is fed to the open spray tower in an aqueous slurry at a molar feed rate of 1.1 moles CaO/mole of SO_2 removed. The SO_2 is absorbed by the reaction with magnesium sulfite, forming magnesium bisulfite. This occurs through the following reactions [20]:

$$SO_2(g) + H_2O(l) \rightarrow H_2SO_3(aq) \rightarrow H^+ + HSO_3^- \tag{4.29}$$

$$H^+ + MgSO_3(s) \rightarrow HSO_3^- + Mg^{++} \tag{4.30}$$

The magnesium sulfite absorbs the H^+ ion and increases the HSO_3^- concentration in Reaction (4.30). This allows the scrubber liquor to absorb more of the SO_2. The absorbed SO_2 reacts with hydrated lime to form solid-phase calcium sulfite. The magnesium sulfite is reformed by the following reactions:

$$Ca(OH)_2(s) + 2HSO_3^- + Mg^{++} \rightarrow Ca^{++}SO_3^{--} + 2H_2O(l) + MgSO_3(s) \tag{4.31}$$

$$Ca^{++}SO_3^{--} + \tfrac{1}{2}H_2O \rightarrow CaSO_3 \cdot \tfrac{1}{2}H_2O(s) \tag{4.32}$$

Inside the absorber, some magnesium sulfite present in the solution is oxidized to sulfate. This sulfite reacts with the lime to form calcium sulfate solids. Calcium sulfite and sulfate solids are the main products of the MagLime process. The calcium sulfite sludge is dewatered using thickener and vacuum filter systems, then fixated using fly ash and lime prior to disposal in a lined landfill. The magnesium remains dissolved in the liquid phase.

Some advantages of the MagLime process compared to the LSFO process are [20]:

- High SO_2 removal efficiency at low liquid-to-gas ratios;
- Lower gas-side pressure drop due to lower liquid-to-gas ratios;
- Reduced potential for scaling, which improves reliability of the system;
- Lower power consumption due to a lower slurry recycle rate;
- Lower capital investment due to smaller reagent handling equipment and no oxidation air compressor; and
- Reduction in fresh water use since the process water may be recycled for the mist eliminator wash.

The three major disadvantages of the process are the expense of the lime reagent compared to the limestone, the use of fresh water for lime slaking, and the difficult dewatering characteristics of the calcium sulfite/sulfate sludge. The sulfite can be oxidized to produce gypsum, but this requires extensive equipment and process control.

4.4.3.2.5 Limestone with dibasic acid

The dibasic acid enhanced limestone process is very similar to the LSFO process. The hot flue gas exits the particulate control device and enters a spray tower where it comes into contact with a diluted limestone slurry. The SO_2 in the flue gas reacts with the limestone and water to form hydrated calcium sulfite

$$SO_2(g) + CaCO_3(s) + \tfrac{1}{2}H_2O(l) \rightarrow CaSO_3 \cdot \tfrac{1}{2}H_2O(s) + CO_2(g) \tag{4.15}$$

This equation is rate limited by the absorption of SO_2 into the scrubbing liquor

$$SO_2(l) + H_2O(l) \leftrightarrow HSO_3^- + H^+ \tag{4.3}$$

The dissolved SO_2 ions then react with the calcium ions to form calcium sulfite. The hydrogen ions in solution are partly responsible for reforming SO_2.

After absorbing the SO_2, the slurry drains from the tower to a recirculation tank. Here the calcium sulfite is oxidized to calcium sulfate dihydrate using oxygen

$$CaSO_3 \cdot \tfrac{1}{2}H_2O(s) + \tfrac{3}{2}H_2O(l) + \tfrac{1}{2}O_2(g) \leftrightarrow CaSO_4 \cdot 2H_2O(s) \tag{4.17}$$

Dibasic acid acts as a buffer by absorbing free hydrogen ions formed by Reaction (4.3). This then shifts the reaction to the right to form more sulfite ions, thus removing more SO_2. Alkaline limestone is added to replace the buffering capabilities of the acid; therefore there is no net consumption of the dibasic acid during SO_2 absorption.

The limestone dissolution rate is increased by increasing the SO_2 removal efficiency at a low slurry pH. This results in a lower reagent consumption due to an increase in calcium carbonate availability in the recirculation tank.

The dibasic acid process offers some advantages compared to the LSFO process [20]:

- Increased SO_2 removal efficiency;
- Reduced liquid-to-gas ratio and the potential to decrease the reagent feed rate. This lowers capital and operating costs for the limestone grinding equipment, slurry handling, and landfill requirements;
- Reduced scaling because of the low pH and reduced gypsum relative saturation levels; and
- Increased system reliability by reducing the maintenance requirements and increasing the flexibility of the system.

Disadvantages of the process include:

- More process capital is needed for the dibasic acid feed equipment;
- There is the potential for corrosion and erosion due to the low system pH;
- Odorous byproducts are produced by the dibasic acid degradation. Although the SO_2 absorption reactions do not consume the dibasic acid, the acid does degrade by carboxylic oxidation into many short chain molecules. One of these molecules is valeric acid which has a musty odor; and
- Control problems may be caused by foaming in the recirculation and oxidation tanks due to the presence of the dibasic acid.

4.4.3.3 Sodium-based scrubbers

Wet sodium-based systems have been in commercial operation since the 1970s. These systems can achieve high SO_2 removal efficiencies while burning coals with medium- to high-sulfur content. A disadvantage of these systems, however, is the production of a waste sludge that requires disposal.

4.4.3.3.1 Lime dual alkali

In the lime dual alkali process, the hot flue gas exits the particulate control device and enters an open spray tower where the gas comes into contact with a sodium sulfite (Na_2SO_3) solution that is sprayed into the tower [20]. An initial charge of sodium carbonate (Na_2CO_3) reacts directly with the SO_2 to form sodium sulfite and CO_2. The sulfite then reacts with more SO_2 and water to form sodium bisulfite ($NaHSO_3$). Some of the sodium sulfite is oxidized by excess oxygen in the flue gas to form sodium sulfate (Na_2SO_4). This does not react with SO_2 and cannot be reformed by the addition of lime to form calcium sulfate. The above process is described by the following reactions:

$$Na_2CO_3(s) + SO_2(g) \rightarrow Na_2SO_3(s) + CO_2(g) \tag{4.33}$$

$$Na_2SO_3(s) + SO_2(g) + H_2O(l) \rightarrow 2NaHSO_3(s) \tag{4.34}$$

$$Na_2SO_3(s) + \tfrac{1}{2}O_2(g) \rightarrow Na_2SO_4(s) \tag{4.35}$$

along with the minor reaction

$$2NaOH(aq) + SO_2(g) \rightarrow Na_2SO_3(s) + H_2O(l) \tag{4.36}$$

The calcium sulfites and sulfates are reformed in a separate regeneration tank and are formed by mixing the soluble sodium salts (bisulfate and sulfate) with slaked lime. The calcium sulfites and sulfates precipitate from the solution in the regeneration tank. The scrubber liquor then has a pH of 6 to 7 and consists of sodium sulfite, sodium bisulfite, sodium sulfate, sodium hydroxide, sodium carbonate, and sodium bicarbonate [20].

The lime dual alkali process has several advantages over the LSFO process, which include [13]:

- The system has a higher availability since there is less potential for scaling and plugging of the soluble absorption reagents and reaction products;
- Corrosion and erosion is prevented with the use of a relatively high pH solution;
- Maintenance labor and materials are lower because of the high reliability of the system;
- The main recirculation pumps are smaller since the absorber liquid/gas feed rate is less;
- Power consumption is lower due to the smaller pump requirements;
- There is not a process blowdown water discharge stream; and
- The highly reactive alkaline compounds in the absorbing solution allow for better turndown and load following capabilities.

There are two main disadvantages of the process compared to the LSFO system. The sodium carbonate reagent is more expensive than limestone and the sludge must be disposed in a lined landfill because of sodium contamination of the calcium sulfite/sulfate sludge.

4.4.3.3.2 Regenerative processes

Regenerative FGD processes regenerate the alkaline reagent and convert the SO_2 to a usable chemical byproduct. Two commercially accepted processes are discussed in this section. The Wellman-Lord process is the most highly demonstrated regenerative technology in the world, while the regenerative magnesia scrubbing process is in commercial service in the United States.

4.4.3.3.2.1 Wellman-lord process The Wellman-Lord process uses sodium sulfite to absorb SO_2, which is then regenerated to release a concentrated stream of SO_2. Most of the sodium sulfite is converted to sodium bisulfite by reaction with SO_2 as in the dual alkali process. Some of the sodium sulfite is oxidized to sodium sulfate. Prescrubbing of the flue gases is necessary to saturate and cool the flue gas to about 130°F. This removes chlorides and remaining fly ash, and avoids excessive evaporation in the absorber. A schematic of the system is shown in Figure 4.3 [13].

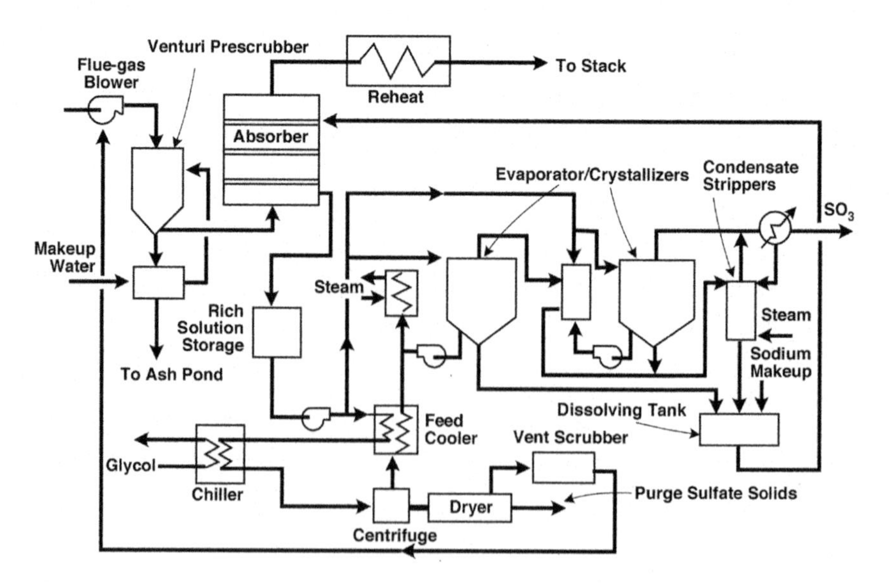

Figure 4.3 Wellman-Lord process.

The basic absorption reaction for the Wellman-Lord process is

$$SO_2(g) + Na_2SO_3(aq) + H_2O(l) \rightarrow 2NaHSO_3(aq) \tag{4.37}$$

The sodium sulfite is regenerated in an evaporator-crystallizer through the application of heat. A concentrated SO_2 stream (i.e. 90%) is produced at the same time. The overall regeneration reaction is

$$2NaHSO_3(aq) + heat \rightarrow Na_2SO_3(s) + H_2O(l) + SO_2(conc.) \tag{4.38}$$

The concentrated SO_2 stream that is produced may be compressed, liquefied, and oxidized to produce sulfuric acid or reduced to elemental sulfur. A small portion of collected SO_2 oxidizes to the sulfate form and is converted in a crystallizer to sodium sulfate solids that are marketed as a salt cake [13].

The advantages of this process include minimal solid wastes production, low alkaline reagent consumption, the use of a slurry rather than a solution, thereby preventing scaling and allowing the production of a marketable byproduct. The disadvantage of the process is the high energy consumption and maintenance due to the complexity of the process, and large area required for the system. Another disadvantage is that the purge stream of about 15% of the scrubbing solution is required to prevent buildup of the sodium sulfate. Thiosulfate must be purged from the regenerated sodium sulfite.

4.4.3.3.2.2 Regenerative magnesia scrubbing In the magnesium oxide process, MgO in the slurry is used in a similar manner to limestone or lime in the lime scrubbing process. The primary difference between the processes is that the magnesium oxide process is regenerative, whereas lime scrubbing is generally a throw-away process.

The magnesium oxide process, shown in Figure 4.4, uses a slurry of slaked magnesium oxide ($Mg(OH)_2$) to remove SO_2 from the flue gas forming magnesium sulfite and sulfate via the basic reactions

$$Mg(OH)_2(s) + SO_2(g) \rightarrow MgSO_3(s) + H_2O(l) \tag{4.39}$$

$$MgSO_3(s) + \tfrac{1}{2}O_2(g) \rightarrow MgSO_4(s) \tag{4.40}$$

A bleed stream of scrubber slurry is centrifuged to form a wet cake containing 75 to 90% solids, which is then dried to form a dry, free-flowing mixture of magnesium sulfite and sulfate. This mixture is heated to decompose most of the magnesium sulfite/sulfate to SO_2 and MgO. A stream of 10 to 15% SO_2 is produced. Coke, or some other reducing agent, is added in the calcination step to reduce any sulfate present. The regenerated MgO is slaked and used in the absorber. The regeneration reactions are

$$MgSO_3(s) + heat \rightarrow MgO(s) + SO_2(g) \tag{4.41}$$

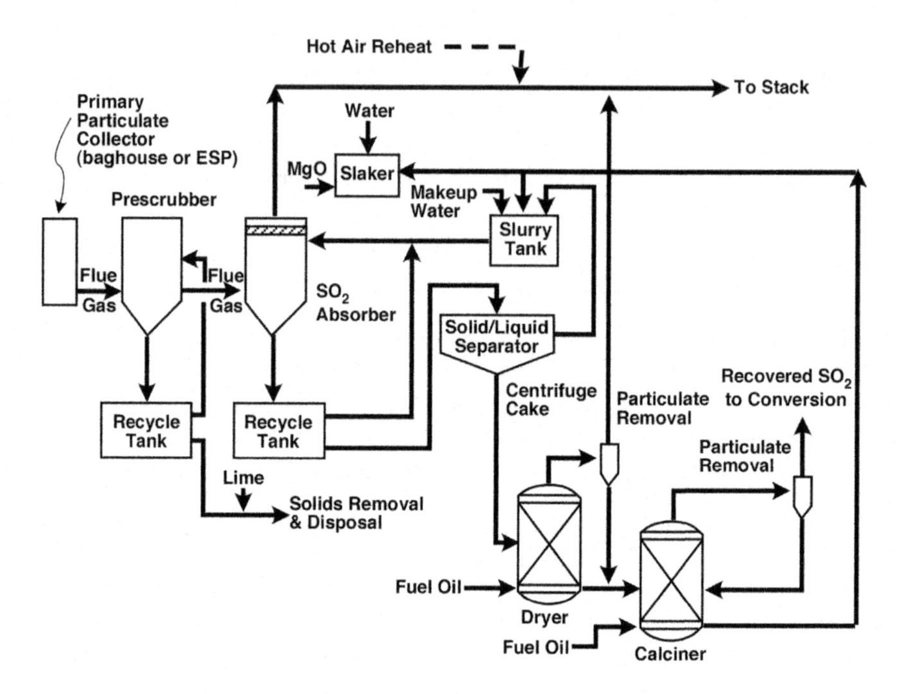

Figure 4.4 Regenerative magnesia scrubbing process.

$$MgSO_4(s) + \tfrac{1}{2}C(s) + \text{heat} \rightarrow MgO(s) + SO_2(g) + \tfrac{1}{2}CO_2(g) \tag{4.42}$$

The SO_2 product gas is generally washed and quenched and fed to a contact sulfuric acid plant to produce concentrated sulfuric acid byproduct. Sulfur production is possible but would be expensive since the SO_2 stream is dilute.

Advantages of this process include high SO_2 removal efficiencies (up to 99%), minimum impact of fluctuations in inlet SO_2 levels on removal efficiency, low chemical scaling potential, the capability to regenerate the sulfate which simplifies waste management, and more favorable economics compared to other available regenerative processes [13]. The main disadvantage of the process is its complexity and the need of a contact sulfuric acid plant to produce a salable byproduct.

4.4.3.4 Seawater scrubbing process

The seawater scrubbing process uses the natural alkalinity of seawater (i.e. the pH of seawater ranges from 7.6 to 8.4 with the inherent alkalinity resulting mainly from the bicarbonate ions (HCO_3^-) and carbonate ions (HCO_3^{2-}) combined in the seawater) to absorb SO_2 [9,21]. The process was developed for and has been applied on small-scale industrial units and oil and coal-fired power plants since 1968; however, by the mid-1990s the process has been applied to 300–700 MWe power plants and the technology is expanding rapidly [9]. Seawater scrubbers have been installed on more than 35 GWe of coal-fired capacity by 2009 with another 4.4 GWe of capacity under construction. Seawater has a large neutralizing capacity and is capable of achieving up to 99% SO_2 removal [9]. The absorbed SO_2 is oxidized to sulfate, which occurs naturally in seawater and therefore can be discharged into the sea.

The seawater FGD process consists of two major systems, the SO_2 absorption system and the seawater treatment plant. The chemistry of the process is similar to the LSFO chemistry except that the limestone comes from being completely dissolved with the seawater. The absorption of the SO_2 occurs in an absorber where seawater and flue gas are brought into close contact in a countercurrent flow. When the flue gas comes into contact with seawater in the absorber, the SO_2 in the flue gas dissolves in water to form bisulfite (HSO_3^-) and some of the bisulfite is converted to sulfite (SO_3^{2-}) [9]:

$$SO_2 + H_2O \rightarrow HSO_3^- + H^+ \tag{4.3}$$

$$HSO_3^- \rightarrow SO_3^{2-} + H^+ \tag{4.43}$$

The bisulfite and sulfite is oxidized to sulfate due to the oxygen in the flue gas and the seawater:

$$HSO_3^- + 1/2O_2 \rightarrow SO_4^{2-} + H^+ \tag{4.44}$$

$$SO_3^{2-} + 1/2O_2 \rightarrow SO_4^{2-} \tag{4.45}$$

As SO_2 dissolves in the seawater to form bisulfite and sulfite, which are in turn oxidized to sulfate, hydrogen ions (H^+) are produced and acidify the seawater, lowering its pH value. The acidified effluent is then neutralized using bicarbonate and carbonate ions from the seawater before discharge into the sea through the reactions:

$$HCO_3^- + H^+ \rightarrow CO_2 + H_2O \tag{4.19}$$

$$CO_3^{2-} + H^+ \rightarrow HCO_3^- \tag{4.46}$$

The neutralizing step is performed by adding more seawater from the cooling system of the power plant to obtain the alkalinity required. Prior to discharge to the sea, air is blown through the seawater effluent to: insure sufficient oxidation of the bisulfite and sulfite ions, strip CO_2 from the effluent to increase the efficiency of the neutralization, and replenish dissolved oxygen in the seawater [9].

The main advantages of the seawater FGD process include: 1) the addition of chemical reagents is not required; 2) no byproduct to handle or dispose; 3) plant design and operation are relatively simple; and 4) high SO_2 removal efficiency (up to 99%) can be achieved [9]. The process, however, is limited to coastal power plants. In addition, the process is limited to fuels containing less than 1.5% sulfur, otherwise additional additives, e.g. sodium hydroxide or magnesium oxide, are needed for the process to neutralize the absorber effluent before its discharge to the sea.

4.4.3.5 Ammonia scrubbing process

The ammonia scrubbing process is similar to the limestone gypsum process except that anhydrous or aqueous ammonia is used as the scrubbing agent to remove SO_2 from the flue gas and the final product is ammonium sulfate, which is used as agricultural fertilizer. Ammonia-based desulfurization technology was developed in the early 1970s in Japan and Italy for producing fertilizer [9]. The high value byproduct is the major advantage of this process. Further development has led to several ammonia scrubbing processes becoming commercially available for the power generation sector and ammonia scrubbing units have been installed on coal- and oil-fired boilers.

The process chemistry consists of several steps. Ammonia, when dissolved in water, forms ammonium hydroxide (NH_4OH). The scrubbing solution is continuously recycled and is circulated in the absorber, thereby the solution also contains ammonium sulfite ($(NH_4)_2SO_3$) and ammonium sulfate ($(NH_4)_2SO_4$). When the flue gas comes into contact with the scrubbing solution, the SO_2 dissolves into the water forming sulfurous acid (HSO_3), which in turn reacts with the ammonia sulfite, sulfate, and hydroxide. The steps are [9]:

$$SO_2 + H_2O \rightarrow H_2SO_3 \quad (4.47)$$

$$H_2SO_3 + 2NH_4OH \rightarrow (NH_4)_2SO_3 + 2H_2O \quad (4.48)$$

$$H_2SO_3 + NH_4OH \rightarrow NH_4HSO_3 + H_2O \quad (4.49)$$

$$H_2SO_3 + (NH_4)_2SO_3 \rightarrow 2NH_4HSO_3 \quad (4.50)$$

$$H_2SO_3 + (NH_4)_2SO_4 \rightarrow NH_4HSO_3 + NH_4HSO_4 \quad (4.51)$$

$$NH_4HSO_3 + NH_4OH \rightarrow (NH_4)_2SO_3 + H_2O \quad (4.52)$$

$$NH_4HSO_4 + NH_4OH \rightarrow (NH_4)_2SO_4 + H_2O \quad (4.53)$$

The pH of the scrubbing solution is maintained at the desired level by the addition of ammonia because the pH is lowered by the formation of sulfurous acid and the acidic intermediate species. A portion of the ammonium sulfite is oxidized into ammonium sulfate by the presence of oxygen in the flue gas. Air is injected into the solution to oxidize the remaining sulfite to sulfate:

$$2(NH_4)_2SO_3 + O_2 \rightarrow 2(NH_4)SO_4 \quad (4.54)$$

The ammonium sulfate solution is then saturated and the ammonium sulfate precipitates from the solution, which can be dried to produce saleable fertilizer.

The ammonia scrubbing process can remove more than 98% of SO_2 even when firing high sulfur fuels [9]. Other advantages of the process include: there is no wastewater to discharge or waste solid to dispose; a high value byproduct is produced; and scaling or blockage problems are not encountered. Some processes have been designed for flexibility such that either limestone or ammonia can be used as the reagent. Disadvantages of the system are: the process does have high capital cost and requires a large footprint; ammonia is both caustic and hazardous to human health; many utility companies are reluctant to own and operate a fertilizer plant on site; and there is a limited market for ammonium sulfate fertilizer in industrialized countries and therefore there is limited application of this process in power plants [9].

4.4.4 Dry flue gas desulfurization technology

Dry flue gas desulfurization (FGD) technology includes lime or limestone spray drying, dry sorbent injection including furnace-, economizer-, duct-, and hybrid methods, and circulating fluidized-bed scrubbers. These processes are characterized with dry waste products that are generally easier to dispose than waste products from wet scrubbers. All dry FGD processes are throwaway types.

4.4.4.1 Spray dry scrubbers

Spray dry scrubbers are the second most widely used method to control SO_2 emissions in utility coal-fired power plants. Prior to 1980, the removal of SO_2 by absorption was usually performed using wet scrubbers. Wet scrubbing requires considerable equipment and alternatives to wet scrubbing were developed including spray dry scrubbers. Lime (CaO) is usually the sorbent used in the spray drying process but hydrated lime ($Ca(OH)_2$) is also used. This technology is also known as semi-dry flue gas desulfurization and is generally used for sources that burn low- to medium-sulfur coal. In the United States, this process has been used in both retrofit applications and new installations on units burning low-sulfur coal [2,22].

In this process, the hot flue gas exits the boiler air heater and enters a reactor vessel. A slurry consisting of lime and recycled solids is atomized/sprayed into the absorber. The slurry is formed by the reaction

$$CaO(s) + H_2O(l) \rightarrow Ca(OH)_2(s) + heat \quad (4.14)$$

The SO_2 in the flue gas is absorbed into the slurry and reacts with the lime and fly ash alkali to form calcium salts:

$$Ca(OH)_2(s) + SO_2(g) \rightarrow CaSO_3 \cdot \tfrac{1}{2} H_2O(s) + \tfrac{1}{2}H_2O(v) \quad (4.55)$$

$$Ca(OH)_2(s) + SO_3(g) + H_2O(v) \rightarrow CaSO_4 \cdot 2H_2O(s) \quad (4.56)$$

Hydrogen chloride (HCl) present in the flue gas is also absorbed into the slurry and reacts with the slaked lime. The water that enters with the slurry is evaporated, which lowers the temperature and raises the moisture content of the scrubbed gas. The scrubbed gas then passes through a particulate control device downstream of the spray drier. Some of the collected reaction product, which contains some unreacted lime, and fly ash is recycled to the slurry feed system while the rest is sent to a landfill for disposal. Factors affecting the absorption chemistry include the flue gas temperature, SO_2 concentration in the flue gas, and the size of the atomized slurry droplets. The residence time in the reactor vessel is typically about 10–12 seconds.

The lime spray dryer process offers a few advantages over the LSFO process [20]. Only a small alkaline stream of scrubbing slurry must be pumped into the spray dryer. This stream contacts the gas entering the dryer instead of the walls of the system. This prevents corrosion of the walls and pipes in the absorber system. The pH of the slurry and dry solids is high, allowing for the use of mild steel materials rather than expensive alloys. The product from the spray dryer is a dry solid that is handled by conventional dry fly ash particulate removal and handling systems, which eliminates the need for dewatering solids handling equipment and reduces associated maintenance and operating requirements. Overall power requirements are decreased since less pumping power is required. The gas exiting the absorber is not saturated and does not require reheat, thereby capital costs and steam consumption is reduced. Chloride concentration increases the SO_2 removal efficiencies (whereas in wet scrubbers increasing chloride concentration decreases efficiency), which allows the use of cooling tower blowdown for slurry dilution after completing the slaking of the lime reagent. The absorption system is less complex, therefore operating, laboratory, and maintenance manpower requirements are lower than those required for a wet scrubbing system.

There are some disadvantages of the lime spray dryer compared to the LSFO system and these, along with the advantages, must be evaluated for specific applications [20]. A major product of the lime spray dryer process is calcium sulfite as only 25% or less oxidizes to calcium sulfate. The solids-handling equipment for the particulate removal device has to have a greater capacity than conventional fly ash removal applications. Fresh water is required in the lime slaking process, which can represent approximately half of the system's water requirement. This differs from the wet scrubbers where cooling tower water can be used for limestone grinding circuits and most other makeup water applications. The lime spray dryer process requires a higher reagent feed ratio than conventional systems to achieve high removal efficiencies. Approximately 1.5 moles CaO/mole of SO_2 removed are needed for 90% removal efficiency. Lime is also more expensive than limestone, so the operating cost is increased. This cost can be reduced if higher coal chloride levels and/or calcium chloride spiking are used since chlorides improve removal efficiency and reduce reagent consumption. A higher inlet flue gas temperature is needed when a higher sulfur coal is used, which in turn reduces the overall boiler efficiency.

Combining spray dry scrubbing with other FGD systems such as furnace or duct sorbent injection and particulate control technology such as a pulse-jet baghouse allows the use of limestone as the sorbent instead of the more costly lime [8]. Sulfur dioxide removal efficiencies can exceed 99% with such a combination.

4.4.4.2 Sorbent injection processes

A number of dry injection processes have been developed to provide moderate SO_2 removal that are easily retrofitted to existing facilities and are low capital cost. There are five basic processes. Two are associated with the furnace – furnace sorbent injection and convective pass (economizer) injection – and three are associated with injection into the ductwork downstream of the air heater – in-duct injection, in-duct spray drying, and hybrid systems. Combinations of these processes are also available. Sorbents include calcium- and sodium-based compounds; however, the use of calcium-based sorbents is more prevalent. Furnace injection has been used in some small plants using low-sulfur coals. Hybrid systems may combine furnace and duct sorbent injection or introduce a humidification step to improve removal efficiency. These systems can achieve as high as 70% removal and are commercially available [8]. Process schematics for dry injection SO_2 control technologies are illustrated in Figure 4.5 (modified from [23]).

Figure 4.6 is a representation of the level of SO_2 removal that the dry calcium-based sorbent injection processes achieve and the temperature regimes in which they operate [23]. The peak at approximately 2,200°F represents furnace sorbent injection, the peak at about 1,000°F is convective pass/economizer injection, and the peak at the low temperature represents all of the processes downstream of the air heater.

Another dry limestone injection technique, LIMB (limestone injection into a multistage burner), was developed in the 1960s to the 1980s but has not been adopted on a commercial scale for utility applications and is not discussed in detail. In this low capital cost process, which is used in some industrial-scale applications where low SO_2 removal is required, limestone is added to the coal stream and fed with the coal directly to the burner. This process gives poor SO_2 removal (typically $\approx$ 15% but in rare cases as much as 50%), experiences dead-burning (i.e. sintering or melting of the sorbent, thereby reducing surface area and lowering sulfur capture), is difficult to introduce in a uniform manner, and can cause operational problems such as tube fouling and impairment of ESP performance because of excessive sorbent addition [13].

Sorbents under development are not discussed; rather this section focuses mainly on commercial applications. Calcium organic salts (e.g. calcium acetate, calcium magnesium acetate, and calcium benzoate), pyrolysis liquor, and other sorbents are under development for use in injection processes.

4.4.4.2.1 Furnace sorbent injection

With the exception of LIMB, furnace sorbent injection (FSI) is the simplest dry sorbent process. In this process, illustrated in Figure 4.5a, pulverized sorbents, most often calcium hydroxide and sometimes limestone, are injected into the upper part of the furnace to react with the SO_2 in the flue gas. The sorbents are distributed

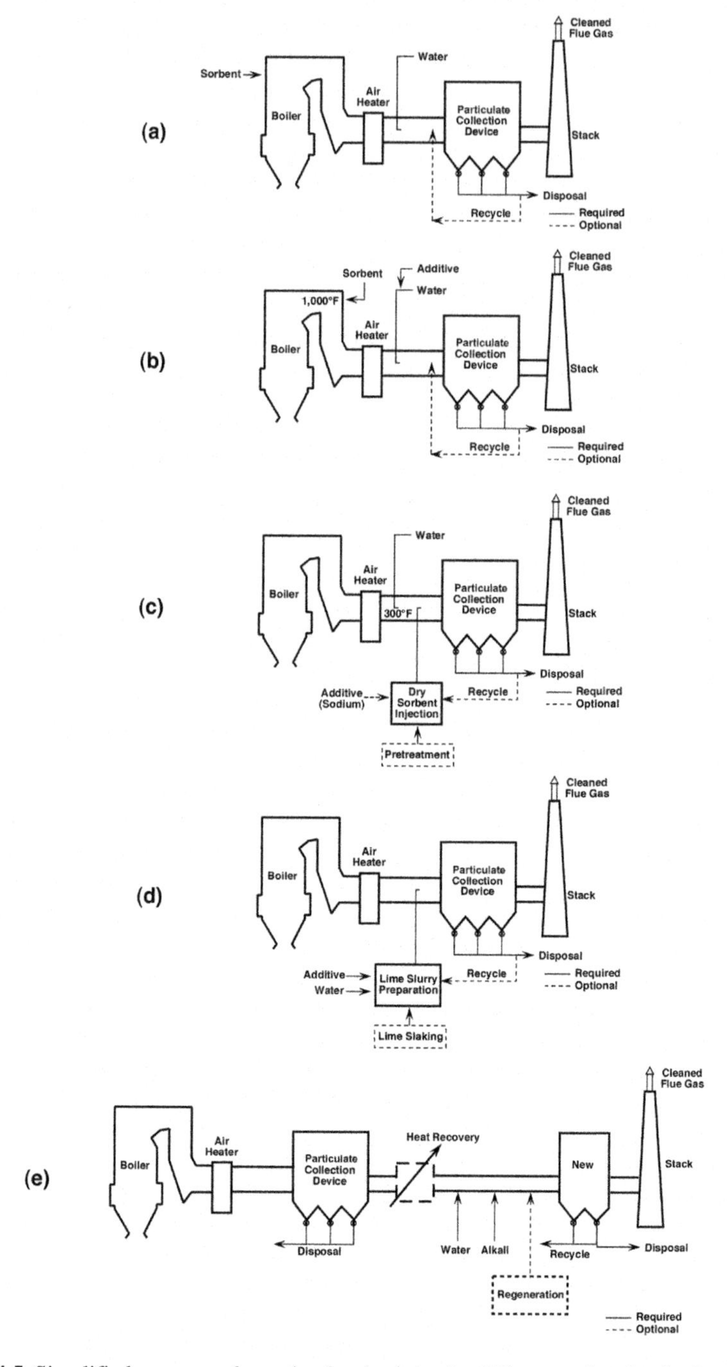

Figure 4.5 Simplified process schematics for dry injection SO_2 control technologies. A. furnace sorbent injection; B. economizer injection; C. duct sorbent injection: dry sorbent injection; D. duct sorbent injection: duct spray drying; E. hybrid system.

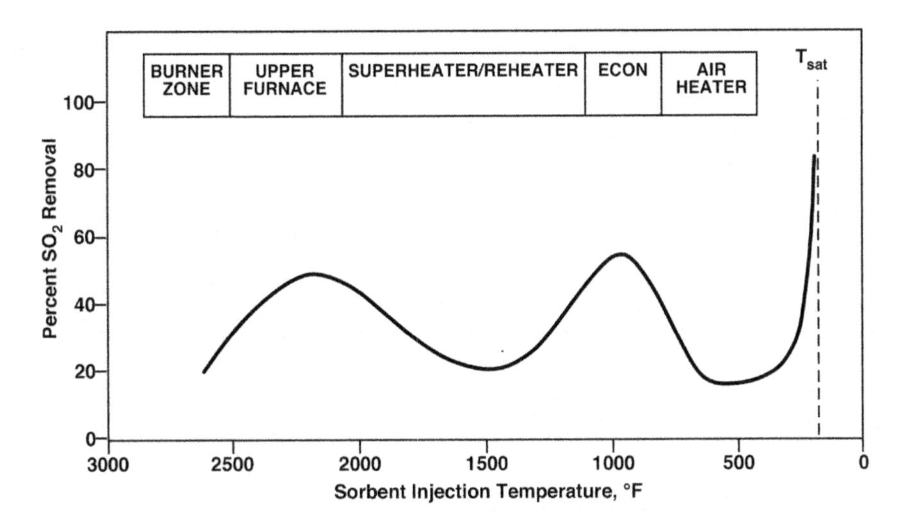

Figure 4.6 SO_2 capture regimes for hydrated calcitic lime at a Ca/S molar ratio of 2.0.

over the entire cross-section of the upper furnace where the temperature is in the range 1,400 to 2,400°F and the residence time for the reactions is 1–2 seconds. The sorbents decompose and become porous solids with high surface area. At temperatures higher than ≈2,300°F, dead-burning or sintering is experienced.

When limestone is used as the sorbent, it is rapidly calcined to quicklime when it enters the furnace

$$CaCO_3(s) + heat \rightarrow CaO(s) + CO_2(g) \tag{4.57}$$

Sulfur dioxide diffuses to the particle surface and heterogeneously reacts with the CaO to form calcium sulfate

$$CaO(s) + SO_2(g) + \tfrac{1}{2}O_2(g) \rightarrow CaSO_4(s) \tag{4.58}$$

Sulfur trioxide, although present at a significantly lower concentration than SO_2, is also captured using calcium-based sorbents

$$CaO(s) + SO_3(g) \rightarrow CaSO_4(s) \tag{4.59}$$

Approximately 15 to 40% SO_2 removal can be achieved using a Ca/S in the flue gas molar ratio of 2.0. The optimum temperature for injecting limestone is ≈1,900 to 2,100°F.

The calcium sulfate that is formed travels through the rest of the boiler flue gas system and is ultimately collected in the existing particulate control device with the fly ash and unreacted sorbent. Some concerns exist regarding increased tube deposits as a result of injecting solids into the boiler and the extent of calcium deposition is influenced by overall ash chemistry, ash loading, and boiler system design.

The following overall reactions occur when using hydrated lime as the sorbent:

$$Ca(OH)_2(s) + heat \rightarrow CaO(s) + H_2O(v) \tag{4.60}$$

$$CaO(s) + SO_2(g) + \frac{1}{2}O_2(g) \rightarrow CaSO_4(s) \tag{4.58}$$

$$CaO(s) + SO_3(g) \rightarrow CaSO_4(s) \tag{4.59}$$

Approximately 50 to 80% SO_2 removal can be achieved using hydrated lime at a Ca/S molar ratio of 2.0. The hydrate is injected at very nearly the same temperature window as limestone and the optimum range is 2,100 to 2,300°F.

The FSI process can be applied to boilers burning low- to high-sulfur coals. The factors that affect the efficiency of the FSI system are flue gas humidification (to condition the flue gas to counter degredation that may occur in ESP performance from the addition of significant quantities of fine, high resistivity sorbent particles), type of sorbent, efficiency of ESP, and temperature and location of the sorbent injection. The process is better suited for large furnaces with lower heat release rates [20]. Systems that use hydrated calcium salts sometimes have problems with scaling; however, this can be prevented by keeping the approach to adiabatic saturation temperature above a minimum threshold.

There are several advantages of the FSI system [20]. One advantage is simplicity of the process. The dry reagent is injected directly into the flow path of the flue gas in the furnace and a separate absorption vessel is not required. The injection of lime in a dry form allows for a less complex reagent handling system. This in turn lowers operating labor and maintenance costs, and eliminates the problems of plugging, scaling, and corrosion found in slurry handling. There are lower power requirements since less equipment is needed. Steam is not required for reheat whereas most LSFO systems require some form of reheat to prevent corrosion of downstream equipment. The sludge dewatering system is eliminated since the FSI process produces a dry solid, which can be removed by conventional fly ash removal systems.

The FSI process has a few disadvantages when compared to the LSFO process [20]. One major disadvantage is that the process only removes up to 40 and 80% SO_2 when using limestone and hydrated lime at Ca/S molar ratios of 2.0, respectively, whereas the LSFO process can remove greater than 90% SO_2 using 1.05 to 1.1 moles CaO/mole SO_2 removed. This is further illustrated in Figure 4.7, which shows calcium utilization (defined as the percent SO_2 removed divided by the Ca/S ratio) of hydrated lime and limestone at various injection temperatures (modified from [23]). Hence, more sorbent is needed in the FSI process and lime, which works better than limestone, is more expensive than limestone. There is a potential for solids deposition and boiler convective pass fouling, which occurs during the humidification step through the impact of solid droplets onto surfaces. Also, there is a potential for corrosion at the point of humidification, the ESP, the downstream ductwork, and the stack. The corrosion at the point of humidification is caused by operating below the acid dewpoint, whereas downstream corrosion is caused by the

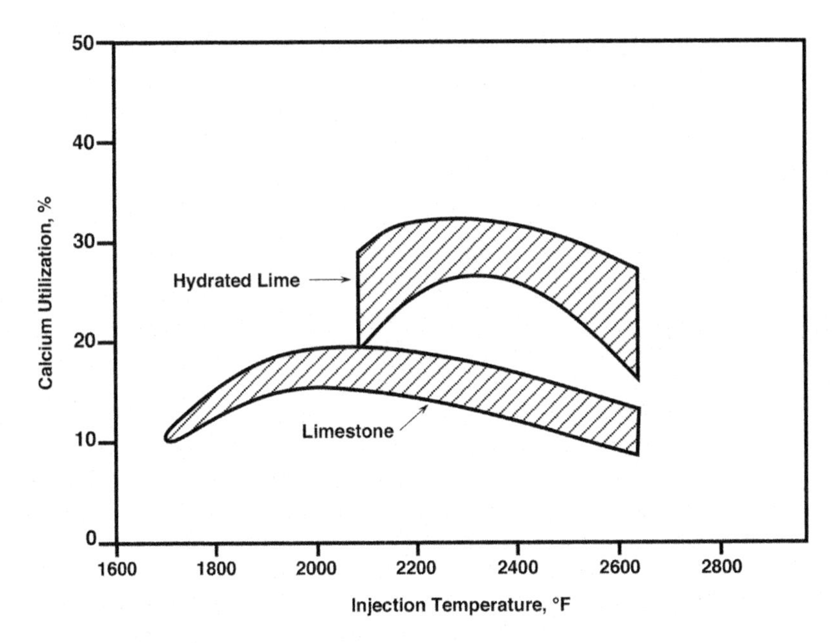

Figure 4.7 Calcium utilization as a function of sorbent injection temperature for furnace sorbent injection.

humidified gas temperature being close to the water saturation temperature. Plugging can also occur, thereby affecting system pressures. The efficiency of an ESP can be reduced by increased particulate loading and changes in the ash resistivity. This can, in turn, lead to the installation of additional particulate collection devices. Sintering of the sorbent is a concern if it is injected at too high a temperature (e.g. >2,300°F for hydrated lime). Multiple injection ports in the furnace wall may be needed to ensure proper mixing and follow boiler load swings and hence shifting temperature zones. Hydration of the free lime in the product may be required. Lime is very reactive when exposed to water and can pose a safety hazard for disposal areas.

4.4.4.2.2 Economizer injection

In an economizer injection process (shown in Figure 4.5b), hydrated lime is injected into the flue gas stream near the economizer inlet where the temperature is between 950–1,050°F. This process is not commercially used at this time but was extensively studied because it was found that the reaction rate and extent of sulfur capture (see Figure 4.6) are comparable to FSI. However, the economizer temperatures are too low for dehydration of the hydrated lime (only about 10% of the hydrated lime forms quicklime) and the hydrate reacts directly with the SO_2 to form calcium sulfite

$$Ca(OH)_2(s) + SO_2(g) \rightarrow CaSO_3(s) + H_2O(v) \tag{4.60}$$

This process is best suited for older units in need of a retrofit process and can be used for low- to high-sulfur coals. The advantages and disadvantages of this system are similar to the FSI process (but will not be discussed in detail since this process is not presently being used in the power industry) with the notable exception that there is no reactive CaO contained in the waste.

4.4.4.2.3 Duct sorbent injection – duct spray drying

Spray dry scrubbers are the second most widely used method to control SO_2 emissions in utility coal-fired power plants. Lime is usually the sorbent used in this technology but sodium carbonate is also used, specifically in the western United States. Spray dryer FGD systems have been installed on over 12,000 MW of total FGD capacity as well as numerous industrial boilers.

The first commercial dry scrubbing system installed on a coal-fired boiler in the United States was in mid-1981 at the Coyote station (jointly owned by Montana-Dakota Utilities, Northern Municipal Power Agency, Northwestern Public Service Company, and Ottertail Power Company) near Beulah, North Dakota. The 425 MW unit burns lignite from a mine-mouth plant and initially used soda ash (Na_2CO_3) as the sulfur removal reagent. The spray dryer was modified about ten years later and the unit currently uses lime as the reagent. The second dry scrubbing system that was installed on a coal-fired utility boiler was on two 440 MW units, that became operational in 1982 and 1983 at the Basin Electric Power Cooperative's Antelope Valley station, also located near Beulah, North Dakota. These units fire mine-mouth lignite and use a slaked lime slurry to remove SO_2 in the spray dryer.

A slaked lime slurry is sprayed directly into the ductwork to remove SO_2 (see Figure 4.5c). The reaction products and fly ash are captured downstream in the particulate removal device. A portion of these solids is recycled and reinjected with the fresh sorbent. Dry spray drying (DSD) is a relatively simple retrofit process capable of 50% SO_2 removal at a Ca/S ratio of 1.5. The concept is the same as conventional spray drying except that the existing ductwork provides the residence time for drying instead of a reaction vessel. The main difference is that the residence time in the duct is much shorter, i.e. 1–2 seconds, compared to 10–12 seconds in a spray drying vessel.

The slaked lime is produced by hydrating raw lime to form calcium hydroxide. This slaked lime is atomized and absorbs the SO_2 in the flue gas. The SO_2 reacts with the slurry droplets as they dry to form equimolar amounts of calcium sulfite and calcium sulfate. The water in the lime slurry improves SO_2 absorption by humidifying the gas. The reaction products, unreacted sorbent, and fly ash are collected in the particulate control device located downstream. Some of the unreacted sorbent may react with a portion of the CO_2 in the flue gas to form calcium carbonate. Also, a little more SO_2 removal is achieved in the particulate control device. The reactions occurring in the process are:

$$CaO(s) + H_2O(g) \rightarrow Ca(OH)_2(s) + heat \tag{4.61}$$

$$Ca(OH)_2(s) + SO_2(g) \rightarrow CaSO_3 \cdot \tfrac{1}{2}H_2O(s) + \tfrac{1}{2}H_2O(v) \tag{4.55}$$

$$Ca(OH)_2(s) + SO_2(g) + \tfrac{1}{2}O_2(g) + H_2O(v) \rightarrow CaSO_4 \cdot 2H_2O(s) \quad (4.62)$$

$$Ca(OH)_2(s) + CO_2(g) \rightarrow CaCO_3(s) + H_2O(v) \quad (4.63)$$

There are two different methods for atomizing the slurry. One method is the use of rotary atomizers, with the ductwork providing the short gas residence time of 1–2 seconds. When using this atomizer, the ductwork must be sufficiently long to allow for drying of the slurry droplets. There must also be no obstructions in the duct.

The second method for atomizing the slurry is the use of dual-fluid atomizers, where compressed air and water are used to atomize the slurry. This is called the Confined Zone Dispersion (CZD) process. The dual-fluid atomizer has been shown to be more controllable due to the adjustable water flow rate. They are also relatively inexpensive and have a long and reliable operating life with little maintenance. The spray is confined in the duct, which allows better mixing with the flue gas rather than impinging on the walls.

There are several advantages of the DSD process compared to wet processes. The DSD process is less complex since the reagent is injected directly into the flow path of the flue gas and a separate absorption vessel is not needed. Less equipment is needed and thereby power requirements are lower. The waste from this process does not contain reactive lime like the FSI process and therefore does not require special handling.

Some of the problems encountered by the DSD system are also common to other dry processes. A main disadvantage of the system includes limited SO_2 removal efficiency (i.e. $\approx 50\%$) and low calcium utilization compared to wet processes. Quicklime is more expensive than limestone. If an ESP is used, there is the potential for reduced efficiency due to changes in fly ash resistivity and the increased dust loading in the flue gas. Additional collection devices may be needed as well as humidification to improve ESP collection efficiency. There must be sufficient length, i.e. residence time, of the ductwork to ensure complete droplet vaporization prior to the particulate collection device. This is necessary for good sulfur capture and to avoid plugging and deposition, which in turn results in an increased pressure drop that the induced draft fans must overcome.

4.4.4.2.4 Duct sorbent injection – dry sorbent injection

Dry sorbent injection (DSI), also referred to as in-duct dry injection, is illustrated in Figure 4.5d. Hydrated lime is the sorbent typically used in this process, especially for power generation facilities. DSI has also been identified as a process to comply with the Boiler MACT regulations while continuing to use coal under normal technological risk. However, sodium-based sorbents have been tested extensively including full-scale utility demonstrations and are used in industrial systems, such as municipal and medical waste incinerators for acid gas control.

When using hydrated lime in this process, it is injected either upstream or downstream of a flue gas humidification zone. In this zone, the flue gas is humidified to within 20°F of the adiabiatic saturation temperature by injecting water into the duct downstream of the air preheater [20]. The SO_2 in the flue gas reacts with the calcium hydroxide to form calcium sulfate and calcium sulfite:

$$Ca(OH)_2(s) + SO_2(g) + \tfrac{1}{2}O_2(g) + H_2O(v) \rightarrow CaSO_4 \cdot 2H_2O(s) \tag{4.62}$$

$$Ca(OH)_2(s) + SO_2(g) \rightarrow CaSO_3 \cdot \tfrac{1}{2}H_2O(s) + \tfrac{1}{2}H_2O(v) \tag{4.55}$$

The water droplets are vaporized before they strike the surface of the wall or enter the particulate control device. The unused sorbent, along with the products and fly ash, are all collected in the particulate control device. About half of the collected material is shipped to a landfill while the other half is recycled for injection with the fresh sorbent into the ducts [20].

The DSI system offers many of the same advantages and disadvantages that other dry systems offer [20]. The process is less complex (i.e. no slurry recycle and handling, no dewatering system, fewer pumps, and no reactor vessel) than a wet system, specifically LSFO. The humidification water and hydrated lime are injected directly into the existing flue gas path. No separate SO_2 absorption vessel is necessary. The handling of the reagent is simpler than in wet systems. The costs for DSI systems are less than in wet systems since there is less equipment to install and, since there is less equipment, operating and maintenance costs are reduced. The waste product is free of reactive lime so that no special handling is required.

Some of the problems encountered by the DSI system and its disadvantages, as compared to the LSFO system are common to other dry processes. Sulfur dioxide removal efficiencies are lower (as is calcium utilization) than wet systems and range from 30 to 70% for a Ca/S ratio of 2.0. Quicklime is more expensive than limestone. When an ESP is used for particulate control, there is the potential for reduced efficiency due to increased fly ash resisitivity and dust loading in the flue gas. Additional collection devices may be required. A sufficient length of ductwork is necessary to ensure a residence time of 1–2 seconds in a straight, unrestricted path. Plugging of the duct can occur if the residence time is insufficient for droplet vaporization leading to increased system pressure drop.

In the dry sodium desulfurization process, a variety of sodium-containing crystalline compounds may be injected directly into the flue gas. The main compounds of interest include [24]:

- Sodium carbonate (Na_2CO_3), a refined product of $\approx$98% purity;
- Sodium bicarbonate ($NaHCO_3$), a refined product of $\approx$98% purity;
- Nacholite ($NaHCO_3$), a natural material of $\approx$76% purity containing high levels of insolubles;
- Sodium sesquicarbonate ($NaHCO_3 \cdot Na_2CO_3 \cdot 2H_2O$), a refined product of $\approx$98% purity; and
- Trona ($NaHCO_3 \cdot Na_2CO_3 \cdot 2H_2O$), a natural material of $\approx$88% purity containing high levels of insolubles.

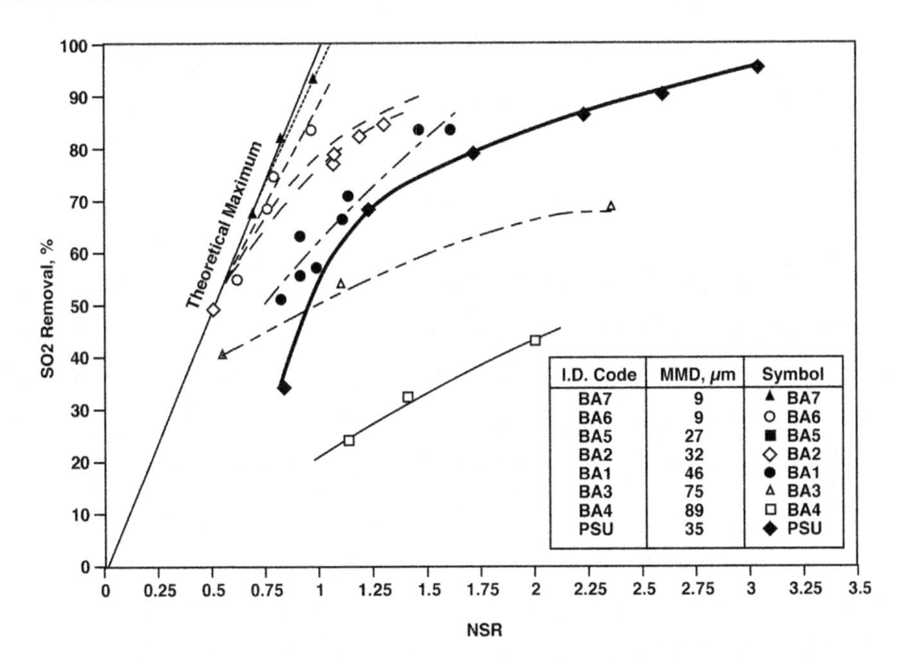

Figure 4.8 Comparison of SO_2 removal as a function of NSR and particle size when using sodium bicarbonate. (Tests identified by the BA code were performed in a coal-fired pilot-scale facility while the tests coded PSU were performed in a coal-fired industrial boiler system. Both facilities were equipped with fabric filter baghouses. The Penn State tests were performed with flue gas temperature of $\approx$380°F while the pilot-scale tests were performed at $\approx$300°F. All tests were performed with sorbent from the same vendor.)

Sodium bicarbonate and sodium sesquicarbonate have been the most extensively tested in pilot, demonstration, and full-scale utility applications due to proven success and commercial availability. In addition, sodium bicarbonate is extensively used in industrial applications for acid gas control.

Of the compounds listed above, sodium bicarbonate has demonstrated the best sulfur capture in coal-fired boiler applications. This is illustrated in Figure 4.8 where SO_2 removal as a function of normalized stoichiometric ratio (NSR) for sodium bicarbonate injection into a coal-fired pilot-scale test facility and industrial boiler equipped with fabric filter baghouses is shown. Note that the NSR represents the molar ratio between the injected sodium compound and the initial SO_2 concentration in the flue gas considering that it takes two moles of sodium to react with only one mole of SO_2 [24,25].

The flue gas stream must be above 240°F for rapid decomposition of the sodium bicarbonate when it is injected or little SO_2 capture will occur. While SO_2 will react directly with the sodium bicarbonate, in the presence of nitric oxide (NO) this reaction is inhibited and does not result in significant sulfur capture. Therefore, for acceptable SO_2 capture to progress rapidly and attain acceptable levels of utilization, the bicarbonate component must begin to decompose. As flue gas temperatures increase into the optimum range of 240 to 320°F, the carbonate is decomposed and the subsequent sulfation reaction occurs [24]. When the bicarbonate component decomposes, carbon dioxide and water vapor are evolved from the particle interior creating a network of void spaces. Sulfur dioxide and NO can diffuse to the fresh sorbent surfaces where the heterogeneous reactions to capture SO_2 (and to a lesser extent NO) take place. The decomposition and sulfation reactions are:

$$2NaHCO_3(s) + heat \rightarrow Na_2CO_3(s) + H_2O(v) + CO_2(g) \tag{4.64}$$

$$SO_2(g) + Na_2CO_3(s) + \tfrac{1}{2}O_2(g) \rightarrow Na_2SO_4(s) + CO_2(g) \tag{4.65}$$

Lower NO_x emissions also result from injection of dry sodium compounds [24]. The mechanism is not well understood but reductions up to 30% have been demonstrated, a function of SO_2 concentration and NSR ratio. There is a side effect of this reduction though. Nitric oxide is oxidized to NO_2 (a reddish-brown gas) and not all of the NO_2 is reacted with the sorbent. As the NO_2 concentration increases in the stack, an undesirable coloration in the plume can be created.

4.4.4.2.5 Hybrid systems

Hybrid sorbent injection processes are typically a combination of FSI and DSI systems with the goal of achieving greater SO_2 removal and sorbent utilization [8]. Various types of configurations have been tested including injecting secondary sorbents, such as sodium compounds, into the ductwork or humidifying the flue gas in a specially designed vessel. Humidification reactivates the unreacted CaO and can increase the SO_2 removal efficiency. Advantages of hybrid processes include high SO_2 removal, low capital and operating costs, less space necessary, thereby lending to easy retrofit, easy operation and maintenance, and no waste water treatment [8].

In some hybrid systems, a new baghouse is installed downstream of an existing particulate removal device (generally an ESP). The existing ESP continues to remove the ash, which can be either sold or disposed. Sulfur dioxide removal is accomplished in a manner similar to in-duct injection, with the sorbent injection upstream of the new baghouse [23].

The potential advantages of this system include the potential for toxic substances control since a baghouse is the last control device (this is further discussed in Chapter 6, "Mercury Emissions Reduction"), easier waste disposal, the potential for sorbent regeneration, separate ash and product streams, and more efficient recycle without ash present [23]. The major issue is the high capital cost of adding a baghouse although the concept of adding one with a high air-to-cloth ratio (3–5 acfm/ft^2) can minimize this cost.

4.4.4.3 Circulating fluidized-bed scrubbers

Circulating fluidized-bed (CFB) scrubbers are the least used commercial option. CFB scrubbers include dry and semi-dry systems [8]. CFB scrubbers are in operation on facilities ranging from 10 to 420 MW in size [11]. Commercial application of the dry CFB is more widespread of the two and this process can achieve SO_2 removal efficiencies of 90–98% at a Ca/S molar ratio of 1.2–1.5. In this system, hydrated lime is injected directly into a CFB reactor along with water to obtain operation close to the adiabatic saturation temperature. The fluidized bed recirculates the material within the reactor to achieve a high retention time. Fresh reactive surface area is exposed by the continuous abrasion of the sorbent particles in the reactor. The flue gas takes $\approx$5 seconds to pass through a deep fluidized bed. The main features of this process include [8]: simple, efficient, and reliable operation with proven high availability; no scaling, plugging, or corrosion problems allowing mild steel construction, which does not require a lining; no moving parts or slurry nozzles; all solids that require handling are dry; water injection is independent of reagent feed; good turndown capability to below 30% maximum continuous rating with a clean flue gas recirculation loop and to around 50% without flue gas recirculation; moderate space requirements; and greater flexibility to handle varying SO_2 and SO_3 concentrations.

The CFB scrubber uses hydrated lime rather than the less expensive limestone commonly used in wet FGD technology processes. Additionally, due to a higher particulate matter concentration downstream of the scrubber, improvements to the particulate removal device, specifically an ESP, may be needed to meet the required particulate emission levels. CFB scrubbers consume 0.3 to 2% of the electric capacity of the plant, largely from the booster induced-draft fans required because of the pressure drop caused by the reactor [11].

4.5 Economics of flue gas desulfurization

The capital, operating, and maintenance costs of an FGD plant are determined by many factors including the technology chosen, the plant size, the SO_2 removal requirements, the fuel used, the costs of the reagent, and waste disposal [9]. For most of the FGD processes commercially available, the overall total costs have been reduced due to significant improvements in process designs because of operating experience accumulated and advances in the technology over the years.

The capital costs of an FGD system depend on many factors including [26]:

- market conditions;
- geographical location;
- preparatory site work required;
- volume of flue gas to be scrubbed;
- concentration of SO_2 in the flue gas;
- extent of SO_2 removal required;
- quality of the products produced;
- process and waste water treatment;
- the need for flue gas reheat;
- the degree of reliability and redundancy required; and
- life of the system.

Operating costs are divided into variable and fixed costs [26]. Variable costs include the costs of the sorbents/reagents, costs associated with disposal or utilization of the byproducts, and steam, power, and water costs. Fixed costs include costs of operating labor, maintenance, and administration. The operating and maintenance costs of an FGD system can be significant. According to Soud [8], operating and maintenance costs for the various subsystems in a pulverized coal-fired power generating facility with state-of-the-art environmental protection are: 78% for the boiler/turbine/generator, 10% for the FGD system, 6% for a selective catalytic reduction system, 4% for wastewater treatment, and 2% for an ESP.

The features and various costs of several FGD processes are compared in Table 4.5, which includes comparisons of some wet and dry FGD systems [9]. In the following sections, the costs associated with different FGD systems, including those in Table 4.5, are further discussed. It must be noted, however, that it is difficult to compare costs (whether it is capital or operating) between systems, since costs have many site-specific factors, are dependent upon the age of the system, are influenced by economies of scale, and are higher for retrofit applications than new installations, and vary between different sources compiling or reporting the data. Therefore, there is a large range in the cost data that is summarized in this section. Also, most of the costs reported are for utility-scale applications. Costs for industrial-sized systems are discussed separately at the end of the section.

The capital costs of wet FGD systems have exhibited a decreasing trend in the United States for most of the last $\approx$40 years. As reported by EIA, the prototype FGD systems of the 1970s cost $400/kW and experienced many problems [27]. With standardization, better chemistry, and improved materials, the cost of FGD systems in the 1980s dropped to $275/kW. Additional developments, such as reduced redundancy, fewer modules, increased competition from foreign vendors, and use of well-engineered packages, dropped capital costs to about $100/kW by 2000. However, recent EIA reports show average installed capital costs increasing over the last several years from $135/kW in 2007 to $250/kW in 2012 [28].

It can be seen from Table 4.5 that the seawater FGD process has lower capital costs compared to most other FGD processes shown. In addition, its operating costs are lower than the other processes listed in Table 4.5.

The ammonia scrubbing process requires the highest capital investment compared to the other FGD processes in Table 4.5. In addition, the ammonia reagent cost is high. However, the income from the sale of the high-value byproduct fertilizer can help offset the high capital and reagent costs. When considering the average 20-year cost of emission control and fuel, the ammonia system is comparable to the limestone gypsum system [9].

The spray dry process generally has lower capital cost but higher and more expensive sorbent use, which is typically lime, than wet processes. This process is

Table 4.5 **Comparison of selected FGD processes and their Costs**

	Ammonia scrubbing	**Seawater scrubbing**	**Sodium scrubbing**	**Limestone gypsum**	**Dry FGD**
Features	High-value byproduct Economics better at high sulfur levels Low operating cost	Low capital cost Simple operation	Low capital cost Simple operation	High efficiency spray zone Low cost reagent Byproduct flexibility	Low capital cost Dry byproduct Small footprint No liquid waste
Reagent	Ammonia	Seawater	Caustic, soda ash	Limestone	Lime
Byproduct	Fertilizer	Treated seawater	Sodium sulfate	Marketable gypsum or landfill	Landfill
SO_2 inlet concentration	High	Low/medium	High	High	Low/medium
SO_2 removal, %	>98	>95	>98	>98	90–95
Capital Cost, $/acfm	35–60	15–25	10–20	25–45	15–25
Reagent Cost, $/acfm	80–105	0	100–130	15–25	60–75
Byproduct Cost, $/ton SO_2 removed	150–250 (sale)	0	N/A	12–20 (disposal) 15 (sale)	12–20

used mostly for small to medium plants firing low- to medium-sulfur coals, and is preferable for retrofits [26].

The sorbent injection process, with a moderate SO_2 removal efficiency, has a relatively low capital cost.

The circulating fluidized-bed processes have a relatively low capital cost, similar to that for the spray dry process. The process also has a low to moderate fixed operating cost but it its variable operating cost is relatively high [26,29].

Regenerative processes generally have a high capital cost and power consumption [26]. The net variable operating cost is moderate because the processes produce saleable byproducts. However, the fixed operating cost is substantially higher than other processes.

Sargent & Lundy performed a study in which they investigated the capital and operating costs for a lime spray dryer FGD (LSD), a circulating dry scrubber (CDS, or fluidized-bed FGD), and a wet limestone FGD with forced oxidation to produce gypsum (LSFO) [29]. They determined costs for both a 400 and 500 MW installation firing three different coals with varying sulfur concentrations – low-sulfur Appalachian coal (2.00 lb/MM Btu), medium-sulfur Appalachian coal (3.00 lb/MM Btu), and a low-sulfur Powder River Basin Coal (1.44 lb/MM Btu). The results of the study are summarized in Table 4.6. These data illustrate the cost dependency on unit size and coal sulfur content as well as new versus retrofit applications.

Sargent & Lundy performed a study for the U.S. EPA in which they determined scrubber costs for LSFO and LSD scrubber types for representative sizes and heat rates [30]. They determined capital costs and variable and fixed operating and maintenance (O&M) costs for power plant capacities ranging from 50 to 1,000 MW and for heat rates varying from 9,000 to 11,000 Btu/kWh. These results are shown in Table 4.7.

Andover Technologies performed a study for the LADCO region states (i.e. upper Midwest) in which they examined the costs of controlling SO_2 emissions from industrial, commercial, and institutional (ICI) boilers [31]. They found several factors that impacted costs. Larger boilers tend to benefit from economics of scale in terms of capital costs. Lower initial SO_2 levels result in higher costs represented in terms of \$/ton of SO_2 removed. Also, some technologies have higher capital cost but may have additional benefits. Limestone wet FGD can result in a marketable byproduct. Dry FGD offers high particulate matter capture. Sodium wet scrubbing is a low-cost approach for many industrial boilers because of its low capital cost, providing that water treatment is available on site and that the sodium reagent remains reasonable in price. Table 4.8 summarizes the costs, \$/ton of SO_2 removed, for control of SO_2 emissions using wet and dry FGD systems and various boiler types and sizes [31].

Table 4.6 **Capital costs and total operating and maintenance costs for 400 and 500 MW new and retrofit power plants as a function of coal sulfur content**

Low sulfur Powder River Basin coal						Low sulfur Appalachian coal						Medium sulfur Appalachian coal					
1.44 lb/MM Btu						2.0 lb/MM Btu						3.0 lb/MM Btu					
LSD[a]		CDS		LSFO		LSD		CDS		LSFO		LSD		CDS		LSFO	
Cap[b]	O&M[c]	Cap.	O&M	Cap.	O&M	Cap.	O&M	Cap.	O&M	Cap.	O&M	Cap.	O&M	Cap.	O&M	Cap.	O&M
400 MW New unit																	
59.1	7.2	60.9	7.2	79.9	6.7	57.7	8.8	59.6	9.1	80.7	7.0	59.5	12.2	61.9	12.6	83.2	7.9
500 MW New unit																	
74.4	8.8	76.4	8.9	91.0	7.8	72.7	10.8	74.5	11.2	91.8	8.2	74.8	15.1	77.2	15.5	94.5	9.2
400 MW Retrofit unit																	
69.7	–[d]	71.4	–	94.5	–	68.1	–	69.8	–	95.7	–	70.0	–	72.4	–	98.6	–
500 MW Retrofit unit																	
88.4	–	90.2	–	107.2	–	86.4	–	88.0	–	108.5	–	88.6	–	91.0	–	111.8	–

[a]LSD = limestone spray dryer; CDS = circulating dry scrubber or circulating fluid bed scrubber; LSFO = limestone forced oxidation.
[b]Capital costs in $million.
[c]Total operating and maintenance costs ($million/year).
[d]Not determined.

Table 4.7 Illustrative scrubber costs (2011$) for representative sizes and heat rates

Scrubber type	Heat rate (Btu/kWh)	Var. O&M (mills/kWh)	Capacity (MW)											
			50		100		300		500		700		1000	
			Capital cost ($/kW)	Fixed O&M ($/kW-yr)	Capital cost ($/kW)	Fixed O&M ($/kW-yr)	Capital cost ($/kW)	Fixed O&M ($/kW-yr)	Capital cost ($/kW)	Fixed O&M ($/kW-yr)	Capital cost ($/kW)	Fixed O&M ($/kW-yr)	Capital cost ($/kW)	Fixed O&M ($/kW-yr)
LSFO	9,000	2.03	819	23.7	819	23.7	600	11.2	519	8.3	471	7.7	426	6.4
	10,000	2.26	860	24.2	860	24.2	629	11.5	544	8.6	495	8.0	447	6.6
	11,000	2.49	899	24.6	899	24.6	658	11.8	569	8.9	517	8.2	467	6.8
LSD	9,000	2.51	854	29.1	701	17.3	513	8.6	444	6.5	422	5.7	422	5.3
	10,000	2.79	894	29.6	734	17.7	538	8.6	564	6.8	442	5.9	442	5.5
	11,000	3.07	933	30.0	766	18.0	561	9.1	485	7.0	461	6.1	461	5.7

Table 4.8 Costs of SO_2 control for wet and dry FGD systems on ICI boilers

Boiler type	FGD technology	Percent reduction	\$/ton SO_2 removed	Year \$
Coal-100 MM Btu/h	Lime Dry FGD	85	1,600–7,700	2008
Coal-250 MM Btu/h			1,500–4,000	
Coal-750 MM Btu/h			1,400–2,400	
Coal-100 MM Btu/h	Lime Wet FGD	85	1,650–7,500	2008
Coal-250 MM Btu/h			1,400–3,800	
Coal-750 MM Btu/h			1,300–2,200	
Coal	Spray Dryer	90	1,700–3,600	2005
Coal		95	1,600–3,400	
Oil		90	1,900–5,200	
Oil		95	1,800–4,900	
Coal	Limestone Wet FGD	90	2,100–3,800	2005
Coal		99	1,900–3,400	
Oil		90	2,200–5,200	
Oil		99	2,000–4,700	
Coal-250 MM Btu/h	Limestone Wet FGD	95	4,400	2006
Coal-100 MM Btu/h			9,500	
Oil-250 MM Btu/h			5,700	
Oil-100 MM Btu/h			12,500	
Coal-250 MM Btu/h	Lime Dry FGD	90	3,700	2006
Coal-100 MM Btu/h			7,900	
No.6-250 MM Btu/h			4,700	
No.6-100 MM Btu/h			10,400	
210 MM Btu/h, 5.75 lb SO_2/ MM Btu	Lime Dry FGD	90	1,000	2009
210 MM Btu/h, 2.59 lb SO_2/ MM Btu			2,000	
210 MM Btu/h, 1.15 lb SO_2/ MM Btu			4,200	
210 MM Btu/h, 0.5 lb SO_2/ MM Btu			9,500	

(*Continued*)

Table 4.8 (Continued)

Boiler type	FGD technology	Percent reduction	\$/ton SO_2 removed	Year \$
420 MM Btu/h, 5.75 lb SO_2/ MM Btu	Lime Dry FGD	90	700	2009
420 MM Btu/h, 2.59 lb SO_2/ MM Btu			1,400	
420 MM Btu/h, 1.15 lb SO_2/ MM Btu			2,900	
420 MM Btu/h, 0.5 lb SO_2/ MM Btu			6,600	
630 MM Btu/h, 5.75 lb SO_2/ MM Btu	Lime Dry FGD	90	600	2009
630 MM Btu/h, 2.59 lb SO_2/ MM Btu			1,200	
630 MM Btu/h, 1.15 lb SO_2/ MM Btu			2,400	
630 MM Btu/h, 0.5 lb SO_2/ MM Btu			5,400	
210 MM Btu/h, 5.75 lb SO_2/ MM Btu	Sodium Wet FGD	90	800	2009
210 MM Btu/h, 2.59 lb SO_2/ MM Btu			1,100	
210 MM Btu/h, 1.15 lb SO_2/ MM Btu			1,800	
210 MM Btu/h, 0.5 lb SO_2/ MM Btu			3,800	
420 MM Btu/h, 5.75 lb SO_2/ MM Btu	Sodium Wet FGD	90	600	2009
420 MM Btu/h, 2.59 lb SO_2/ MM Btu			900	
420 MM Btu/h, 1.15 lb SO_2/ MM Btu			1,400	
420 MM Btu/h, 0.5 lb SO_2/ MM Btu			2,900	
630 MM Btu/h, 5.75 lb SO_2/ MM Btu	Sodium Wet FGD	90	600	2009
630 MM Btu/h, 2.59 lb SO_2/ MM Btu			800	
630 MM Btu/h, 1.15 lb SO_2/ MM Btu			1,300	
630 MM Btu/h, 0.5 lb SO_2/ MM Btu			2,600	

References

[1] Davis WT, editor. Air pollution engineering manual. 2nd ed. New York: John Wiley & Sons, Inc; 2000.

[2] Wark K, Warner CF, Davis WT. Air pollution: its origin and control. 3rd ed. Menlo Park, California: Addison Wesley Longman, Inc; 1998.

[3] EPA (United States Environmental Protection Agency). Latest findings on national air quality: 1997 status and trends. Washington, DC: Office of Air Quality Planning and Standards; 1998.

[4] EPA (United States Environmental Protection Agency). National Emissions Inventory (NEI), 1970–2013 Average annual emissions, all criteria pollutants, www.epa.gov/ttnchie1/trends/; February 27, 2014.

[5] EIA (United States Energy Information Administration). Annual energy review 2011. U.S. Department of Energy; 2012.

[6] EPA (United States Environmental Protection Agency). Clean Air Interstate Rule, Acid Rain Program, and Former NO_x Budget Trading Program 2012 Progress Report, 2013.

[7] EPA (United States Environmental Protection Agency). Acid rain and related programs: 2008 emission, compliance, and market analysis. Washington, DC: Office of Air Quality Planning and Standards, U.S. Government Printing Office; 2008.

[8] Soud HN. Developments in FGD. London: IEA Coal Research; 2010.

[9] Zhu Q. Non-calcium Desulphurization Technologies. London: IEA Clean Coal Centre; 2010.

[10] Moretti AL, and Jones CS. Advanced emissions control techniques for coal-fired power plants, Technical Paper BR-1886, Babcock & Wilcox Power Generation Group, October 2012.

[11] Carpenter AM. Advances in multi-pollutant control. London: IEA Coal Research; 2013.

[12] Macedonia J. Summary of EPA Power Sector Regulation: Air Toxics Standards, Bipartisan Policy Center, January 24, 2012.

[13] Elliot TC, editor. Standard handbook of powerplant engineering. New York: McGraw-Hill Publishing Company; 1989.

[14] EIA (United States Energy Information Agency). U.S. coal reserves: 1997 Update. (U. S. Department of Energy, Office of Coal, Nuclear, Electric and Alternate Fuels, U.S. Government Printing Office, Washington, D.C., February 1999).

[15] Nowling U. Utility options for leveraging natural gas. Power, pp., 48–54, October 23.

[16] Age C. Upcoming, recent coal-fired power unit retirements, pp. 18–19, April 2014. Available from: http://www.coalage.com/departments/market-watch/3570-upcoming-recent-coal-fired-power-unit-retirements.html#.VPo6j-FEIk9

[17] Harrison CD. Fuel options to mitigate emissions reduction costs. Proceedings of the 28th International Technical Conference on Coal Utilization & Fuel Systems, Coal & Slurry Technology Association, Washington, D.C., 2003.

[18] Stultz SC, Kitto JB, editors. Steam: its generation and use. 40th ed. The Babcock and Wilcox Company; 1992.

[19] Kitto JB, Stultz SC, editors. Steam: its generation and use. 41st ed. The Babcock and Wilcox Company; 2005.

[20] Radcliffe PT. Economic evaluation of flue gas desulfurization systems. Palo Alto, California: Electric Power Research Institute; 1991.

[21] Srivastava RK. Controlling SO_2 emissions: a review of technologies. U.S. Environmental Protection Agency, November 2000.
[22] Srivastava RK, Singer C, and Jozewicz W. SO_2 Scrubbing technologies: a review. Proceedings of the AWMA 2000 Annual Conference and Exhibition, 2000.
[23] Rhudy R, McElroy M, and Offen G. Status of calcium-based dry sorbent injection SO_2 control. Proceedings of the Tenth Symposium on Flue Gas Desulfurization, November 17–21, 1986, pp.9–69 to 9–84.
[24] Bland VV, and Martin CE. Full-scale demonstration of additives for NO_2 reduction with dry sodium desulfurization. Electric Power Research Institute, EPRI GS-6852, June 1990.
[25] Miller BG, Boehman AL, Hatcher P, Knicker H, Krishnan A, McConnie J, et al. The development of coal-based technologies for department of defense facilities phase II final report. Prepared for the U.S. Department of Energy Federal Energy Technology Center, Pittsburgh, Pennsylvania, July 31, 2000, DE-FC22-92PC92162, 784 pages.
[26] Wu Z. Air pollution control costs for coal-fired power stations. London: IEA Coal Research; 2001.
[27] Smith DJ. Cost of SO_2 scrubbers down to $100/kW. Power Engineering, September 2001, pp. 63–68.
[28] EIA (United States Energy Information Administration). Electric Power Annual 2012. U.S. Department of Energy; 2013.
[29] Sargent & Lundy. Flue gas desulfurization technology evaluation, Dry Lime vs. Wet Limestone FGD, March 2007.
[30] EPA (United States Environmental Protection Agency). EPA's power sector modeling platform v.5.13, emission control technologies. www.epa.gov/airmarkets/progsregs/epa-ipm; November 2013.
[31] Staudt JE. Candidate SO_2 control measures for industrial sources in the LADCO region, Lake Michigan Air Directors Consortium, January 24, 2012. Available from: http://dnr.wi.gov/topic/AirQuality/documents/LADCO_SO2_Final.pdf

Nitrogen oxides formation and control

5

5.1 Introduction

There are seven oxides of nitrogen that are present in ambient air [1]. These include nitric oxide (NO), nitrogen dioxide (NO_2), nitrous oxide (N_2O), NO_3, N_2O_3, N_2O_4, and N_2O_5. Nitric oxide and nitrogen dioxide are collectively referred to as NO_x due to their interconvertibility in photochemical smog reactions. The term NO_y is often used to represent the sum of the reactive oxides of nitrogen and all other compounds that are atmospheric products of NO_x. NO_y includes compounds such as nitric acid (HNO_3), nitrous acid (HNO_2), nitrate radical (NO_3), dinitrogen pentoxide (N_2O_5), and peroxyacetyl nitrate (PAN). It excludes N_2O and ammonia (NH_3) because they are not normally the products of NO_x reactions [1].

Nitrogen oxide emissions from coal combustion are produced from three sources – thermal NO_x, fuel NO_x, and prompt NO_x. Nitrogen oxides are primarily produced as a result of the fixation of atmospheric nitrogen at high temperatures (thermal NO_x) and the oxidation of coal nitrogen compounds (fuel NO_x). Prompt NO_x is formed when hydrocarbon radical fragments in the flame zone react with nitrogen to form nitrogen atoms, which then form NO. The majority of the oxide species produced is NO with NO_2 accounting for less than 5% of the total, which is usually referred to as NO_x [2].

Both NO_x and NO_y (i.e. HNO_3) have been shown to accelerate damage to materials in the ambient air. NO_x affects dyes and fabrics resulting in fading, discoloration of archival and artistic materials and textile fibers, and lost of textile fabric strength [2]. NO_2 absorbs visible light and at a concentration of 0.25 ppmv will cause appreciable reduction in visibility. NO_2 affects vegetation as studies have shown suppressed growth of pinto beans and tomatoes and reduced yields of oranges. In combination with unburned hydrocarbons (i.e. volatile organic compounds, VOCs, which are emitted primarily from motor vehicles but also from chemical plants, refineries, factories, consumer and commercial products, and other industrial sources), nitrogen oxides react in the presence of sunlight to form photochemical smog.

Nitrogen oxides also contribute to the formation of acid rain. NO and NO_2 in the ambient air can react with moisture to form NO_3^- and H^+ in the aqueous phase (i.e. nitric acid), which can cause considerable corrosion of metal surfaces. The kinetics of nitric acid formation are not as well understood as those for the formation of sulfuric acid discussed in Chapter 4. Nitrogen oxides contribute to changes in the composition and competition of some species of vegetation in wetland and terrestrial systems, acidification of freshwater bodies, eutrophication (i.e. explosive algae

Fossil Fuel Emissions Control Technologies. DOI: http://dx.doi.org/10.1016/B978-0-12-801566-7.00005-1

growth leading to depletion of oxygen in the water) of estuarine and coastal waters, and increase in levels of toxins harmful to fish and other aquatic life [3].

Nitrogen dioxide acts as an acute irritant and in equal concentrations is more injurious than NO. However, at concentrations found in the atmosphere, NO_2 is only potentially irritating and potentially related to chronic obstructive pulmonary disease [1]. EPRI has shown from their ARIES study that the nitrate components of coal combustion were not the cause for adverse health effects [4]. The United States EPA reports that short-term exposures (e.g. less than 3 hours) to current NO_2 concentrations may lead to changes in airway responsiveness and lung function in individuals with preexisting respiratory illnesses and increases in respiratory illnesses in children from 5 to 12 years of age [3]. EPA also reports that long-term exposures to NO_2 may lead to increased susceptibility to respiratory infection and may cause alterations in the lung. Atmospheric transformation of NO_x can lead to the formation of ozone and nitrogen-bearing particles (most notably in some western United States urban areas), which are associated with adverse health effects [3].

Nitrogen oxide emissions totaled about 13 million short tons in 2013 (see Table 2.20), with stationary fuel combustion accounting for approximately 29% of the total anthropogenic emissions (i.e. 3.7 million short tons). Of the 3.7 million short tons, electric generating utilities are the largest source of NO_x emissions compared to industry with electric utilities accounting for 50% or 1.8 million short tons of all NO_x emissions. Nitrogen oxide emissions have, however, decreased approximately 72% for the period 1990 to 2013 [5,6]. Reductions in NO_x emissions and concentrations since 1990 are primarily due to controls implemented under EPA's Acid Rain and CAIR NO_x programs. In 2012, 797 coal-fired units with controls (combustion, postcombustion, or other) generated all but 1% of annual coal-fired generation and 637 combined cycle units (firing gas or oil) with controls generated all but 1% of annual combined cycle generation [6].

NO_x formation mechanisms are reviewed in this section, followed by technologies used to control NO_x emissions. Similar to the discussion on SO_2 control technologies, NO_x control technologies will focus on commercially available, commercially used systems.

5.2 NO_x formation mechanisms

The majority of nitrogen oxides emitted from stationary combustion systems are in the form of nitric oxide (NO), with only a small fraction as nitrogen dioxide (NO_2) and nitrous oxide (N_2O). NO originates from the fuel-bound nitrogen and nitrogen in the air used in the combustion process and is produced through three mechanisms: thermal NO, prompt NO, and fuel NO. Fuel-bound nitrogen accounts for 50 to 95% of the total NO generated, which is a function of fuel type, while thermal and prompt NO accounts for the balance with prompt NO being no more than 5% of the total NO [7]. The reason for the fuel NO dominance is because the N-H and N-C bonds, common in fuel-bound nitrogen, are weaker than the triple bond in molecular nitrogen, which must be dissociated to produce thermal NO.

The major source of NO_x emissions from coal and oil is the conversion of the organically-bound nitrogen in the fuel [8]. Fuel NO_x contributes approximately 50% of the total uncontrolled emissions when firing residual fuel oil and 75 to 95% when firing coal [8,9]. Residual fuel oil typically contains approximately a few tenths of a percent (by weight) nitrogen, while coal can contain up to a couple of percent of nitrogen [10]. Although it is a major factor in NO_x emissions, only 20 to 30% of the fuel-bound nitrogen is converted to NO. Natural gas and other high-quality fuels such as propane and distillate fuel oils are generally low or devoid of fuel-bound nitrogen so there is no to a minimal amount of fuel NO_x generated. Thermal NO_x is the primary source of NO_x formation from natural gas and distillate oils.

Coal combustion generally produces the highest NO_x emissions. Oil combustion generates less NO_x emissions while gas firing produces even less. When firing fuel oils, a reduction in fuel nitrogen content results in reduced NO_x emissions. However, a similar correlation is not observed with coal as other factors in coal chemistry (e.g. volatile species, oxygen, and moisture content) strongly influence the formation of NO_x [8].

The factors that influence NO_x emissions in pulverized coal-fired boilers can be generally categorized as boiler design, boiler operation, and coal properties [7]. However, NO_x formation is complex and many parameters influence its production [11,12]. Boiler design factors include boiler type, capacity, burner type, number and capacity of the burners, burner zone heat release, residence times, and presence of overfire air ports. Similarly, boiler operation factors include load, mills in operation, excess air level (i.e. stoichiometry), burner tilt, and burner operation. Coal properties that influence NO_x production include volatiles release and nitrogen partitioning, ratio of combustibles-to-volatile matter, heating value, rank, and nitrogen content.

5.2.1 Thermal NO

Thermal NO formation involves the high temperature (>2,370°F) reaction of oxygen and nitrogen from the combustion air and the production of thermal NO is a function of the combustion temperature and fuel-to-air ratio and increases exponentially with temperature above 2,650°F [7,13]. Thermal NO can be predicted by the following equation [13]:

$$[NO] = K_1 e^{-K_2/T}[N_2][O_2]^{1/2}t \tag{5.1}$$

where T is temperature, t is time, K_1 and K_2 are constants, and $[N_2]$ and $[O_2]$ are concentrations in moles [7]. Accordingly, thermal NO can be decreased by reducing the time, temperature, and concentration of N_2 and O_2. The principal reaction governing the formation of NO is the reaction of oxygen atoms formed from the dissociation of O_2 with nitrogen. These reactions, referred to as the Zeldovich mechanism, are:

$$N_2 + O^{\bullet} \leftrightarrow NO + N^{\bullet} \tag{5.2}$$

$$N^{\bullet} + O_2 \leftrightarrow NO + O^{\bullet} \tag{5.3}$$

These reactions are sensitive to temperature, local stoichiometry, and residence time. High temperature is required for the dissociation of oxygen and to overcome the high activation energy for breaking the triple bond of the nitrogen molecule. These reactions dominate in fuel-lean high-temperature conditions. Under fuel-rich conditions, there are increased hydroxyl and hydrogen radical concentrations, which initiate the oxidation of the nitrogen radicals and at least one additional step should be included in this mechanism:

$$N^{\bullet} + OH^{\bullet} \leftrightarrow NO + H^{\bullet} \tag{5.4}$$

Reactions (5.2)–(5.4) are usually referred to as the extended Zeldovich mechanism. In addition, the following reactions can occur in fuel-rich conditions [7]:

$$H^{\bullet} + N_2 \leftrightarrow N_2H \tag{5.5}$$

$$N_2H + O^{\bullet} \leftrightarrow NO + NH^{\bullet} \tag{5.6}$$

Thermal NO is of greater significance in the post-flame region than within the flame. Consequently, there are several technologies that have been developed for reducing thermal NO by lowering the peak temperature in the flame, minimizing the residence time in the region of the highest temperature, and controlling the excess air levels.

5.2.2 Prompt NO

Prompt NO is the fixation of atmospheric (molecular) nitrogen by hydrocarbon fragments in the reducing atmosphere in the flame front [7]. Prompt NO is most significant in fuel-rich flames where the concentration of radicals such as O and OH can exceed equilibrium values, thereby enhancing the rate of NO formation. Prompt NO occurs by the collision of hydrocarbons with molecular nitrogen in the fuel-rich flames to form HCN (hydrogen cyanide) and N. The HCN is then converted to NO by a series of reactions between H, O, OH, NH, and N. The amount of prompt NO generated is proportional to the concentration of N_2 and the number of carbon atoms present in the gas phase but the total amount produced is low in comparison to the total thermal and fuel NO in coal combustion. The proposed mechanism is [12]:

$$CH^{\bullet} + N_2 \leftrightarrow HCN + N^{\bullet} \tag{5.7}$$

$$HCN + O^{\bullet} \leftrightarrow NH^{\bullet} + CO \tag{5.8}$$

$$NH^{\bullet} + O^{\bullet} \leftrightarrow NO^{\bullet} + H^{\bullet} \tag{5.9}$$

The main reaction product of hydrocarbon radicals with N_2 is HCN with the amount of NO formed governed by the reactions of the nitrogen atoms with available radical species. In fuel-rich environments, therefore, the formation of N_2 is favored due to the reduced concentrations of hydroxyl and oxygen radical concentrations [7].

5.2.3 Fuel NO

Nitrogen in the coal, which typically ranges from 0.5 to 2.0 weight percent, occurs mainly as organically-bound heteroatoms in aromatic rings or clusters [7]. Pyrrolic (5-membered ring) nitrogen is the most abundant form and contributes 50 to 60% of the total nitrogen. Pyridinic (6-membered ring) nitrogen comprises about 20 to 40% of the total nitrogen. The remaining 0 to 20% nitrogen is thought to be amine or quaternary nitrogen forms.

Coal nitrogen is first released during volatilization in the coal flame as an element in aromatic compounds referred to as tar. The tar undergoes pyrolysis to convert most of the nitrogen to HCN as well as some NH_3 and NH. Some nitrogen is expelled from the char as HCN and occasionally NH_3; however, this is at a much slower rate than evolution from the volatiles. The partitioning of nitrogen between volatiles and char is important in NO_x formation.

NO formation proceeds along two paths [7]. The nitrogen from the char reacts with oxygen to form NO. The NH_3 and NH released from the volatile matter, and to a lesser extent the coal, reacts with oxygen atoms, forming NO. HCN is converted to NO via a pathway of hydrogen abstraction to form ammonia species and subsequently NO. Volatile nitrogen species can also be converted to nitrogen atoms through a series of fuel-rich pyrolysis reactions. Also, reactions between NO and volatile nitrogen species and carbon particles can result in the formation of nitrogen molecules:

$$C + NO \leftrightarrow \tfrac{1}{2}N_2 + CO \tag{5.10}$$

$$CH^{\bullet} + NO \leftrightarrow HCN + O^{\bullet} \tag{5.11}$$

In a fuel-rich environment, the main product of the reaction of NO with hydrocarbon radicals is HCN, which is then converted to N_2 in an oxygen-deficient environment. This is the basis for reburning, discussed later in this chapter, where a secondary hydrocarbon fuel is injected into combustion products containing NO. Unlike thermal NO, the production of fuel NO is relatively insensitive to temperature over the range found in pulverized coal flames and more sensitive to the air-to-fuel ration [13,14].

5.2.4 Nitrogen dioxide (NO_2) and Nitrous oxide (N_2O)

Small amounts of NO_2 and N_2O are formed during coal combustion but they comprise less than 5% of the total NO_x production. The oxygen levels are too low and the residence times are too short in high-temperature coal flames for much of the

NO to be oxidized to NO_2. Nitrous oxide, however, can be formed in the early part of fuel-lean flames by gas phase reaction by the reactions [7]:

$$O^{\bullet} + N_2 \leftrightarrow N_2O \tag{5.12}$$

$$NH^{\bullet} + NO \leftrightarrow N_2O + H^{\bullet} \tag{5.13}$$

$$NCO^{\bullet} + NO \leftrightarrow N_2O + CO \tag{5.14}$$

5.3 NO_x control technologies

Unlike sulfur and mineral matter constituents in fuels, nitrogen species contained in the fuel cannot be easily removed or reduced. Therefore the most common precombustion option for reducing NO_x levels is to switch to a fuel with lower nitrogen content. Typically this means switching from coal to fuel oil or natural gas or from fuel oil to natural gas.

Technologies for control of NO_x emissions from fossil fuel-fired stationary heat and power plants can be divided into two groups: 1) combustion modifications where the NO_x production is reduced during the combustion process; and 2) post-combustion, i.e. flue gas, treatment, which removes the NO_x from flue gas following its formation. Sometimes the practice of injecting reducing agents to reduce NO_x to molecular nitrogen (N_2) is classified separately; however, in this chapter it is included as a flue gas treatment. Specifics of the various control technologies are discussed later in the chapter.

Table 5.1 lists various NO_x control technologies with a summary of their attributes [15]. This summary essentially covers technologies applicable for all fossil fuels. The abatement or emission control principles for these various control methods include reducing peak flame temperatures, reducing the residence time at peak flame temperatures, chemically reducing NO_x, oxidizing NO_x with subsequent absorption, removing nitrogen, using a sorbent, or a combination of these methods.

Table 5.2 lists the common NO_x control options for coal-fired boilers [modified from 16]. The coal-fired boiler types that the control options are typically applied to are listed in Table 5.2 as are their potential NO_x reduction levels.

In boilers fired on crude oil or residual oil, the control of fuel NO_x is important because fuel NO_x typically accounts for 60 to 80% of the total NO_x formed [16]. Fuel nitrogen conversion to NO_x is highly dependent upon the fuel-to-air ratio in the combustion zone and, in contract to thermal NO_x formation, is relatively insensitive to small changes in combustion zone temperature. The most common combustion modification technique is to suppress combustion air levels below the theoretical amount required for complete combustion. The lack of combustion creates reducing conditions, which, given sufficient time at high temperatures, cause volatile fuel nitrogen to convert to N_2 rather than to NO. Combustion controls are

Table 5.1 NO_x control technologies

Technique	Description	Advantages	Disadvantages	Impacts	Applicability
Less Excess Air (LEA)	Reduces oxygen availability	Easy modification	Low NO_x reduction	High CO Flame length Flame stability	All fuels
Off Stoichiometric a. Burners Out of Service (BOOS) b. Overfire Air (OFA)	Staged combustion	Low cost No capital cost for BOOS	a. Higher air flow for CO reduction b. High capital cost	Flame length Fan capacity Header pressure	All fuels Multiple burners required for BOOS
Low NO_x Burner	Internal staged combustion	Low operating cost Compatible with FGR	Moderately high capital cost	Flame length Fan capacity Turndown capability	All fuels
Flue Gas Recirculation (FGR)	<30% flue gas recirculated with air, decreasing temperature	High NO_x reduction potential for low nitrogen fuels	Moderately high capital and operating costs Affects heat transfer and system pressures	Fan capacity Furnace pressure Burner pressure drop Turndown stability	All fuels
Water/Steam Injection	Reduces flame temperature	Moderate capital cost NO_x reduction similar to FGR	Efficiency penalty Fan power higher	Flame stability Efficiency penalty	All fuels
Reduced Air Preheat	Air not preheated, reduces flame temperature	High NO_x reduction potential	Significant efficiency loss (1% / 40°F)	Fan capacity Efficiency penalty	All fuels

(*Continued*)

Table 5.1 (Continued)

Technique	Description	Advantages	Disadvantages	Impacts	Applicability
Selective Catalytic Reduction (SCR)	Catalyst located in air flow, promotes reaction between ammonia and NO_x	High NO_x removal	Very high capital cost High operating cost Catalyst siting Increased pressure drop Possible water wash required	Space requirements Ammonia slip Hazardous materials Disposal	All fuels
Selective Non-Catalytic Reduction (SNCR) a. Urea b. Ammonia	Inject reagent to react with NO_x	a. Low capital cost Moderate NO_x removal Nontoxic chemical b. Low operating cost Moderate NO_x removal	a. Temperature dependent NO_x reduction less at lower loads b. Moderately high capital cost Ammonia storage, handling, injection system	a. Furnace geometry Temperature profile b. Furnace geometry Temperature profile	All fuels
Fuel Reburning	Inject fuel to react with NO_x	Moderate cost Moderate NO_x removal	Extends residence time	Furnace temperature profile	All fuels (pulverized solid)
Combustion Optimization	Change efficiency of primary combustion	Minimal cost	Extends residence time	Furnace temperature profile	All fuels
Inject Oxidant	Chemical oxidant injected into flow	Moderate cost	Nitric acid removal	Add-on	All fuels

Oxygen instead of Air	Uses oxygen as oxidizer	Moderate to high cost Intense combustion Eliminate thermal NO_x	Eliminate prompt NO_x Furnace alteration	Equipment to handle oxygen	All fuels
Ultra-Low Nitrogen Fuel	Uses low-nitrogen fuel	Eliminates fuel NO_x No capital cost	Possible rise in operating cost	Minimal change	All ultra-low nitrogen fuels
Sorbent Injection a. Combustion b. Duct to baghouse c. Duct to ESP	Uses a chemical to absorb NO_x, or an adsorber to capture it, or to reduce it	Can control other pollutants as well as NO_x Moderate operating cost	Cost of sorbent	Add-on Space for the sorbent storage and handling	All fuels
Air Staging	Admit air in separated stages	Reduce peak combustion temperature	Extend combustion to a longer residence time at lower temperature	Add ducts and dampers to control air Furnace modification	All fuels
Fuel Staging	Admits fuel in separated stages	Reduce peak combustion temperature	Extend combustion to a longer residence time at lower temperature	Adds fuel injectors to other locations Furnace modification	All fuels

Table 5.2 NO_x control options for coal-fired boilers

Control technique[a]	Description of technology	Applicable boiler types	NO_x reduction potential (%)	Comments
Combustion modifications				
Load reduction	Reduction of coal and air	Stokers	Minimal	Applicable to stokers that can reduce load without increasing excess air; may cause reduction in boiler efficiency; NO_x reduction varies with percent load reduction.
Operational modifications (BOOS, LEA, BF, or combination)	Rearrangement of air or fuel in the main combustion zone	Pulverized coal boilers (some designs); stokers (LEA only)	10–20	Must have sufficient operational flexibility to achieve NO_x reduction potential without sacrificing boiler performance.
Overfire air	Injection of air above main combustion zone	Pulverized coal boilers and stokers	20–30	Must have sufficient furnace height above top row of burners in order to retrofit this technology to existing boilers.
Low NO_x Burners	New burner designs controlling air-fuel mixing	Pulverized coal boilers	35–55	Available in new boiler designs and can be retrofit in existing boilers.
LNB with OFA	Combination of new burner designs and injection of air above main combustion zone	Pulverized coal boilers	30–70	Available in new boiler designs and can be retrofit in existing boilers with sufficient furnace height above top row of burners.
Reburn	Injection of reburn fuel and completion air above main combustion zone	Pulverized coal boilers, cyclone furnaces	25–70	Reburn fuel can be natural gas, fuel oil, or pulverized coal; must have sufficient furnace height to retrofit this technology to existing boilers.

Post-Combustion Modifications				
SNCR	Injection of NH_3 or urea in the convective pass	Pulverized coal boilers, cyclone furnaces, stokers, and fluidized-bed boilers	20–60	Applicable to new boilers or as retrofit technology; must have sufficient residence time at correct temperature; possible load restrictions on boiler; possible air preheater fouling
SCR	Injection of NH_3 in combination with catalyst material	Pulverized coal boilers, cyclone furnaces	75–85	Applicable to new boilers or as a retrofit technology provided there is sufficient space; hot-side SCR best on low-sulfur fuel and low fly ash applications; cold-side SCR can be used on high-sulfur/high-ash applications if equipped with an upstream FGD system
LNB with SNCR	Combination of new burner designs and injection of NH_3 or urea	Pulverized coal boilers	50–80	Same as LNB and SNCR alone
LNB with OFA and SCR	Combination of new burner design, injection of air above combustion zone, and injection of NH_3 or urea	Pulverized coal boilers	85–95	Same as LNB, OFA, and SCR alone

[a]BOOS = burners out of service; LEA = lean excess air; BF = biased burner firing; OFA = overfire air; LNB = low NO_x burner.

the most widely used method of controlling NO_x emissions in all types of boilers and include low excess air (LEA), burners out of service (BOOS), biased-burner firing (BF), flue gas recirculation (FGR), overfire air (OFA), and low-NO_x burners (LNB) [16]. Postcombustion controls are also used including selective catalytic reduction (SCR) and selective noncatalytic reduction (SNCR), which can be used separately or in combination.

In natural gas-fired boilers, the two most prevalent control techniques used to reduce NO_x emissions are flue gas recirculation and low-NO_x burners [16]. When low-NO_x burners and FGR are used in combination, NO_x emissions can be reduced by 60 to 90%. Other combustion control techniques used to reduce NO_x emissions included staged combustion (i.e. BOOS and OFA) and gas reburning. Postcombustion technologies include SCR and SNCR.

In stationary gas turbines, there are three generic types of emission controls for NO_x control – wet controls using steam or water injection to reduce combustion temperatures, dry controls using advanced combustor design to suppress NO_x formation and/or promote CO burnout, and postcombustion catalytic control to selectively reduce NO_x and/or oxidize CO emissions from the turbine [16]. NO_x emissions can be reduced by 60% or higher using water or steam injection. This is accompanied by an efficiency penalty of typically 2 to 3% but an increase in power output of typically 5 to 6%.

Controls for reducing NO_x emissions have been developed for firetube and watertube boilers firing liquefied petroleum gas (LPG), i.e. propane or butane [16]. These systems use a combination of low-NO_x burners and FGR. Some systems use water or steam injection into the flame zone as a trimming technique for NO_x emissions.

The remainder of the chapter will discuss the primary NO_x reduction techniques in some detail. The discussions will focus on coal combustion since NO_x emissions are the highest from coal firing and require the most control of any fossil-fuel feedstock.

Reducing combustion temperature is accomplished by operating at nonstoichiometric conditions to dilute the available heat with an excess of fuel, air, flue gas, or steam [15]. The combustion temperature is reduced by: using fuel-rich mixtures to limit the availability of oxygen; using fuel-lean mixtures to dilute energy input; injecting cooled oxygen-depleted flue gas into the combustion air to dilute energy; injecting cooled flue gas with the fuel; or injecting water or steam.

Reducing residence time at high combustion temperature is accomplished by restricting the flame to a short region to limit the nitrogen from becoming ionized. Fuel, steam, more combustion air, or recirculated flue gas is then injected immediately after this region.

Chemically reducing NO_x removes oxygen from the nitrogen oxides. This is accomplished by reducing the valence level of nitrogen to zero after the valence has become higher.

Oxidizing NO_x intentionally raises the valence of the nitrogen ion to allow water to absorb to it. This is accomplished by using a catalyst, injecting hydrogen peroxide, creating ozone within the air flow, or injecting ozone into the air flow.

Removing nitrogen from combustion is accomplished by removing nitrogen as a reactant either by using low nitrogen content fuels or using oxygen instead of air. The ability to vary coal nitrogen contents, however, is limited.

Treatment of flue gas by injection sorbents such as ammonia, limestone, aluminum oxide, or carbon can remove NO_x and other pollutants. This type of treatment has been applied in the combustion chamber, flue gas, and particulate control device.

Many of these methods can be combined to achieve a lower NO_x concentration than can be achieved alone by any one method. In some cases, technologies that are used to control other pollutants, such as SO_2, can also reduce NO_x.

5.3.1 Combustion modifications

Primary NO_x control technologies involve modifying the combustion process. Several technologies have been developed and applied commercially and include:

- low-NO_x burners;
- furnace air staging;
- flue gas recirculation;
- fuel staging (i.e. reburn); and
- process optimization.

Options to control NO_x during combustion and their effects are different for new and existing boilers. For new boilers, combustion modifications are easily made during construction whereas for existing boilers, viable alternatives are more limited. Modifications can be complicated and unforeseen problems may arise. When combustion modifications are made, it is important to avoid adverse impacts on boiler operation and also the formation of other pollutants such as N_2O or CO. Issues pertaining to low-NO_x operation include:

- Safe operation (e.g. stable ignition over the desired load range);
- Reliable operation to prevent corrosion, erosion, deposition, and uniform heating of the tubes;
- Complete combustion to limit formation of other pollutants such as CO, polyorganic matter, or N_2O;
- Minimal adverse impact on the flue gas cleaning equipment; and
- Low maintenance costs.

Combustion modification technologies redistribute the fuel and air to slow mixing, reduce the availability of oxygen in the critical NO_x formation zones, and decrease the amount of fuel burned at peak flame temperatures. In addition, reburning chemically destroys the NO_x formed by hydrocarbon radicals during the combustion process. The commercially applied technologies are discussed in detail in the following sections. One technology listed in Table 5.1, low excess air (LEA), is the simplest of the combustion control strategies but it is not discussed in detail as it has limited success in coal-fired applications (i.e. 1 to 15%). In this technique, excess air levels are reduced until there are adverse impacts on

CO formation and flame length and stability. Similarly, a technique termed burners out of service (BOOS) has limited success with coal and is not discussed in detail. In this technique, the fuel flow to the selected burner is stopped but airflow is maintained to create staged combustion in the furnace. The remaining burners operate fuel-rich, which limits oxygen availability, lowering peak flame temperatures, and reducing NO_x formation. The unreacted products combine with the air from the burners out of service to complete burnout before exiting the furnace.

5.3.1.1 Low-NO_x burners

Prior to concern with NO_x emissions in the early 1970s, coal burners were designed to provide highly turbulent mixing and combustion at peak flame temperature to ensure high combustion efficiency, a condition that is ideal for NO_x formation [10]. In 1971, industry began developing low-NO_x burners for coal-fired boilers with the promulgation of New Source Performance Standards. By the mid-1970s, low-NO_x burners were being demonstrated and commercial operation started in the late 1970s [17]. They have undergone considerable improvements in design spurred by the 1990 Clean Air Act Amendments Title IV, Phase II Acid Rain regulations and Title I Ozone regulations [18,19]. The technology is well proven for NO_x control in both wall- and tangentially fired boilers and is commercially available, with a significant number of them installed worldwide.

Low-NO_x burners work under the principle of staging the combustion air within the burner to reduce NO_x formation. Rapid devolatilization of the coal particles occurs near the burner in a fuel-rich, oxygen-starved environment to produce NO. NO_x formation is suppressed because oxygen molecules are not available to react with the nitrogen released from the coal and present in the air, and the flame temperature is reduced. Hydrocarbon radicals that are generated under the substoichiometric conditions then reduce the NO that is formed to N_2. The air required to complete the burnout of the coal is added after the primary combustion zone where the temperature is sufficiently low so that additional NO_x formation is minimized.

Larger and more branched flames are produced by staging the air [7]. This flame structure limits coal and air mixing during the initial devolatilization stage while maximizing the release of volatiles from the coal. The more volatile nitrogen that is released with the volatiles and the longer the residence time in the fuel-rich zone, the lower the amount of fuel NO that is produced. An oxygen rich layer is produced around the flame that aids in carbon burnout.

An example of this concept is shown in Figure 5.1, which is a schematic of a low-NO_x burner (i.e. Ahlstom Power's Radially Stratified Fuel Core burner) illustrating a typical flow field emanating from it [20]. A photograph of the burner, which is a 20 million Btu/h prototype used for developmental work at Penn State prior to its commercialization and worldwide deployment, is shown in Figure 5.2 depicting the various dampers and air scoops for channeling and controlling the quantity and degree of swirl of the various air streams [21].

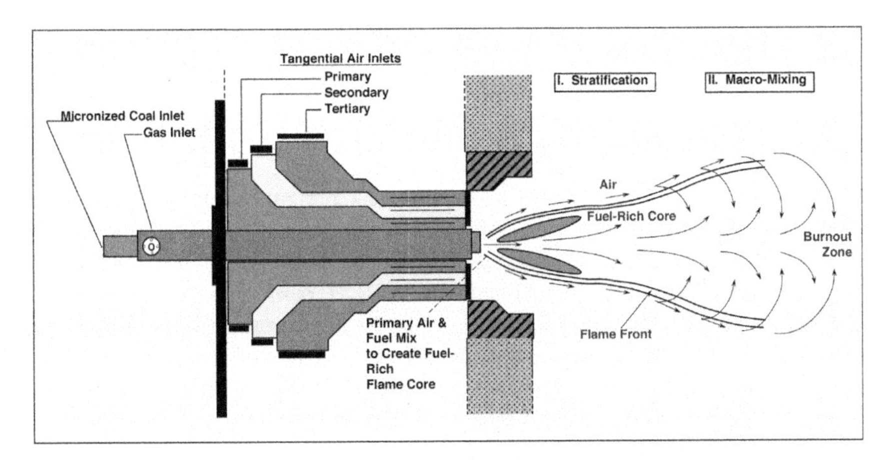

Figure 5.1 Schematic diagram of Alstom's RSFC burner depicting flow fields.

Figure 5.2 Photograph of the RSFC burner showing internal components.

Low-NO_x burners are designed to accomplish the following [22]:

- Maximize the rate of volatiles evolution and total volatile yield from the fuel with the fuel nitrogen evolving in the reducing part of the flame;
- Provide an oxygen-deficient zone where the fuel nitrogen is evolved to minimize its conversion to NO_x but ensuring there is sufficient oxygen to maintain a stable flame;

- Optimize the residence time and temperature in the reducing zone to minimize the fuel nitrogen conversion to NO_x;
- Maximize the char residence time under fuel rich conditions to reduce the potential for NO_x formation from the nitrogen remaining in the char after devolatilization; and
- Add sufficient air to complete combustion.

All low-NO_x burners employ the air-staging principle but the designs vary widely between manufacturers. All of the major boiler manufacturers have one or more versions of low-NO_x burners employed in boilers throughout the world. Mitchell [7] reported that there were over 370 units worldwide fitted with low-NO_x burners with a total generating capacity of more than 125 GW prior to 1998. By 2008, the number of generating units that installed low-NO_x burners in the U.S. alone increased significantly as DOE reported that low-NO_x burners were found on more than 75% of United States coal-fired power capacity [23], where 1,470 coal-fired steam-electric generators with nameplate capacity of 336 GW were producing more than 1,994 billion kWh of electricity [24]. Low-NO_x burners are commonplace now on not only utility boilers but industrial and commercial boilers as well.

Low-NO_x burners, based on air-staging alone, are capable of achieving 30 to 60% NO_x reduction. In addition, they should perform in such ways that [22]:

- the overall combustion efficiency is not significantly reduced;
- flame stability and turndown limits are not impaired;
- the flame has an oxidizing envelope to minimize the potential for high temperature corrosion at the furnace walls;
- flame length is compatible with furnace dimensions; and
- the performance should be acceptable for a wide range of coals.

The major concern with low-NO_x burners is the potential for reducing combustion efficiency and thereby increasing the unburned carbon level in the fly ash. An increase in the unburned carbon level will lower the fly ash resistivity, which can reduce the efficiency of an ESP. In addition, it may also affect the sale of the ash. Some operating parameters that can be adjusted to mitigate the impact of the unburned carbon are [22]:

- fire coal with high reactivity and high volatile matter content;
- reduce the size of the coal particles;
- balance coal distribution to the burners; and
- use advanced combustion control systems.

5.3.1.2 Furnace air staging

A technique to stage combustion is done by installing secondary and even tertiary overfire air (OFA) ports above the main combustion zone. This is a well-proven, commercially available technology for NO_x reduction at coal-fired power plants and is applicable to both wall- and tangentially-fired boilers [22].

When OFA is employed, 70 to 90% of the combustion air is supplied to the burners with the coal (i.e. primary air) with the balance introduced to the furnace above the burners (i.e. overfire air). The primary air and coal produce a relatively low-temperature, oxygen-deficient, fuel-rich environment near the burner, which reduces the formation of fuel-NO_x. The overfire air is injected above the primary combustion zone producing a relatively low-temperature secondary combustion zone that limits the formation of thermal NO_x.

Overfire air in combustion with low-NO_x burners can reduce NO_x emissions by 30 to 70%. Advanced OFA systems such as separated overfire air (SOFA), where the overfire air is introduced some distance above the burners, and close-coupled overfire air (CCOFA), where the overfire air nozzles are immediately above the burners, can achieve higher NO_x reduction efficiency [7]. Mitchell [7] reports that furnace air staging is used in approximately 300 pulverized coal-fired units with a total generating capacity of over 100 GW. There are a number of advanced overfire air systems commercially available with designs varying between suppliers [22].

Furnace air staging can increase unburned carbon levels in the ash by 35 to 50% with the degree of increase being dependent on the reactivity of the coal used [22]. In addition, operational problems can be experienced including waterwall corrosion, changes in slagging and fouling patterns, and a loss in steam temperature.

5.3.1.3 Flue gas recirculation

Flue gas recirculation (FGR) involves recirculating part of the flue gas back into the furnace or the burners to modify conditions in the combustion zone by lowering the peak flame temperature and reducing the oxygen concentration, thereby reducing thermal NO_x formation. FGR has been used commercially for many years at coal-fired units. However, unlike gas- and oil-fired boilers, which can achieve high NO_x reduction, coal-fired boilers typically realize less than 20% NO_x reduction due to a relatively low contribution of thermal NO_x to total NO_x.

In conventional FGR applications, 20 to 30% of the flue gas is extracted from the boiler outlet duct upstream of the air heater (at $\approx$570 to 750°F) and is mixed with the combustion air. This process reduces thermal NO_x formation without any significant effect on fuel NO_x.

A major consideration of FGR is the impact on boiler thermal performance [22]. The reduced flame temperature lowers heat transfer, potentially limiting the maximum heating capacity of the unit, which results in a reduction in steam-generating capacity.

5.3.1.4 Fuel staging (reburn)

Reburn is a comparatively new technology, which combines the principles of air and fuel staging. In this technology, a reburn fuel, which can be coal, oil, gas, orimulsion, biomass, coal-water mixtures, etc., is used as a reducing agent to convert NO_x to N_2. The process does not require modifications to the existing main combustion system and can be used on wall-, tangential-, and cyclone-fired boilers.

Reburn is a combustion hardware modification in which the NO_x produced in the main combustion zone is reduced downstream in a second combustion zone (i.e. the reburn zone). This, in turn, is followed by a zone where overfire air is introduced to complete burnout. This is illustrated in Figure 5.3 [25].

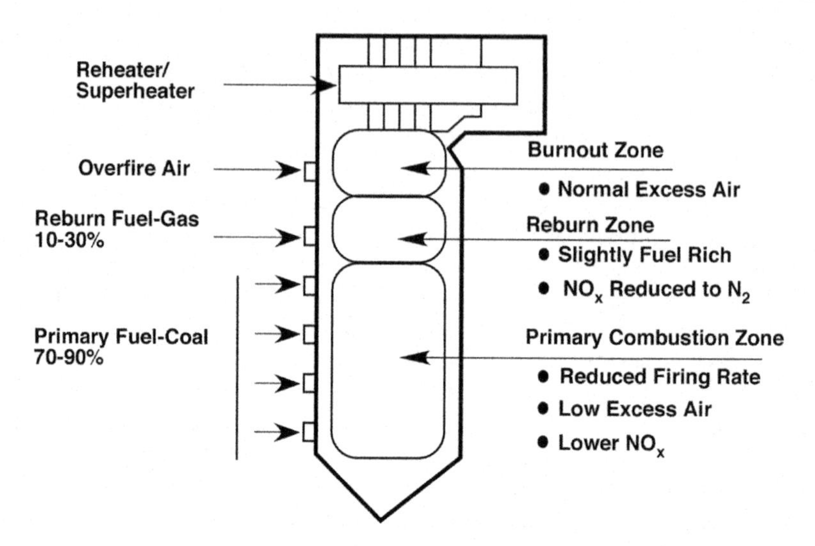

Figure 5.3 Schematic of the reburn process.

In the primary combustion zone, the burners are operated at a reduced firing rate with low excess air (stoichiometry of 0.9 to 1.1) to produce lower fuel and thermal NO_x levels. The reburn fuel, which can be 10 to 30% of the total fuel input, on a heat input basis, is injected above the main combustion zone to create a fuel-rich zone (stoichiometry of 0.85–0.95) [25]. In this zone, most of the NO_x reduction occurs with hydrocarbon radicals formed in the reburn zone reacting with the NO_x forming N_2 and water vapor. The temperature in this zone must be greater than 1,800°F. The remaining combustion air is injected above the reburn zone to produce a fuel-lean burnout zone.

While reburn technology is considered relatively new and numerous pilot-scale tests and full-scale demonstrations have been conducted, the concept was proposed in the late 1960s [26]. The concept was based on the principle of Myerson, et al. [27] that CH fragments can react with NO.

The major chemical reactions for the reburn process are [25]:

$$\text{Hydrocarbon fuel} \xrightarrow[\text{Heat}O_2\text{deficiency}]{} {}^{\bullet}CH_2 \tag{5.15}$$

where hydrocarbon radicals are produced due to the pyrolysis of the fuel in the oxygen-deficient, high-temperature reburn zone. The hydrocarbon radicals then mix with the combustion gases from the primary combustion zone:

$$ {}^{\bullet}CH_3 + NO \rightarrow HCN + H_2O \tag{5.16}$$

$$N_2 + {}^{\bullet}CH_2 \rightarrow {}^{\bullet}NH_2 + HCN \tag{5.17}$$

$$ {}^{\bullet}H + HCN \rightarrow {}^{\bullet}CN + H_2 \tag{5.18}$$

The radicals then react with the NO to form molecular nitrogen:

$$NO + {}^{\bullet}NH_2 \rightarrow N_2 + H_2O \quad (5.19)$$

$$NO + {}^{\bullet}CN \rightarrow N_2 + CO \quad (5.20)$$

$$2NO + 2CO \rightarrow N_2 + 2CO_2 \quad (5.21)$$

An oxygen deficient atmosphere is critical for Reactions (5.16)–(5.18) to occur. If oxygen levels are high, the NO_x reduction reactions will not occur and the following will predominate:

$$CN + O_2 \rightarrow CO + NO \quad (5.22)$$

$$NH_2 + O_2 \rightarrow H_2O + NO \quad (5.23)$$

To complete the combustion process, air is introduced above the reburn zone. Some NO_x is formed from conversion of HCN and ammonia compounds; however, the net effect is to significantly reduce the total quantity of NO_x emitted from the boiler. The reactions with HCN and ammonia are:

$$HCN + \tfrac{5}{4}O_2 \rightarrow NO + CO + \tfrac{1}{2}H_2O \quad (5.24)$$

$$NH_3 + \tfrac{5}{4}O_2 \rightarrow NO + \tfrac{3}{2}H_2O \quad (5.25)$$

$$HCN + \tfrac{3}{4}O_2 \rightarrow \tfrac{1}{2}N_2 + CO + \tfrac{1}{2}H_2O \quad (5.26)$$

$$NH_3 + \tfrac{3}{4}O_2 \rightarrow \tfrac{1}{2}N_2 + \tfrac{3}{2}H_2O \quad (5.27)$$

Reburn offers the advantages of being able to operate over a wide range of NO_x reduction levels using a variety of reburn fuels. A reburn system can be varied from relatively low levels of reduction, 25–30%, using an overfire air system without any reburn fuel, to ≈70% reduction when reburn fuel is added [7,22]. This allows for fine-tuning to meet emissions limits.

Concerns regarding using reburn technology are similar to those for other combustion modification processes. This includes concerns about incomplete combustion (i.e. CO and hydrocarbon production, and unburned carbon in the fly ash), changes in slagging and fouling characteristics, different ash characteristics and fly ash loadings, corrosion of boiler tubes in reducing atmospheres, higher fan power consumption, and pulverizer constraints (if pulverized coal is used as the reburn fuel).

5.3.1.5 Cofiring

Cofiring is the practice of firing a supplementary fuel, such as coal-water mixtures (CWF) or biomass, with a primary fuel (i.e. coal) in the same burner or separately but into the main combustion zone. This technology was originally developed to utilize opportunity fuels; however, various levels of NO_x reduction were achieved and provide an option for NO_x reduction without investing in a postcombustion system when the emissions are near the regulatory requirements. This technique is not currently used as a commercial means for NO_x reduction; however, it is briefly discussed in this section because several demonstrations of this technology have been conducted, with a few still ongoing. In addition, it is considered a viable option for NO_x trimming especially if used in conjunction with legislation that mandates a percentage of electricity be generated from renewable/sustainable sources. Such legislation has been seriously discussed in the United States and has been included in congressional bills although they have not yet passed.

CWF technology was originally developed as a fuel oil replacement with considerable research and development from the late 1970s to the late 1980s. During the late 1980s and early 1990s, coal suppliers and coal-fired utilities began to evaluate the production of CWF using bituminous coal fines from coal-cleaning circuits in an effort to reduce dewatering/drying costs and/or to recover and utilize low-cost impounded coal fines [28,29]. This marked a philosophical change in the driving force behind utilizing CWF in the United States as well as the CWF characteristics of these two fuel types since cofire CWFs are quite different from fuel oil replacement CWFs, i.e. low solids content (50%) and no additive package to wet the coal, provide stability, and modify rheology compared to high solids content ($\approx$70%) with an expensive additive package, respectively. Extensive testing was performed by several companies and universities culminating with waste impoundment characterizations and several utility demonstrations in pulverized coal-fired boilers (both wall- and tangentially-fired units) and cyclone-fired boilers. Funding for these demonstrations was provided by industry, United States DOE, Electric Power Research Institute, and state agencies. Penn State provided fuel support in all but one of these demonstrations, which is summarized in a CWF preparation and operation manual prepared by Penn State [30], where the CWFs were being developed to provide coal preparation plants a means for utilizing hard to dewater fines, cleaning up waste coal impoundments thereby reducing coal mine liability, and supplying utilities with a low-cost fuel that also serves as a low-cost NO_x reduction technology. NO_x reductions varying from approximately 11% in cyclone-fired boilers [31] to approximately 30% in wall-fired boilers [32,33] to approximately 35% in tangentially-fired boilers [34] were achieved. Several mechanisms were responsible for the NO_x reduction including lower flame temperature from the addition of the water, staged combustion from cofiring in low-NO_x burners, and the CWF acting as a reburn fuel when injected in upper level burners.

Biomass cofiring has been demonstrated and deployed at a number of power plants in the United States and Western Europe using a variety of materials including sawdust, urban wood waste, switchgrass, straw and other similar materials [35]. Biomass fuels have been cofired with all ranks of coal – bituminous and subbituminous coals and lignites. The benefits of biomass cofiring include reduced NO_x, fossil CO_2, SO_2, and mercury emissions.

Cofiring biomass, particularly sawdust and urban wood waste but also switchgrass to a lesser extent, in large-scale pulverized coal- and cyclone-fired units has been demonstrated at several utilities with seven commercial installations in the United States [36,37]. Many of the demonstrations were conducted to achieve NO_x reduction, which can vary significantly but can be as high as approximately 35%. Tillman [35] noted that the dominant mechanism for NO_x reduction is to support deeper staging of combustion when staging has not been particularly extensive. When biomass can introduce or accentuate staging by early release of volatile matter, then NO_x reduction can be significant [35,38]. A secondary mechanism for NO_x reduction is the influence of cofiring on furnace exit gas temperature (FEGT). Data indicates that cofiring has minimal impact on flame temperatures but can have a pronounced impact on FEGT, thereby reducing NO_x emissions. A third influence is the reduction in fuel nitrogen content when a low-nitrogen fuel such as sawdust is used.

5.3.1.6 Process optimization

Several software packages have been developed that apply optimization procedures to the distributed control system of the boiler to provide tighter control of plant operation parameters [22]. Artificial intelligence (AI)-based optimization software has been used to improve fossil fuel-fired steam power plant boiler operations for more than 15 years, with power plant boiler optimization initially focused on reducing NO_x emissions [39]. Using AI-based technologies such as neural networks, design of experiments, model predictive control, and rule-based optimization, combustion optimizers extract knowledge about the combustion process and determine the optimal balance of fuel and air flows into the furnace [40]. These technologies are summarized in Table 5.3 [39].

The combustion process is optimized resulting in lower NO_x emissions and improved boiler efficiency while maintaining safe, reliable, and consistent unit operation. Also, combustion optimization approaches have been developed where advanced computational and experimental approaches are used to make design and operational modifications to the process equipment and boiler as a whole [41]. Combustion optimization systems optimize the distribution of fuel and air in boilers by biasing control system settings to those that best meet a set of objectives and constraints [39]. To determine the optimal biases, real-time and historical plant data are used to model relationships between variables that impact combustion quality, such as damper, burner tilt, and pulverizer settings, and optimization objectives, such as reducing NO_x emissions, improving boiler efficiency, and controlling carbon monoxide. SCR systems and other NO_x reducing hardware are being combined with boiler optimization software and the comparatively inexpensive boiler optimization solutions reduce NO_x at the source and allow power generators to lower their SCR-related operating costs, better manage the interactions between combustion and postcombustion systems, and potentially run their SCR systems less aggressively [40]. The use of these packages has resulted in NO_x reductions of 10 to 40%, reduced unburned carbon levels by 25 to 50%, increased boiler efficiencies by 1 to 3%, and increased heat rates by 0.5 to 5% [22,41].

Table 5.3 Optimization and advanced control technologies used for boiler optimization

Technology	Description	Example of use
Neural Network Model-Based Optimization	Nonlinear, multivariable steady-state models used to identify the best combinations of variables under varying conditions. The type of model is flexible and able to adapt its own structure based on experience.	Combustion process manipulated variables that can be adjusted over time to balance unit operations (e.g. damper biasing)
Model Predictive Control (MPC)	Dynamic models used to predict changes over the next several minutes and anticipate the effects of disturbances and future moves.	Optimizing the major gross air controls that must respond quickly to plant conditions (e.g. reheat and superheat temperature control)
Rule-Based Optimization	Heuristic models that capture and codify human expertise as rules that can be systematically applied by an inference engine to rank a set of possible actions given a set of conditions.	Comparing a set of proposed sootblowing actions until the optimal action is found
Design of Experiments/ Direct Search	A form of direct optimization in which the optimizer makes a small change and, if that change improves the process, makes another small change in the same direction. This trial and error approach is repeated to continually search for improved operating performance.	Neural network model training

5.3.2 *Flue gas treatment*

Flue gas treatment technologies are postcombustion processes to convert NO_x to molecular nitrogen or nitrates. The two primary strategies that have been developed for postcombustion control and that are commercially available are selective catalytic reduction (SCR) and selective noncatalytic reduction (SNCR). There are additional concepts under development including combining SCR and SNCR technologies (known as hybrid SCR/SNCR) and rich reagent injection; however, these are not extensively used at this time. Of these technologies, SCR is being identified by utilities as the strategy to meet stringent NO_x requirements. These technologies are discussed in the following sections, with an emphasis on SCR.

5.3.2.1 Selective catalytic reduction

Selective catalytic reduction (SCR) of NO_x using ammonia (NH_3) as the reducing gas was patented in the United States by Englehard Corporation in 1957 [42]. This technology can achieve NO_x reductions in excess of 90% and is widely used in commercial applications in Western Europe and Japan, which have stringent NO_x regulations, and is becoming the postcombustion technology of choice in the United States and China. Stringent NO_x regulations in Western Europe essentially mandate the installation of secondary NO_x reduction systems, the majority of which is SCR with only a few boilers using the selective noncatalytic reduction process [43]. Japan applied SCR technology to thermal plants in the late 1970s and Germany led widespread application in Europe beginning in 1986. United States utilities initially deployed SCR for coal-fired units for new and retrofit applications in 1991 and 1993, respectively [44]. As of early 2009, more than 200 SCR were installed on fossil fuel-fired power generation facilities with overall capacity greater than 100 GW. In the U.S. in 2012, 241 coal-fired power units (i.e. 54% of coal-fired MWh generation capacity) and 425 natural gas or oil-fired combined cycle units (i.e. 78% of gas- or oil-fired MWh generation) were equipped with SCR systems [6].

The SCR process uses a catalyst at approximately 570–750°F to facilitate a heterogeneous reaction between NO_x and an injected reagent, vaporized ammonia, to produce nitrogen and water vapor as depicted schematically in Figure 5.4 [8]. Ammonia chemisorbs onto the active sites on the catalyst. The NO_x in the flue gas reacts with the adsorbed ammonia to produce nitrogen and water vapor. The principal reactions are [45]:

$$4NO + 4NH_3 + O_2 \rightarrow 4N_2 + 6H_2O \tag{5.28}$$

$$2NO_2 + 4NH_3 + O_2 \rightarrow 3N_2 + 6H_2O \tag{5.29}$$

A small fraction of the sulfur dioxide is oxidized to sulfur trioxide over the SCR catalyst. In addition, side reactions may produce the undesirable byproducts

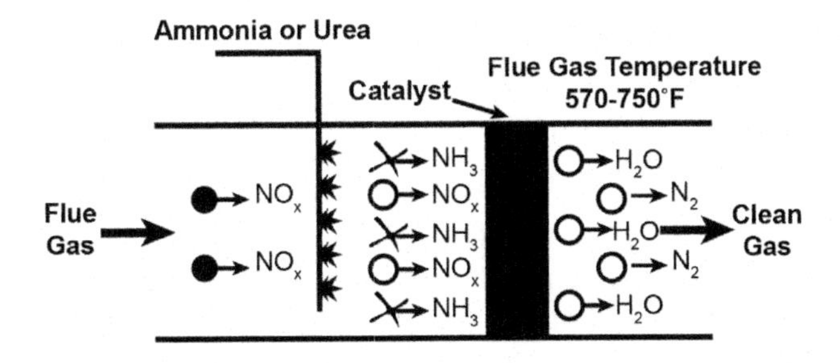

Figure 5.4 Schematic representation of an SCR system.

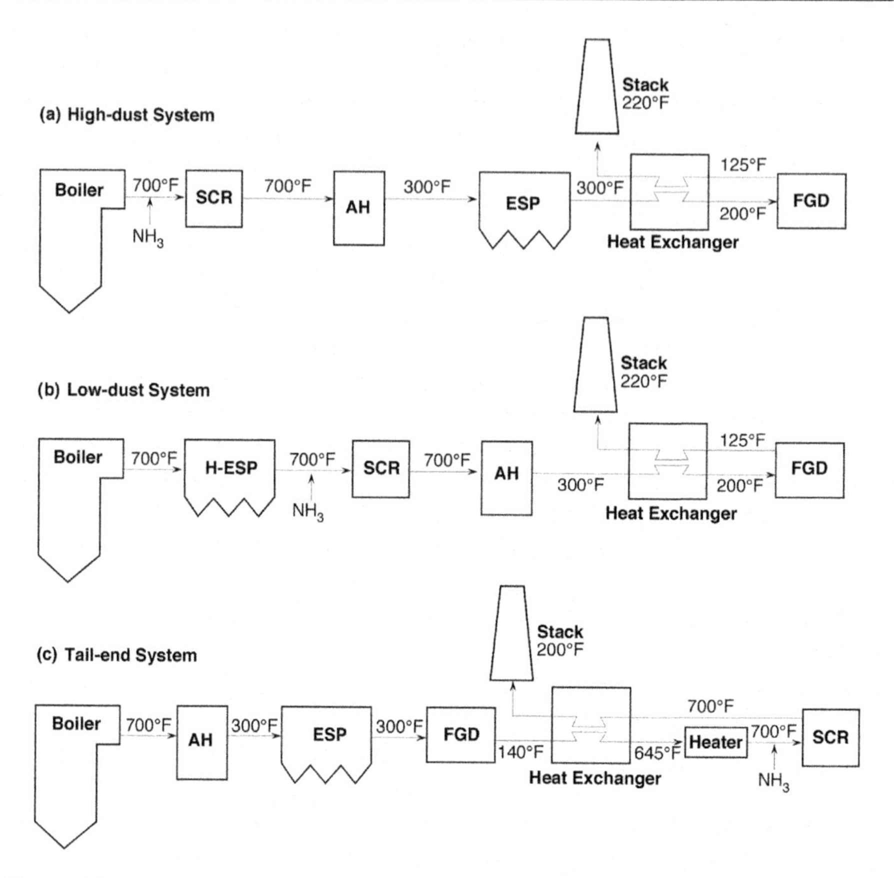

Figure 5.5 SCR configurations with typical system temperatures [45].

ammonium sulfate ($(NH_4)_2SO_4$) and ammonium bisulfate (NH_4HSO_4), which can cause plugging and corrosion of downstream equipment. These side reactions are [42]:

$$SO_2 + \tfrac{1}{2}O_2 \rightarrow SO_3 \tag{5.30}$$

$$2NH_3 + SO_3 + H_2O \rightarrow (NH_4)_2SO_4 \tag{5.31}$$

$$NH_3 + SO_3 + H_2O \rightarrow NH_4HSO_4 \tag{5.32}$$

There are three SCR system configurations for coal-fired boilers and they are known as high-dust, low-dust, and tail-end systems. These are shown schematically in Figure 5.5 [45]. In a high-dust configuration, the SCR reactor is placed upstream of the particulate removal device between the economizer and the air preheater. This configuration (also referred to as hot side, high dust) is the most commonly used, particularly with dry-bottom boilers [24] and is the principal type planned for the United States installations [43]. In this configuration, the catalyst is exposed to the fly ash and chemical compounds present in the flue gas that have the potential to degrade the catalyst by ash erosion and chemical reactions (i.e. poisoning). However, these can be addressed by proper design as evidenced by the extensive use of this configuration.

In a low-dust installation, the SCR reactor is located downstream of the particulate removal device. This configuration (also referred to as hot side, low dust) reduces the degradation of the catalyst by fly ash erosion. However, this configuration requires a costly hot-side ESP or a flue gas reheating system to maintain the optimum operating temperature.

In tail-end systems (also referred to as cold side, low dust), the SCR reactor is installed downstream of the FGD unit. It may be used mainly in wet-bottom boilers and also on retrofit installations with space limitations and lack of outage duration for the retrofit [22]. However, this configuration is typically more expensive than the high-dust configuration due to flue gas reheating requirements. This configuration does have the advantage of longer catalyst life and use of more active catalyst formulations to reduce overall catalyst cost, especially when firing fuels that contain rapid catalyst deactivation components. Also, industrial boiler applications may have a greater need for this arrangement. Package boilers may require a duct burner to achieve the required gas temperature [8].

There are several issues that need to be considered in the design and operation of SCR systems including coal characteristics, catalyst and reagent selections, process conditions, ammonia injection, catalyst cleaning and regeneration, low-load operation, and process optimization [22]. Coals with high sulfur in combination with significant quantities of alkaline, alkaline earth, arsenic, or phosphorus in the ash can severely deactivate a catalyst and reduce its service life. In addition, the SO_3 can react with residual ammonia resulting in ammonium sulfate deposition in the air preheater and loss of performance.

Catalyst formulation, pitch (plate centerline or cell spacing dimensions), and sizing are important for successful SCR applications. Catalyst formulations are proprietary among manufacturers but the most common components used in stationary heat and power applications are titanium dioxide with small amounts of vanadium, molybdenum, tungsten, or a combination of other reagents [8]. The catalyst commonly consists of a vanadium pentoxide active material on a titanium dioxide substrate.

Catalyst types generally consist of three categories – base metal, zeolite, and precious metal – with base metal catalysts the most common in stationary heat and power applications [8]. Base metal catalysts are typically titania-vanadia compositions and can be provided on three geometries: plate type, honeycomb type, or corrugated fiber type. The honeycomb form usually is an extruded ceramic with the catalyst either incorporated throughout the structure (homogenous) or coated on the substrate. In the plate geometry, the support material is generally coated with catalyst by paste pressing the catalyst into the substrate to form a catalyst plate. Corrugated catalyst consists of a corrugated glass fiber substrate coated with titanium dioxide with the base impregnated with the active components.

Zeolite catalysts are less common than the base metal catalysts [8]. These catalysts are aluminosilicate materials which function similar to base metal catalysts but can be operated at higher temperatures (e.g. >1,000°F). They will also oxidize SO_2 to SO_3 but are more sensitive to fly ash-laden flue gas. Therefore, they tend to be used in clean flue gas applications such as gas turbines.

Precious metal catalysts are manufactured from platinum and rhodium but are not widely used in steam-generating applications because of their high costs [8]. These catalysts can be used to convert CO to CO_2 so some manufacturers have combined the precious metal and base metal catalysts to allow both NO_x reduction and CO oxidation within one catalyst chamber.

Catalyst performance depends on the surface area velocity, which influences the NO_x reduction reactions to occur. The term that relates catalyst volume with a specific surface area is space velocity, S_v, which is expressed as

$$S_v = \text{flue gas flow (standard ft}^3/h)/\text{catalyst volume (ft}^3) \tag{5.33}$$

The greater the space velocity, the less catalyst required. Space velocity is dependent upon many factors including required NO_x reduction, flue gas flow, ammonia slip, gas velocity, flue gas constituents, required catalyst life, and distribution profiles of gas temperature, velocity, NO_x, and NH_3/NO_x molar ratio.

SCR reactor performance is influenced by the distribution of velocity, temperature, and reacting gas components through the catalyst bed [8]. The degree of a given profile uniformity is described by the coefficient of variance (C_v), which is reported as the percentage of the profile standard deviation from the arithmetic mean [8]:

$$C_v = \frac{\sigma}{\overline{x}} 100\% \tag{5.34}$$

where

$$\sigma = \sqrt{\frac{1}{(n-1)} \sum_{i=1}^{n} (x_i - \overline{x})^2} \tag{5.35}$$

$$\overline{x} = \frac{1}{n} \sum_{i=1}^{n} x_i \tag{5.36}$$

where σ is standard deviation and $\overline{x}$ equals arithmetic mean.

SCR reactor distribution goals for flow and molar ratio are commonly specified in this coefficient of variation format. Temperature distribution criteria are typically expressed as a minimum and maximum deviation about the arithmetic mean. The objective is to achieve the most uniform profile as possible at the SCR catalyst inlet for NO_x concentration, NH_3/NO_x molar ratio, velocity, and temperature.

For optimum SCR performance, the reagent must be well mixed with the flue gas and in direct proportion to the amount of NO_x reaching the catalyst. Anhydrous ammonia has been commonly used as reagent, accounting for over 90% of current world SCR applications [22]. It dominates planned installations in the United States although numerous aqueous systems will be installed. Recently, urea-based processes are being developed to address utilizing anhydrous ammonia, which is a hazardous and toxic chemical. When urea $CO(NH_2)_2$ is used, it produces ammonia, which is the active reducing agent, by the following reactions:

$$NH_2 - CO - NH_2 \rightarrow NH_3 + HNCO \quad (5.37)$$

$$HNCO + H_2O \rightarrow NH_3 + CO_2 \quad (5.38)$$

Catalyst is not consumed during operation but does deteriorate. During the operation of the SCR, the catalyst is deactivated by fly ash plugging, catalyst poisoning, and/or the formation of blinding layers and is most pronounced in the high-dust arrangement. Deterioration occurs when the active sites within the catalyst micropore structure become contaminated and inactive. Moisture in liquid form can accelerate deterioration. Operation at temperatures greater than 800°F can also accelerate deterioration due to sintering of the pore structure [8]. Arsenic in coal ash will poison the catalyst.

The most common method of catalyst cleaning has been the installation of steam soot blowers although acoustic cleaners have been successfully tested. Once the catalyst has been severely deactivated, it is conventional practice to add additional catalyst or replace it; however, several regeneration techniques have evolved over the last few years providing extended service life for catalysts [22].

Low-load boiler operation can be problematic with SCR operation, specifically with high-sulfur coals. There is a minimum temperature below which the SCR should not be operated; therefore, system modifications such as economizer bypass to raise the SCR temperature during low-load operation may be required [22].

5.3.2.2 Selective non-catalytic reduction

Selective non-catalytic reduction (SNCR) is a proven, commercially available technology that has been applied since 1974 with over 300 systems installed worldwide on various combustion sources including coal-fired utility applications [22]. In the U.S. in 2012, 132 coal-fired power units (i.e. 8% of coal-fired MWh generation capacity) were equipped with SNCR systems [6]. No SNCR systems were installed on natural gas or oil-fired combined cycle units. The SNCR process involves injecting nitrogen-containing chemicals into the upper furnace or convective pass of a boiler within a specific temperature window without the use of an expensive catalyst. Figure 5.6 is a schematic representation of the SNCR process [8]. There are different chemicals that can be used that selectively react with NO in the presence of oxygen to form molecular nitrogen and water, but the two most common are ammonia and urea. Other chemicals that have been tested in research include amines, amides, amine salts, and cyanuric acid. In recent years, urea-based reagents such as dry urea, molten urea, or urea solution have been increasingly used, replacing ammonia at many plants because anhydrous ammonia is the most toxic and requires strict transportation, storage, and handling procedures [22]. The main reactions when using ammonia or urea are, respectively:

$$4NO + 4NH_3 + O_2 \rightarrow 4N_2 + 6H_2O \tag{5.28}$$

$$4NO + 2CO(NH_2)_2 + O_2 \rightarrow 4N_2 + 2CO_2 + 6H_2O \tag{5.29}$$

A critical issue is finding an injection location with the proper temperature window for all operating conditions and boiler loads. The chemicals then need to be adequately mixed with the flue gases to ensure maximum NO_x reduction without producing too much ammonia. Ammonia slip from an SNCR can affect downstream equipment by forming ammonium sulfates.

The temperature window varies for most of the reducing chemicals used, but generally it is between 1,650–2,100°F. Ammonia can be formed at temperatures below the temperature window and the reducing chemicals can actually form more NO_x above the temperature window. Ammonia has a lower operating temperature than urea – 1,560–1,920°F compared to 1,830–2,100°F, respectively. Enhancers such as hydrogen, carbon monoxide, hydrogen peroxide (H_2O_2), ethane (C_2H_6), light alkanes, and alcohols have been used in combination with urea to reduce the temperature window [46]. Several processes use proprietary additives with urea in order to reduce NO_x emissions [47].

Most of the SNCR applications in coal-fired units are on stoker-fired and fluidized-bed boilers where the appropriate temperature range and residence time are available. In large utility boilers, the proper temperature range occurs in the convective pass cavities, making application more challenging [8]. For these applications, overall control may be limited to the 20 to 30% reduction range.

The efficiency of reagent utilization is significantly less with SNCR than with SCR. In commercial SNCR systems, the utilization is typically between 20 and 60%; consequently, usually three to four times as much reagent is required with SNCR to achieve NO_x reduction similar to that of SCR. SNCR processes typically achieve 20 to 50% NO_x reduction with stoichiometric ratios of 1.0 to 2.0.

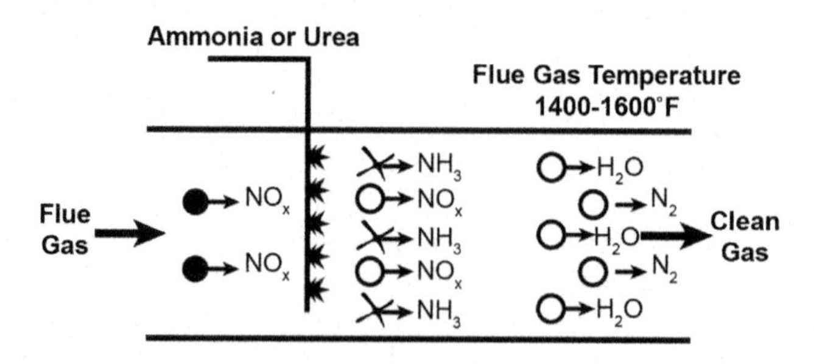

Figure 5.6 Schematic representation of the SNCR process.

The major operational impacts of SNCR include air preheater fouling, ash contamination, N_2O emissions, and minor increases in heat rate. A major plant impact of SNCR is on the air preheater where residual ammonia reacts with the SO_3 in the flue gas to form ammonium sulfate and bisulfate (see Reactions (5.30)–(5.32)) causing plugging and downstream corrosion. High levels of ammonia slip can contaminate the fly ash and reduce its sale or disposal. Significant quantities of N_2O can be formed when the reagent is injected into areas of the boiler that are below the SNCR optimum operating temperature range. Urea injection tends to produce a higher level of N_2O compared to ammonia. The unit heat rate is increased slightly due to the latent heat losses from vaporization of injected liquids and/or increased power requirements for high-energy injection systems. The overall efficiency and power losses normally range from 0.3 to 0.8% [22].

5.3.2.3 Hybrid SNCR/SCR

SCR generally represents a relatively high capital requirement whereas SNCR has a high reagent cost. A hybrid SNCR/SCR system balances these costs over the life cycle for a specific NO_x reduction level, provides improvements in reagent utilization, and increases overall NO_x reduction [22]. However, there is limited experience with these hybrid systems as full-scale power plant operation to date has only been in demonstrations. They are discussed here since they have demonstrated NO_x reductions as high as 60 to 70%.

In a hybrid SNCR/SCR system, the SNCR operates at lower temperatures than stand-alone SNCRs, resulting in greater NO_x reduction but also higher ammonia slip. The residual ammonia feeds a smaller-sized SCR reactor, which removes the ammonia slip and decreases NO_x emissions further. The SCR component may achieve only 10 to 30% NO_x reduction with reagent utilization as high as 60 to 80% [22]. Hybrid SNCR/SCR systems can be installed in different configurations including [22]:

- SNCR with conventional reactor-housed SCR;
- SNCR with in-duct SCR which uses catalysts in existing or expanded flue gas ductwork;
- SNCR with catalyzed air preheater, where catalytically active heat transfer elements are used; and
- SNCR with a combination of in-duct SCR and catalyzed air heater.

5.3.2.4 Rich reagent injection

Cyclone burners, with their turbulent and high-temperature environment, are conducive for NO_x production. Lower cost methods than installing SCRs to reduce NO_x production in cyclone-fired boilers have been tested, such as CWSF or biomass cofiring, while others are under development. One such process currently under development is the rich reagent injection (RRI) process. It involves injection of amine reagents in the fuel-rich zone above the main combustion zone at temperatures of 2,370 to 3,100°F. NO_x in the flue gas is converted to molecular nitrogen and reductions of 30% have been achieved. The capital costs for an RRI system are consistent with those of SNCR; however, the operating costs are expected to be 2 to 3 times that of SNCR due to increased reagent usage.

5.3.3 NO_x control in fluidized-bed combustion

The fluidized-bed combustion (FBC) process inherently produces lower NO_x emissions due to its lower operating temperature (i.e. bed temperature of $\approx$1,450 to 1,600°F). Also, the bed is a reducing region where available oxygen is consumed by carbon, thereby reducing ionization of nitrogen. Additional combustion modifications or flue gas treatment for NO_x control, discussed previously in this chapter, can also be employed. Techniques currently used for FBC include reducing the peak temperature by flue gas recirculation (FGR), natural gas reburning (NGR), overfire air (OFA), low excess air (LEA), and reduced air preheat [15]. Postcombustion control is also used including SCR, SNCR, and fuel reburning, which achieve 35 to 90% NO_x reduction. Also, low nitrogen fuel can be used (e.g. sawdust) thereby reducing the amount of fuel nitrogen available. Injecting sorbents into the combustion chamber or in the ducts can reduce NO_x by 60 to 90% [15].

5.3.4 NO_x control in stoker-fired boilers

NO_x control in stokers, specifically traveling grate and spreader stokers, include abatement methods to reduce the peak temperature, reduce the residence time at peak temperature, chemically reduce the NO_x, use low-nitrogen fuels, and inject a sorbent [15]. In traveling grate stokers, the peak temperature can be reduced by FGR, NGR, combustion optimization, OFA, LEA, water or steam injection, and reduced air preheat, thereby achieving 35 to 50% NO_x reduction. Air or fuel staging, which reduces the residence time at peak temperature, can achieve 50 to 70% NO_x reduction while using SCR, SNCR, or fuel reburning technologies can achieve 55 to 80% NO_x reduction. Sorbent injection, which can achieve 60 to 90% NO_x reduction, and using fuels with low nitrogen content are technologies also employed. NO_x technologies used for spreader stokers are similar to traveling grate stokers with slightly different results. FGR, natural gas reburning, low-NO_x burners, combustion optimization, OFA, LEA, water or steam injection, and reduced air preheat temperature are control options to reduce peak temperatures and can achieve 50 to 65% NO_x reduction. Air or fuel staging or steam injection, which reduces the residence time at peak temperature, can achieve 50 to 65% NO_x reduction while using SCR, SNCR, or fuel reburning technologies achieves 35 to 80% NO_x reduction. Additional NO_x reduction technologies include sorbent injection, which can achieve 60 to 90% reduction, and using lower nitrogen fuels.

5.4 Economics of NO_x reduction/removal

The costs for NO_x reduction/removal techniques are site and performance specific, thereby making it difficult for comparing generalized system costs. These techniques depend on several factors including whether the installation is new or a retrofit application, the degree of retrofit difficulty when retrofitting an existing boiler system, unit size, uncontrolled NO_x levels, and required NO_x reduction [22].

Table 5.4 NO_x **combustion control costs (2011$) for coal-fired boilers 300 MW in size**

Boiler type	Technology	Costs		
		Capital ($/kW)	Fixed O&M ($/kW-year)	Variable O&M (mills/kWh)
Dry bottom, wall-fired	Low NO_x burner without overfire air	48	0.3	0.07
	Low-NO_x burner with overfire air	65	0.5	0.09
Tangentially fired	Low NO_x coal and air nozzles with close-coupled overfire air	26	0.2	0.00
	Low NO_x coal and air nozzles with separated overfire air	35	0.2	0.03
	Low NO_x coal and air nozzles with close-coupled and separated overfire air	41	0.3	0.03
Vertically fired	NO_x combustion control	31	0.2	0.06

This section will summarize costs for the various systems using information published over the last several years. Even in the case of data published several years ago, the costs illustrate the relative differences between various control technologies. Also, most of the costs reported in this section are for utility applications. Costs for industrial-sized boilers are discussed separately at the end of the section.

Sargent & Lundy performed a study for the U.S. EPA in which they determined costs for various combustion control options [48]. They determined capital costs and variable and fixed operating and maintenance (O&M) costs, in 2011 dollars, for three types of 300 MW boilers firing coal. These results are provided in Table 5.4. The capital costs for these options range from $31/kW to $65/kW. Fixed O&M and variable O&M ranged from $0.2/kW-year to $0.5/kW-year and from 0.00 mills/kWh to 0.09 mills/kWh, respectively.

Wu [22] reports that the capital cost for a low-NO_x burner is in the range of $650 to 8,300/MM Btu. The operating cost can range from $340–$1,500/MM Btu. The levelized cost can vary from $240–$4,300/ton of NO_x removed with the average cost closer to the lower end of the range [49].

The costs for furnace air staging are similar to those for low-NO_x burners [22]. The capital cost is ≈$8–$23/kW and the levelized cost ranges from $110–$210/short ton of NO_x removed. If furnace air staging is combined with low-NO_x burners, the capital cost will increase to $15–$30/kW while the levelized cost remains relatively unchanged. Retrofits of furnace air staging in tangentially fired boilers are generally more expensive than those in wall-fired boilers, \$5–\$11/kW and \$11–\$23/kW, respectively.

The capital costs for conventional fuel gas recirculation (FGR) is similar to that for low-NO_x burners and OFA – \$8–\$35/kW [49]. However, capital costs of induced FGR, a design derivative of conventional forced flue gas desulfurization, have been reduced to \$1–\$3/kW.

The capital cost for reburn technology depends on the size of the unit, ease of retrofit, control system upgrade requirements, and, for natural gas reburn, availability of natural gas at the plant [22]. The retrofit costs are typically about \$15–\$20/kW for natural gas, coal, or oil reburn, excluding the cost of any natural gas pipeline. The operating cost for a reburn retrofit is mainly due to the differential cost of the reburn fuel over the main fuel. For coal reburn this is zero while reburn fuels like natural gas or oil are usually more expensive than the main fuel. This differential, however, can be offset by reductions in SO_2 emissions, ash remediation and disposal, and pulverizer power. The levelized cost for reburn is ≈\$110–210/short ton of NO_x removed [50].

Cofiring of CWF is not commercially used at this time. Biomass cofiring on the other hand has been demonstrated at several plants with commercial operations being performed at several utilities. The capital costs for biomass cofiring ranges from \$175–\$250/kW [36].

The total turnkey installation cost for an advanced combustion control system ranges from \$150,000 to \$500,000 [22]. It is possible to achieve moderate cost reductions on a per unit basis for similar units at the same power plant site. The size of the unit typically has little impact on the cost of a system.

The capital costs for an SCR system depend on the level of NO_x removal and other site-specific conditions, such as inlet NO_x concentration, unit size and ease of retrofit, and in the case of retrofits, have a wide range due to the uniqueness of each retrofit [51]. The capital costs of an SCR system include [22]:

- catalyst and reactor system;
- flow control skid and valving system;
- ammonia injection grid;
- ammonia storage;
- piping;
- ducts, expansion joints, and dampers;
- fan upgrades/booster fans;
- air preheater changes;
- foundations, structural steel, and electrical work; and
- installation.

The operating costs for an SCR include:

- ammonia usage;
- pressure drop changes;
- excess air change;
- unburned carbon change;

- ash disposal;
- catalyst replacement;
- vaporization/injection energy requirements; and
- other auxiliary power usage.

Sargent & Lundy performed a study for the U.S. EPA in which they determined costs for postcombustion NO_x control options [48]. They determined capital costs and variable and fixed operating and maintenance (O&M) costs, in 2011 dollars, for coal-fired power plant capacities ranging from 100 to 1,000 MW and for heat rates varying from 9,000 to 11,000 Btu/kWh. These results are provided in Table 5.5.

Sargent & Lundy also determined SCR costs for oil and gas-fired steam-generating plants [48]. They reported capital costs of $80/kW and fixed and variable O&M of $1.16/kW-year and $0.13/MWh, respectively, for a system removing 80% of the NO_x.

SNCR is less capital-intensive than SCR. Two SNCR scenarios were studied by Sargent & Lundy and they are reported in Table 5.5. In addition, Wu reported that the cost of an SNCR retrofit is $10–$20/kW, whereas incorporating SNCR into a new boiler is $5–$10/kW [22]. The difference is due to the cost associated with modifying the existing boiler to install the reagent injection ports. The operating costs associated with the reagent, auxiliary power, and potential adverse plant impacts are of the order 1–2 mills/kWh. The levelized costs average ≈$1,000/short ton of NO_x removed.

A combination of flue gas treatment with combustion modification is being increasingly used. This technology provides higher overall NO_x reduction and can be more cost effective than stand-alone technology for the same level of NO_x control [22].

The costs of SCR can be reduced when it is used in combination with combustion modifications such as low-NO_x burners and overfire air [22]. Capital costs are lowered because combustion modifications lower the inlet NO_x concentration, which reduces the catalyst volume, support systems, and installation cost of SCR. In addition, operating costs are lowered due to reductions in catalyst replacement and reagent consumption.

SNCR can be combined with low-NO_x burners or gas reburn. SNCR and gas reburn have comparable economics at the same level of NO_x reduction; however, combining the two technologies considerably lowers cost while achieving a slightly higher NO_x reduction. An example of annual costs, reported by Wu [22], are ≈$1,140, $1,120, and $730 per short ton NO_x removed, respectively, for urea SNCR, gas reburn, and urea SNCR/gas reburn.

The U.S. EPA developed costs for SCR and SNCR NO_x control for industrial boilers and gas turbines [52,53]. These results are shown in Table 5.6. The SCR costs are based on coal- and oil-fired boilers with firing rates of 350 MM Btu/h and gas turbines with generating capacities of 75 MW and 5 MW for the large and small units, respectively. SNCR costs are based on coal-fired boilers with firing rates greater than 100 MM Btu/h.

Table 5.5 **Illustrative postcombustion NO_x control costs (2011$) for coal-fired power plants for representative sizes and heat rates**

Control type	Heat rate (Btu/kWh)	Var. O&M (mills/Kwh)	Capacity (MW)									
			100		300		500		700		1,000	
			Capital cost ($/kW)	Fixed O&M ($/kW-yr)	Capital cost ($/kW)	Fixed O&M ($/kW-yr)	Capital cost ($/kW)	Fixed O&M ($/kW-yr)	Capital cost ($/kW)	Fixed O&M ($/kW-yr)	Capital cost ($/kW)	Fixed O&M ($/kW-yr)
SCR	9,000	1.23	321	1.76	263	0.76	243	0.64	232	0.58	222	0.53
	10,000	1.32	349	1.86	287	0.81	266	0.69	255	0.63	244	0.57
	11,000	1.41	377	1.96	311	0.87	289	0.73	277	0.67	265	0.62
SNCR-Tangent.	9,000	1.04	55	0.48	30	0.26	22	0.20	18	0.16	15	0.13
	10,000	1.15	56	0.50	30	0.27	23	0.20	19	0.17	15	0.14
	11,000	1.27	57	0.51	31	0.27	23	0.21	19	0.17	15	0.14
SNCR-Fluidized Bed	9,000	1.04	41	0.36	22	0.20	17	0.15	14	0.12	11	0.10
	10,000	1.15	42	0.37	23	0.20	17	0.15	14	0.12	12	0.10
	11,000	1.27	43	0.38	23	0.21	17	0.15	14	0.12	12	0.10

Table 5.6 Capital and operating costs for industrial boilers and gas turbines

Unit type	Capital cost ($/MM Btu)	O&M cost ($/MM Btu)	Annual cost ($/MM Btu)	Cost per short ton of NO_x removed ($/short ton)
SCR				
Industrial boiler -coal	10,000–15,000	300	1,600	2,000–5,000
Industrial boiler -oil, gas	4,000–6,000	450	700	1,000–3,000
Large gas turbine	5,000–7,500	3,500	8,500	3,000–6,000
Small gas turbine	17,000–35,000	1,500	3,000	2,000–10,000
SNCR				
Industrial boiler >100 MM Btu/h	900–2,500 (9,000–25,000 $/MW)	100–500 (1,000–5,000 $/MW)	300–1,000 (3,000–10,000 $/MW)	400–2,500

References

[1] Wark K, Warner CF, Davis WT. Air pollution: its origin and control. 3rd ed. Menlo Park, California: Addison Wesley Longman, Inc; 1998.
[2] Wall TF. Principles of combustion engineering for boilers. London: Harcourt Brace Jovanovich; 1987. pp. 197–294
[3] EPA (United States Environmental Protection Agency). Latest findings on national air quality: 1997 status and trends. Washington, DC: Office of Air Quality Planning and Standards; 1998.
[4] EPRI (Electric Power Research Institute). Air pollution and health effects research at epri: the aries program, strategic overview fact sheet. California: EPRI, Hillview; 2002.
[5] EPA (United States Environmental Protection Agency). National Emissions Inventory (NEI), 1970–2013 Average Annual Emissions. All Criteria Pollutants, www.epa.gov/ttnchie1/trends/; February 27, 2014.
[6] EPA (United States Environmental Protection Agency). Clean Air Interstate Rule, Acid Rain Program, and Former NO_x Budget Trading Program 2012 Progress Report, 2013.
[7] Mitchell SC. NO_x in pulverized coal combustion. London: IEA Coal Research; 1998.
[8] Kitto JB, Stultz SC, editors. Steam: its generation and use. 41st ed. The Babcock and Wilcox Company; 2005.
[9] Baukal CE, Schwartz RE, editors. The John Zink combustion handbook. CRC Press LLC; 2001.

[10] Lawn CJ, editor. Principles of combustion engineering for boilers. London, England: Academic Press; 1987.

[11] Davidson RM. How coal properties influence emissions. London: IEA Coal Research; 2000.

[12] Moreea-Taha R. NO_x modelling and prediction. London: IEA Coal Research; 2000.

[13] Singer JG, editor. Combustion: fossil power systems. Windsor, Connecticut: Combustion Engineering, Inc; 1981.

[14] Wood SC. Select the right NO_x control technology. (Chemical Engineering Progress, Volume 90, No. 1, 1994), pp. 32–38.

[15] (NO_x) Why and how they are controlled. (Office of Air Quality Planning and Standards, U.S. Government Printing Office, Washington, D.C., November 1999).

[16] AP-42. Emission Factors, Fifth Edition, Chapter 1: External Combustion Sources and Chapter 3: Stationary Internal Combustion, EPA Office of Air Quality Planning and Standards and Office of Air and Radiation, 1995 with latest revisions in 2010.

[17] Tsiou C, Lin H, Laux S, and Grusha J. Operating results from foster wheeler's new vortex series low-NO_x burners. Proceedings of Power Gen 2000, 2000.

[18] Steitz TH, and Cole RW. Field experience in over 30,000 MW of wall fired low NO_x installations. Proceedings of Power Gen 1996, 1996.

[19] Steitz TH, Grusha J, and Cole R. Wall fired low NO_x burner evolution for global NO_x compliance. Proceedings of the 23rd International Technical Conference on Coal Utilization & Fuel Systems, Coal & Slurry Technology Association, 1998.

[20] Patel RL, Thornock DE, Borio RW, Miller BG, and Scaroni AW. Firing micronized coal with a low NO_x RSFC burner in an industrial boiler designed for oil and gas. Proceedings of the Thirteenth Annual International Pittsburgh Coal Conference, 1996.

[21] Borio RW, Patel RL, Thornock DE, Miller BG, Scaroni AW, and McGowan JG. Task 5 – final report: one thousand hour demonstration test in the penn state boiler. Prepared for U.S. Department of Energy, Federal Energy Technology Center, No.DE-AC22-91PC91160, March 1998.

[22] Wu Z. NO_x control for pulverized coal fired power stations. London: IEA Coal Research; 2002.

[23] DOE (United States Department of Energy). Innovation for existing plants. Washington, DC: Office of Fossil Energy; 2007.

[24] EIA (United States Energy Information Agency). Electric power monthly. (U.S. Department of Energy, Office of Coal, Nuclear, Electric and Alternate Fuels, U.S. Government Printing Office, Washington, D.C., September 11, 2009).

[25] EPA (United States Environmental Protection Agency). Control of NO_x emissions by reburning. Washington, DC: Office of Research and Development, U.S. Government Printing Office; 1996.

[26] Wendt JOL, Sternling CV, and Matovich MA. Reduction of sulfur trioxide and nitrogen oxides by secondary fuel injection. Proceedings of the 14th Symposium (International) on Combustion, Combustion Institute, 1973, pp. 897–904.

[27] Myerson AL, Taylor FR, and Faunce BG. Ignition limits and products of the multistage flames of propane-nitrogen dioxide mixtures, 6th Symposium (International) on Combustion. The Combustion Institute, pp. 154–163, 1957.

[28] Stoesssner RD, and Zawadzki E. Coal water slurry dual firing project for homer city station – phase I test results, 16th International Technical Conference on Coal Utilization & Fuel Systems, Coal & Slurry Technology Association, 1991, pp. 599–608.

[29] Falcone Miller S, Miller BG, Scaroni AW, Britton SA, Clark D, Kinneman WP, et al. Coal-water slurry fuel combustion program. Pennsylvania Electric Power Company, 1993, 98 pages.
[30] Morrison JL, Miller BG, and Scaroni AW. Determining coal slurryability: a UCIG/ Penn state initiative. Electric Power Research Institute, WO3852-06, January 1998.
[31] Ashworth RA, and Sommer TM. Economical use of coal water slurry fuels produced from impounded coal fines. Proceedings of Effects of Coal Quality on Power Plants, Electric Power Institute, 1997.
[32] Falcone Miller S, Morrison JL, and Scaroni AW. The effect of cofiring coal-water slurry fuel formulated from waste coal fines with pulverized coal on NO_x emissions. Proceedings of the 21st International Technical Conference on Coal Utilization & Fuel Systems, Coal & Slurry Technology Association, Washington, D.C., 1996.
[33] Miller BG, Falcone Miller S, Morrison JL, and Scaroni AW. Cofiring coal-water slurry fuel with pulverized coal as a NO_x reduction strategy. Proceedings of the Fourteenth International Pittsburgh Coal Conference, 1997.
[34] Battista JJ. personal communication, 1997.
[35] Tillman DA. NO_x Reduction achieved through biomass cofiring. Proceedings of the 20th Annual International Pittsburgh Coal Conference, September 2003.
[36] Tillman DA, Foster Wheeler NA. personal communication, November 2003.
[37] Tillman DA, Harding NS. Fuels of opportunity: characteristics and uses in combustion systems. London: Elsevier; 2004.
[38] Tillman DA, Miller BG, and Johnson D. Analyzing opportunity fuels for firing in coal-fired boilers. Proceedings of the 20th Annual International Pittsburgh Coal Conference, September 2003.
[39] James R, and Spinney P. The boiler optimization journey. Power Engineering, May 2011.
[40] James R, and Spinney P. Boiler optimization and SCR systems: reducing NO_x, Managing tradeoffs. Power Engineering, July 2008.
[41] Vasquez ER, Gadalla H, McQuistan K, Iman F, and Sears RE. NO_x Control in coal-fired cyclone boilers using smartburn combustion technology. Proceedings of the EPRI-DOE-EPA Combined Power Plant Air Pollution Control MEGA Symposium, 2003.
[42] DOE (United States Department of Energy). Clean coal technology, control of nitrogen oxide emissions: Selective Catalytic Reduction (SCR), Topical Report Number 9. (The United States Department of Energy, Washington, D.C., July 1997).
[43] McIlvaine RW, Weiler H, and Ellison W. SCR operating experience of german power-plant owners as applied to challenging, U.S., high-sulfur service. Proceedings of the EPRI-DOE-EPA Combined Power Plant Air Pollution Control MEGA Symposium, 2003.
[44] Cichanowicz JE, Smith LL, Muzio LJ, and Marchetti J. 100 GW of SCR: installation status and implications of operating performance on compliance strategies. Proceedings of the EPRI-DOE-EPA Combined Power Plant Air Pollution Control MEGA Symposium, 2003.
[45] EPA (United States Environmental Protection Agency). Performance of selective catalytic reduction on coal-fired steam generating units. Washington, DC: Office of Air and Radiation, U.S. Government Printing Office; 1997.
[46] Lodder P, Lefers JB. Effect of natural gas, C_2H_6, and CO on the homogenous gas phase reduction of NO_x by NH_3. Chem Eng J 1985;Volume 30(No. 3):161.
[47] Ciarlante V, Zoccola MA. Conectiv energy successfully using SNCR for NO_x control. Power Engineering 2001;Volume 105(Number 6):61–2.

[48] EPA (United States Environmental Protection Agency). EPA's power sector modeling platform v.5.13. Emission Control Technologies, www.epa.gov/airmarkets/progsregs/epa-ipm; November 2013.
[49] Frederick N, Agrawai RK, Wood SC. NO_x control on a budget: induced flue gas recirculation. Power Engineering 2003;Volume 107(Number 7):28–32.
[50] Wu Z. Air pollution control costs for coal-fired power stations. IEA Coal Research; 2001.
[51] Hoskins B. Uniqueness of SCR retrofits translates into broad cost variations. Power Engineering 2003;Volume 107(Number 5):25–30.
[52] EPA (United States Environmental Protection Agency). Air pollution control technology fact sheet – Selective Catalytic Reduction (SCR). Technology Transfer Network; 2007.
[53] EPA (United States Environmental Protection Agency). Air pollution control technology fact sheet – Selective Non-Catalytic Reduction (SNCR). Technology Transfer Network; 2007.

Mercury emissions reduction

6

6.1 Introduction

Mercury exists in trace amounts in fossil fuels, vegetation, crustal material, and waste products [1]. Mercury vapor can be released to the atmosphere through combustion or natural processes where it can drift for a year or more, spreading over the globe. An estimated 5,000–8,000 metric tons of mercury were emitted or re-emitted globally in 2010 from both natural and anthropogenic sources [2]. Total anthropogenic emissions of mercury to the atmosphere in 2010 are estimated at 1,960 metric tons with coal burning emitting approximately 475 metric tons [3]. Estimates of U.S. mercury emissions in 2008 are 61 short tons, of which coal-fired utility boilers and industrial-commercial-institutional boilers and process heaters are estimated to have emitted 29.5 and 4.5 short tons, respectively [4,5]. Reductions in coal-fired boiler mercury emissions have not been as significant as reductions from other industries, which is one reason mercury legislation has been recently passed (see Chapter 2 for emissions details and regulations). The 29.5 short tons of mercury emitted from utility coal-fired boilers in 2008 were from units greater than 25 MW in size.

Mercury is a liquid silvery metal that is considered toxic and, when in the form of methyl mercury, is extremely toxic [6,7]. The mercury directly emitted from power plants is measured as three forms – elemental (Hg^0), oxidized (Hg^{+2}), and condensed on ash particles (Hg_p). In the natural environment, mercury can go through a series of chemical transformations to convert mercury to a highly toxic form, methylmercury (CH_3Hg) that is concentrated in fish and birds [8]. Methylation rates in the ecosystems are a function of mercury availability, bacterial population, nutrient load, acidity and oxidizing conditions, sediment load, and sedimentation rates. Methylmercury enters the food chain, particularly in aquatic organisms, and bioaccumulates.

Volatile elements that are emitted from power plants, such as mercury, are mostly found either in gaseous form or enriched on the surface of fine particles and physically should be available for uptake by plants [9]. There is evidence, however, that almost no mercury from the soil is taken into the shoots of the plants and hence, this appears to be an important barrier against entry of mercury into the above-ground ecosystem, even if accumulation in the soil has occurred.

As previously mentioned, mercury exits in trace amounts in fossil fuels, including coal, vegetation, crustal material, and waste products. Through combustion or natural processes, mercury vapor can be released to the atmosphere, where it can drift for a year or more, spreading with air currents over vast regions of the world [1]. Research indicates that mercury poses adverse human health effects with fish consumption the primary pathway for human and wildlife exposure. Mercury bioaccumulates in fish as methylmercury (CH_3Hg) and poses a serious health hazard for humans. Other

Fossil Fuel Emissions Control Technologies. DOI: http://dx.doi.org/10.1016/B978-0-12-801566-7.00006-3

research suggests that other forms of mercury may be harmful as well [10,11]. Ingested mercury in elemental, organic, and inorganic form is converted to mercuric mercury, which is slowly eliminated from the kidneys but remains fixed in the brain indefinitely [11]. Exposure to high levels of metallic, inorganic, or organic mercury can permanently damage the brain, kidneys, and developing fetus [10]. Documented associations between low-dose prenatal exposure to methylmercury and neurodevelopmental effects on attention, motor function, language, visual-spatial, and verbal abilities have been made [12].

Loss of sight has been associated with cases of extreme mercury ingestion [13]. Chronic thallium poisoning in the Guizhou Province, China has been reported where vegetables are grown on mercury/thallium-rich mining slag. Most symptoms that have been reported, such as hair loss, are typical of thallium poisoning. However, many patients from this region have lost their vision, which is being attributed to mercury poisoning since the mercury concentration of this coal is 55 ppm or about 200 times the average mercury concentration in United States coals.

There are several technologies under development or being demonstrated that involve removal of mercury from the flue gas. They include sorbent injection, particulate collection systems, catalysts, or chemical additives to promote the oxidation of elemental mercury and facilitate its capture in particulate and sulfur dioxide control systems, and fixed structures in flue gas ducts that adsorb mercury. These are discussed in this chapter, which begins with an overview of mercury in fossil fuels, a discussion of mercury chemistry, and a review of mercury emissions from stationary heat and power systems. The chapter concludes with a discussion of cost estimates to control mercury emissions.

6.2 Mercury in fossil fuels

6.2.1 U.S. coals

Prior to 1999, anthropogenic mercury emissions were unknown. The U.S. EPA conducted an Information Collection Request (ICR) in 1999 to assess mercury emissions from power plants. The data collection was conducted in three phases. In Phase I, information was collected on the fuels, boiler types, and air pollution control devices used at all coal-fired utility boilers in the United States, and emission testing was performed on 84 coal-fired units in Phase III. Prior to the emission testing, coal data were collected and analyzed by the utility industry for 1,140 coal-fired and three integrated gasification combined cycle (IGCC) electric power generating units in Phase II. Each coal sample was analyzed for mercury content, chlorine content, sulfur content, moisture content, ash content, and calorific value.

Over 40,000 fuel samples (of which more than 39,000 were coal samples) were analyzed as part of the ICR and a summary of the ICR coal data, by point of origin for six regions and corresponding coal rank, is provided in Table 6.1 [14]. Appalachian bituminous coal and Western subbituminous coal accounted for approximately 75% of United States coal production in 1999 and over 80% of the mercury entering

Table 6.1 Summary of ICR data on mercury in coal

Coal rank:	Bituminous			Subbituminous	Lignite		Totals
Region:	**Appalachian**	**Interior**	**Western**	**Western**	**Fort union**	**Gulf coast**	
No. of samples	**19,530**	**3,763**	**1,471**	**7,989**	**424**	**623**	
Average ICR coal analysis (dry basis)							
Hg, ppm	0.126	0.09	0.049	0.068	0.09	0.119	
Cl, ppm	948	1,348	215	124	139	221	
S, %	1.7	2.5	0.6	0.5	1.2	1.4	
Ash, %	11.7	10.4	10.5	7.9	13.4	23.6	
Btu/lb	13,275	13,001	12,614	11,971	10,585	9,646	
Other coal related factors							
Ca, ppm, dry basis	2,700	6,100	7,000	14,000	32,000	33,000	
Fe, ppm, dry basis	16,000	23,000	4,200	10,000	12,000	20,000	
Moisture, % as received	2.5	6.6	4.2	19.4	37.3	34.5	
Typical heat rate, Btu/kWh	10,002	10,067	10,047	10,276	10,805	10,769	
Regional coal production for utility use							
Million short tons, as rec'd.	342	67	75	336	23	57	900
Million short tons, dry coal	333	63	72	271	14	37	790
Mercury in coal used by utilities							
Short tons of Hg	42.1	5.4	3.5	18.4	1.3	4.5	75.1
Pounds of Hg/10^{12} Btu	9.5	6.6	3.9	5.7	8.3	12.5	
Pounds of Hg/GWh	0.0951	0.07	0.039	0.0584	0.09	0.134	

Table 6.2 **Mercury in U.S. coals by rank**

Coal type	Number of analyses	lb Hg/trillion Btu (based on HHV)		
		Low	High	Median
Bituminous	27,884	0.04	103.81	8.59
Subbituminous	8,193	0.39	71.08	5.74
Lignite	1,047	0.93	75.08	10.54

coal-fired power plants. The composition of these coals is quite different which can affect their mercury emissions. Appalachian coals typically have high mercury, chlorine, and sulfur contents and low calcium content, resulting in a high percentage of oxidized mercury (i.e. Hg^{2+}) whereas Western subbituminous coals typically have low concentrations of mercury, chlorine, and sulfur contents and high calcium content, resulting in a high percentage of elemental mercury (i.e. Hg^0).

The average mercury content data in the ICR database for the different coal regions ranged from approximately 4 to 12.5 lb/trillion Btu (based on higher heating value; HHV) with most of the mercury contents varying from approximately 6 to 10 lb Hg/trillion Btu. While there are modest differences between the ranks of U.S. coals, the variation in mercury content among coals within a given rank is much larger. This is illustrated in Table 6.2 [15,16].

The United States Geological Society (USGS) compiled a nationwide coal information database of over 7,000 coal samples for different coal-producing regions of the U.S. [17]. These results are provided in Table 6.3 and list mercury concentration and calorific value for a more detailed regional breakdown than that reported in Table 6.1. On a lb Hg/trillion Btu basis (see Figure 6.1), these data indicate that the Gulf Coast lignite may have the highest potential for mercury emissions, and the Green River coal from western Wyoming may have the lowest mercury emissions on an equivalent Btu basis [17]. Of the two major bituminous coal-producing regions, samples from the Appalachian region contain higher mercury levels than those from the Eastern Interior. Samples from the Powder River Basin are slightly higher in mercury levels than the subbituminous coals of the San Juan River Basin.

6.2.2 Petroleum, natural gas liquids, and refined products

A number of mercury species have been identified in crude oil and natural gas condensates [18–20]. These include:

- Elemental mercury (Hg^0) – Elemental mercury is soluble in crude oil and hydrocarbon liquids in atomic form to a few ppm.
- Dissolved organic mercury, which can be present as dialkylmercury (R-Hg-R, where R is an organic group like methyl ($-CH_3$) or R-Hg-X, where R is an organic group and X is an inorganic ion like Cl^-) – Dissolved organic mercury compounds are highly soluble in crude oil and gas condensates.

Table 6.3 Median and mean values for mercury concentrations and calorific values on an as-received, whole coal basis for selected coal-producing regions in the U.S.

Coal-producing region	Mercury concentration (ppm)			Calorific value (Btu/lb)		
	Median	Mean	No. of samples	Median	Mean	No. of samples
Appalachian, northern	0.19	0.24	1,613	12,570	12,440	1,506
Appalachian, central	0.10	0.15	1,747	13,360	13,210	1,648
Appalachian, southern	0.18	0.21	975	12,850	12,760	969
Eastern Interior	0.07	0.10	289	11,150	11,450	255
Fort Union	0.08	0.10	300	6,280	6,360	277
Green River	0.06	0.09	388	9,940	9,560	264
Gulf Coast	0.13	0.16	141	6,440	6,470	110
Pennsylvania, anthracite	0.10	0.10	51	12,860	12,520	39
Powder River	0.06	0.08	612	8,050	8,090	489
Raton Mesa	0.05	0.09	40	12,500	12,300	34
San Juan River	0.04	0.08	192	9,340	9,610	173
Uinta	0.04	0.07	253	11,280	10,810	226
Western Interior	0.14	0.18	286	11,320	11,420	261
Wind River	0.08	0.15	42	9,580	9,560	42

- Inorganic (ionic) mercury salts ($Hg^{2+}X$ or $Hg^{2+}X_2$, where X is an organic ion) – Mercury salts, mostly halides, are soluble in oil and gas condensate but preferentially partition to the water phase in primary separations.
- Complexed mercury (HgK or HgK_2) – Mercury can exist in hydrocarbons as a complex, where K is a ligand such as an organic acid, porphyrin, or thiol.
- Suspended mercury compounds – The most common examples are mercuric sulfide (HgS) and selenide (HgSe), which are insoluble in water and oil but may be present as suspended solid particles of very small particle size.

Another category of mercury association in crude oil is asphaltenes mercury. This is mercury that can be present in a combination of forms. Of these mercury species, elemental mercury is typically the major mercury component, dialkylmercury is typically a trace component, mercury associated with asphaltenes can be a major component in some oils, and complexed ionic mercury species are present in some oils as a minor component [18,20].

Organo-mercury compounds are easily broken down when heated in a reducing atmosphere and elemental mercury is released during the hydrotreating and catalytic reforming stages of a refinery. This has the result that much of the mercury in the crude oil feed to the refinery ends up as elemental mercury in the C_3 to C_6

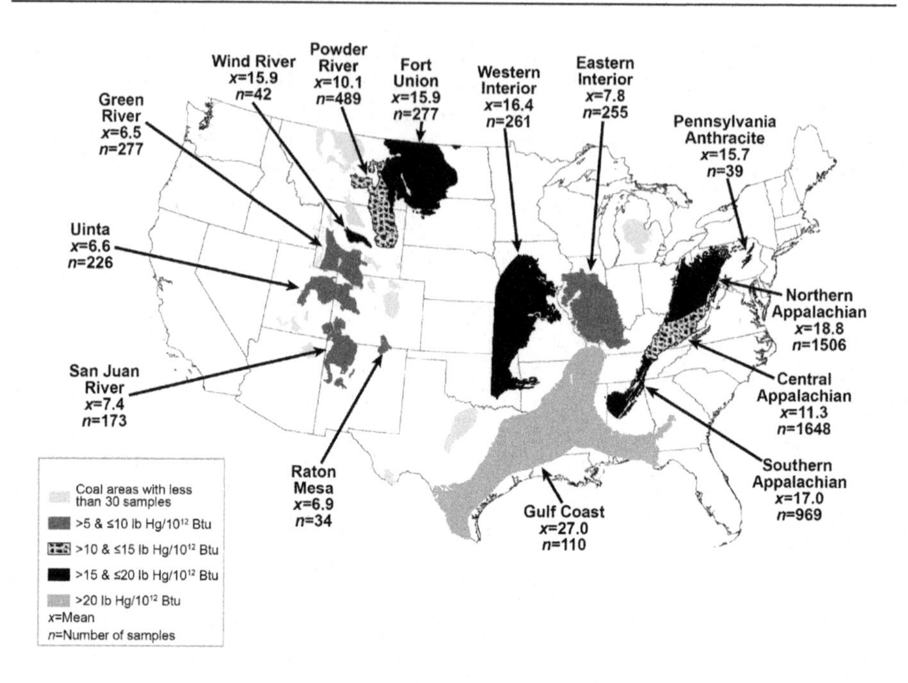

Figure 6.1 Mercury concentrations (lb Hg/trillion Btu) of in-ground coal for selected U.S. coal-producing regions.

product streams, namely liquefied petroleum gas (LPG), gasoline, and naphtha [21]. Studies that have been made on the distribution of mercury in gas processing plants show that approximately half of the mercury present in the raw gas ends up in the sales gas and natural gas liquids (NGLs) and the rest is removed in the acid gas removal plant and the driers [21].

Numerous studies have reported mercury concentrations in crude oils, NGLs, and refinery products and the results vary significantly. Some of these results are reported in this section showing the variability of the concentrations.

The International Petroleum Industry Environmental Conservation Association (IPIECA) developed a database of mercury concentrations in 446 global crude oils and condensates [19]. These results are summarized in Table 6.4, which shows the range of mercury levels and the percent of samples that fell into each range. Table 6.5 shows the regional breakdowns of the 446 assays.

Monthly crude oil imports into the U.S., as of August 2014 totaled approximately 9.5 million barrels per day (bpd) or about 52% of the total 18.1 million bpd of crude oil consumed in the U.S. [22]. The top eight countries that exported oil to the U.S., which was approximately 7 million bpd, are listed in Table 6.6 along with average mercury concentrations by country [18,23]. Canada provided nearly one-third of the U.S.'s imported crude oil, of which nearly half of the total crude oil was crude oil derived from oils sands with the balance conventional crude oil [24]. Alberta Research Council reported that the average mercury content of 88 samples of Alberta crude oil sampled near the wellhead ranged from 1.6 to 399 ppb with an

Table 6.4 Range of mercury levels in global crude oils and condensates

Concentration range (ppb)	Number of samples	Percent
≤2	284	64
2–5	68	15
5–15	42	10
15–50	33	7
50–100	6	1
>100	13	3
Total	446	100

average concentration of 50.9 ppb [25]. In contrast, the Petroleum Alliance of Canada (PTAC) reported that the average mercury content of Western Canadian crude was 1.6 ppb. In Alberta, diluted bitumen and synthetic crude feeds to a refinery contained 5 to 11 ppb mercury and 0.4 to 1.6 ppb mercury, respectively.

Concentrations of mercury in crude oil produced in the United States also vary significantly. Various sources report average levels of total mercury in U.S. crude oils – 78 to 30,000 ppb from various locations in California [20–22]; 23 ppb in Louisiana [21,22]; 77 ppb in Wyoming [21,22]; and 4.3 ppb for a U.S. average [18].

Similarly, mercury levels in refined liquid products vary significantly as shown in Table 6.7. Ranges and average mercury concentrations for a variety of refined products have been compiled from several sources and reported in Table 6.7 [26–30]. As a refresher, Figure 1.22, which depicts the various fractions from a distillation tower, has been modified with the addition of the mercury species that have been measured in different product fractions and the result is shown in Figure 6.2 [31].

6.2.3 Natural gas

Mercury has been detected in natural gas on all five continents and in the Pacific [32]. The concentration of mercury in natural gas varies widely from 450–5,000 $\mu g/Nm^3$ in some fields in Germany to less than 0.01 $\mu g/Nm^3$ in some parts of the U.S. and Africa [21]. To compare these concentrations to previously reported concentrations in parts per billion (ppb), note that 1 $\mu g/Nm^3$ is equal to 1.2 parts per billion by weight for natural gas with a molecular weight of 20 [32]. Table 6.8 lists mercury levels measured by UOP LLC (formerly knowns as Universal Oil Products) in 13 natural gas plants, most of which are in the United States [32]. Mercury must be removed below detectable levels, or 0.1 to avoid mercury-caused damage in equipment, hence, mercury concentrations in natural gas are very low when used in combustion systems. PTAC reports that mercury in North American natural gas ranges from 0.005 to 0.04 $\mu g/Nm^3$ while most Alberta Canada natural gas contains less than 0.08 $\mu g/Nm^3$ [25].

Table 6.5 **Regional breakdown of mercury in crude oil and condensates**

Region	No. of samples	Median Hg level (ppb)	Percentage of crude oils and condensates in specific ranges of mercury levels (ppb Hg)					
			<2 ppb	2–5 ppb	5–15 ppb	15–50 ppb	50–100 ppb	>100 ppb
Africa	90	1.0	72	15	9	3	1	—
Eurasia	95	1.2	74	10	6	4	1	5
Middle East	34	1.0	79	18	3	—[a]	—	—
North America	95	1.2	64	21	9	6	—	—
Pacific and Indian Ocean	93	3.0	41	13	16	18	4	8
South America	39	1.4	69	12	8	8	—	3
Condensates	51	2.4	48	14	14	12	8	4
Non-condensates	395	1.3	65	15	9	7	1	3

[a]Not detected.

Table 6.6 Mercury in crude oil imported to the U.S. by country (thousand barrels per day as of August 2014)

Country	Thousand bpd	Mercury concentration (ppb)
Canada	2,961	2.1
Saudi Arabia	840	0.9
Venezuela	800	Not Reported
Iraq	714	0.7
Mexico	553	1.3
Kuwait	520	0.8
Ecuador	293	1.8
Colombia	247	3.4
Total	6,958	–

Table 6.7 Mercury concentrations in refined petroleum products

Product type	Range (ppb)	Average (ppb)	Reference
LPG	NR[a]	<10	[21]
Gasoline	0.22–1.43	0.7	[26]
	0.72–3.2	1.5	[26]
Naphtha	3–40	15	[27]
	8–60	40	[28]
Kerosene	NR	0.4	[26]
Diesel	NR	0.4	[26]
	NR	2.97	[26]
Heating oil	NR	0.59	[26]
Light distillates	NR	1.32	[29]
Utility fuel oil	NR	0.67	[29]
	NR	0.59	[29]
	60–120	90	[20]
	NR	1	[29]
	NR	10	[30]
Residual fuel oil #6	2–6	4	[30]
Distillate fuel oil #2	NR	<120	[29]
Asphalt	NR	0.27	[29]
Petroleum coke	0–250	50	[28]

[a]Not reported.

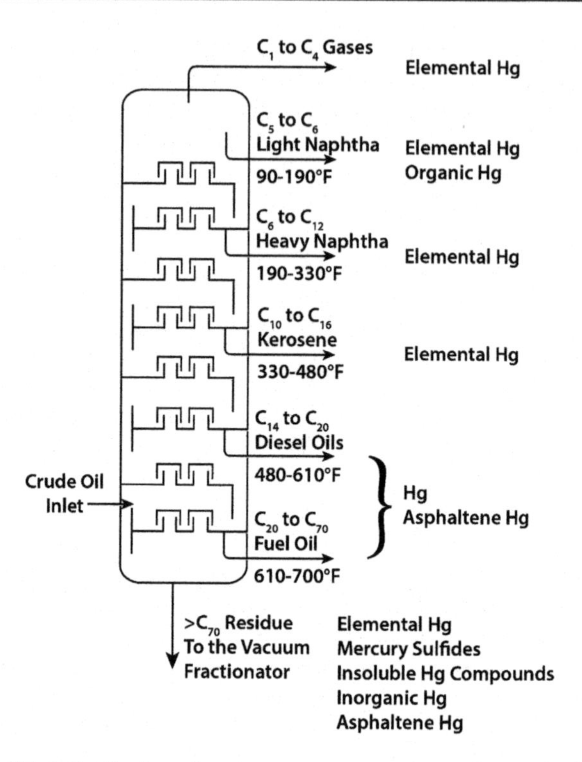

Figure 6.2 Simplified distribution of mercury components in a refinery distillation column.

Table 6.8 Mercury levels in natural gas for selected plants

Plant location	Mercury level (μg/Nm^3)
U.S.	5.0
U.S.	<0.5
U.S.	13
U.S.	0.6
U.S.	8
U.S.	0.6
U.S.	2.1
U.S.	5.5
U.S.	5.1
U.S.	120
U.S.	0.5
Far East	50
Africa	<0.5

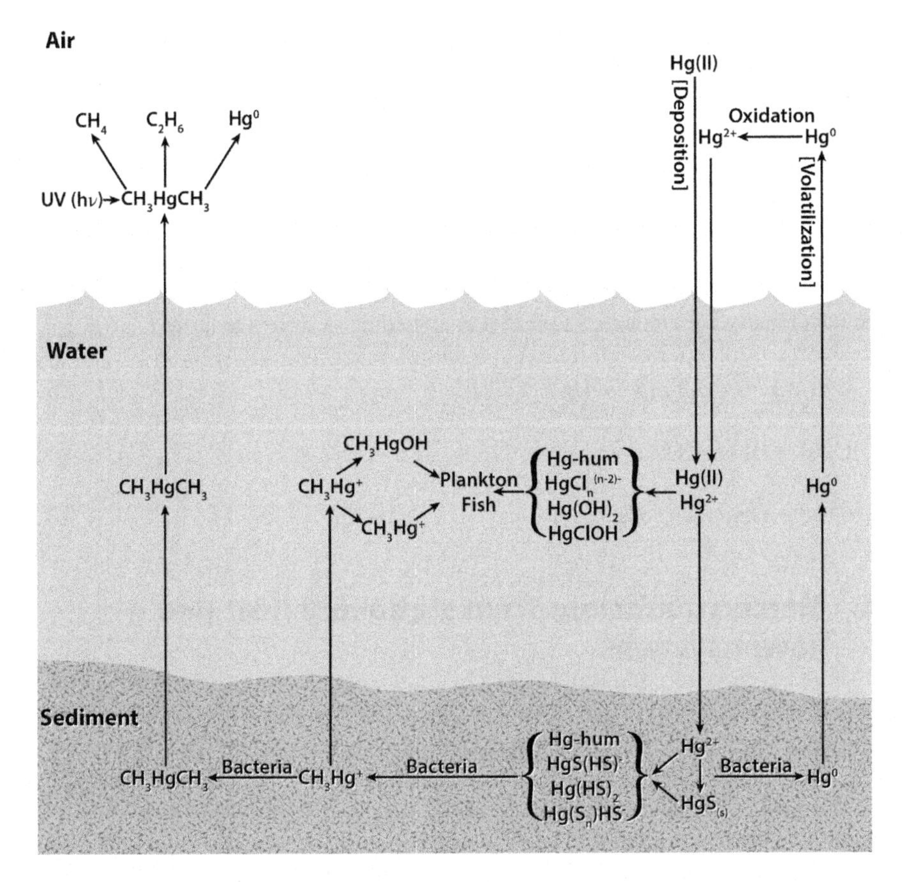

Figure 6.3 Mercury cycling.

6.3 Mercury chemistry

Metallic mercury is quite stable under ambient conditions; however, in the soil and water, biochemical reactions create organic mercury compounds such as salts of monomethylmercury (CH_3Hg^+) that can be transported across membranes and stored in fat tissues [33]. Figure 6.3 summarizes the formation and fate of mercury compounds in an aqueous environment [modified from 33–35]. If mercury from anthropogenic sources or major natural sources, like a volcano eruption, enter aqueous systems, natural biological processes convert mercury into the neurotoxin dimethylmercury, $(CH_3)_2Hg$ [33]. When dimethylmercury is produced faster than natural organisms can degrade it, it accumulates in the flesh of fish and shellfish.

Thermodynamically the divalent mercury in surface waters, Hg(II) is not present as the free ion Hg^{2+} but should be complexed in variable amounts to hydroxide ions ($Hg(OH)^+$, $Hg(OH)_2$, $Hg(OH)_3^-$) and to chloride ions ($HgCl^+$, $HgClOH$, $HgCl_2$, $HgCl_3^-$,

$HgCl_4^{2-}$) depending on the pH and the chloride concentration [34]. It is also possible that some of the Hg(II) might be bound to sulfides (S^{2-} and HS^-).

Transformation of inorganic mercury compounds is catalyzed by microorganisms. Aerobes oxidize materials such as HgS to form sulfites and sulfates that form water soluble Hg^{2+}. Enzymes present in some bacteria reduce Hg^{2+} to elemental mercury, Hg^0. Elemental mercury has a high vapor pressure and can be transferred from sediment and the aqueous environment into the atmosphere. The water soluble Hg^{2+} is converted into mono- and dimethylmercury. Dimethylmercury is volatile and is photolyzed by UV in the atmosphere to yield Hg^0 plus the methyl free radicals $(CH_3)^{\bullet}$, which eventually are converted to methane and ethane via the reactions:

$$UV(h\nu) + (CH_3)_2Hg \rightarrow Hg^0 + 2CH_3^{\bullet} \tag{6.1}$$

$$CH_3^{\bullet} + H^{\bullet} \rightarrow CH_4 \tag{6.2}$$

$$CH_3^{\bullet} + CH_3^{\bullet} \rightarrow C_2H_6 \tag{6.3}$$

6.4 Mercury emissions from stationary heat and power systems

As previously discussed, Title III of the Clean Air Act (CAA) regulates stationary sources that emit hazardous air pollutants (HAPs). Section 112 in Title III was extensively amended in 1990. Under the amended CAA Section 112(b), Congress listed specific chemicals, compounds, and groups of chemicals as HAPs. Mercury is one of the chemicals included on this list. The EPA is directed by Section 112 to regulate HAP emissions from stationary sources by establishing national emission standards for hazardous air pollutants [16]. The 1990 CAA amendments to Section 112 also directed EPA to perform additional studies that include analyses of mercury emissions from electric steam generating units. One result of these studies was that in EPA's 1998 report to Congress on HAP emissions from electric utility steam generating units, EPA identified additional information needed to gain a better understanding of the risks, impacts, and control of mercury emissions from coal-fired steam generating units [16]. As part of EPA's effort to gather this information, EPA conducted an information collection project beginning in late 1998 to survey all coal-fired steam generating units meeting the CAA Section 112(a) definition that were operating in the United States. This information collection provided EPA with data on mercury content and amount of coal burned by these units during the 1999 calendar year. In addition, EPA selected a subset of the coal-fired electric utility steam generating units, representing the various boiler types, where field testing was performed to obtain mercury emission data for various air pollution control devices. The results from this comprehensive data collection are presented in this section. In addition, results from the Electric Power Research Institute's (EPRI's) updated HAPs emissions estimates published in 2009 are also discussed as are the sources of the largest mercury emitters from coal-fired power plants. Mercury emissions from industrial boilers and oil- and natural gas-fired stationary heat and power systems are presented in this section.

Table 6.9 Distribution of electric utility units by coal-firing configuration in the U.S. for the year 1999 as reported in the ICR data

Coal-firing configuration	Total number of units	Percent of total	Percent of electricity generating capacity
Pulverized coal-fired furnace	979	85.6	90.1
Cyclone furnace	87	7.6	7.6
Fluidized-bed combustor	42	3.7	1.3
Stoker-fired furnace	32	2.8	1.0
Integrated gasification combined cycle	3	0.3	<0.1
TOTAL	1,143	100	100

6.4.1 *Emissions from existing control technologies from coal-fired power plants*

6.4.1.1 *Electric generating units*

6.4.1.1.1 Data from the 1999 mercury information collect request

The nationwide distribution of the 1,143 coal-fired electric utility boilers listed in the EPA Information Collection Request (ICR) data by coal-firing configuration is shown in Table 6.9 [36]. Pulverized coal-fired designs account for the majority of the coal-fired electric utility boilers both in terms of total number of units and U.S. generating capacity comprising approximately 86% of the total units and 90% of the electric generating capacity. Cyclone furnace design comprised about 8% of the total units as well as 8% of the total U.S. electric generating capacity. Fluidized-bed combustors, stoker-fired furnaces, and integrated gasification combined cycle units combined comprise approximately 6% of the total units and 2% of the total electric generating capacity.

Of the 1,143 boiler units in operation in 1999, EPA selected a subset of 81 units for field testing of mercury emissions. A summary of these 81 boiler and coal type configurations is given in Table 6.10 [36]. Of these boiler units, 65 were pulverized coal-fired boilers, since they represent the majority of the coal-fired electric utility boilers in the U.S. as shown in Table 6.9.

A summary of the flue gas cleaning devices installed on the pulverized coal-fired boilers is given in Table 6.11 as a function of fuel burned in each unit in 1999 [36]. These data show that:

- A total of 28 boilers were equipped with a cold-side electrostatic precipitator (CS-ESP) (14 units), hot-side electrostatic precipitator (HS-ESP) (8 units), or a fabric filter (FF) (6 units).
- The 11 dry flue gas desulfurization (FGD) units were equipped with either a spray dryer absorber (SDA)/ESP (3 units) or a SDA/FF (8 units).
- The 20 wet FGD units were equipped with a particle scrubber (PS) + wet FGD (6 units), CS-ESP + wet FGD (6 units), HS-ESP + wet FGD (6 units), or FF + wet FGD (2 units).
- Two units were equipped with a CS-ESP + FF.
- One unit was equipped with a PS.

Table 6.10 Distribution of ICR mercury emission stack testing by boiler/coal type configurations

Boiler unit type	Number of boiler units tested				
	Fuel burned in boiler unit				Total number of units tested
	Bituminous coal	Subbituminous coal	Lignite	Other[a]	
Pulverized coal-fired	26	29	9	1	65
Cyclone-fired	3	2	2	0	7
Fluidized-bed fired	1	0	2	2	5
Stoker fired	2	0	0	0	2
IGCC[b]	2	0	0	0	2
Total number of units	34	31	13	3	81

[a]Some units used coal wastes or a blend of fuels.
[b]Integrated gasification combined cycle.

The analysis results of more than 39,000 coal samples were reported as part of the ICR data [36]. These results include the mercury content of as-fired coals and supplemental fuels burned in electric utility boilers in 1999. The mercury content of the bituminous coal, subbituminous coal, and lignite (the three most commonly used fuels) samples was generally less than 15 lb/10^{12} Btu. Statistical information on each type of fuel burned in coal-fired utility boilers in 1999 is presented in Table 6.12 [36] and with regional results shown in Table 6.1 [14].

Estimates for mercury emissions from coal-fired power plants with various control technologies, based on the 1999 ICR data, are given in Tables 6.13 and 6.14 [14,36]. The results shown in Table 6.13 are from data from all of the boilers tested while the results in Table 6.14 are from pulverized coal-fired units. The data in Table 6.13 show that mercury emissions were estimated at $\approx$49 short tons in 1999. Because this estimate is based on a small subset out of more than 1,100 units in the United States and, since there are questions on bias based on the number of samples from Eastern versus Western coal-fired boilers, various estimates of mercury emissions ranged from 40 to 52 short tons/year [14]. The ICR data indicate that the speciation of mercury exiting the stack of the boilers is primarily gas-phase oxidized (i.e. 43%) or elemental (i.e. 54%) mercury with some particulate-bound (i.e. 3%) mercury present [1]. Variations in mercury speciation are further illustrated in Table 6.15 [36].

Tables 6.13 and 6.14 provide information on the influence of various existing air pollution control devices (APCDs) on mercury removal; however, mercury capture across the APCDs can vary significantly based on coal properties, fly ash properties, including unburned carbon, specific APCD configurations, and other factors [1]. Mercury removals across cold-side ESPs averaged 27%, compared to 4% for hot side ESPs [14]. Removals for fabric filters were higher, averaging 58%, owing to

Table 6.11 Distribution of ICR mercury emission test data for pulverized coal-fired boilers by post-combustion emission control device configuration

Post-combustion control strategy	Post-combustion emissions control device configuration[a]	Number of boiler units tested				
		Fuels burned in boiler unit				Total
		Bituminous coal	Subbituminous coal	Lignite	Other	
Particulate matter control	CS-ESP	7	5	1	1	14
	HS-ESP	4	4	0	0	8
	FF	4	2	0	0	6
	CS-ESP + FF	0	0	2	0	2
	PS	0	1	0	0	1
Particulate matter control and dry SO_2 scrubber system	SDA + CS-ESP	0	3	0	0	3
	SDA + FF	3	3	2	0	8
	DI + CS-ESP	1	0	0	0	1
Particulate matter control and wet SO_2 scrubber system	PS + wet FGD	1	4	1	0	6
	CS-ESP + wet FGD	1	3	2	0	6
	HS-ESP + wet FGD	1	5	0	0	6
	FF + wet FGD	2	0	0	0	2
Other control device configuration		2	0	0	0	2
Number of units tested:		27	29	8	1	65

[a]Abbreviations:
Particulate Matter Controls
CS-ESP = cold-side electrostatic precipitator
HS-ESP = hot-side electrostatic precipitator
FF = fabric filter
PS = particle scrubber

SO_2 Controls
DI = dry injection
FGD = flue gas desulfurization system
SDA = spray dryer absorber system

Table 6.12 Comparison of mercury content normalized by heating value in as-fired coals and supplemental fuels for electric utility boilers in 1999

Fuel type	Number of analyses	Mercury content (lb Hg per 10^{12} Btu)			
		Range	Mean	Median	Standard deviation
Anthracite	114	5.02–35.19	15.25	13.37	6.23
Bituminous coal	27,884	0.04–103.81	8.59	7.05	6.69
South American bituminous coal	270	0.70–66.81	5.94	4.91	5.28
Subbituminous coal	8,193	0.39–71.08	5.74	5.00	3.59
Indonesian subbituminous coal	78	0.79–4.61	2.51	2.39	0.86
Lignite	1,047	0.93–75.06	10.54	7.94	9.05
Waste anthracite refuse	377	2.49–73.02	29.31	27.77	11.94
Waste bituminous coal	575	2.47–172.92	60.50	53.32	44.35
Waste subbituminous coal	53	5.81–30.35	11.42	10.79	4.66
Petroleum coke	1,149	0.06–32.16	23.18	2.16	3.18
Tire-Derived fuel	149	0.38–19.89	3.58	2.79	2.78

additional gas-solid contact time for oxidation. Both wet and dry FGD systems removed 80 to 90% of the gaseous oxidized mercury, but elemental mercury was not affected. High mercury removals (i.e. 86%) in fluidized-bed combustors with fabric filters were attributed to mercury capture on high carbon content fly ash.

Differences in mercury emissions as a function of coal rank are one of the most significant findings of the ICR and subsequent DOE testing, as can be observed in Table 6.16 [1]. Table 6.16 lists the ranges (note that values less than zero in the ICR data, due to mercury measurement limitations, have been changed to zero for averaging purposes) and average co-benefit mercury capture by coal rank and APCD configuration [85]. Specifically, it is the fact that units burning subbituminous coal and lignite frequently demonstrate worse mercury capture than similarly equipped bituminous coal-fired plants. An exception to this is the case where a subbituminous coal is fired in a boiler containing an SCR and cold-side ESP. The SCR is oxidizing the elemental mercury, which is enhancing the ESP's effectiveness in capturing the mercury. These data also demonstrate the improved mercury capture effectiveness of wet FGD systems when an upstream SCR system is in service. For example, average mercury capture for bituminous coal-fired plants equipped with a cold-side ESP and wet FGD increased from 69 to 86% with the addition of an

Table 6.13 Estimated mercury removal by various control technologies

Control technology	Short tons of mercury entering	No. of U.S. power plants	Number of ICR part III test sites	Estimated mercury removals (%)[a]	Hg emission calculation, EPRI ICR (short tons)
ESP cold	39.4	674	18	27	28.8
ESP cold + FGD wet	16.8	117	11	49	8.6
ESP hot	5.5	120	9	4	5.3
Fabric filter	2.9	58	9	58	1.2
Venturi particulate scrubber	2.2	32	9	18	1.8
Spray dryer + fabric filter	1.6	47	10	38	1
ESP hot + FGD wet	1.6	20	6	26	1.2
Fabric filter + FGD wet	1.5	14	2	88	0.2
Spray dryer + ESP cold	0.3	5	3	18	0.2
FBC + fabric filter	3.4	39	5	86	0.5
Integrated gasification combined cycle	0.07	2	2	4	0.1
FBC + ESP cold	0.02	1	1	–	0.1
Totals	75.3	1,128	84		48.8

[a]Removals as percentage of mercury in coal calculated by EPRI.

SCR. FGD systems are more efficient at mercury removal with upstream SCR systems; however, FGD systems also remove significant mercury quantities without an SCR system. The data indicate that for pulverized coal-fired units, the greatest co-benefit for mercury control is obtained for bituminous coal-fired units equipped with a fabric filter for particulate matter control and either a wet FGD or spray dryer absorber for sulfur dioxide control. In these cases, average mercury capture of 98% was observed for both cases. The worst performing pulverized bituminous coal-fired units were those equipped only with a hot-side ESP [1].

The rank-dependency on mercury removal is due to the speciation of the mercury in the flue gas, which can vary significantly between power plants depending on coal properties. Power plants that burn bituminous coal typically have higher levels of oxidized mercury than power plants that burn subbituminous coal or lignite, which is attributed to the higher chlorine and sulfur content of the bituminous coal. The oxidized mercury, as well as the particulate mercury, can be effectively captured in some conventional control devices such as an ESP, fabric filter, or FGD system while elemental mercury is not as readily captured. The oxidized mercury

Table 6.14 Average mercury capture by existing post-combustion control configurations used for pulverized coal-fired boilers

Post-combustion control strategy	Post-combustion emission control device configuration	Average mercury capture		
		Bituminous coal	subbituminous coal	Lignite
PM control only	CS-ESP	36	3	−4
	HS-ESP	9	6	Not tested
	FF	90	72	Not tested
	PS	Not tested	9	Not tested
PM control and spray dryer absorber	SDA + ESP	Not tested	35	Not tested
	SDA + FF	98	24	0
	SDA + FF + SCR	98	Not tested	Not tested
PM control and wet FGD SO_2 control	PS + wet FGD	12	−8	33
	CS-ESP + wet FGD	74	29	44
	HS-ESP + wet FGD	50	29	Not tested
	FF + wet FGD	98	Not tested	Not tested

Notes:
PM = particulate matter; PS = particle scrubber;
CS-ESP = cold-side electrostatic precipitator; SDA = spray dryer absorber;
HS-ESP = hot-side electrostatic precipitator; FGD = flue gas desulfurization;
FF = fabric filter.

can be more readily adsorbed onto fly ash particles and collected with the ash in either an ESP or fabric filter. Also, because the most likely form of oxidized mercury present in the flue gas, mercuric chloride ($HgCl_2$), is water soluble, it is more readily absorbed in the scrubbing slurry of plants equipped with wet FGD systems compared to elemental mercury, which is not water soluble [1].

6.4.1.1.2 Updated emissions estimates

The Electric Power Research Institute (EPRI) performed an assessment of hazardous air pollutant (HAP) emissions (including mercury) and the consequent human health risks by inhalation for all U.S. coal-fired electric generation units [36]. EPRI performed this study because there had not been a comprehensive study of the emissions of and risks due to HAPs from coal-fired electric utilities since the mid-1990s and those studies used data from 1990 and earlier. In the mid-1990s, EPRI started conducting various data analyses and emissions correlations and worked extensively with the 1999 ICR dataset for mercury and chlorine and examined mercury emissions in more detail. Since that time, additional full-scale emissions information has been generated through various test programs sponsored by organizations such as EPRI, the U.S. Department of Energy, and individual utilities. There were approximately 150 coal-fired units in which baseline mercury emissions data were collected. These data were collected during periods in which no mercury

Table 6.15 **Effects of coal and control technology inlet and outlet speciation factor and capture for pulverized coal-fired boilers**

Coal/Control device	Inlet			Outlet			% Reduction total Hg
	SPF_p	SPF^{2+}	SPF^0	SPF_p	SPF^{2+}	SPF^0	
Bituminous coal							
CS-ESP	0.35	0.58	0.07	0.02	0.78	0.20	36
SNCR + CS-ESP	0.92	0.03	0.05	0.20	0.35	0.45	91
HS-ESP	0.09	0.53	0.37	0.04	0.59	0.37	9
FF	0.92	0.04	0.04	0.01	0.52	0.47	90
SDA + FF	0.59	0.28	0.15	0.01	0.22	0.77	98
SCR + SDA + FF	0.82	0.17	0.01	0.05	0.46	0.48	98
Subbituminous coal							
CS-ESP	0.05	0.25	0.70	0.00	0.31	0.69	3
HS-ESP	0.02	0.15	0.83	0.00	0.17	0.83	6
FF	0.33	0.23	0.44	0.01	0.87	0.12	72
SDA + ESP	0.13	0.26	0.61	0.00	0.05	0.94	35
SDA + FF	0.01	0.06	0.84	0.01	0.05	0.94	24
North Dakota lignite							
CS-ESP	0.01	0.01	0.98	0.00	0.04	0.96	4
SDA + FF	0.03	0.04	0.93	0.00	0.03	0.947	0

(Continued)

Table 6.15 (Continued)

Coal/Control device	Inlet			Outlet			% Reduction total Hg
	SPF_p	SPF^{2+}	SPF^0	SPF_p	SPF^{2+}	SPF^0	
Texas lignite							
CS-ESP + FF	0.09	0.31	0.60	0.00	0.70	0.30	NA
CS-ESP + wet FGD	0.00	0.52	0.47	0.01	0.14	0.85	44

Notes:

Speciation Factors – inlet sampling trainis

$SPF_p = Hg_p$ (inlet)/Hg_T (inlet)
$SPF^{2+} = Hg^{2+}$ (inlet)/Hg_T (inlet)
$SPF^0 = Hg^0$ (inlet)/Hg_T (inlet)

Speciation Factors – outlet sampling train

$SPF_p = Hg_p$ (outlet)/Hg_T (outlet)
$SPF^{2+} = Hg^{2+}$ (outlet)/Hg_T (outlet)
$SPF^0 = Hg^0$ (outlet)/Hg_T (outlet)

CS-ESP = cold-side electrostatic precipitator
SNCR = selective non-catalytic reduction
HS-ESP = hot-side electrostatic precipitator
FF = fabric filter
SDA = spray dryer absorber
SCR = selective catalytic reactor
FGD = flue gas desulfurization

Table 6.16 Average co-benefit mercury capture by coal rank and APCD configurations

APCD configuration	Average percentage mercury capture (Average and range of mercury capture, %)				
	Bituminous coal		Subbituminous coal		Lignite
	w/o SCR	w/ SCR	w/o SCR	w/ SCR	w/o SCR
CS-ESP	28 (0–92)	8 (0–18)	13 (0–61)	69 (58–79)	8 (0–18)
CS-ESP + Wet FGD	69 (41–91)	85 (70–97)	29 (2–60)	N/A	44 (21–56)
HS-ESP	15 (0–43)	N/A	7 (0–27)	N/A	N/A
HS-ESP + Wet FGD	49 (38–59)	N/A	29 (0–49)	N/A	N/A
FF	90 (84–93)	N/A	72 (53–87)	N/A	N/A
FF + Wet FGD	98 (97–99)	N/A	N/A	N/A	N/A
SDA + FF	98 (97–99)	95 (89–99)	19 (0–47)	N/A	4 (0–8)
SDA + CS-ESP	N/A	N/A	38 (0–63)	N/A	N/A
PS	N/A	N/A	9 (5–14)	N/A	N/A
PS + Wet FGD	32 (7–58)	91 (88–93)	10 (0–74)	N/A	33 (9–51)

Note: CS-ESP, cold-side ESP; HS-ESP, hot-side ESP; PS, particulate scrubber; SDA, spray dryer absorber; FF, fabric filter; SCR, selective catalytic reduction; FGD, flue gas desulfurization.

controls were used. In 2008, EPRI started a project to identify and compile these new sources of data and subsequently use these data to supplement and update work previously conducted by EPRI. One objective of this program was to prepare revised emission estimates for the current fleet of coal-fired electric generating units for 2007, which was the base year selected for the study.

A detailed discussion of the methodology used and emission estimates generated are provided in the final report published by EPRI in 2009 [36]. Using unit configuration and operational characteristics, plant measurements, and fuel analyses, a procedure was developed for estimating power plant emissions of HAPs species. This procedure integrated information from plant databases, data on trace materials in utility fuels, and emission estimating correlations and factors derived from updated field test datasets. Estimates were developed for total mercury as well as for speciated mercury – elemental, oxidized, and particulate bound. In summary, the calculation of emission estimates for coal-fired units involved the following steps:

- The unit characteristic databases were used to determine the operations of the individual units for the 2007 base year.
- Blended coal characteristics were assigned to coal-fired units based on coal composition research and 2007 fuel purchase records.
- Stack particulate emission rates for each coal-fired unit were defined.
- Initial trace substance emissions for coal-fired units were calculated.
- Input parameters and plant characteristics information were reviewed/updated, and trace substance emission estimates were finalized for all units >25 MW that sell power to the grid.

Table 6.17 summarizes estimated normalized emission rates for mercury for all coal-fired units within a specified control class [36]. Speciated forms of mercury are presented as a percentage of total mercury emissions in each control class (35 control

Table 6.17 **Summary of unit-level mercury emissions by control technology class**

Control class[a]	MW	No. of units	T Btu input	Coal Hg (lb/yr)	Stack Hg (lb/yr)	Percent reduction[b]	Stack Hg (lb/TBtu)[c]	Percent elemental	Percent oxidized	Percent particulate
ESPc	76,970	367	4,414	34,367	25,775	25	5.8	44.2	52.3	3.5
ESPc con	30,254	113	1,930	14,805	7,402	50	3.8	45.8	50.2	4.0
ESPc ACI	109	1	5	29	3	90	0.6	54.0	42.5	3.5
ESPc FGDd	1,512	4	112	683	649	5	5.8	94.0	5.6	0.4
ESPc FGDw	34,259	76	2,416	23,926	10,006	58	4.1	87.8	11.6	0.7
ESPh	13,434	77	794	5,458	5,294	3	6.7	62.6	34.9	2.5
ESPh FGDw	6,925	16	504	2,740	2,572	6	5.1	91.0	6.4	2.6
FF	18,012	126	1,216	13,548	5,731	58	4.7	23.0	76.2	0.8
FF ACI	270	3	20	92	9	90	0.5	23.0	76.2	0.8
FF FBC	1,235	14	57	462	65	86	1.1	56.0	42.0	2.0
FF FGDd	7,583	33	543	3,352	2,342	30	4.3	92.9	4.3	2.8
FF FGDw	6,422	13	476	2,295	321	86	0.7	74.0	21.0	5.0
VS FGDw	5,922	17	443	2,660	2,075	22	4.7	94.0	5.0	1.0
SCR ESPc	23,103	51	1,337	9,557	5,409	43	4.0	20.9	78.1	0.9
SCR ESPc con	21,787	37	1,351	10,871	5,435	50	4.0	39.8	56.2	4.0
SCR ESPc FBC	80	1	9	53	7	86	0.9	56.0	42.0	2.0
SCR ESPc FGDw	40,055	69	2,590	27,146	3,206	88	1.2	59.0	40.2	0.8
SCR ESPc FGDw con	1,080	1	68	481	241	50	3.5	29.7	66.3	4.0
SCR ESPh	8,496	14	500	5,944	5,766	3	11.5	20.0	77.5	2.5
SCR ESPh FGDw	3,035	7	194	1,499	436	71	2.2	91	6.4	2.6
SCR FF	2,681	4	171	1,043	685	34	4.0	30.0	69.2	0.8
SCR FF ACI	922	1	90	546	55	90	0.6	23.0	76.2	0.8

SCR FF FGDd	2,822	18	184	1,413	551	61	3.0	30.0	69.2	0.8
SCR FF FGDw	880	2	68	425	59	86	0.9	74.0	21.0	5.0
SCR VS FGDw	3,270	4	192	1,590	731	54	3.8	56.0	43.0	1.0
SNCR ESPc	8,911	42	512	5,351	1,498	72	2.9	20.0	76.5	3.5
SNCR ESPc ACI	74	1	5	27	3	90	0.6	20.0	76.5	3.5
SNCR ESPc FGDw	5,049	18	305	3,553	440	88	1.4	59.0	40.2	0.8
SNCR ESPh	1,294	9	72	400	388	3	5.4	20.0	77.5	2.5
SNCR FF	2,682	25	169	3,438	629	82	3.7	23.0	76.2	0.8
SNCR FF FBC	240	2	14	89	12	86	0.9	56.0	42.0	2.0
SNCR FF FGDd	863	5	54	2,113	681	68	12.7	76.5	20.7	2.8
SNCR FF FGDw	848	2	58	336	47	86	0.8	74.0	21.0	5.0
SNCR VS FGDw	187	1	8	61	38	38	4.8	75.0	24.0	1.0
IGCC	633	2	27	151	145	4	5.3	96.0	3.5	0.5
Total	331,634	1,173	20,908	180,504	88,706	51	4.2	50.4	47.1	2.5

[a]Key: EPSc = cold-side electrostatic precipitator; FF = fabric filter; Con = flue gas conditioning; FBC = fluidized-bed combustion; ACI = activated carbon injection; VS = venturi scrubber; FGDd = dry flue gas desulfurization; SCR = selective catalytic reduction; FGDw = wet flue gas desulfurization; SNCR = selective non-catalytic reduction; ESPh = hot-side electrostatic precipitator.
[b]Calculated based on the total mercury input (lb/yr) and the total stack mercury emissions (lb/yr) for a given control class category.
[c]Calculated based on the total trillion Btu input and the total stack mercury emissions (lb/hr) for a given control class category.

Table 6.18 Annual mercury emissions at the station level

Mercury component	Annual emissions (lb/year)			
	Median	Mean	Maximum	Minimum
Total (Hg_T)	93	192	3,300	0.47
Elemental (Hg^0)	41	97	1,740	0.1
Oxidized (Hg^{2+})	40	91	2,350	0
Particulate (Hg_p)	2	5	119	0

technology categories were used in the study). The mean normalized emission rate for mercury across all units is 4.2 lb/trillion (T) Btu with an overall coal-to-stack reduction of 51%. The total estimated mercury emissions for all units were approximately 44 short tons. The impact of using an SCR, which was discussed earlier, in combination with wet FGD is evident. For example, the combination of SCR, cold-side ESP, and wet FGD shows a mean emission factor of 1.2 lb Hg/T Btu, which is an 88% reduction compared to 4.1 lb Hg/T Btu (58% reduction) when using a cold-side ESP and wet FGD.

Table 6.18 shows the station level mercury emissions, which includes total mercury and speciated mercury components [36]. The emissions of elemental mercury (Hg^0) and oxidized mercury (Hg^{2+}) are nearly equal in amount and comprise 98% of the total mercury emitted, with particulate-bound mercury (H_p) contributing about 2% of the total mercury emitted.

6.4.1.1.3 Survey of largest mercury emitters from electric generating units

In the previous section, it was shown that EPRI estimated that the mercury emissions from electric generating units were approximately 44 short tons in 2007. Data compiled from EPA's Toxics Release Inventory (TRI) for power plants in 2010 indicate that approximately 33 short tons of mercury were emitted [37]. Mercury emissions will continue to decrease further as more facilities are installing mercury controls as a result of the Clean Air Mercury Rule (which was discussed in Chapter 2). In 2010, the top ten states with power plant mercury emissions are listed in Table 6.19 while the top 25 mercury-emitting power plants are listed in Table 6.20 [37].

6.4.1.2 Industrial boilers

Initially, estimates of mercury emissions from industrial and commercial boilers were reported to be as high as 28.4 tons/year with coal-fired boilers contributing 20.7 tons/year (or 73% of the total) and oil-fired boilers contributing 7.7 tons/year (or 27% of the total) [6]. These data were reported in 1997 and no emissions were reported from natural gas-fired boilers. In data collected by EPA to promulgate the industrial boiler MACTs in 2012 (see Chapter 2), the mercury emissions from all major source boilers (i.e. boilers that emit $\geq$10 tons/yr of a single HAP or $\geq$25 tons/yr of any combination of HAPs) was reported at 6.66 tons/year [38]. According to EPA there are 14,111 major source boilers of which approximately 12% (or 1,753 boilers) need to meet emission limits to minimize air toxics [39].

Table 6.19 Top 10 states with power plant mercury emissions in 2010

Rank	State	Mercury emissions (lb)
1	Texas	11,127
2	Ohio	4,218
3	Pennsylvania	3,964
4	Missouri	3,835
5	Indiana	3,175
6	Alabama	3,002
7	West Virginia	2,495
8	North Dakota	2,363
9	Kentucky	2,287
10	Michigan	2,253

A breakdown of the 14,111 boilers, fuels they fire, and their baseline (i.e. no mercury controls) mercury emissions are provided in Table 6.21. The 14,111 boilers emit 6.66 tons of mercury per year. The subset of 1,753 boilers emit approximately 5 tons of mercury per year or about 75% of the total mercury emitted. The breakdown of the 1,753 boilers with firing rates greater than 10 million Btu/hour is [40]:

- 487 boilers firing biomass
- 601 boilers firing coal
- 293 boilers firing heavy liquids
- 252 boiler firing light liquids
- 42 non-continental liquid boilers
- 78 process gas boilers

The 601 coal-fired industrial boilers, comprising 34% of the 1,753 boilers, emit 2.55 tons of mercury per year, or approximately 51% of the 5 tons of mercury per year.

Table 6.22 lists the 20 states/non-continental territories that reported the highest levels of mercury emissions, which was 4.06 tons of mercury per year. This was approximately 61% of the 6.66 tons of mercury emitted per year [40].

6.4.2 Mercury emissions from stationary heat and power systems firing fuel oils and gases

There are 665 boilers firing liquid fuels or natural gas that are subject to the industrial boiler MACT, which comprise 38% of the 1,753 boilers. These boilers contributed 1.78 tons of mercury per year of the approximately 5 tons of mercury emitted per year or approximately 36% of the total mercury emissions.

6.5 Technologies for mercury control

Many research organizations, federal agencies, technology vendors, and utility and industrial companies have been actively identifying, developing, and demonstrating

Table 6.20 **Top 25 power plants emitting mercury in 2010**

Rank	Facility	Mercury emissions (lb)	State
1	Big Brown Steam Electric Station	1,610	Texas
2	Ameren Missouri Labadie Energy Center	1,527	Missouri
3	Martin Lake Steam Electric Station	1,420	Texas
4	Limestone Electric Generating Station	1,150	Texas
5	H.W. Pirkey Power Plant	1,070	Texas
6	Miller Steam Plant	1,037	Alabama
7	Monticello Steam Electric Station	1,005	Texas
8	Big Cajun 2	850	Louisiana
9	Gavin Plant	829	Ohio
10	W.A. Parish Electric Generating Station	820	Texas
11	Coal Creek Station	779	North Dakota
12	Shawville Station	702	Pennsylvania
13	Grand River Dam Authority Coal Fired Complex	670	Oklahoma
14	Detroit Edison Monroe Power Plant	660	Michigan
15	Independence Steam Electric Station	601	Arkansas
16	American Electric Power Amos Plant	585	West Virginia
17	Conemaugh Power Plant	576	Pennsylvania
18	IPL Petersburg	568	Indiana
19	Salt River Project Navajo Generating Station	566	Arizona
20	White Bluff Generating Plant	559	Arkansas
21	EME Homer City Generation LP	547	Pennsylvania
22	Gaston Steam Plant	545	Alabama
23	Milton R. Young Station	544	North Dakota
24	Calaveras Power Station	540	Texas
25	Jeffrey Energy Center	524	Kansas

Table 6.21 **Mercury emissions from industrial boilers**

Fuel	Number of boilers	Mercury emissions (tons/year)
Biomass	487	0.67
Coal	601	2.55
Liquid	777	1.22
Non-continental liquid	44	0.13
Gas	10,883	0.87
Gas-metal furnace	698	0.07
Process Gas	125	0.46
Limited Use	479	0.02
TOTAL	14,111	6.66

Table 6.22 Top 20 States/non-continental territories reporting industrial boiler mercury emissions

Ranking	State	Emissions (lb/yr)
1	Indiana	956
2	Ohio	797
3	Iowa	650
4	Pennsylvania	526
5	West Virginia	431
6	North Carolina	426
7	Tennessee	403
8	Michigan	393
9	Alabama	384
10	South Caroline	381
11	Georgia	356
12	Virginia	340
13	Florida	331
14	Illinois	311
15	Minnesota	277
16	Wyoming	275
17	Wisconsin	250
18	Maryland	232
19	Virgin Islands	209
20	New York	188
Total		8,116

cost-effective mercury control technologies for the electric utility and industrial sectors. There are many technology options at various levels of testing, demonstration, and commercialization but based upon the current state of development, fuel switching/blending, coal treatment/combustion modifications, sorbent injection, and FGD enhancement/oxidation represent the best potential for reducing mercury emissions and meeting mercury regulations [41–43]. This section begins with a discussion of capture mechanisms and then discusses the various technology options for reducing mercury emissions.

6.5.1 Capture mechanisms

The two major capture mechanisms for mercury control are absorption and adsorption. The absorption process was presented in Chapter 4 as part of the sulfur dioxide capture discussion and some comments are provided in this section, which specifically pertain to mercury. When using sorbent injection for mercury control, adsorption is the mechanism of capture and it is discussed in this section.

6.5.1.1 Absorption

Gaseous elemental mercury (Hg^0) is insoluble in water and therefore does not absorb in the aqueous slurry of a wet FGD system. Gaseous compounds of oxidized mercury (Hg^{2+}) are water-soluble and do absorb in a wet FGD system. When these compounds are absorbed in the liquid slurry, the dissolved species react with dissolved sulfides to form mercuric sulfide (HgS), which precipitates from the liquid solution as a sludge [16]. When there are insufficient sulfides in the liquid solution, a competing reaction that reduces/converts dissolved Hg^{2+} to Hg^0 is believed to occur [16]. When this conversion takes place, the newly formed insoluble Hg^0 is transferred to the flue gas passing through the wet FGD unit. The transferred Hg^0 increases the concentration of Hg^0 in the flue gas passing through the wet FGD unit. Transition metals in the slurry that originate from the flue gas are thought to play an active role in the conversion reaction since they can act as catalysts and/or reactants for reducing oxidized species.

6.5.1.2 Adsorption

Adsorption is a separation process based on the ability of solids to preferentially remove gaseous components from a flow stream like flue gas [33,44–46]. The pollutant gas or vapor molecules present in the waste stream collect on the surface of the solid material. The solid adsorbing medium is the adsorbent or sorbent while the gas or vapor adsorbed is called the adsorbate. Locations on the adsorbent surface where pollutant molecules adhere are called active sites.

The adsorption process is classified as either physical adsorption or chemisorption, and in both cases the process is exothermic. In physical adsorption, the gas molecules adhere to the surface of the solid adsorbent as a result of intermolecular attractive forces (van der Waals forces) between them. If a chemical bond occurs between the adsorbent and adsorbate, the process is called chemisorption. The bonding force associated with this type of adsorption is much stronger than that for physical adsorption. Consequently, the heat liberated during chemisorption is much larger than that liberated during physical adsorption.

Adsorption occurs by a series of steps [45]. In the first step, the contaminant diffuses from the gaseous waste stream to the external surface of the adsorbent particle. In the second step, the contaminant molecule migrates from the relatively small are of the external surface to the pores within each adsorbent particle. In the third step, the contaminant molecule adheres to the surface of the pore.

Adsorbents are very porous materials containing countless minuscule pores. Particle diameters may range from 0.5 inches to as small as 200 μm or less. A large surface area per unit weight is essential. Surface area refers to area occupied by adsorbing molecules in monolayer coverage on the surfaces within an adsorbent. The total surface area per unit mass of adsorbent is enormous −100 to as high as 3,000 m^2/g. Pore size distribution is related to the fraction of space within an adsorbent particle occupied by micropores (diameter $<$20 Angstroms (note 1 Angstrom (Å) = 10^{-10} m), mesopores (diameter between 20 and 500 Å), and macropores (diameter $>$500 Å). Pore dimensions influence the capacity and kinetics of adsorption.

Several materials are used effectively as adsorbing agents. The most common adsorbents are activated carbon, activated alumina ($Al_2O_3 \cdot nH_2O$), silica (SiO_2 as a gel or porous borosilicate glass), and zeolites (aluminosilicates, which are stoichiometric compounds of silica and alumina), which are also referred to as molecular sieves. Activated carbons and zeolites (molecular sieves) have been developed for mercury removal from gas streams.

An important thermodynamic concept governing the extent of adsorption of a single component at equilibrium is the adsorption isotherm. The isotherm is a plot of the adsorbent capacity as a function of the partial pressure of the adsorbate at a constant temperature. Adsorbent capacity is usually given in weight percent expressed as gram of adsorbate per 100 grams of adsorbent.

A variety of theoretical equations are used to describe pure-component isotherms – Langmuir isotherm, Brunauer-Emmett-Tellar isotherm, Redlich-Peterson isotherm, etc. [33]. It is difficult to develop generalized equations that can predict adsorption equilibrium from physical data because adsorption isotherms take many shapes depending on the forces involved. One of the most useful mathematical models to describe adsorption equilibria is the Langmuir isotherm [33]. It is based on the assumptions that the adsorbed phase is a unimolecular layer and at equilibrium the rate of adsorption is equal to the rate of desorption from the surface. The rate of adsorption, r_a, is proportional to the partial pressure of the adsorbate, p, and to the fraction of the solid surface area available for adsorption, $1-f$, where f is the fraction of the total surface occupied by adsorbate molecules:

$$r_a = C_a p(1-f) \tag{6.4}$$

where C_a is a constant. The rate of desportion, r_a, however is proportional to the fraction of the surface area occupied by the adsorbate:

$$r_d = C_d f \tag{6.5}$$

where C_d is a constant. At equilibrium, the rate of adsorption is equal to the rate of desportion. The fraction of the surface covered is then given by:

$$f = \frac{C_a p}{C_d + C_a p} \tag{6.6}$$

Because the adsorbed phase is a unimolecular layer, the mass of adsorbate per unit mass of adsorbent, m, is also proportional to the surface covered:

$$m = C_m f \tag{6.7}$$

where C_m is a constant. Combining equations (6.6) and (6.7) gives:

$$m = \frac{k_1 p}{k_2 p + 1} \tag{6.8}$$

where $k_1 = C_a C_m / C_d$, and $k_2 = C_a / C_d$. Equation (6.8) is known as the Langmuir isotherm.

Equation (6.8) can be rearranged to give:

$$\frac{p}{m} = \frac{1}{k_1} + \frac{k_2}{k_1}p \tag{6.9}$$

This equation is a straight line when plotted as p/m versus p.

At very low adsorbate equilibrium partial pressure k_2p is nearly zero and equation (6.8) becomes

$$m = k_1 p \tag{6.10}$$

At high equilibrium partial pressure:

$$m = \frac{k_1}{k_2} \tag{6.11}$$

Over an intermediate range of partial pressures:

$$m = kp^n \tag{6.12}$$

where k is a constant and n is a constant with a value between 0 and 1. Equation (6.12) is known as the Freundlich isotherm. The values of k and n are obtained from adsorbent manufacturers or experimental data [33].

6.5.2 Fuel switching/blending

One of the simplest ways of reducing mercury emissions from coal combustion is to burn less coal or a coal with lower mercury content. This can be accomplished through fuel switching or fuel blending. Fuel switching includes options such as switching to a mercury-compliant coal (if coal firing is still desired), repowering with natural gas or residual or distillate oils, or switching to other solid fuels such as biomass. Cofiring coal with a fuel that contains lower mercury content such as natural gas or fuel oils is a control option. Other options include blending coal with other solid fuels such as biomass, petroleum coke, or tire-derived fuels. Also, blending coals to reduce the mercury in the coal is an option. Low-rank coals, like subbituminous coal, tend to have low halogen content and produce low carbon content in the fly ash so blending them with bituminous coal can improve mercury capture by existing pollution control devices. The effect of halogens on mercury capture is discussed later in this chapter.

6.5.3 Coal cleaning

Coal cleaning is an option for removing mercury from the coal prior to utilization. Of the approximately 1 billion short tons of coal mined each year in the United States, about 600 to 650 million short tons are processed to some degree [47,48]. According to *Coal Age*'s 2013 Prep Plant Census, the U.S. employs 268 coal preparation plants to beneficiate coal. The majority of the plants are located in West Virginia (80), Kentucky (56), Pennsylvania (44), Virginia (18), Illinois (16), and Indiana (14). Coal cleaning removes pyritic sulfur and ash. Mercury tends to have a strong inorganic association (i.e. it is associated with the pyrite), especially for Eastern bituminous coals, but mercury removal efficiencies reported for physical coal cleaning vary considerably. Physical coal cleaning is effective in reducing the concentration of many trace elements, especially if they are present in the coal in relatively high concentrations. The degree of reduction achieved is coal specific, relating in part to the degree of mineral association of the specific trace element and the degree of liberation of the trace element-bearing mineral. High levels of mercury removal (up to approximately 80%) have been demonstrated with advanced cleaning techniques such as column flotation and selective agglomeration [49] while conventional cleaning methods, such as heavy media cyclone, combined water-only cyclone/spiral concentrators, and froth flotation have been shown to remove up to 62% of the mercury [50]. The average mercury removal with conventional coal washing is around 20 to 30% [43]. In both the conventional and advanced cleaning techniques, the results varied widely and were coal dependent.

6.5.4 Sorbent injection

Sorbents can be used to capture mercury on particles, which are then caught in a particulate control device. The U.S. Department of Energy (DOE) performed a three-phase research program on mercury control at coal-fired power plants during the 1990s and early 2000s. Phase I of the program focused on activated carbon injection (ACI) and the improvement of mercury capture in FGD systems while Phase II concentrated on chemically-treated ACI, sorbent enhancement additives, and sorbents designed to preserve fly ash quality [43]. DOE carried out full-scale field tests in both phases at nearly 50 coal-fired plants. Phase III had the goal of developing advanced control technologies that can achieve >90% mercury capture at cost of 50–75% less than $60,000/lb of mercury removed.

In a typical configuration, powdered activated carbon (PAC) is injected downstream of the power plant's air heater and upstream of the particulate control device – either an ESP or a fabric filter. The PAC adsorbs the mercury from the combustion flue gas and is subsequently captured along with the fly ash in the particulate control device. A variation of this concept is the TOXECON™ process, where a separate baghouse is installed after the primary particulate collector (especially when it is a hot-side ESP) and air heater and PAC is injected prior to the TOXECON unit (i.e. TOXECON I™). This concept allows for separate treatment or disposal of fly ash collected in the primary particulate control device. A variation of the process is the injection of PAC into a downstream ESP collection field to eliminate the requirement of a retrofit fabric filter and allow for potential sorbent recycling (i.e. TOXECON II™ configuration).

6.5.4.1 Overview of powdered activated carbon injection for mercury control

The performance of PAC in capturing mercury is influenced by the flue gas characteristics, which is determined by factors such as coal type, APCD configuration, and additions to the flue gas including SO_3 for flue gas conditioning [51]. Research has shown that HCl and sulfur species (i.e. SO_2 and SO_3) in the flue gas significantly impact the adsorption capacity of fly ash and activated carbon for mercury. Specifically, it has been found that [41]:

- HCl and H_2SO_4 accumulate on the surface on the carbon;
- HCl increases the mercury removal effectiveness of activated carbon and fly ash for mercury, particularly as the flue gas concentration increases from 1 to 10 ppm. The relative enhancement in mercury removal performance is not as great above 10 ppm HCl. Other strong Brønsted acids such as the hydrogen halides – HCl, HBr, or HI – should have a similar effect. Halogens such as Cl_2 and Br_2 should also be effective at enhancing mercury removal effectiveness, but this may be the result of the halogens reacting directly with mercury rather than the halides thereby promoting the effectiveness of the activated carbon; and
- SO_2 and SO_3 reduce the equilibrium capacity of activated carbon and fly ash for mercury. Activated carbon catalyzes SO_2 to H_2SO_4 in the flue gas. Because the concentration of SO_2 is much higher than mercury in the flue gas, the overall adsorption capacity of mercury is likely dependent on the SO_2 and SO_3 concentrations in the gas as these compete for the adsorption sites and form H_2SO_4 on the surface of the carbon.

Figure 6.4 illustrates some results from conventional PAC injection tests performed through numerous DOE test programs [41]. Conventional PAC injection was the focus of initial field testing (which was in 2001–2002) and served as the benchmark for all field PAC injection tests. This work showed that a maximum of approximately 65% mercury capture could be achieved when firing a subbituminous coal in a power plant using an ESP. Also, when using conventional PAC costs of 50¢/lb, it was found that the cost to remove 70% mercury from a bituminous coal-fired plant with an ESP would cost approximately $70,000/lb of mercury removed, which is higher than the targets of $30,000 to $45,000/lb of mercury removed (in 1999 dollars) [52].

The conventional PAC testing was followed by work with chemically-treated PACs, which were developed for low-rank coal applications due the low mercury capture results in the initial testing. Some results from the chemically-treated PAC testing are shown in Figure 6.5 [41]. With PAC injection at 1 lb/MMacf (million actual cubic feet of flue gas), mercury removal ranges from 70 to 95% for low-rank coals. For 90% mercury removal, costs are estimated at ≈$6,000/lb mercury removed for a subbituminous coal/SDA-FF combination to <$20,000/lb mercury

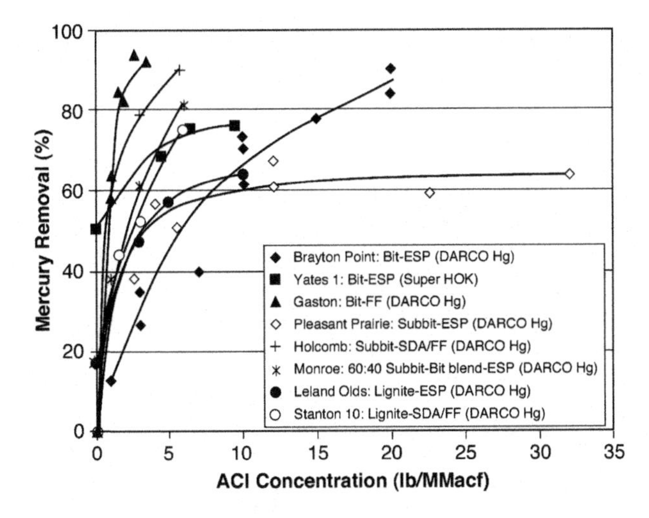

Figure 6.4 Mercury removal as a function of activated carbon injection concentration [86]. Note: legend lists power plant name: coal rank-APCD configuration (PAC type).

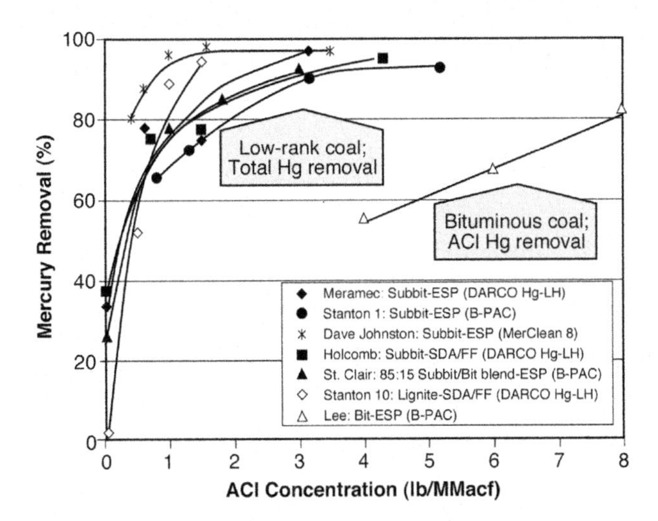

Figure 6.5 Mercury removal as a function of chemically-treated activated carbon injection concentration. Note: legend lists power plant name: coal rank-APCD configuration (PAC type).

removed for a subbituminous coal/ESP combination assuming chemically-treated PAC costs of \$0.75 to \$1.00/lb [41].

Results from testing using the TOXECON technology are given in Table 6.23 (fabric filter configuration) and Figure 6.6 (ESP configuration) [51]. In the fabric filter configuration, mercury removals of 70 to 90% have been achieved using low-rank coals while removal efficiencies of 50 to 90% are expected when using

Table 6.23 Demonstrated mercury removal at 2 or 5 lb/MMacf

	ESP (5 lb/MMacf)	ESP with SO_3[a] (5 lb/MMacf)	TOXECON (2 lb/MMacf)	SDA + FF (2 lb/MMacf)
Low S, very low Cl (PRB or North Dakota lignite)	78–95% with brominated PAC	40–91% with brominated PAC	7–90%	90–95% with brominated PAC
Low S, >50 ppm Cl (Some Texas lignites)	78–95%	40–91%	70–90%	90–95% with brominated PAC
Low S, bituminous coal	55–75%	40–91%	ND[b]	ND
Low S, bituminous coal with SCR	15–70%	NA[c]	ND	ND
Low S, bituminous coal	<15%	NA	NA	NA

[a]SO_3 from SO_3 injection.
[b]Not available.
[c]Not applicable or configuration unlikely.

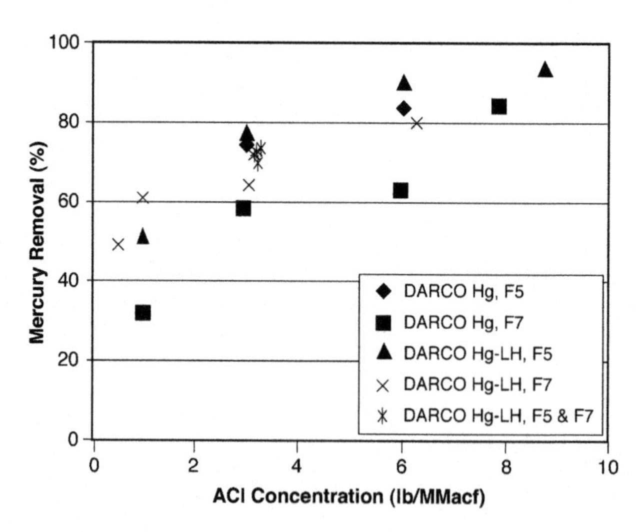

Figure 6.6 Mercury removal as a function of activated carbon injection concentration using the TOXECON II™ configuration at the Independence Station; Note: DARCO Hg is a conventional PAC while DARCO Hg-LH is a brominated PAC.

Table 6.24 Summary of mercury removal performance for sorbent injection at units firing subbituminous coals

APCD	Type of sorbent injection	Range of observed mercury removal			Factors influencing performance
		1 lb/ MMacf	5 lb/ MMacf	10 lb/ MMacf	
Fabric Filter or TOXECON	ACI	70–95%	>95%	–	Temperature, improvement with BAC plus ACI
	Brominated ACI	80–95%	>95%	–	Temperature
ESP (No flue gas conditioning)	ACI	–	40–95%	50–95%	Improvement with BAC plus ACI; temperature effect not demonstrated, but suspected
	Brominated ACI	65 to >95%	80 to >95%	–	Temperature effect not demonstrated, but suspected
Spray Dryer	ACI	25–60%	45–90%	60–90%	Higher removals possible with BAC plus ACI or with upstream SCR
	Brominated ACI	60–95%	85 to >95%	–	–

low-sulfur bituminous coals [51]. From Figure 6.6, 50 to 80% mercury reduction was achieved when injecting chemically-treated PAC at 4–5 lb/MMacf into the next-to-last ESP field (i.e. F5) and last ESP field (i.e. F7) [41].

Additional results from full-scale testing using brominated activated carbon (BAC) are provided in Tables 6.24 and 6.25 for tests firing Powder River Basin subbituminous coal and eastern bituminous coals, respectively [53]. Flue gas associated with western coals is relatively low in chloride and high in elemental mercury; therefore, using brominated PACs, which can capture elemental mercury in a low-halogen flue gas, or increasing flue gas oxidized mercury by adding bromide directly into the boiler in conjunction with PAC typically improves performance over activated carbon injection (ACI) for western coals. The flue gas associated with eastern bituminous coals is relatively high in chloride and oxidized mercury; therefore, the use of BAC to increase flue gas oxidized mercury does little to improve performance over ACI for bituminous coals.

The improved mercury capture efficiency of the advanced chemically-treated sorbent injection systems has given U.S. coal-fired power plant operators the

Table 6.25 **Summary of mercury removal performance for sorbent injection at units firing eastern bituminous coals**

APCD	Coal	Type of sorbent injection	Range of observed mercury removal			Factors influencing performance
			1 lb/MMacf	5 lb/MMacf	10 lb/MMacf	
Fabric Filter or TOXECON	Low sulfur	ACI	75–90%	>90%	–	Temperature; air-to-cloth ratio; no improvement with BAC
ESP (No flue gas conditioning)	Low sulfur	ACI	20–60%	20–70%	20–75%	Temperature
		Brominated ACI	30–40%	35–60%	60 to >80%	Temperature
ESP	Medium or high sulfur	ACI	0–35%	0–70%	5–80%	Temperature; SO_3 concentration; co-injection of SO_3 sorbent with ACI; use of SO_3-tolerant sorbents
		Brominated ACI	0–35%	0–70%	10 to >90%	Temperature; SO_3 concentration

confidence to deploy the technology. In 2011, prior to the MATS legislation, around 155 coal-fired boilers totaling 85 GW of power had awarded contracts for sorbent-based mercury control systems [43]. This number was expected to triple between the issuance of the MATS legislation and the compliance date. ACI is expected to be the primary technology of choice for controlling mercury emissions.

6.5.4.2 Balance-of-plant issues

One concern with PAC injection is the effect it may have on fly ash. The presence of carbon in the fly ahs affects the performance of concrete and may result in the loss of fly ash sales for utilities that sell their fly ash to the cement industry. Minimizing the impact of carbon in the fly ash led to the development of the TOXECON technology. In addition, companies are now developing sorbents that do not result in loss of fly ash sales [43]. These include amended silicates, calcium-based sorbents, and others.

No impact of PAC injection upstream of the air preheater has been observed. In general, no balance-of-plant issues including increases in particulate emissions and changes in ESP operation have been identified as a result of PAC injection in the ductwork upstream of the APCDs. One exception was an increase in the arc rate at low boiler load conditions as compared to baseline arcing, during PAC injection [51]. Particulate measurements collected at the outlet of the ESP when injecting PAC in the TOXECON II configuration indicated that the PAC injection did not impact particulate emissions [51]. However, stack opacity and ESP outlet particulate spikes associated with PAC injection following fourth field raps were observed.

Two significant balance-of-plant issues were discovered during TOXECON I testing at Presque Isle [51]. The first was that PAC can self ignite in the hoppers of the baghouse when it is allowed to accumulate and when exposed to external heating from hopper heaters. The second was that the ash/PAC mixture becomes stickier than ash alone and harder to remove from the hoppers and transport with the ash removal system when it is heated. The first issue was addressed by frequently evacuating the hoppers and controlling the maximum temperature of the heating elements to less than 300°F. The second issue was addressed through equipment design.

No balance-of-plant problems have been noted at sites using spray dryer absorbers during PAC injection. This includes opacity or changes in spray dryer absorber or fabric filter operation.

Balance-of-plant impacts from using bromine compounds are being investigated [53]. This includes the effect of bromine on fly ash for concrete use, leachability of bromine from fly ash, bromine corrosion potential in the boiler and wet FGD, effect of bromine on mercury partitioning between wet FGD liquor and solids, effect of bromine of mercury re-emissions from wet FGD.

6.5.5 Wet flue gas desulfurization

Although PAC injection has shown much promise as a near-term mercury control technology, wet FGD systems for SO_2 control can also be extremely effective at reducing mercury emissions. Oxidized mercury is soluble and is therefore captured in FGD solutions to be removed from the plant in the liquid or solid waste. Maximizing the proportion of mercury in the oxidized form maximizes its capture in the FGD. Mercury capture in wet FGD systems can be up to and greater than 90% [43,54]. FGD-related technologies under development include: 1) coal and flue gas chemical additives with fixed-bed catalysts to increase levels of oxidized mercury in the flue gas; and 2) wet FGD chemical additives to promote mercury capture and prevent re-emissions of previously captured mercury from the FGD absorber vessel.

Wet FGD systems, especially those associated with bituminous coal-fired power plants equipped with SCR systems, are good candidates for capturing mercury. The catalysts used in SCR systems for NO_x reduction have the co-benefit effect of converting some of the elemental mercury into the oxidized form, thereby making it easier to capture in FGD systems. The combination of SCR and FGD systems can achieve over 90% mercury control [54].

6.6 Cost estimates to control mercury emissions

Cost estimates to control mercury emissions are quite variable and it is difficult to compare various control technologies on the same bases using published data. Cost estimates represent "snapshots" in time based on many assumptions and operating conditions. As a consequence, the economics are plant- and condition-specific and are based on relatively small data sets [55]. This section will present cost estimate data primarily to illustrate comparative information between different control strategies and current data on the leading control technology at this time, i.e. PAC injection.

6.6.1 Coal cleaning

The costs of cleaning Eastern bituminous coals for mercury removal range from no additional cost (for coals already washed for sulfur removal) to a cost of $33,000 per pound of mercury removed [56]. The costs for cleaning Powder River Basin subbituminous coals are higher and approach $58,000 per pound of mercury removed [14]. However, mercury reductions from washing methods currently being applied are already built into the ICR mercury data for delivered coal; consequently, to realize a benefit from coal cleaning, higher levels of coal cleaning must be employed. Advanced cleaning methods can remove additional mercury, but they are generally not economical.

6.6.2 General economic comparisons

As previously mentioned, cost estimates for mercury compliance are very approximate and these estimates vary from $5,000 to $70,000 per pound of mercury removed, from 0.03 to 0.8 cents/kWh, and from $1.7 to $7 billion annually for the total national cost depending on technical advances [14]. A breakdown of costs by various technology options is provided in Table 6.26 [57]. The costs are dated, i.e. from 2000, but are provided to compare costs between technologies.

6.6.3 Activated carbon injection

Activated carbon injection (ACI) is the leading technology for mercury control in industrial and electric generating power plants as there has been a significant amount of research, development, and demonstrations performed. There has been an improvement in both the cost and performance of mercury control during full-scale tests with chemically-treated (or brominated) ACI.

There are three scenarios for evaluating the cost of mercury control with sorbents such as PAC: 1) using only activated carbon; 2) using activated carbon with an additive, e.g. brominated PAC; and 3) installation of a baghouse to handle the additional particulate load of the PAC [58]. The third scenario is the most costly but if a baghouse is installed for the carbon capture, the quality of the fly ash

Table 6.26 Estimates of current and projected annualized operating costs for mercury emissions control technology

Coal		Existing controls	Retrofit control[a]	Current cost (mills/kWh)	Projected cost (mills/kWh)
Type[b]	S, %				
Bit	3	ESP cold + FGD	PAC	0.727–1.197	0.436–0.718
Bit	3	Fabric filter + FGD	PAC	0.305–0.502	0.183–0.301
Bit	3	ESP hot + FGD	PAC + PFF	1.501–NA[c]	0.901–NA
Bit	0.6	ESP cold	SC + PAC	1.017–1.793	0.610–1.076
Bit	0.6	Fabric filter	SC + PAC	0.427–0.753	0.256–0.452
Bit	0.6	ESP hot	SC + PAC + PFF	1.817–3.783	1.090–2.270
Subbit	0.5	ESP cold	SC + PAC	1.150–1.915	0.690–1.149
Subbit	0.5	Fabric filter	SC + PAC	0.423–1.120	0.254–0.672
Subbit	0.5	ESP hot	Sc + PAC + PFF	1.419–2.723	0.851–1.634

[a]PAC = powdered activated carbon; SC = spray cooling; PFF = polishing fabric filter.
[b]Bit = bituminous coal; subbit = subbituminous coal.
[c]NA = not available.

does not change significantly and has a minimal effect on fly ash sales if the fly ash was being sold.

The capital and operating costs for installing a PAC injection system is dependent upon the scale of the power plant. Table 6.27 illustrates ACI costs generated by Sargent & Lundy for EPA for representative sizes and heat rates for three options – ACI with an existing ESP, ACI with an existing baghouse, and ACI with an additional baghouse (i.e. TOXECON) [59]. The data were developed assuming a carbon feed rate of 5 lb/million acfm of flue gas to achieve 90% mercury reduction.

DOE has published economic data on mercury control at the various field-testing sites where they have funded testing [60]. These data provide plant-specific cost estimates and were performed so DOE could assess its ability to achieve a target of reducing the baseline (1999) mercury control costs estimate of $60,000/lb mercury removed by 25 to 50%. Figure 6.7 shows the 20-year levelized cost estimates for the incremental cost of mercury control for 90% ACI mercury removal (using chemically-treated ACI) at seven of DOE's field testing sites [60]. DOE found the results encouraging especially for the lower rank Powder River Basin coals and lignite. Figure 6.7 also shows an assessment of the potential for ACI to negatively impact the sale and disposal of the fly ash and therefore impact the overall cost of mercury control. Although it was demonstrated that ACI systems have the potential to remove more than 90% of the mercury in many applications at a cost estimate below $10,000/lb mercury removed, DOE pointed out that only through experience gained during long-term continuous operation of these technologies in a range of commercial applications will their actual costs and performance be determined.

Brominated activated carbon costs (FOB manufacturing plant) about $1.00/lb compared to $0.70/lb for untreated activated carbon [53]. However, the performance of BAC is significantly better than that of PAC for western coal applications, making it a more cost-effective approach for these coals. Figures 6.8 and 6.9 show that annual mercury control costs for a 500 MW plant firing western PRB coals or lignite, and low-sulfur eastern bituminous coal, respectively [modified from 53]. The estimates (in 2008 dollars) assume a 500 MW power plant with a flue gas flow rate of 2 million acfm, 0.65 capacity factor, no ash sales, and a mercury flue gas concentration of 10 $\mu g/Nm^3$, resulting in mercury emissions of approximately 275 lb/year. Equipment costs are amortized using a capital recovery factor of 0.15.

In another scenario, DOE estimated the capital cost for a calcium bromide ($CaBr_2$) injection operation at a 500 MW plant to be $78,000 (in 2008 dollars) [61]. DOE estimated the cost per pound of mercury removed to be from $2,800 (with an SCR) to $7,000 (with a cold-side ESP and wet FGD) for a low-rank coal on a 20-year level cost basis. The cost included the sale of fly ash.

Table 6.27 Illustrative activated carbon injection costs (2011$) for representative sizes and heat rates

Control type	Heat rate (Btu/kWh)	Var. O&M (mills/Kwh)	Capacity (MW)									
			100		300		500		700		1,000	
			Capital cost ($/kW)	Fixed O&M ($/kW-yr)	Capital cost ($/kW)	Fixed O&M ($/kW-yr)	Capital cost ($/kW)	Fixed O&M ($/kW-yr)	Capital cost ($/kW)	Fixed O&M ($/kW-yr)	Capital cost ($/kW)	Fixed O&M ($/kW-yr)
ACI system with an existing ESP and sorbent injection rate of 5 lb/million acfm	9,000 10,000 11,000	2.19 2.43 2.68	37.89 38.51 39.07	0.32 0.32 0.33	14.90 15.14 15.35	0.13 0.13 0.13	9.65 9.81 9.95	0.08 0.08 0.08	7.25 7.36 7.47	0.06 0.06 0.06	5.35 5.44 5.52	0.04 0.05 0.06
ACI system with an existing baghouse with sorbent injection rate of 5 lb/million acfm	9,000 10,000 11,000	1.57 1.75 1.92	33.03 33.54 34.02	0.28 0.28 0.29	12.98 13.18 13.38	0.11 0.11 0.11	8.41 8.54 8.68	0.07 0.07 0.07	6.32 6.42 6.51	0.05 0.05 0.06	4.66 4.74 4.81	0.04 0.04 0.04
TOXECON with sorbent injection rate of 5 lb/million acfm	9,000 10,000 11,000	0.47 0.52 0.57	291.26 314.32 336.91	1.02 1.10 1.18	219.74 238.18 256.26	0.77 0.83 0.90	195.35 212.02 228.37	0.68 0.74 0.80	181.36 196.97 212.28	0.63 0.69 0.74	167.98 182.55 196.83	0.59 0.64 0.69

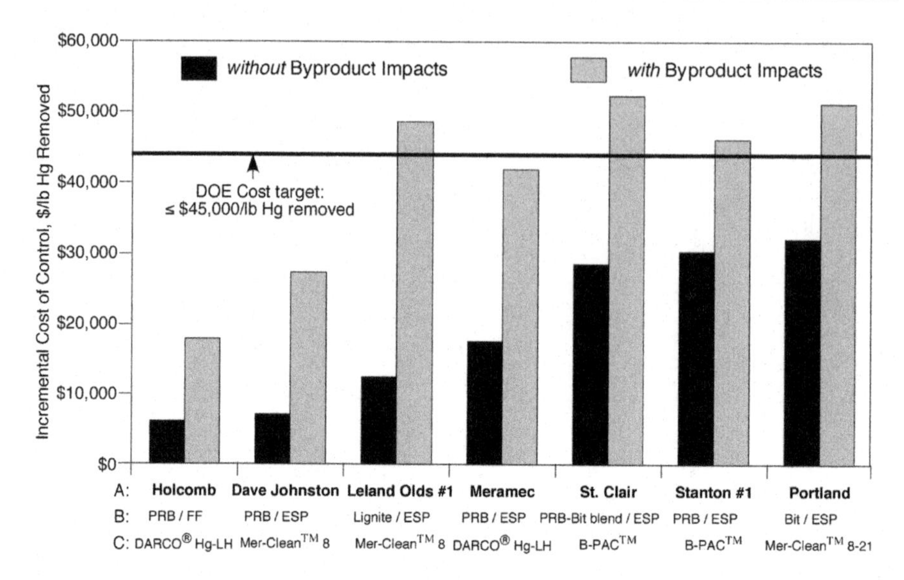

Figure 6.7 20-year levelized incremental cost of 90% ACI mercury control. A, Power Plant Name; B, Coal Rank/Particulate Control Device; C, Sorbent Name.

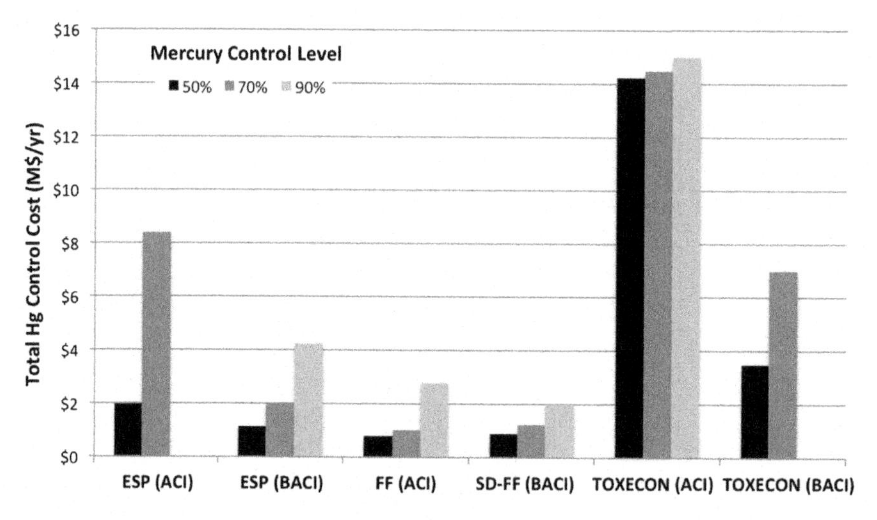

Figure 6.8 Total annual cost of mercury control for Powder River Basin subbituminous coal and North Dakota lignite for a 500 MW power plant.

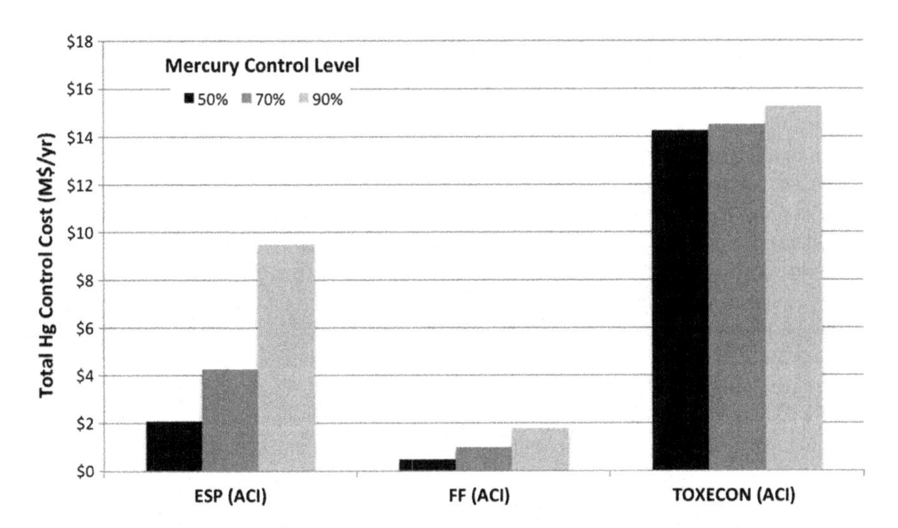

Figure 6.9 Total annual cost of mercury control for low-sulfur eastern bituminous coal for a 500 MW power plant.

References

[1] Feeley TJ, Murphy J, Hoffmann J, Renninger SA. A review of DOE/NETL's mercury control technology R&D program for coal-fired power plants. Pittsburgh, Pennsylvania: U.S. Department of Energy; 2003.

[2] EPA (United States Environmental Protection Agency). Mercury emissions: the global context, <http://www2.epa.gov/international-cooperation/mercury-emissions-global-context#learn>; 2014.

[3] UNEP (United Nations Environment Programme). Global mercury assessment 2013: sources, emissions, releases and environmental transport; 2013.

[4] EPA (United States Environmental Protection Agency). Our nation's air status and trends through 2010; 2012.

[5] EPA (United States Environmental Protection Agency). Controlling power plant emissions: emissions progress, <http://www.epa.gov/mercury/control_emissions/emisisons.htm>; 2009.

[6] EPA (United States Environmental Protection Agency). Mercury study report to congress. (Office of Air Quality Planning & Standards, U.S. Government Printing Office, Washington, D.C., 1997).

[7] Emsley J. The elements. Oxford: Clarendon Press; 1989.

[8] USGS (United States Geological Survey). Mercury in U.S. coal – abundance, distribution, and modes of occurrence. (USGS Fact Sheet FS-095-01, U.S. Government Printing Office, Washington, D.C., 2001).

[9] Clarke LE, Sloss LL. Trace elements – emissions from coal combustion and gasification. London: IEA Coal Research; 1992.

[10] CERHR (Center for the Evaluation of Risks to Human Reproduction). CERHR: mercury (5/16/02), <http://cerhr.niehs.nih.gov/genpub/topics/mercury2-ccae.html>; 2002.

[11] Aposhian HV, Aposhian MM. Elemental, mercuric, and methylmercury: biological interactions and dilemmas. In: Proceedings of the air quality II: mercury, trace elements, and particulate matter conference. (Grand Forks, North Dakota): University of North Dakota; 2000.

[12] National Research Council. Toxicological effects of methyl mercury. Washington, DC: National Academy Press; 2000.

[13] Finkelman RB, Orem W, Castranova V, Tatu CA, Belkin HE, Zheng B, et al. Health impacts of coal and coal use: possible solution. Int J Coal Geol 2002;50:425–43.

[14] Pavlish JJ, Sondreal EA, Mann MD, Olson ES, Galbreath KC, Laudal DL, et al. Status review of mercury control options for coal-fired power plants. Fuel Process Technol 2003;82:89–1653.

[15] Kitto JB, Stultz SC, editors. Steam: its generation and use. 41st ed. The Babcock and Wilcox Company; 2005.

[16] EPA (United States Environmental Protection Agency). Control of mercury emissions from coal-fired electric utility boilers: interim report including errata dated 3-21-02; 2002.

[17] United States Geological Survey. Mercury in U.S. coal – abundance, distribution, and modes of occurrence, USGS fact sheet FS-095-01; 2001.

[18] Wilhelm SM, Liang L, Cussen D, Kirchgessner DA. Mercury in crude oil processed in the United States (2004). Environ Sci Technol 2007;41(13):4509–14.

[19] IPIECA (International Petroleum Environmental Conservation Association). Mercury management in petroleum refining: an IPIECA good practice guide; 2014.

[20] EPA (United States Environmental Protection Agency). Mercury in petroleum and natural gas: estimation of emissions from production, processing, and combustion; 2001.

[21] Foster A, Carnel P. Mercury emissions control. Hydrocarbon engineering; 2006.

[22] EIA (U.S. Energy Information Administration). September 2014 monthly energy review, <www.eia.gov/mer>; 2014.

[23] EIA (U.S. Energy Information Administration). Petroleum & other liquids, <www.eia.gov/petroleum/imports/companylevel/>; 2014.

[24] Bipartisan Policy Center. U.S. imports of Canadian oil sands have doubled since 2005, <http://bipartisonaplicy.org>; 2014.

[25] Alberta Research Council. Potential release of heavy metals and mercury from the UOG industry into the ambient environment – literature review, prepared for petroleum technology alliance Canada; 2009.

[26] Liang L, Horvat M, Danilchik P. A novel analytical method for determination of picogram levels of total mercy in gasoline and other petroleum based products. Sci Total Environ 1996;187:57.

[27] Olsen S, Westerlund S, Visser R. Analysis of metals in condensates and naphthas by ICP-MS. Analyst 1997;122:1229.

[28] Tao H, Mukakami T, Tominaga M, Miyazaki A. Mercury speciation in natural gas condensate by gas chromatography-inductively coupled plasma mass spectrometry. J Anal At Spectrom 1998;13:1085.

[29] Bloom NS. Analysis and stability of mercury speciation in petroleum hydrocarbons. Fresenius J Anal Chem 2000;366(5):438.

[30] EPA (United States Environmental Protection Agency). Locating and estimate air emissions from sources of mercury and mercury compounds, office of air quality planning and standards; 1997.

[31] Littlepage T. Mercury in crude oils, 2013 MCA (Marine Chemist Association) spring seminar series; 2013.

[32] Markovs J, Corvini J. Mercury removal from natural as & liquid streams, gas liquid streams John Markovs adsorption solution, <pdfmanual4.com>, 2014.
[33] Heinsohn RJ, Kabel RL. Sources and control of air pollution. Prentice Hall; 1999.
[34] Morel FMM, Kraepiel AML, Amyot M. The chemical cycle and bioaccumulation of mercury. Annu Rev Ecol Syst 1998;29:543–66.
[35] Department of Geosciences, Princeton University, Mercy Cycling and Methylation; 2012.
[36] Chu P, Levin L. Updated Hazardous Air Pollutants (HAPs) emissions estimates and inhalation human health risk assessment for U.S. coal-fired electric generating units, Electric Power Research Institute; 2009.
[37] Madsen T, Randall L. America's biggest mercury polluters, Environment America Research & Policy Center; 2011.
[38] EPA (United States Environmental Protection Agency). Boiler MACT/Impacts memo & appendices, <www.epa.gov/ttn/atw/boilers/boilerpg.html>; 2011.
[39] EPA (United States Environmental Protection Agency). Reconsideration proposal for boilers at area sources, boilers/process heaters at major sources, and commercial/industrial solid waste incinerators and proposed definition of "non-hazardous solid waste"; 2011.
[40] Smith, Brian. The toxic air burden from industrial power plants, EarthJustice, 2012.
[41] Feeley TJ. U.S. DOE's Hg control technology RD&D program – significant progress, but more work to be done! mercury 2006 – conference on mercury as a global pollutant, Madison, WI; 2006.
[42] Pavlish JH, Hamre LL, Zhuang Y. Mercury control technologies for coal combustion and gasificaiton systems. Fuel 2010;89:838–47.
[43] Sloss L. Legislation, standards and methods for mercury emissions control. London: IEA Clean Coal Centre; 2014.
[44] Wark K, Warner CF, Davis WT. Air pollution: its origin and control. 3rd ed. Menlo Park, California: Addison Wesley Longman, Inc.; 1998.
[45] Davis WT, editor. Air pollution engineering manual. 2nd ed. New York: John Wiley & Sons, Inc.; 2000.
[46] Flagan RC, Seinfeld HH. Fundamentals of air pollution engineering. Englewood Cliffs, New Jersey: Prentice-Hall Inc.; 1988.
[47] National Research Council. Coal waste impoundments: risks, responses, and alternatives. Washington, DC: National Academy Press; 2002.
[48] Fiscor S. Prep plants hold steady in uncertain market. Coal Age 2014;119(10):.
[49] Jha MC, Smit FJ, Shields GL, Moro N. Engineering development of advanced physical fine coal cleaning for premium fuel applications project final report, DOE Contract No. DE-AC22-92PC92208; 1997.
[50] Akers DJ, Raleigh CE. The mechanism of trace element removal during coal cleaning. Coal Prep 1998;19(3):257–69.
[51] Sjostrom S, Campbell T, Bustard J, Stewart R. Activated carbon injection for mercury control: overview, The 32nd international technical conference on coal utilization & fuel systems, Clearwater, Florida; 2007.
[52] Jones AP, Hoffmann JW, Smith DN, Feeley TJ III, Murphy JT. DOE/NETI's phase II mercury control technology field testing program updated economic analysis of activated carbon injection, <http://www.netl.doe.gov/technologies/coalpower/ewr/index.html>; 2007.
[53] Chang R, Dombrowski K, Senior C. Near and long term options for controlling mercury emissions from power plants, URS power technical paper; 2010.
[54] Moretti AL, Jones CS. Advanced emissions control technologies for coal-fired power plants, Power-Gen Asia; 2012.

[55] Jones AP, Hoffman JW, Smith DN, Feeley TJ, Murphy JT. DOE/NETL's phase II mercury control technology field testing program: preliminary economic analysis of activated carbon injection. Environ Sci Technol 2007;41(4):1365–71.

[56] Akers DJ, Toole-O'Neil B. Coal cleaning for HAP control: cost and performance. In: Proceedings of the 23rd international technical conference on coal utilization & fuel systems, Coal & Slurry Technology Association, Washington, DC; 1998.

[57] Kilgroe JD, Srivastava RK. Technical memorandum: control of mercury emissions from coal-fired electric utility boilers United States environmental protection agency; 2000.

[58] Sloss LL. Economics of mercury control. London: IEA Coal Research; 2008.

[59] EPA (United States Environmental Protection Agency). EPA's power sector modeling platform v.5.13, emission control technologies, <www.epa.gov/airmarkets/progsregs/epa-ipm>; 2013.

[60] Feeley TJ, Jones AP. An update on DOE/NETL's mercury control technology field test program, U.S. Department of Energy White Paper. (Office of Fossil Energy, Washington, DC; 2008).

[61] Fielding WR. Calcium bromide technology for mercury emissions reductions. Great Lakes Solutions; 2012.

Formation and control of acid gases, and organic and inorganic hazardous air pollutants

7

7.1 Introduction

This chapter presents a discussion of a suite of pollutants including hazardous air pollutants (HAPs), both organic and inorganic, and acid gases. Mercury, which was discussed in detail in Chapter 6, is not included in detail in this chapter although at times there will be reference to it. Similarly, many inorganic trace elements were discussed in Chapter 3, "Particulate Formation and Control Technologies," and reference will be made to Chapter 3 on occasion. This chapter begins with a discussion of the environmental and health effects of organic and inorganic hazardous air pollutants and acid gases, followed by discussions on the occurrence of trace elements, organic compounds, and halogens in fossil fuels, and the formation and emissions of these pollutants. It concludes with discussions on the control technologies and their costs.

7.2 Environmental and health effects of organic and inorganic hazardous air pollutants

7.2.1 Environmental and health effects of organic compounds

Organic compounds are unburned gaseous combustibles that are emitted from fossil fuel-fired boilers but generally in very small amounts [1]. However, there are brief periods where unburned combustible emissions may increase significantly such as during system startup or upsets. The organic emissions from natural gas-fired units and fuel oil-fired boilers are lower than coal-fired units with emissions from pulverized coal-fired or cyclone-fired units lower than from smaller stoker-fired boilers where operating conditions are not as well controlled [2].

Organic emissions are either due to constituents present in the fuel or are formed as products of incomplete combustion. Polycyclic organic matter (POM) has also been referred to as polynuclear or polycyclic aromatic compounds (PACs). Nine major categories of POM have been identified by the United States EPA [3]. The most common organic compounds in the flue gas of fossil fuel-fired boilers are polycyclic aromatic hydrocarbons (PAHs). Hydrocarbons as a class are not listed as criteria pollutants although a large number of specific hydrocarbon compounds are listed among the 187 hazardous air pollutants under Title III of the Clean Air Act Amendments of 1990 (see Chapter 2 for a discussion of regulations).

Fossil Fuel Emissions Control Technologies. DOI: http://dx.doi.org/10.1016/B978-0-12-801566-7.00007-5

Furthermore, EPA has specified 16 PAH compounds as priority pollutants. These are naphthalene, acenaphthylene, acenaphthene, fluorene, phenanthrene, fluoranthene, pyrene, chrysene, benz[a]anthracene, benzo[b]fluoranthene, benzo[a]pyrene, indeno [1,2,3−cd]pyrene, benzo[ghi]perylene, and dibenz[ah]anthracene.

Gaseous hydrocarbons as a broad class do not appear to cause any appreciable corrosive damage to materials [2]. Of all the hydrocarbons, only ethylene has adverse effects on plants at known ambient concentrations, which includes inhibiting plant growth and injury to orchids and cotton.

Studies of the effects of ambient air concentrations of many of the gaseous hydrocarbons have not demonstrated direct adverse effects upon human health [2]. Certain airborne PAHs, however, are known carcinogens. Also, studies of carcinogenicity of certain classes of hydrocarbons indicate that some cancers appear to be caused by exposure to aromatic hydrocarbons found in soot and tars. An extreme example is the occurrence of highly elevated incidences of lung cancer in China from PAH exposure [4]. PAHs are released during unvented coal combustion of "smoky" coal in homes, resulting in lung cancer mortality that is five times the national average in China.

7.2.2 Environmental and health effects of carbon monoxide

Carbon monoxide is a colorless, odorless gas that is very stable and has a life of 2 to 4 months in the atmosphere [2]. Similar to organic compounds, it is formed when fuel is not burned completely.

Carbon monoxide appears to have no detrimental effects on material surfaces [2]. At ambient concentrations, experiments have not shown CO to produce any harmful effects on plant life. Carbon monoxide has been found to be a minor participant in photochemical reactions leading to ozone formation.

High concentrations of CO can cause physiological and pathological changes and ultimately death. Carbon monoxide enters the bloodstream through the lungs and reduces oxygen delivery to the body's organs and tissues [5]. The health threat from lower levels of CO is most serious for those who suffer from cardiovascular disease, such as angina pectoris. At much higher levels of exposure, CO can be poisonous and even healthy individuals can be affected. Visual impairment, reduced work capacity, reduced manual dexterity, poor learning ability, and difficulty in performing complex tasks are all associated with exposure to elevated CO levels.

7.2.3 Environmental and health effects of trace elements

Heavy fuel oils and coals contain small concentrations of trace elements. Trace elements enter the atmosphere through natural processes and sources of trace elements include soil, seawater, and volcanic eruptions. Human activities, such as power generation and industrial and commercial sectors also lead to emissions of some elements.

As previously discussed in Chapter 3, Title III of the United States Clean Air Act Amendments (CAAA) of 1990 designates 187 hazardous air pollutants (HAPs). Included in the list are 11 trace elements: antimony (Sb), arsenic (As), beryllium (Be), cadmium (Cd), chromium (Cr), cobalt (Co), lead (Pb), manganese (Mn), mercury (Hg) (discussed in detail in Chapter 6), nickel (Ni), and selenium (Se). In addition, barium (Ba) is regulated by the Resources Conservation and Recovery Act, and boron (B) and molybdenum (Mo) are regulated by Irrigation Water Standards [52]. Vanadium (V) is regulated based on its oxidation state and vanadium pentoxide (V_2O_5) is a highly toxic regulated compound. Other elements, such as fluorine (F) and chlorine (Cl), which produce acid gases (i.e. HF and HCl) upon combustion, and radionuclides such as radon (Rn), thorium (Th) and uranium (U) are also of interest.

The U.S. National Research Council (NRC) conducted a study to classify HAPs from coal-fired electric generating units that are of concern based on known adverse health effects or because of their abundances in coal. The NRC classifications are as follows [6]:

- Major concern: arsenic (As), boron (B), cadmium (Cd), lead (Pb), mercury (Hg), molybdenum (Mo), and selenium (Se).
- Minor concern: chromium (Cr), vanadium (V), copper (Cu), zinc (Zn), nickel (Ni), and fluorine (F).
- Minor concern: barium (Ba), strontium (Sr), sodium (Na), manganese (Mn), cobalt (Co), antimony (Sb), lithium (Li), chlorine (Cl), bromine (Br), and germanium (Ge). These elements are of little environmental concern but were classified mainly on the basis that they are present in residues.
- Elements of concern but with negligible concentrations: beryllium (Be), thallium (Tl), silver (Ag), and tellurium (Te). These elements have known documented relationships to health but the low levels present are considered to have negligible impact.
- Radioactive elements: uranium (U) and thorium (Th).

Partitioning of the trace elements in the bottom ash, ash collected in the air pollution control device, and fly ash and gaseous constituents emitted into the atmosphere depends on many factors including the volatility of the elements, temperature profiles across the system, pollution control devices, and operating conditions [7,8]. The classification scheme for trace elements is discussed in detail in Chapter 3.

The environmental effects of trace elements are a function of the chemical and physical form in which they are found [8]. Environmental effects may occur due to the element itself or as a result of a combination of the element and other compounds. Linking a specific environmental effect to an individual element is difficult as is determining the contribution from human activities since the trace elements also occur naturally.

Some trace elements may have an immediate effect in the atmosphere. Trace element metals such as Mn (II) and Fe (III) may contribute to acid rain by promoting oxidation of sulfur dioxide to sulfate in water droplets [8]. Trace elements may also be involved in the complex atmospheric chemistry that forms photochemical smog, and may affect cloud formation.

Soils may contain high concentrations of certain trace elements due to natural minerals and ores. In addition, deposition of trace elements downwind from power plants can lead to high concentrations in the soils and uptake by plants. Some elements found in coal are major plant nutrients, specifically calcium, magnesium, and potassium [8]. These elements are not considered trace elements since they occur in quantities greater than 1,000 ppm (i.e. >0.1% by weight). Other elements (both major and trace) are considered minor plant nutrients such as iron, manganese, copper, zinc, molybdenum, cobalt, and selenium. Elements such as aluminum, sodium, and vanadium are considered essential for some species while others are potentially toxic including chromium, nickel, lead, arsenic, and cadmium.

7.2.3.1 Environmental effects

7.2.3.1.1 Cadmium (Cd)

Cadmium is a silvery metal and is both toxic and carcinogenic [9]. The correlation between soil concentration of cadmium and plant uptake is not clear; consequently, there is concern about cadmium concentrations in the environment and ingestion of plant-based foods by the general population.

7.2.3.1.2 Lead (Pb)

Lead is the only metal currently listed as a criteria pollutant. Lead is a gray metal with a low melting point that is soft, malleable, ductile, resistant to corrosion, and a relatively poor electrical conductor [2]. For these reasons, lead has been used for over 4,000 years for plates and cups, food storage vessels, paints, piping, roofing, storage containers for corrosive materials, radiation shields, lead-acid batteries, and as an organolead additive in gasoline. As result, lead can be found throughout the world, including trace amounts in Antarctica and the Arctic [2]. In the past, automotive sources were the major contributor of lead emissions to the atmosphere [5]. With EPA's regulatory efforts to remove lead from gasoline, along with banning lead from paint pigments and solder, a decline in lead emissions has been observed. The highest concentrations of lead are found in the vicinity of nonferrous and ferrous smelters and battery manufacturers. Lead can be deposited on the leaves of plants, presenting a hazard to grazing animals.

7.2.3.1.3 Selenium (Se)

Crops, including animal forages are sensitive to the addition of small amounts of selenium in the soil [8]. Selenium is a silvery metallic allotrope or red amorphous powder [9]. Selenium is toxic to plants at low concentrations and has been shown to cause stunting and brown spots in some varieties of beans, and reduce germination and cause stunting in cereals and cotton.

7.2.3.1.4 Other trace elements

Clarke and Sloss [8] also listed examples where trace elements from fossil fuel, specifically coal, combustion may have beneficial effects in some areas. Boron is a micronutrient and is required in trace amounts by many plants and animals and the amount of boron released from coal-fired power plants is likely to be beneficial to local agriculture. Copper, iron, manganese, and zinc are necessary for normal growth of plants. It is recognized, however, that excessive concentrations of copper and zinc lead to damage to root formation and growth, but the quantities being deposited around coal-fired power stations are likely to be beneficial to local soils. Also, manganese can be harmful in large doses, especially in acidic soils. Fluorine has an adverse effect on forage [6].

7.2.3.2 *Health effects*

Trace element emissions can potentially cause a number of harmful effects on human health. While there is no evidence that most trace elements from coal-fired power plants are causing health effects due to their low ambient air concentration, there is concern that pollutants may become accumulated through the food chain. This is especially true of mercury, which was discussed in detail in Chapter 6. The health impacts of various trace elements are discussed.

7.2.3.2.1 Arsenic (As)

The combustion of most fossil fuels is unlikely to contribute toxic amounts of arsenic to air [11]. There is some concern, however, that arsenic in fly ash disposal and coal cleaning wastes may be leached into the groundwater. Arsenic, which is gray, metallic, soft, and brittle, may be considered essential; however, it is toxic in small doses [9]. Arsenic can cause anemia, gastric disturbance, renal symptoms, ulceration, and skin and lung cancer [8]. In addition, arsenic can damage peripheral nerves and blood vessels, and is a suspected teratogen (i.e. causes damage to embryos and fetuses). The chemical form of arsenic can affect its toxicity with organic forms of arsenic being more toxic than elemental arsenic.

An extreme example of chronic arsenic poisoning is occurring in the Gizhou Province of China [4,10]. In this province, the villagers bring their chili pepper harvest indoors in the autumn to dry. They hang their peppers over open-burning stoves where arsenic-rich coal, up to 35,000 ppm, is used to heat and cook. Chili peppers, which normally contain <1 ppm arsenic, can contain as much as 500 ppm arsenic after drying. About 3,000 people are exhibiting typical symptoms of arsenic poisoning including hyperpigmentation (flushed appearance, freckles), hyperkeratosis (scaly lesions on the skin, generally concentrated on the hands and feet), Bowen's disease (dark, horny, precancerous lesions of the skin), and squamous cell carcinoma.

7.2.3.2.2 Boron (B)

Boron is similar to arsenic in that the concentration emitted to the atmosphere is small and is unlikely to cause problems as an airborne pollutant [11]. However, boron in the fly ash can become soluble in ash disposal sites. Boron, a dark powder, is essential for plants but can be toxic in excess [9].

7.2.3.2.3 Beryllium (Be)

Beryllium, a silvery, lustrous, relatively soft metal, is toxic and carcinogenic [9]. It can cause respiratory disease and lymphatic, liver, spleen, and kidney effects [8].

7.2.3.2.4 Cadmium (Cd)

Cadmium has no known biological function and is therefore not a nutritional requirement [8]. Cadmium is toxic, carcinogenic, and teratogenic [9]. It can cause emphysema, fibrosis of the lung, renal injury, and possibly cardiovascular disease.

7.2.3.2.5 Chromium (Cr)

Chromium, a hard, blue-white metal, is an essential trace element; however, its chromates are toxic and carcinogenic [9]. Chromium, which is ingested by humans through food and drink, can be toxic when it accumulates in the liver and spleen [8]. The oxidation state of chromium affects its mobility and toxicity. Chromium (III) is nontoxic and has a tendency to absorb to clays, sediments, and organic matter and therefore is not very mobile. Chromium (IV), however, is more mobile and toxic and may make up approximately 5% of the total chromium particles emitted from power plants.

7.2.3.2.6 Fluorine (F)

Fluorine is a pale yellow gas and is the most reactive of all elements [9]. It is an essential element and is commonly used for protecting the enamel of teeth but excess fluoride is toxic. Fluorosis includes mottling of tooth enamel (dental fluorosis) and various forms of skeletal damage including osteosclerosis, limited movement of the joints, and outward manifestations such as knock-knees, bowlegs, and spinal curvature [10]. Fluorosis, combined with nutritional deficiencies in children, can result in severe bone deformation. An extreme example of this is exhibited in China where the health problems caused by fluorine volatilized during domestic coal use are far more extensive than those caused by arsenic [4,10]. More than 10 million people in Guizhou Province and surrounding areas suffer from various forms of fluorosis where corn is dried over unvented ovens burning high (>200 ppm) fluorine coal.

7.2.3.2.7 Manganese (Mn)

Manganese is unlikely to cause health problems as an airborne-pollutant from the combustion of most fossil fuels but leaching of ash may be a concern [11]. Manganese, a hard, brittle, silvery metal, is considered an essential nutrient, nontoxic, and a suspected carcinogen [9]. It is also reported to cause respiratory problems [8].

7.2.3.2.8 Molybdenum (Mo)

Molybdenum, a lustrous, silvery, and fairly soft metal, is an essential nutrient, moderately toxic, and a teratogenic [9]. Under some conditions, molybdenosis may occur in animals, notably ruminants, through eating vegetation with relatively high concentrations of molybdenum [11].

7.2.3.2.9 Nickel (Ni)

Nickel, a silvery, lustrous, malleable, and ductile metal, has no known biological role and nickel and nickel oxide are carcinogenic [9]. Nickel can cause dermatitis and intestinal disorders [8].

7.2.3.2.10 Lead (Pb)

Exposure to lead occurs mainly through inhalation of air and ingestion of lead in food, water, soil, or dust. It accumulates in the blood, bones, and soft tissues [5]. Lead can adversely affect the kidneys, liver, nervous system, and other organs. Excessive exposure to lead may cause neurological impairments, such as seizures, mental retardation, and behavioral disorders. Even at low doses, lead exposure is associated with damage to the nervous systems of fetuses and young children. Lead may be a factor in high blood pressure and heart disease.

7.2.3.2.11 Selenium (Se)

Selenium is considered an essential element but is toxic in excess of dietary requirements and is also a carcinogen [9]. Livestock consuming plants with excessive amounts of selenium can suffer two diseases known as alkali disease and blind staggers, as well as experience infertility and cirrhosis of the liver, and in extreme cases, death [8]. In humans, selenium can cause gastrointestinal disturbance, liver and spleen damage, and anemia, and is a suspected teratogen. Symptoms of selenium poisoning include hair and nail loss. Selenosis has been reported in southwest China where selenium-rich carbonaceous shales, locally known as "stone coals," are used for home heating and cooking [10]. The ash from his selenium-rich coal (as much as 8,400 ppm Se) is then used as a soil amendment, thereby introducing high concentrations of selenium into the soil that is subsequently uptaken by crops.

7.2.3.2.12 Vanadium (V)

Vanadium, a shiny, silvery, soft metal, is an essential trace element although some compounds, specifically vanadium pentoxide (V_2O_5), are quite toxic [8]. Health effects associated with vanadium include acute and chronic respiratory dysfunction [8].

7.2.3.2.13 Radionuclides

Radionuclides are listed generically as a 1990 CAAA HAP. Radioactivity arises mainly from isotopes of lead, radium, radon, thorium, and uranium [1,8,11,12]. Health effects from radiation are well documented including various forms of cancer; however, radionuclide emissions from power plants are quite low [1,8]. During coal combustion, most of the uranium, thorium, and their decay products are released from the original coal matrix and are distributed between the gas phase and solid combustion products [12]. Virtually 100% of the radon gas present in the coal feed is transferred to the gas phase and is lost in stack emissions. In contrast, less volatile elements such as thorium and uranium, and the majority of their decay products are retained in the solid combustion wastes [1,12]. Fly ash is commonly used as an additive to concrete building products, but the radioactivity of typical fly ash is not significantly different from that of more conventional concrete additives or other building materials such as granite or red brick [12].

7.3 Trace elements/hazardous organic compounds/ halogens in fossil fuels

7.3.1 Coal

7.3.1.1 Trace elements

The concentrations of trace elements in coals vary considerably between ranks of coal and geographic regions, which has been influenced by their depositional environment and the geologic processes occurring during the formation of the coal beds. The concentrations of trace elements in coals are provided in Tables 7.1–7.4.

Table 7.1 Concentrations of trace elements in 1500 bituminous and subbituminous coal samples from the U.S., UK, and Australia (Arithmetic mean values, ppm)

Element	U.S.	U.K.	Aus.	For most coals	Element	U.S.	U.K.	Aus.	For most coals
>50 ppm					**1–10 ppm**				
Barium (Ba)	150	70–300	70–300	20–1000	Antimony (Sb)	1.1–1.3	0.5–3.1	0.5	0.05–10
Boron (B)	50–102	30–60	30–60	5–400	Beryllium (Be)	1.6–2.0	1.5–1.8	1.5	0.1–15
Fluorine (F)	61–74	114–150	150	20–500	Cadmium (Cd)	1.3–2.5	0.08–0.4	0.08	0.1–3.0
Manganese (Mn)	49–100	84–130	130	5–300	Cesium (Cs)	–	1.3	1.3	0.3–5.0
Phosphorus (P)	71	–	–	10–3000	Cobalt (Co)	7–9.6	4.0	4.0	0.5–30
Strontium (Sr)	37–100	100	100	15–500	Gallium (Ga)	3.1	4.0	4.0	1.0–20
Titanium (Ti)	700–800	63–900	900	10–2000	Germanium (GE)	6.6	5.1–6.8	6.0	0.5–50
Zinc (Z)	39–272	25	25	5–300	Iodine (I)	2.0	–	–	–
10–50 ppm					Lanthanum (La)	6.9	16	16.0	–
Arsenic (As)	14–15	1.5–18	1.5	0.5–80	Molybdenum (Mo)	3.0–7.5	1.5 – <2.0	1.5	0.1–10
Cerium (Ce)	11	–	–	–	Niobium (Nb)	3.0	–	–	1.0–20
Bromine (Br)	15	–	–	–	Scandium (Sc)	2.4–3.0	4.0	4.0	1.0–10
Chlorine (Cl)	–	150	150	50–200	Selenium (Se)	2.1–4.1	0.8–2.8	0.8	0.2–4.0
Chromium (Cr)	14–15	6–34	6	0.5–60	Thallium (Tl)	–	–	–	<0.2–1.0
Copper (Cu)	15–19	15–48	15	0.5–50	Thorium (Th)	2.0	2.7–3.9	2.7	0.5–10
Lead (Pb)	16–35	10–48	10	2–80	Uranium (U)	1.6–1.8	1.3–2.0	2.0	0.5–10
Lithium (Li)	20	20	20	1–80	**<1 ppm**				
Nickel (Ni)	15–21	15–28	15	0.5–50	Mercury (Hg)	0.18–2.0	0.1–0.2	0.1	0.02–1.0
Rubidium (Rb)	14	–	–	2–50	Silver (Ag)	0.2	<0.1	<0.1	0.02–2.0
Vanadium (V)	20–33	20–76	20	2–100	Tantalum (Ta)	0.15	–	–	–
Zirconium (Zr)	30–72	100	100	5–200					

Table 7.2 Concentrations of trace elements in some bituminous coals (in ppm, dry coal)

Element	Country					
	U.S.	Australia	Canada	South Africa	U.K.	Germany
As, arsenic	<1–170	<0.1–16	0.2–240	0.3–10	2–73	1.5–50
B, boron	<0.6–160	1–300	9–360	11–109	0.5–160	2–236
Ba, barium	4–570	<40–1000	10–1000	86–474	<6–500	45–350
Be, beryllium	0.2–14	<0.4–8	–	–	<0.6–17	0.9–3.5
Cd, cadmium	<0.004–9	0.05–0.36	–	0.1–0.8	0.02–5	0.02–21
Cl, chlorine	130–2400	20–1000	30–1160	19–341	200–9100	1350–2450
Co, cobalt	0.8–90	<0.6–30	0.2–21	2–14	0.4–60	7–30
Cr, chromium	2–84	<1.5–80	2.1–95	12–70	1–45	4–80
Cu, copper	3–160	2.5–70	0.2–52	4.2–29	7–240	10–60
F, fluorine	<13–1900	15–458	31–890	50–465	27–202	20–370
Hg, mercury	0.01–1.8	0.026–0.40	0.02–1.3	0.08–7.0	0.03–2.0	0.1–1.4
Mn, manganese	1–1400	2.5–900	2–600	11–272	1–600	55–68
Mo, molybdenum	<0.1–16	<0.3–6	0.4–13	<1–2.7	0.1–20	6–30
Ni, nickel	1.4–130	0.8–70	2–38	6.9–32	3–60	15–95
P, phosphorus	1–1200	30–4000	45–5200	20–2490	<10–1000	40–1240
Pb, lead	<1–62	1.5–60	1.8–53	1.9–25	1–900	0.1–390
Sb, antimony	<0.1–7.4	<0.1–1.7	0.1–16	0.24–0.38	1–10	0.14–5.0
Se, selenium	<0.6–20	0.21–2.5	<0.1–8.0	<0.4–0.9	0.3–5.1	0.6–5.5
Sn, tin	<0.3–3	0.6–15	2–15	<1–11	0.3–75	3.6–3.9
Th, thorium	<3–26	<0.2–8	0.1–9	2.7–21	0.7–6.7	1.6–4.4
Tl, thallium	<0.3–13	<0.05–1.6	–	–	0.6–1.7	0.01–0.72
U, uranium	<0.1–15	0.4–5	0.4–12	1.2–7.3	1.1–3.0	0.3–2.2
V, vanadium	2–120	4–100	3.4–200	17–132	3–150	31–179
Zn, zinc	1–1600	9–500	2.0–62	3.2–19	3–1700	14–1742

Table 7.3 Concentrations of trace elements in some low-ranks coals (in ppm, dry coal)

Coal rank:	Brown coal	Lignite	Subbituminous coal		
Country:	Australia	U.S.	U.S.	Canada	Australia
Element					
As, arsenic	<0.01–1.3	<0.1–31	0.1–420	4–53	<0.1–36
B, boron	2–70	0.8–500	–	5–32	1–14
Be, beryllium	<0.05–2.0	<0.1–15	0.05–32	–	<0.5–12
Cd, cadmium	<0.01–0.10	<0.1–5.5	0.03–3.7	–	–
Co, cobalt	<0.1–10	0.1–43	0.06–70	1.1–31	<5–321
Cr, chromium	0.08–19	0.3–87	0.54–70	5.7–64	<5–117
Cu, copper	0.2–30	0.3–433	0.16–120	14–80	<1–747
Hg, mercury	<0.01–1.3	<0.01–12	0.01–8.0	0.02–0.44	–
Mn, manganese	0.45–200	1–1075	1.4–3500	1.6–1000	<2–667
Mo, molybdenum	<0.02–0.9	<0.3–280	0.13–41	<0.8–9	<0.1–6
Ni, nickel	0.1–18	0.52–84	0.32–69	–	2–161
Pb, lead	<0.2–15	0.3–129	0.7–76	1.9–19	<1–595
Sb, antimony	<0.01–0.5	0.1–5.2	0.04–43	0.1–2.4	–
Se, selenium	<0.02–1.5	0.10–18	0.10–16	0.5–2.0	–
Th, thorium	<0.01–3.5	0.28–28	0.08–54	<0.1–9	<0.4–26
U, uranium	0.01–1.8	<0.1–17	0.06–76	0.2–2.4	<1–58
V, vanadium	0.1–24	0.1–140	0.14–370	23–300	2–279
Zn, zinc	0.5–129	1.0–486	0.88–910	7.6–83	<2–273

Different sources of data provide generalized concentrations by country, generalized concentrations by rank, and specific concentrations by state (in the U.S.), rank, and seam. In addition, the various sources report the concentrations for different combinations of trace elements. Tables 7.1–7.4 provide an overview of trace element concentrations in coals.

Table 7.1 lists the concentrations of trace elements in 1500 bituminous and subbituminous coal samples from the U.S., United Kingdom, and Australia [6]. These data are based on several studies and are grouped into ranges of arithmetic mean values. The concentrations from both bituminous and subbituminous coals are reported in the same range of values.

Tables 7.2 and 7.3 list the concentrations of trace elements in some bituminous coals and low-rank coals, respectively, by various countries [13]. These data have been compiled from various sources.

Table 7.4 lists the concentrations of selected trace elements in U.S. coals for the range of ranks of coal found in the U.S. The examples provided in Table 7.4 include anthracite, semi-anthracite, several ranks of bituminous coal – low volatile, medium volatile, high volatile A, high volatile B, and high volatile C – subbituminous B coal, and lignite. These data have been compiled from Penn State's Coal Sample Bank [14,15].

Table 7.4 Concentrations of selected trace elements in U.S. coals (ppm whole coal, dry)

Rank	Seam	State	Trace elements (ppm whole coal, dry)						
			Chlorine Cl	Chromium Cr	Mercury Hg	Nickel Ni	Lead Pb	Strontium Sr	Vanadium V
Anthracite	Mammoth	PA	1300	41	–	21	–	82	58
	Buck Mtn.	PA	100	44	0.05	27	11	17	24
	#8 Leader	PA	100	27	0.08	15	<3	11	22
Semi-anthracite	PA #8	PA	400	71	–	53	19	5	76
	Gunnison	CO	100	4	–	4	–	178	11
	L. Spadra	AR	100	16	0.03	29	10	116	20
Low Vol. bituminous	L. Kittanning	PA	1300	26	–	10	–	203	37
	Elk Lick	WV	700	32	–	15	8	152	38
Medium Vol. bituminous	U. Freeport	PA	100	19	0.16	20	13	172	26
	Dutch Creek	CO	700	3	–	<1	–	200	6
	Sewell	WV	2600	11	–	11	–	104	11
High Vol. A bituminous	Pittsburgh #8	PA	2300	21	–	11	–	111	7
	L. Kittanning	WV	600	31	–	18	–	39	28
	U. Kittanning	WV	3700	18	–	6	–	62	7
	Blind Canyon	UT	1200	6	–	<0.4	–	85	9
	L. Elkhorn	KY	900	21	0.08	15	7	73	25
High Vol. B bituminous	Hazard #5	KY	1100	31	–	24	10	297	34
	Hiawatha	UT	300	19	–	24	8	262	8
	Kentucky #9	KY	1600	23	–	8	–	21	19
	Illinois #9	IL	–	–	–	–	–	–	–

(Continued)

Table 7.4 (Continued)

Rank	Seam	State	Trace elements (ppm whole coal, dry)						
			Chlorine Cl	Chromium Cr	Mercury Hg	Nickel Ni	Lead Pb	Strontium Sr	Vanadium V
High Vol. C bituminous	Wadge	CO	100	4	–	<2	–	175	7
	Illinois #5	IL	800	10	–	13	–	17	14
	Indiana #6	IN	500	14	–	6	–	24	19
Subbituminous B	Wyodak	WY	–	3	0.12	2	5	164	11
	Rosebud	MT	200	7	–	7	5	225	5
Lignite A	Beulah	ND	800	5	–	<0.4	–	772	14
Lignite	L. Wilcox	TX	110	17	–	5	–	14	34

7.3.1.2 Hazardous organic compounds

Coal does not consist of organic hazardous compounds per se. Coal consists of aromatic clusters with 1 to 10 fused aromatic rings linked together by aliphatic bridges [15]. The aromatic structures provide the backbone of the fuel. These structures include functional groups and heteroatoms linked either to the aromatic structures or to the bridges. The number of fused aromatic rings provides critical insights into the rank of the coal. Lignites typically contain single aromatic rings linked by open aromatic structures. Subbituminous coals and bituminous coals contain 2 to 4 fused aromatic rings on average. Anthracites contain many fused aromatic rings and far fewer bridges, functional groups, and heteroatoms. Upon combustion these hydrocarbon structures are not released as hazardous air pollutants but are converted to water and carbon dioxide provided there is complete combustion. This is discussed later in the chapter.

7.3.1.3 Halogens

The halogens chlorine, fluorine, bromine, and iodine are found in coal with chorine being the element of most interest because of its role in corrosion and deposition and its presence in coal in higher concentrations. Some examples of concentrations of all four halogens are provided in Table 7.1, chlorine and fluorine concentrations are provided in Table 7.2, and chlorine concentrations are listed in Table 7.4. Because of chlorine's contribution to a number of combustion phenomena (corrosion, deposition, mercury capture), more detailed chlorine concentrations are provided in Table 7.5 (chlorine concentrations in world-wide coals), Table 7.6 (chlorine concentrations in U.S. coals by state/region), Table 7.7 (chlorine concentrations in selected Eastern bituminous coals), and Table 7.8 (chlorine concentrations in selected western U.S. coals) [16].

Chlorine occurs in coal in several forms: chlorine in saline coals (e.g. as NaCl), true organic chlorine covalently bonded in the coal macromolecule, and as organically associated chlorine in the form of anion chlorine adsorbed on the coal organic surface in pores, and being surrounded by pore moisture [16]. Chloride anions adsorbed on the pore walls are probably the dominant form of chlorine in coal. The mode of occurrence of fluorine is primarily association with minerals [17]. Bromine, like chlorine, is generally associated with the organic matter of the coal [17].

7.3.2 Petroleum/fuel oils

7.3.2.1 Trace elements

Similar to coal, the concentrations of trace elements in crude oils vary considerably, which is influenced by their depositional environment and source rock type. Unlike coals, which are utilized as mined (recognizing that many coals undergo processing to reduce mineral matter content), crude oils undergo refining to produce various products for utilization. In Chapter 6 it was mentioned that mercury is removed during crude oil processing because of its detrimental effect on refinery catalysts. However, other trace elements are found in petroleum products after refining. As with

Table 7.5 Chlorine concentrations in some world-wide coals (ppm, dry basis)

Country	Coal	Chlorine concentration (ppm)
Bulgaria	Maritza West	150
	Sofia	80
	Elhovo	90
	Maritza East	200
	Bobov Dol	360
	Balkan	150
Australia	Ebenezer	370
	Wambo	360
	Blair Athol	440
	Lithow	480
	Moura	710
U.S.	Usibelli	90
	Black Thunder	200
	Illinois	750
Japan	Taiheiyo	1,090
	Akabira	110
	Sunagawa	200
	Takashima	230
Canada	Coal Valley	140
	Fording	280
South Africa	Ermelo	260
China	Datong	210
Ukraine	Donbass	500

coal studies, the types of trace elements reported in crude oils and refined products varies between studies. While most of the coal consumed in the U.S. is mined from U.S. coalfields, the U.S. imports oil from around the world. Therefore, the concentrations of trace elements in crude oils produced from various regions of the world are provided in this section. This section also includes concentrations of refined products.

Table 7.9 lists the ranges of concentration of nickel and vanadium from 76 crude oil samples from six geographic regions [18]. These regions are Abu Dhabi, Suez, North Sea, China, Indonesia, and the Gippland Basin in Australia. Nickel and vanadium are the two most abundant metals in petroleum.

The concentrations of 17 metals in 14 crude oils from Saudi Arabia are given in Table 7.10 [19]. The most abundant trace metals were vanadium and nickel with concentrations ranging from 2 to 60 ppm and 0.5 to 17 ppm, respectively.

The concentrations of 22 trace elements in 86 conventional crude oils from Alberta, Canada were determined and the results for 13 of the trace elements

Table 7.6 Concentrations of chlorine in U.S. coals by state or selected region (ppm, dry coal basis)

Province/Region	Arithmetic mean	Maximum	Standard deviation
Eastern	730	8,760	680
Interior	540	3,000	600
Pacific Coast	180	560	120
Alaska	150	4,900	470
Rocky Mountain	150	3,400	300
Gulf	120	900	110
Northern Great Plains	100	1,370	100
Appalachian	730	8,760	680
Green River	180	3,400	420
Powder River	100	1,370	120
Unita	170	2,100	230
Texas Lignite	120	900	110
Fort Union	90	350	50
Pennsylvania	160	360	90

Table 7.7 Chlorine concentrations in selected eastern U.S. bituminous coals (ppm, as received)

State	Coal region	Avg. Cl	Max. Cl	2nd Max. Cl[a]
Alabama	Southern Appalachian – Cahaba	369	1,500	1,200
Alabama	Southern Appalachian – Warrior	283	3,300	2,900
Georgia	Southern Appalachian	872	1,600	1,500
Kentucky	Central Appalachian	1,148	8,800	4,300
Maryland	Northern Appalachian – Castleman	800	1,100	1,100
Ohio	Northern Appalachian	730	3,300	2,500
Pennsylvania	Northern Appalachian/Main	956	2,600	2,400
Virginia	Central Appalachian	502	2,200	2,000
West Virginia	Central Appalachian	1,262	8,200	2,500
West Virginia	Northern Appalachian	949	1,600	1,500

[a]To ensure that the maximum chlorine content was not an outlier, the second highest chlorine content in the population of coal samples is also reported.

are provided in Table 7.11. The crude oils were grouped into three families based on previous analyses of sulfur content and type and concentration of hydrocarbons present (e.g. aromatic, aliphatic). The minimum–maximum range concentrations reported in Table 7.11 are for all three families combined into one range [20]. Each family reported an average concentration and the range of the three averages is reported in Table 7.11.

Examples of selected trace metal contents of in U.S. crude oils are shown in Table 7.12 [21]. The ranges of values of these elements in oil are 0.0002 to 106 ppm vanadium, 0.03 to 35 ppm nickel, 0.03 to 1.7 ppm copper, and 0.00004 to 0.013 ppm uranium.

Table 7.8 Chlorine concentrations in selected western U.S. coals (ppm, as received)

State	Coal region	Coal type	Avg. Cl	Max. Cl	2nd Max. Cl[a]
Alaska	Beluga	Subbituminous	100	100	N/A
Alaska	Nenana	Subbituminous	70	N/A[b]	N/A
Colorado	Green River	Bituminous	129	1,000	550
Colorado	Uinta	Bituminous	100	200	100
Montana	Fort Union	Lignite	77	200	100
Montana	PRB	Subbituminous	88	400	300
N. Dakota	Fort Union	Lignite	110	350	300
Utah	Uinta	Bituminous	101	300	200
Wyoming	Hanna Basin	Subbituminous	75	140	100
Wyoming	Rock Springs	Subbituminous	149	1,200	400
Wyoming	PRB–Powder River	Subbituminous	73	100	90
Wyoming	PRB–Sheridan	Subbituminous	79	100	93

[a]To ensure that the maximum chlorine content was not an outlier, the second highest chlorine content in the population of coal samples is also reported.
[b]Not applicable.

Table 7.9 Nickel and vanadium concentrations for 76 crude oils from six geographic regions

Region	Number of oils	Nickel (Ni), ppm	Vanadium (V), ppm
Abu Dhabi	6	4–28	8–92
Suez	9	4–148	10–190
North Sea	24	1–22	<2–26
China	15	<1–21	<1–5
Indonesia	12	0.12–3	<2
Australia	10	<1	<1

Ultimately, it is the concentration of the trace elements in the refined products that are utilized in combustion systems that are of concern. Table 7.13 lists the concentrations of 18 different trace elements from two crude oil samples (western Canada and Venezuela) and four heating and residual oils [22]. The concentrations of selected trace metals for 102 heating oil samples and 16 residual fuel oil samples utilized in northeastern U.S. are provided in Table 7.14 [23].

7.3.2.2 Hazardous organic compounds

Unlike coal, distillate and residual fuel oils used for combustion do contain volatile organic compounds and polycyclic aromatic hydrocarbon (PAHs) as part of their overall composition [24,25]. Volatile organic compounds include benzene, toluene, ethylbenzene, and xylenes while PAHs include phenanthrene, 1- and 2-methylphenanthrene, fluoranthene, pyrene, benz[a]anthracene, chrysene, triphenylene, benzo[a]pyrene, benzo[e]pyrene, perylene, and others. These compounds have

Table 7.10 Concentrations of trace metals in crude oils from Saudi Arabia

Crude field	Metals (ppm)															
	Ag	Al	Ca	Cd	Cr	Cu	Fe	K	Mg	Mn	Na	Ni	Pb	Si	Sn	V	Zn
1	0.31	1.02	0.38	0.57	0.59	0.21	1.48	<0.05	0.33	<0.05	<0.05	0.55	0.45	0.56	<0.05	2.23	0.51
2	0.39	0.89	1.15	0.43	0.90	0.88	3.44	0.31	2.43	<0.05	13.39	1.61	0.40	1.25	1.74	6.28	1.12
3	0.06	1.10	0.40	0.77	0.49	0.12	0.24	<0.05	0.15	<0.05	0.23	3.94	0.48	0.36	<0.05	17.17	0.23
4	0.15	0.75	0.93	1.76	0.78	0.31	0.55	0.25	0.24	<0.05	6.95	4.80	1.35	0.56	<0.05	18.44	0.55
5	<0.05	0.89	0.50	0.40	<0.05	<0.05	2.40	0.05	0.10	<0.05	2.11	5.20	<0.05	0.05	<0.05	20.30	0.14
6	0.05	1.22	2.05	1.10	0.69	0.69	0.96	1.10	0.24	0.05	21.03	6.24	1.06	4.08	<0.05	17.70	0.88
7	0.15	1.40	1.15	2.37	1.18	0.21	0.65	0.40	0.70	0.05	2.00	15.90	0.87	0.36	<0.05	48.80	1.82
8	<0.05	1.10	1.26	0.14	0.10	0.12	<0.05	0.75	0.15	0.05	8.66	16.90	0.00	0.26	<0.05	52.50	0.23
9	0.15	1.20	4.00	0.64	0.59	0.31	6.64	1.60	1.24	0.15	34.88	18.30	0.97	0.66	<0.05	54.60	2.55
10	<0.05	0.40	<0.05	0.19	0.51	0.03	3.21	<0.05	<0.05	<0.05	0.75	6.18	1.57	0.89	<0.05	17.19	0.11
11	0.05	1.10	0.40	0.52	0.52	0.15	0.41	0.06	0.25	<0.05	2.20	0.55	0.25	0.40	<0.05	2.20	0.12
12	<0.05	1.3	0.40	0.60	0.57	0.20	1.06	<0.05	0.10	<0.05	1.20	3.21	0.50	0.41	<0.05	15.53	0.25
13	0.10	1.7	1.20	0.95	0.62	0.30	0.78	<0.05	0.26	<0.05	2.50	8.24	0.36	0.50	<0.05	27.89	0.11
14	0.15	1.5	1.30	2.20	1.05	0.21	0.95	0.15	0.50	<0.05	3.00	16.70	0.65	0.40	<0.05	57.90	0.18

Fields:
1. Fadhili.
2. Abqaiq.
3. Ghawar.
4. Abu-Safah.
5. Khursaniyah.
6. Zuluf.
7. Safaniya.
8. Marjan Well.
9. Marjan Lower.
10. Bahrain.
11. Arabian Berri.
12. Arabian Light.
13. Arabian Medium.
14. Arabian Heavy.

Table 7.11 Concentrations of trace elements in Alberta, Canada crude oils

Element	Number of samples reporting element	Concentration (ppb)	
		Minimum – Maximum	Average
Zn, zinc	56	11.51–5921.0	272–573
Hg, mercury	35	1.34–398.6	27.6–85.6
As, arsenic	64	0.020–1988.0	11.4–196
Sb, antimony	21	0.10–34.8	0.343–6.15
V, vanadium	75	3–138,800	834–28,100
Se, selenium	53	3.74–252.3	12.9–78.7
Cr, chromium	37	3.63–205.8	8.39–41.6
Cl, chlorine	58	0.50–1014.0	9.48–77.5
Br, bromine	72	1.75–12,470.0	32.3–923
I, iodine	45	7.771–9005.0	410–1077
Mn, manganese	70	1.0–176.6	47.1–54.9
Co, cobalt	75	0.261–202.6	8.05–22.7
Ni, nickel	61	0.022–57.48	0.939–11.5

Table 7.12 Concentrations of trace metals in selected U.S. crude oils

State	Field	Metal concentration (ppm)			
		Vanadium	Nickel	Copper	Uranium
Arkansas	Schuler	15	10	0.34	0.0025
	Stevens-Smart	1.8	2.1	1.4	0.00048
California	Wilmington	35	35	0.22	0.00088
Colorado	Gramps (Dakota)	0.42	0.76	0.03	0.00017
	Gramps (Morrison)	0.96	1.6	0.03	0.00050
Kansas	Brewster	2.6	1.7	0.32	0.00028
	Coffeyville	3.2	1.0	0.04	0.00024
	Iola	16	9.0	0.09	0.00260
Montana	Big Wall	24	13	0.12	0.00012
N. Mexico	Table Mesa	0.002	0.03	1.7	0.00054
Oklahoma	Kendrick	0.84	0.22	1.3	0.00032
	Laffoon	44	20	0.18	0.0020
Utah	Roosevelt	0.16	3.0	0.05	0.00004
Wyoming	Elk Basin	38	9.2	0.16	0.00030
	Grass Creek	106	21	0.38	0.00076
	Halfmoon (Embar)	99	28	1.5	0.0017
	Halfmoon (Tensleep)	51	0.23	0.23	0.00069
	Hamilton Dome (Curtis)	106	24	0.38	0.0015
	Hamilton Dome (Embar)	55	9.1	0.19	0.00036
	Hamilton Dome (Madison)	106	27	0.27	0.0011
	Lost Soldier	0.56	0.72	0.03	0.00024
	Oregon Basin, N. (Embar)	60	11	0.05	0.0065
	Oregon Basin, N. (Tensleep)	72	15	0.40	0.013
	Oregon Basin, N. (Madison)	77	22	0.11	0.0077

Table 7.13 **Concentrations of trace elements in selected petroleum products (in ppb)**

Element	Crude oils		Heating oils[a]			Fuel Oil No. 6
	Canada	Venezuela	Exxon	Shell	Texaco	
As, arsenic	6.8	6.8	0.64	0.36	0.68	37
Br, bromine	40	190	9.3	4.9	6.6	260
Cl, chlorine	140	6640	360	800	9400	200,000
Co, cobalt	29	48	<20	<20	<20	70
Fe, iron	21,000	<5000	<10,000	<10,000	<10,000	<10,000
I, iodine	24	7	3.0	1.1	4.9	70
Mn, manganese	<0.5	<0.5	1.2	2.5	1.6	100
Mo, molybdenum	<10	<10	<5	<5	<5	170
Na, sodium	62	1930	550	81	410	31,200
Ni, nickel	5000	19,000	<1500	<1500	<1500	<1500
Se, selenium	<50	150	<50	<50	<50	<50
V, vanadium	3730	72,000	<30	47	<60	128,000

[a]Pre-merger companies.

been measured in distillate and residual fuels oils in concentrations ranging from less than one to approximately 8,000 ppm. However, as with coal, upon combustion these hydrocarbon structures are not released as hazardous air pollutants but are converted to water and carbon dioxide provided there is complete combustion (which will be discussed later in the chapter); hence, detailed compositional data for VOCs and PAHs are not given in this section.

7.3.2.3 Halogens

Halogens are naturally occurring elements in most crude oils with the concentrations varying among different sources (see Tables 7.11 and 7.13) [20]. Chlorides exist in crude oil in two forms: organic chloro-hydrocarbons and inorganic chloride such as compounds of magnesium, sodium, calcium [26]. Crude oils have varying salt distributions and a typical distribution is 75% NaCl, 15% $MgCl_2$, and 10% $CaCl_2$ [26]. Typically, chlorine is the most abundant halogen that is found in fuel oils and the sources are natural as well as contamination from brackish water (i.e. water containing chloride salts such as $MgCl_2$, $CaCl_2$, and NaCl) from the wells, transportation in ocean vessels (from ballasting of tankers with seawater), and processing of crude oils in refineries (i.e. from fluidized catalytic cracker catalysts) [27].

7.4 Formation and emissions of acid gases

In this section, acid gases refer specifically to halogens and to a lesser extent sulfur trioxide (SO_3). Nitrogen oxides and sulfur dioxide are also considered acid gases in that they contribute to acid rain; however, these gases were discussed in previous chapters.

Table 7.14 **Concentrations of trace metals in home heating oil/diesel and residual fuel oil (in ppb)**

Fuel		Metals concentration (ppb)									
Type		**Ni**	**Zn**	**As**	**Se**	**Hg**[a]	**V**	**Mn**	**Co**	**Sb**	**Pb**
Light distillate	Range	<3–9	<6–66	<1–10	<5–11	<1–13	<4–20	<5–114	<6–ND[c]	<10–ND	<6–144
	Average	3.2	14.8	2.1	4.3	1.9–2.0	NR[b]	NR	NR	NR	NR
Residual	Range	11,900–22,600	813–4960	<20–523	<125–197	1–2	849–8940	1820–4870	697–1650	3880–16,900	<10–ND
	Average	16,988	1963	172	119	1.3	2967	2851	1113	8873	188

[a]Ranges are from two different analytical techniques.
[b]Not reported.
[c]Not determined.

7.4.1 Formation of acid gases

Sulfur trioxide (SO_3) is formed directly in the combustion of sulfur-containing fuels, e.g. coals and fuel oils, and indirectly by the conversion of small quantities of flue gas sulfur dioxide (SO_2) to SO_3 in the presence of iron, some ash constituents like vanadium, and some selective catalytic reduction (SCR) catalysts [28]. Under normal operating conditions in coal-fired boilers without SCR systems, about 0.75% of the SO_2 is typically converted to SO_3. High vanadium containing ashes can increase this amount to 0.7 to 1.5% of the SO_2 concentration but conversions as high as 5% of the SO_2 concentration have been reported in coal-fired boilers [29]. Boiler oxidation rates forming SO_3 can run as high as 10% of the SO_2 concentration for residual oil-fired boilers [29].

The oxidation of SO_2 to SO_3 occurs via two distinct mechanisms – one involving homogeneous gas phase reactions and the other involving surface heterogeneous processes [30]. The main homogeneous reactions are:

$$SO_2 + O_2 \leftrightarrow SO_3 + O \tag{7.1}$$

$$SO_2 + O + M \leftrightarrow SO_3 + M \tag{7.2}$$

where M is a general third molecule. Reaction (7.1) is dominant at temperatures below 1650°F. At higher temperatures, reaction (7.2) becomes more significant.

As the flue gas temperature falls below 1000°F, SO_3 begins to react with water in the flue gas to form sulfuric acid vapor (H_2SO_4). Sulfur acid can condense on metal surfaces causing corrosion and if the concentration is high can result in a visible blue plume or opacity issues. Also, in units equipped with SCR systems, excess ammonia (NH_3) leaving the SCR can react with SO_3 to form sticky, acidic particles of ammonium bisulfate (NH_4HSO_4), which can foul air heaters.

Halogens are released from coal and fuel oil combustion primarily in the form of the acid gases hydrogen chloride (HCl) and hydrogen fluoride (HF). Free halogens are not common as they quickly react with hydrogen donors to form HCl, HF, and to a much lesser extent HBr and HI. After SO_2 and NO_x, HCl emissions are probably the major manmade contribution to atmospheric acidity [31]. HCl is acidic because it dissolves in water to produce the H^+ ion.

7.4.2 Emissions of acid gases

Sulfur trioxide concentrations in the flue gas prior to any control technology generally range from a few ppm for low-sulfur coals to 30–70 ppm or more for very high sulfur coals [31]. The ability to remove SO_3 varies considerably with control technology and ranges from 30% to in excess of 90%, thereby resulting in emissions ranging from a few ppm to 20–40 ppm for both coal and residual fuel oil combustion. Control technologies are discussed later in the chapter.

As discussed in detail in Chapter 6, the Electric Power Research Institute (EPRI) performed an assessment of hazardous air pollutant (HAP) emissions and the consequent health risks by inhalation for all U.S. coal-fired electric generation units [33]. A detailed discussion of the methodology used and emission estimates generated are provided in the final report published by EPRI in 2009 [33] with a summary of the methodology discussed in Chapter 6. Table 7.15 summarizes estimated emission rates for acid gas HAPs substances from coal-fired units normalized on a lb/T (trillion) Btu heat input basis. Results are presented for all units, for units with particulate matter control only, and for units with particulate matter and SO_2 controls. Significant differences are seen in the normalized emission rates for HCl, Cl_2, and HF for units with particulate matter only controls compared to units with particulate matter and SO_2 controls.

EPRI also summarized emissions estimates on an annual basis at the station level for 462 operating stations (808 stack emission points) [33]. These results are given in Table 7.16.

EPRI also evaluated the HCl emissions by fuel type and flue gas desulfurization (FGD) design [34]. This evaluation of the 130 best performing units is summarized in Table 7.17. The overall average HCl emissions for this group of best performing units is 260 lb/T Btu input. Eighty-five of the 87 bituminous coal-fired facilities used either a wet or dry FGD system. Ten low chloride subbituminous coal-fired units without FGD controls appear in the group of 130 best performing units.

Table 7.15 Emission rate summary for acid gas HAPs for coal-fired units (lb/T Btu heat input)

	HCl	Cl_2	HF
All Units			
Median	3,020	1,471	1,678
Mean	22,289	2,357	2,410
Maximum	123,973	32,435	7,277
Minimum	34.1	16.1	20.9
Particulate Control Only			
Median	13,669	1,979	3,130
Mean	29,945	2,975	3,147
Maximum	123,973	32,435	7,277
Minimum	1073	58.4	948
Particulate and SO_2 Controls			
Median	553	551	308
Mean	662	656	285
Maximum	2,202	2,202	731
Minimum	34.1	16.1	20.9

Table 7.16 Annual acid gas emission summary at the station level (lb/year)

	Annual emissions at station level (lb/year)			
	Median	Mean	Maximum	Minimum
HCl	98,600	762,000	16,500,000	252
CL_2	33,100	95,700	2,790,000	14
HF	39,700	85,200	1,180,000	109

Table 7.17 HCl emissions by fuel type and control type for the 130 best performing coal-fired units

Coal type	FGD Type	Average (lb/T Btu)	Number of units
All coals	–	260	130
Bituminous	None	670	2
Eastern bituminous	Dry FGD	99	14
Western bituminous	Dry FGD	220	6
Bituminous[a]	Wet FGD	310	65
Lignite	Wet FGD	94	1
Subbituminous	None	420	10
Subbituminous	Dry FGD	210	11
Subbituminous	Wet FGD	150	11
Bituminous	FBC	100	4
Lignite	FBC	220	4
Waste coal	FGC	160	2

[a]5 of the 65 bituminous units were located in CO and AZ, the rest of the units were located in the east.

Electric utilities are the top industrial source of HCl emissions into the air, accounting for 80% of all HCl air emissions [35]. According to EIP using U.S. EPA Toxics Release Inventory, the electric utilities released approximately 82,400 short tons of HCl into the air in 2010 [35]. In 2010, 479 power plants reported HCl emissions. Table 7.18 lists the top 20 producers of HCl emissions, which account for approximately 33% or 27,260 short tons of HCl emissions.

As previously discussed in Chapter 6, according to EPA there are 14,111 major source boilers of which approximately 12% (or 1,753 boilers) need to meet emission limits to minimize air toxics [36]. A breakdown of the 14,111 boilers, fuels they fire, and their baseline (i.e. no HCl controls) HCl emissions are provided in Table 7.19. The 14,111 boilers emit 54,227 short tons of HCl per year. The subset of 1,753 boilers emit approximately 50,000 short tons of HCl per year or about 92% of the total HCl emitted. The breakdown of the 1,753 boilers with firing rates greater than 10 million Btu/hour is [37]:

- 487 boilers firing biomass
- 601 boilers firing coal

Table 7.18 Top 20 power plants emitting HCl in 2010

Rank	Facility	HCl (lb)	State
1	Muskingum River Plant	5,300,000	OH
2	Big Sandy Plant	4,900,00	KY
3	Monroe Power Plant	4,500,000	MI
4	EME Homer City Generation LP	4,100,105	PA
5	Crystal River Energy Complex	4,100,000	FL
6	Branch Steam Electric Generating Plant	3,800,000	GA
7	Keystone Power Plant	3,500,000	PA
8	Yates Steam Electric Generating Plant	3,100,000	GA
9	Cardinal Plant	2,900,000	OH
10	Johnsonville Fossil Plant	2,700,000	TN
11	Beckjord Generating Station	2,460,000	OH
12	Plant Watson	2,400,000	MS
13	State Line Generating Plant	2,360,000	IN
14	Shawville Station	2,300,005	PA
15	Indian River Generating Station	2,300,000	DE
16	LV Sutton Electric Plant	2,100,000	NC
17	Philip Sporn Plant	2,100,000	WV
18	Gallagher Generating Station	2,010,000	IN
19	Portland Power Plant	2,000,005	PA
20	Amos Plant	2,000,000	WV

Table 7.19 HCl emissions from industrial boilers

Fuel	Number of boilers	HCl emissions (short tons/year)
Biomass	487	2,541
Coal	601	46,830
Liquid	777	2,337
Non-continental liquid	44	202
Gas	10,883	2,063
Gas-metal furnace	698	42
Process gas	125	131
Limited use	479	81
TOTAL	14,111	54,227

- 293 boilers firing heavy liquid
- 252 boiler firing light liquids
- 42 non-continental liquid boilers
- 78 process gas boilers

The 601 coal-fired industrial boilers, comprising 34% of the 1,753 boilers, emit 46,830 short tons of HCl per year, or approximately 86% of the 54,227 short tons of HCl per year.

Table 7.20 Top 20 states/non-continental territories reporting industrial boiler HCl emissions

Ranking	State	Emissions (short tons/yr)
1	North Caroline	4,932.4
2	Michigan	4,345.8
3	Iowa	4,149.2
4	West Virginia	3,621.3
5	Ohio	3,368.0
6	Illinois	2,733.9
7	Wisconsin	2,569.6
8	Minnesota	2,235.3
9	Indiana	2,162.1
10	Virginia	2,129.9
11	Tennessee	2,077.6
12	Wyoming	1,987.8
13	Pennsylvania	1,912.5
14	South Carolina	1,866.7
15	Missouri	1,796.2
16	Alabama	1,594.8
17	Georgia	1,422.3
18	New York	1,202.9
19	Florida	683.5
20	Colorado	631.3
Total		47,422.9

Table 7.20 lists the 20 states/non-continental territories that reported the highest levels of HCl emissions, which were 47,423 short tons of HCl per year. This was approximately 87% of the 54,227 short tons of HCl emitted per year [37].

There are 665 boilers firing liquid fuels or natural gas that are subject to the industrial boiler MACT, which comprise 38% of the 1,753 boilers. These boilers contributed 4,856 short tons of HCl per year of the approximately 47,423 short tons of HCl emitted per year or approximately 10% of the total HCl emissions.

7.5 Formation and emissions of organic compounds

7.5.1 Formation of organic compounds

Organic compounds from combustion processes take on many different forms. By U.S. federal legislation, volatile organic compounds (VOCs) means any compound of carbon, excluding carbon monoxide (CO), carbon dioxide (CO_2), carbonic acid, metallic carbides or carbonates, and ammonium carbonate, that participates in atmospheric photochemical reactions to form ozone and other photochemical oxidants [38]. It should be noted that methane, a nonreactive compound, is not a VOC. Hazardous air pollutants are defined by a list of chemicals that are known or suspected of causing cancer or other serious health effects. Organic HAPs are contained in the list of 187 HAPs provided in Appendix C. As discussed at the beginning of the chapter, polycyclic organic matter (POM) has also been referred to as polynuclear or polycyclic compounds (PACs). Nine major categories of POM have been identified by the U.S. EPA with the most common organic compounds in the flue gas of fossil fuel-fired boilers being polycyclic aromatic hydrocarbons (PAHs) [3]. EPA has specified 16 PAH compounds as priority pollutants and they are listed in the beginning of the chapter. Other organic emissions of interest are polychlorinated dibenzo-p-dioxins and polychlorinated dibenzofurans (PCDD/PCDF).

VOCs and CO are unburned gaseous combustibles that are emitted from fossil fuel-fired boilers but generally in quite small amounts. However, during startups, boiler upsets, or other conditions preventing complete combustion, unburned combustible emissions may increase significantly. The application of low-NO_x burners and flue gas recirculation decreases combustion efficiency, in some cases resulting in higher CO and VOC emissions and, in the case of natural gas firing, most of the VOC emissions are formaldehyde or benzene, which are present in natural gas in trace amounts [28]. When firing fuel oils with incomplete combustion, organic compounds present in the flue gas include aliphatic and aromatic hydrocarbons, esters, ethers, alcohols, carbonyl compounds, phenols, and POM (organic matter having two or more benzene rings). Some of the VOCs are classified as HAPs. Formaldehyde is formed during the combustion of fuel oil and is subject to oxidation/decomposition at the high temperatures in the combustion zone [28]. Larger units with more efficient combustion emit less formaldehyde than smaller, less efficient units. When firing coal, VOC and CO emissions are generally lower from pulverized coal or cyclone-fired furnaces than from smaller stoker units where operating conditions are not so well controlled. Emissions include alkanes, aldehydes, alcohols, benzene, toluene, xylene, and ethyl benzene [28]. Polychlorinated dibenzo-p-dioxins and polychlorinated dibenzofurans can form downstream from reactions in air pollution control devices depending on the flue gas temperature with the maximum potential in the temperature window of 450–650°F. Essentially, CO, VOCs, and other organic compounds found in the flue gas of fossil fuel-fired units are primarily the products of incomplete combustion. Figure 7.1 shows the structure of some of these organic compounds, which are important in air pollution.

7.5.2 Emissions of organic compounds

National CO emissions totals are provided in Table 2.20. Electrical utilities and industrial combustion sources account for 1.1 and 1.3% of total CO emissions, respectively [39]. The contribution of VOC emissions from stationary fuel combustion is even lower with electric utilities and industrial combustion sources responsible for 0.23 and 0.61% of total CO emissions, respectively.

In the EPRI HAPs assessment for coal-fired utility boilers, organic compounds that were detected in the flue gas at one or more sampling sites were reported [33]. Eighty-eight compounds were detected and they are listed in Table 7.21 along with the mean emissions value in lb/trillion Btu heat input.

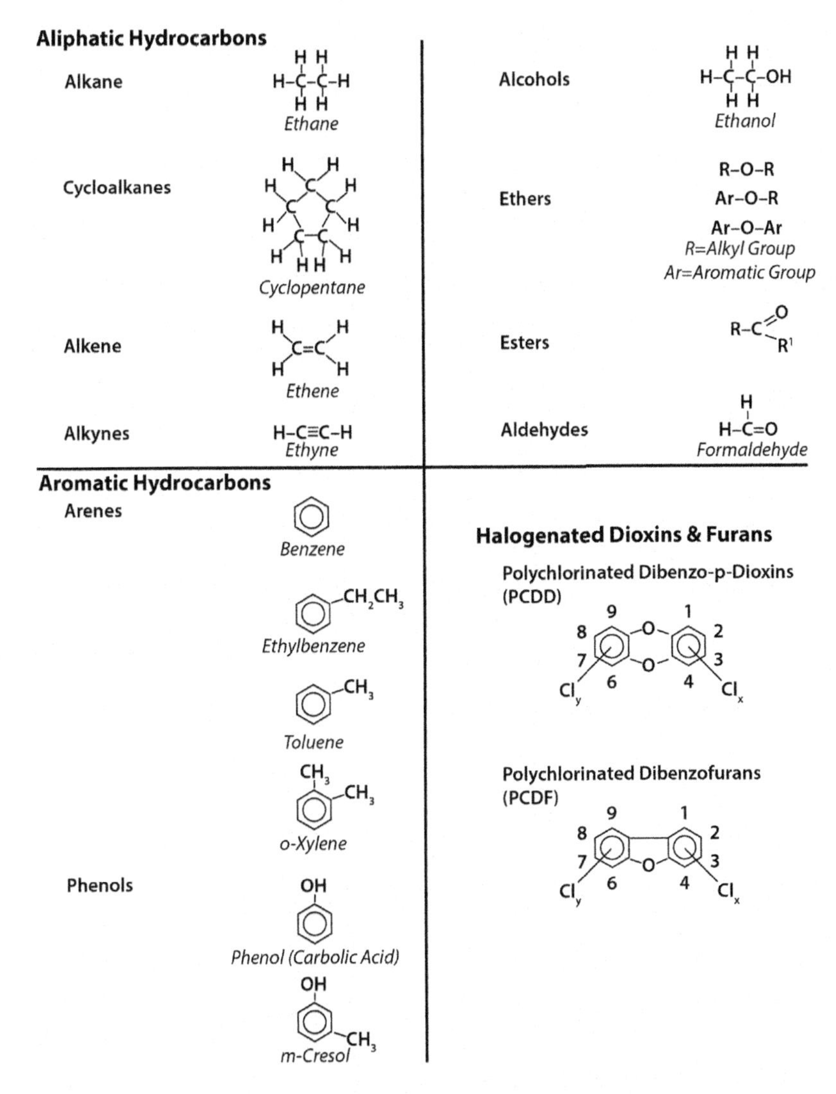

Figure 7.1 Classes and examples of organic compounds.

EPRI summarized the emissions estimates on an annual basis at the station level for 462 operating electric generating stations (808 stack emission points) [33]. Table 7.22 summarizes selected organic compounds or compound groups.

A breakdown of the 14,111 major source industrial boilers, according to the U.S. EPA, the fuels they fire, and their baseline organic compounds emissions is provided in Table 7.23 [36]. The organic compounds are reported as dioxin/furan TEQs (toxic equivalents), total hydrocarbons (THC), and volatile organic compounds (VOCs).

Table 7.21 **Organic substances detected in the flue gas**

Chemical substance	Mean (lb/T Btu)	Chemical substance	Mean (lb/T Btu)
1,1-Dichloroethane	0.68	Chlorobenzene	0.14
1,2,4-Trichlorobenzene	1.1	Chloroethane	0.43
1,2-Dibromethane	2.6	Chloroform	0.64
1,3,5-Trimethylbenzene	0.95	Chloromethane	1.8
1,3-Dichlorobenzene	0.75	Cis-1,3-Dichloropropene	0.59
1,4-Dichlorobenzene	0.79	Cumene	0.21
1-Naphthylamine	0.011	Dibenzo(a,h)anthracene	0.00098
2,3,7,8-TCDD equiv.[a]	0.00000141	Dibenzo(a,j)acridine	0.001
2,4-Dinitrotoluene	0.2	Dibenzofuran	0.61
2,5-Dimethylbenzaldehyde	14	Dibutylphthalate	0.11
2,6-Dinitrotoluene	0.11	Diethylphthalate	0.2
2-Butanone	2.4	Dimethylphthalate	0.09
2-Chloronaphthalene	0.0005	Ethylbenzene	0.65
2-Hexanone	2.1	Ethylene dibromide	0.07
2-Methylnaphthalene	0.042	Fluoranthene	0.13
3-Chloropropylene	2.9	Fluorene	0.13
4-Ethyltoluene	2.8	Formaldehyde	2.4
4-Methyl-2-pentanone	1.4	Hexaldehyde	5.7
4-Methylphenol	1.1	Indeno(1,2,3-c,d)pyrene	0.0018
5-Methylchrysene	0.0006	Iodomethane	2
Acenaphthene	0.021	Isophorone	1.2
Acenaphthylene	0.0073	m/p-Tolualdehyde	3.2
Acetaldehyde	2.6	m/p-Xylene	0.7
Acetone	1.0	1,1,1-Triclhloroethane	0.44
Acetophenone	1.2	Methylmethacrylate	1.1
Acrolein	1.9	Methylene chloride	3.1
Anthracene	0.011	Naphthalene	0.9
Benzo(a)pyrene equiv.	0.00336	n-Butyraldehyde	8.3
Benzaldehyde	4.2	n-Hexane	0.48
Benzene	3.5	o-Tolualdehyde	2.9
Benzo(a)anthracene	0.0066	o-Xylene	0.37
Chrysene	0.0049	Perylene	0.0033
Benzo(a)pyrene	0.0019	Phenanthrene	0.40
Benzo(b)fluoranthrene	0.0083	Phenol	3.3
Benzo(e)pyrene	0.0031	Propionaldehyde	1.9
Benzo(g,h,i)perylene	0.0016	Pyrene	0.055
Benzoic acid	22	Styrene	0.59
Benzyl alcohol	2	Tetrachloroethylene	0.35
Benzylchloride	0.28	Toluene	1.7
Biphenyl	0.16	Trichlorofluoromethane	0.72
Bis(2-Ethylhexyl)phthalate	3.6	Valeraldehyde	7.6
Bromomethane	1.1	Vinyl acetate	0.25
Butylbenxylphthalate	0.3	Vinyl chloride	0.58
Carbon disulphide	1.0	HCN	13.3

[a]equivalents for the family of dioxins and furans.

Table 7.22 **Annual emission summary for selected organic compounds or compound groups at the station level**

	Annual emissions at station level (lb/year)			
	Median	Mean	Maximum	Minimum
Benzene	99	158	919	0.28
Toluene	48	77	446	0.14
Formaldehyde	68	108	630	0.19
Benzo(a)pyrene equiv.	0.095	0.15	0.88	0.00027
2,3,7,8-TCDD equiv.	0.00004	0.000064	0.00037	0.00000011

Table 7.23 **Organic compound emissions from industrial boilers**

Fuel	Number of boilers	Emissions in short tons/year		
		Dioxin/ Furan TEQ	Total hydrocarbons	Volatile organic compounds
Biomass	487	0.00000462	17,615	10,110
Coal	601	0.0000319	9,129	5,240
Liquid	777	0.00000188	4,338	2,518
Non-cont. liquid	44	0.000000455	718	412
Gas	10,883	0.00000811	14,872	8,536
Gas-metal furnace	698	0.0000028	4,059	2,330
Process gas	125	0.000000391	291	167
Limited use	479	0.0000000543	107	62
TOTAL	14,111	0.0000502	51,179	29,375

7.6 Characteristics and emissions of trace elements

7.6.1 Characteristics/formation of inorganic trace elements

The characteristics and formation of inorganic trace elements were provided in detail in Chapter 3, "Particulate Formation and Control Technologies," as part of the overall particulate matter discussion with mercury presented in detail in Chapter 6, "Mercury Emissions Reduction."

The inorganic trace elements specifically identified by the U.S. EPA in the 1990 Clean Air Act Amendments as hazardous air pollutants were compounds of antimony,

Table 7.24 Emission rate summary for toxic metal HAPs in coal-fired units (lb/trillion Btu heat input)

	As	Be	Cd	Co	Cr	Mn	Ni	Pb	Sb	Se
All units										
Median	3.6	0.3	0.6	1.2	6.4	11.9	5.4	3.4	0.3	51.5
Mean	5.5	0.4	0.7	1.3	7.1	13.3	6.1	4.1	0.4	57.3
Max	35.8	2.7	4.9	4.8	24.4	58.3	43.8	19.2	1.8	262
Min	0.1	0.024	0.049	0.2	0.9	1.3	1.1	0.3	0.0	0.0
PM control only										
Median	3.9	0.3	0.6	1.3	6.6	12.6	5.6	3.8	0.3	63.5
Mean	5.8	0.4	0.7	1.4	7.4	14.0	6.3	4.3	0.4	63.8
Max	35.8	2.7	4.9	4.8	24.4	47.8	39.4	19.2	1.8	262
Min	0.1	0.027	0.073	0.2	1.0	1.6	1.1	0.4	0.0	0.0
PM and SO_2 controls										
Median	2.8	0.3	0.5	1.0	5.2	9.6	4.7	2.8	0.3	40.2
Mean	4.7	0.4	0.6	1.1	6.2	11.3	5.6	3.5	0.3	39.4
Max	33.5	2.2	3.4	3.8	20.9	58.3	43.8	14.5	1.8	180
Min	0.3	0.024	0.049	0.2	0.9	1.3	1.1	0.3	0.0	0.4

arsenic including arsine, beryllium, cadmium, chromium, cobalt, lead, manganese, mercury, nickel, and selenium. Radionuclides were also included. Trace elements can be generally classified into three classes [6]:

- Class 1: elements that are approximately equally concentrated in the fly ash and bottom ash, or show little or no small particle enrichment. Examples include manganese, beryllium, cobalt, nickel, and chromium.
- Class 2: elements that are enriched in the fly ash relative to the bottom ash, or show increasing enrichment with decreasing particle size. Examples include arsenic, cadmium, lead, and antimony.
- Class 3: elements that are emitted in the gas phase. Examples include mercury (discussed in more detail in Chapter 6) and, in some cases, selenium.

7.6.2 Emissions of inorganic trace elements

Table 7.24 is a summary of coal-fired emission rates determined by EPRI as part of their update on HAPs emissions, which has been discussed in detail earlier [33]. Table 7.24 summarizes estimated emission rates for toxic metal (except for mercury) HAPs from coal-fired units normalized on a lb/T (trillion) Btu heat input basis.

Table 7.25 **Annual emission summary for toxic metals at the station level (lb/year)**

	Annual emissions at station level (lb/year)			
	Median	**Mean**	**Maximum**	**Minimum**
As, arsenic	104	214	2,950	1.0
Be, beryllium	9	16	185	0.1
Cd, cadmium	17	27	253	0.1
Co, cobalt	34	54	472	0.1
Cr, chromium	178	293	2,270	1.0
Mn, manganese	358	575	3,960	2.0
Ni, nickel	157	249	3,330	1.0
Pb, lead	100	164	1,280	1.0
Sb, antimony	9	15	148	0.04
Se, selenium	1,360	2,490	30,000	0.0

In EPRI's updated HAPs emissions study they summarized the toxic metal emissions estimates on an annual basis at the station level for 462 operating electric generating stations (808 stack emission points) [33]. Table 7.25 summarizes selected toxic metals HAPs at the station level.

Data compiled from EPA's Toxics Release Inventory (TRI) for power plants in 2010 indicate overall toxic power plant emissions are declining, when compared to data from 2007; however, data also show that some power plants are still emitting high levels of toxics with certain toxics increasing in some states [35]. Emissions of toxic metals in 2010 were approximately:

- Arsenic – 31.5 short tons
- Chromium – 67.5 short tons
- Lead – 99.5 short tons
- Mercury – 33 short tons
- Nickel – 122 short tons
- Selenium – 125 short tons

The electric utility sector is the top emitter of arsenic, mercury, and selenium of all industry sectors, and the utility sector is the second highest emitter of chromium, cobalt, and nickel of all industry sectors. Table 7.26 lists the top 15 states for emissions of HAPs by their average rank based on each state's ranking for the chemical listed [35].

As previously mentioned, there are 1,753 industrial boilers, out of the 14,111 major source boilers, that will need to meet emissions limits to minimize air toxics. The baseline emissions of selected metals, according to EPA data, are 5 short tons of mercury, 107.5 short tons of lead, and 61.5 short tons of chromium [37]. The top 20 states with industrial boilers emitting mercury, lead, and chromium are listed in Table 7.27.

Table 7.26 State averages of eight hazardous air pollutant rankings based on 2010 TRI data

State	Rank of each HAP								
	As	Cr	Co	HCl	Pb	Hg	Ni	Se	Average[a]
PA	1	7	1	2	1	3	6	3	3.0
OH	4	6	2	1	4	2	5	2	3.3
IN	5	4	4	3	3	5	4	11	4.9
KY	2	5	6	5	2	9	8	10	5.9
TX	13	3	9	13	6	1	9	1	6.9
WV	9	11	3	9	10	7	10	4	7.9
GA	6	8	5	8	7	22	12	9	9.6
MI	16	2	21	4	5	10	13	12	10.4
FL	12	20	10	7	9	15	7	8	11
AL	8	16	11	14	16	6	17	6	11.8
NC	11	17	14	6	14	24	11	5	12.8
ND	3	12	7	34	13	8	14	–	13.0
MO	17	18	17	20	15	4	19	–	15.7
SC	10	15	23	10	12	29	22	15	17.0
WY	7	13	8	43	28	14	21	7	17.6

[a]Average of HAP rankings.

Table 7.27 Top 20 states with baseline mercury, lead, and chromium emissions from industrial boilers

Rank	Mercury		Lead		Chromium	
	State	lb/yr	State	Short tons/yr	State	Short tons/yr
1	IN	956	OH	7.72	OH	6.53
2	OH	797	GA	7.41	IN	4.25
3	IA	650	NC	7.22	IA	3.30
4	PA	526	LA	6.93	NC	3.23
5	WV	431	IN	6.61	WI	2.89
6	NC	426	SC	6.02	PA	2.82
7	TN	403	AL	5.77	TN	2.81
8	MI	393	PA	3.95	SC	2.69
9	AL	384	IA	3.80	WV	2.37
10	SC	381	WA	2.965	VA	2.31
11	GA	356	TN	2.96	IL	2.16
12	VA	340	VA	2.90	MN	1.72
13	FL	331	FL	2.86	Virgin Is.	1.70
14	IL	311	WV	2.73	GA	1.64
15	MN	277	MN	2.69	AL	1.62
16	WY	275	MI	2.67	MI	1.52
17	WI	250	IL	2.64	NY	1.48
18	MD	232	AR	2.61	FL	1.42
19	Virgin Is.	209	NY	2.60	MO	1.27
20	NY	188	Virgin Is.	2.58	NJ	1.19
Total		8,116		85.65		48.92
Total from 1,753 power plants		10,000		107.50		61.50

Table 7.28 **Total metals emissions from industrial boilers**

Fuel	Number of boilers	Metals emissions (short tons/year)
Biomass	487	354
Coal	601	379
Liquid	777	3,069
Non-continental liquid	44	352
Gas	10,883	85
Gas-metal furnace	698	2
Process gas	125	7
Limited use	479	45
TOTAL	14,111	4,294

A breakdown of the 14,111 major source industrial boilers, according to the U.S. EPA, the fuels they fire, and their baseline total metals emissions is provided in Table 7.28 [36]. Total metals include antimony, arsenic, chromium, cobalt, lead, manganese, nickel, and selenium.

7.7 Control technologies

Many of the control technologies for acid gases and inorganic trace elements are the same as those presented in earlier chapters discussing sulfur dioxide and particulate matter control. This section will briefly discuss options for controlling acid gases, organic compounds, inorganic trace elements.

7.7.1 Control technologies for acid gases

7.7.1.1 SO_3

As previously discussed, SO_3 in the flue gas hydrates to form H_2SO_4 vapor. As the flue gas temperature cools, the H_2SO_4 vapor condenses to form a submicron aerosol (i.e. an acid mist). Due to the small size of these aerosol particles (e.g. less than 0.5 μm), the sulfuric acid mist is difficult to capture or collect as it passes through the air pollution control equipment (i.e. ESPs, fabric filters, and FGD systems) [40].

There are different techniques used to control SO_3 emissions including [28,29,40,41]:

- Fuel switching to a lower sulfur fuel
- Furnace injection of alkalis
- Alkali injection into the economizer outlet or SCR outlet duct
- Flue gas humidification (with or without alkali injection) upstream of a dry ESP
- Wet ESP

7.7.1.1.1 Sorbent injection

There are several sorbents that can be used to reduce SO_3 emissions and they typically utilize alkaline-based reagents that contain calcium, magnesium, sodium, or ammonia [40]. Detailed discussions of the various sorbent injection types were provided in Chapter 4. The reagents used for SO_3 control include magnesium oxide (MgO), calcium hydroxide ($Ca(OH)_2$), sodium carbonate (Na_2CO_3), sodium bicarbonate ($NaHCO_3$), and sodium sesquicarbonate, i.e. trona ($Na_2CO_3 \cdot NaHCO_3 \cdot 2H_2O$). Depending on the reagent used, these reagents can either be injected dry or wet, and either in aqueous or slurry form. Sorbent injection into the furnace, downstream of the furnace (upstream or downstream of the air heater), SCR, ESP, or scrubber can effectively reduce SO_3 emissions. SO_3/H_2SO_4 removals from 80 to over 99% can be achieved.

When magnesium oxide is used as the reagent, the SO_3 is captured via the reaction

$$MgO + SO_3 \rightarrow MgSO_4 \tag{7.3}$$

Sulfur trioxide is captured via the following reaction when using calcium hydroxide:

$$Ca(OH)_2 + SO_3 \rightarrow CaSO_4 + H_2O \tag{7.4}$$

When sodium bicarbonate or trona are used, they first decompose to sodium carbonate, water, and carbon dioxide. Sodium carbonate reacts with SO_3 to form sodium sulfate by the following reactions when the reagent is in excess of the amount of SO_3:

$$Na_2CO_3 + SO_3 \rightarrow Na_2SO_4 + CO_2 \tag{7.5}$$

If the amount of reagent injected is much less than the amount of SO_3, then the sodium sulfate will continue to react with SO_3 to form sodium bisulfate [32]:

$$Na_2SO_4 + SO_3 + H_2O \rightarrow 2NaHSO_4 \tag{7.6}$$

Magnesium-based sorbents have an advantage over calcium-based ones for furnace injection since the former do not react with SO_2 at furnace temperatures and therefore SO_2 does not compete with SO_3 for reactive sites on the sorbent particle [42]. A high ratio of injected calcium to SO_3 would be required to capture SO_3 with calcium-based sorbents.

Injecting calcium- or sodium-based or other alkaline sorbents, either wet or dry, into the duct is performed commercially to remove SO_3/H_2SO_4. Duct sorbent injection removes SO_3/H_2SO_4 more efficiently than wet FGD scrubbers and also captures HCl and HF (which is discussed later). If the sorbent is injected downstream of the SCR then both furnace and SCR-generated SO_3 will be removed.

Dry hydrated lime is commercially utilized to lower SO_3 emissions. Over 80% reduction has been achieved when it was injected before the ESP at a Ca:SO_3 molar ratio of 4.2:1 [42]. Injecting the hydrated lime after the air heater can lower SO_3 concentrations by 98%.

Around 60 to 70% of the SO_3 can be removed when dry magnesium hydroxide is injected between the SCR and air heater at Mg:SO_3 molar ratios of 1:1 to 2:1 with removals greater than 80% also achieved [42].

Sodium-based sorbents are more reactive than limestone and hydrated lime. Dry trona injection can capture over 90% of the SO_3 at a stoichiometric ratio of approximately 1.5 [42]. SO_3 removals in excess of 99% have been achieved [32].

Ammonia (NH_3) is also used for SO_3 control through duct injection. Best results are obtained of ammonia is injected in the temperature range 300–390°F (150–200°C) with a NH_3:SO_3 molar ratio in the range of 1:1 to 2:1 [42]. A molar ratio of 1:1 produces mostly ammonium bisulfate

$$NH_3 + SO_3 + H_2O \rightarrow NH_4HSO_4 \tag{7.7}$$

whereas a 2:1 ratio yields largely ammonium sulfate:

$$2NH_3 + SO_3 + H_2O \rightarrow (NH_4)_2SO_4 \tag{7.8}$$

Injecting ammonia between the air heater and ESP can remove 90 to 95% of the SO_3 at molar ratios in the range of 1.5:1 to 2:1 [42].

7.7.1.1.2 Humidification

Cooling the flue gas by humidification upstream of an existing dry ESP will allow the dry ESP to collect some sulfuric acid. It is possible to operate a dry ESP at 20°F below the acid dew point without encountering excessive corrosion because the acid mist is absorbed on the fly ash surface rather than on the metal surfaces [28].

7.7.1.1.3 Wet ESPs

The wet ESP is also an effective device to remove SO_3, especially when ultra-low outlet concentrations are required [28,40]. Wet ESP technology can be used after the wet FGD as a final polishing stage to remove very fine particulate, sulfuric acid, and any other mist. Sulfuric acid removals in the range of 30 to 40% can be achieved. Acid mist removal efficiencies of 80 to 90% can be achieved.

7.7.1.2 Halogens

As previously discussed, fluorine is associated with the mineral, ash-forming part of the coal (of the fossil fuels, coal contains the highest concentrations of halogens). Consequently, as coal is cleaned to reduce its ash content, reductions in coal fluorine content have been reported [17]. The amount of removal is a function of the coal and the level of coal cleaning. Fluorine reductions from 34 to 80% have been reported. Chlorine, on the other hand, can be associated with the organic content of the coal and its concentration is higher in the fuels that are burned; hence, chlorine emissions are of greater concern.

The removal of halogen emissions, specifically HCl and HF, is achieved through the same technologies used to control SO_2 and SO_3. When designing a system for

Table 7.29 **Typical maximum removal efficiencies**

Pollution control technology	Maximum removal efficiency (%)	
	HCl	HF
Fabric filter/wet FGD/ wet ESP	99+	99+
ESP or fabric filter/ wet FGD	99+	99+
Circulating dry scrubber/ fabric filter	98+	98+
Spray drier absorber/ fabric filter	98+	98+
Duct sorbent injection/ fabric filter	80 to 95+	80 to 95+

SO_2/SO_3 removal, there is the beneficial effect of also removing HCl and HF emissions. HCl and HF emissions can be reduced by approximately 25% without specifically designing for their removal [42]. When a control system is designed to remove all of the acid gases, maximum removal efficiencies can exceed 98% for removing HCl and HF. Maximum removal efficiencies for HCl and HF using various control technologies are shown in Table 7.29 [40].

When using calcium-based sorbents for duct injection or a spray drier absorber with fabric filter, the reactions for capturing HCl and HF are

$$Ca(OH)_2 + 2HCl \rightarrow CaCl_2 + 2H_2O \tag{7.9}$$

and

$$Ca(OH)_2 + 2HF \rightarrow CaF_2 + H_2O \tag{7.10}$$

Sodium-based sorbents are used for capturing HCl. The reaction of sodium carbonate with HCl forms sodium chloride:

$$Na_2CO_3 + 2HCl \rightarrow 2NaCl + H_2O + CO_2 \tag{7.11}$$

7.7.2 Control technologies for organic compounds

There are control technologies available specifically for the control of organic compounds. However, these are used in industries where there are fugitive VOC emissions or in the hazardous waste incineration sector. Generally, the emissions of organic compounds from boilers are controlled through proper combustion controls/ operations so the organic compounds are not formed. Dioxins, furans, and other gas-phase organic carbon-based compounds can be adsorbed onto the surface of activated carbon and subsequently removed by particulate matter controls [43,44]. Activated carbon injection for mercury control can have the co-benefit of organic compound capture. Wet scrubbers can also remove VOCs present in the flue gas.

7.7.3 Control technologies for trace elements

The control of most trace elements is accomplished through the use of particulate matter control technologies such as ESPs, baghouses/fabric filters, and cyclones. These are discussed in detail in Chapter 3. The more volatile trace elements like mercury are controlled through technologies discussed in Chapter 6. Other trace elements such as the halogens are discussed earlier in this chapter.

7.8 Economics of control technologies

The economics of control technologies are not discussed in this chapter because they were presented in previous chapters on particulate matter control (Chapter 3), sulfur dioxide control (Chapter 4), and mercury control (Chapter 6).

References

[1] Miller SJ, Ness SR, Weber GF, Erickson TA, Hassett DJ, Hawthorne SB, et al. A comprehensive assessment of toxic emissions from coal-fired power plants: phase I Results from the U. S. Department of Energy Study Final Report, 1996.
[2] Wark K, Warner CF, Davis WT. Air pollution: its origin and control. 3rd ed. Menlo Park, California: Addison Wesley Longman, Inc; 1998.
[3] Brooks AW. Estimating air toxic emissions from coal and oil combustion sources, Report No. EPA-450/2-89-001. (U.S. EPA, North Carolina, 1989).
[4] Finkelman RB, Skinner HCW, Plumlee GS, Bunnell JE. Medical geology. Alexandria, Virginia: Geosciences and Human Health, American Geological Institute; 2001. pp. 20–23.
[5] EPA (United States Environmental Protection Agency). Latest findings on national air quality: 1997 status and trends. Washington, DC: Office of Air Quality Planning and Standards; 1998.
[6] Nalbandian H. Trace element emissions from coal. London: IEA Coal Research; 2012.
[7] Davis WT, editor. Air pollution engineering manual. 2nd ed. John Wiley & Sons Inc; 2000.
[8] Clarke LE, Sloss LL. Trace elements – emissions from coal combustion and gasification. London: IEA Coal Research; 1992.
[9] Emsley J. The elements. Oxford: Clarendon Press; 1989.
[10] Finkelman RB, Orem W, Castranova V, Tatu CA, Belkin HE, Zheng B, et al. Health impacts of coal and coal use: possible solution, Volume 50. International Journal of Coal Geology, Elsevier Science; 2002. pp. 425–443.
[11] Swaine DJ, Goodarzi F, editors. Environmental aspects of trace elements in coal. Dordrecht, The Netherlands: Kluwer Academic Publishers; 1995.
[12] USGS (United States Geological Survey). Radioactive elements in coal and fly ash: abundance, forms, and environmental significance. (USGS Fact Sheet FS-163-97; U.S. Government Printing Office, Washington, D.C., 1997).
[13] Swaine DJ, Goodarzi F. Environmental aspects of trace elements in coal, Volume 2. Dordrecht, The Netherlands: Kluwer Academic Publishers; 1995.

[14] Penn State (2015). Coal sample bank. Earth and Mineral Sciences Energy Institute, Available at www.energy.psu.edu/copl/index.html.

[15] Miller BG, Tillman DA, editors. Combustion engineering issues for solid fuel systems. Oxford, United Kingdom: Academic Press; 2008.

[16] Tillman DA, Duong D, Miller BG. Chlorine in solid fuels fired in pulverized fuel boilers – sources, forms, reactions, and consequences: a literature review. Energy & Fuels 2009; 23:3379–91.

[17] Davidson RM. Chlorine and Other Halogens in Coal. London: IEA Coal Research; 1996.

[18] Barwise AJG. Role of nickel and vanadium in petroleum classification. Energy & Fuels 1990;4:647–52.

[19] Trace metals in crude oils from Saudi Arabia. Ind. Eng. Chem. Prod. Res. Dev., Vol. 22, pp. 691–694, 1983.

[20] Hitchon B, Filby RH. Geochemical studies – 1, trace elements in alberta crude oils. Alberta Research Council Open File Report 1983-02, 1983.

[21] Ball JS, Wenger WJ. Metal content of twenty-four petroleums. J Chem Eng Data 1960;5(4):553–7.

[22] Bergerious C, Zikovsky L. Determination of 18 trace elements. J Radioanal Chem 1978; 46:277–84.

[23] Graham J. Select trace elemental composition of fuel oil used in the Northeastern United States. Air & Waste Manage Assoc 2010;16–22.

[24] Wang Z, Hollebone BP, Fingas M, et al. Characteristics of spilled oils, fuels, and petroleum products: 1. Composition and properties of selected oils. National Exposure Research Laboratory, U.S. Environmental Protection Agency, July 2003.

[25] International Agency for Research on Cancer. IARC monographs on the evaluation of carcinogenic risks to humans, occupational exposures in petroleum refining; Crude Oil and Major Petroleum Fuels, Volume 45, 1989.

[26] Oil & Gas Journal. Gas Oil Desalting Reduces Chlorides in Crude, www.ogi.com, Volume 98, Issue 42, October 16, 2000.

[27] Harkov R, Ross J. Hydrogen Chloride (HCl) emissions from maryland utility boilers. Maryland Department of Natural Resources, June 1999.

[28] Kitto JB, Stultz SC, editors. Steam: its generation and use. 41st ed. The Babcock and Wilcox Company; 2005.

[29] Farthing WE, Walsh PM, Gooch JP, McCain JD, Hinton WS, Heapy RF. Identification of (and responses to) potential effects of SCR and wet scrubbers on submicron particulate emissions and plume characteristics. U.S. Environmental Protection Agency, August 2004.

[30] Fernando R. SO_3 issues for coal-fired plant. London: IEA Coal Research; 2003.

[31] Sloss LL. Halogen emissions from coal combustion. London: IEA Coal Research; 1992.

[32] Gray S, Harpenau M, Copsey P. Ultra-High SO_3 removal: SBS injection as an alternative to W-ESP, URS, August 28, 2008.

[33] Chu P, Levin L. Updated Hazardous Air Pollutants (HAPs) emissions estimates and inhalation human health risk assessment for U.S. coal-fired electric generating units. Electric Power Research Institute, 2009.

[34] Chu P, Goodman N. White paper: EPRI's preliminary evaluation of the available HAPs ICR data. January 7, 2011.

[35] EIP (Environmental Integrity Project). America's top power plant toxic air polluters. December 2011.

[36] EPA (United States Environmental Protection Agency). Reconsideration proposal for boilers at area sources, boilers/process heaters at major sources, and commercial/ industrial solid waste incinerators and proposed definition of "non-hazardous solid waste". December 2011.
[37] EarthJustice. The toxic air burden from industrial power plants. February 2012. Available from http://earthjustice.org/features/campaigns/boilers
[38] EPA (United States Environmental Protection Agency). Title 40: part 51.100(s) – requirements for preparation, adoption, and submittal of implementation plans, subpart f – procedural requirements. www.e-CFR, ecfr.gov, as of October 30, 2014.
[39] EPA (United States Environmental Protection Agency). National Emissions Inventory (NEI), 1970–2013 Average Annual Emissions, All Criteria Pollutants, www.epa.gov/ttnchie1/trends/, February 27, 2014.
[40] Moretti AL, Jones CS. Advanced emissions control technologies for coal-fired power plants. Babcock and Wilcox Power Generation Group, Inc., Technical Paper BR-1886, 2012.
[41] Nalbanian H. Economics of retrofit air pollution control technologies. London: IEA Coal Research; 2006.
[42] Carpenter AM, Low-Water FGD. Technologies. London: IEA Coal Research; 2012.
[43] Andover Technology Partners. Control technologies to reduce conventional and hazardous air pollutants from coal-fired power plants, 2011.
[44] Environmental Health & Engineering, Inc., Emissions of hazardous air pollutants from coal-fired power plants, 2011.

Greenhouse gas – carbon dioxide emissions reduction technologies

8

8.1 Introduction

Many chemical compounds found in the earth's atmosphere act as greenhouse gases as discussed previously in Chapter 2. When sunlight strikes the earth's surface, some of it is reradiated back toward space as infrared radiation (i.e. heat). Greenhouse gases absorb this infrared radiation and trap its heat in the atmosphere.

There are several major greenhouse gases that are emitted as a result of human activity. These include [1,2]:

- Carbon dioxide (CO_2), which comes from the combustion of coal, petroleum, natural gas, solid waste, trees, and wood products and also as a result of certain chemical reactions (e.g. manufacture of cement)
- Methane (CH_4), which comes from municipal landfills, petroleum and natural gas systems, coal mines, and agriculture
- Nitrous oxide (N_2O), which comes from the use of nitrogen fertilizers, burning of fossil fuels, and from certain industrial and waste management processes (e.g. nitric acid production and adipic acid production)
- High global warming potential (GWP) gases (see Chapter 2 for a discussion on GWPs), which are manmade fluorinated industrial gases including hydrofluorocarbons (HFCs), perfluorocarbons (PFCs), and sulfur hexafluoride (SF_6)

This chapter discusses the environmental and health effects of the greenhouse gases and emissions of the greenhouse gases, with an emphasis on those emitted from fossil fuel combustion, specifically CO_2. This is followed by discussions of CO_2 capture technologies, transport and storage of CO_2, and the economics of CO_2 capture and sequestration. The chapter concludes with a discussion of the permanence, monitoring, mitigation, and verification of CO_2 capture and storage.

8.2 Environmental and health effects

Climate change impacts society and ecosystems in several ways [3–6]. Sectors impacted by climate change include:

- Agriculture and food supply
- Coastal areas
- Ecosystems
- Energy demand and supply
- Forest growth and productivity

Fossil Fuel Emissions Control Technologies. DOI: http://dx.doi.org/10.1016/B978-0-12-801566-7.00008-7

- Transportation
- Water resources
- Human health

Agriculture and fisheries are highly dependent on specific climate conditions. Warmer temperatures may make many crops grow more quickly, but warmer temperatures could also reduce yields because with some crops (such as grains) faster growth reduces the amount of time that seeds have to grow and mature. Higher CO_2 levels can increase yields for some crops; however, factors such as increased temperature may counteract these yields. More extreme temperature and precipitation can prevent crops from growing. Many weeds, pests, and fungi thrive under warmer temperatures, wetter climates, and increased CO_2 levels.

Climate change could affect livestock production both directly and indirectly. Heat waves threaten livestock. Drought may threaten pasture and feed supplies. The prevalence of parasites and diseases that affect livestock may increase.

Fisheries could be impacted by temperature changes by changing the ranges of many fish and shellfish. Aquatic species that can find colder areas of streams or lakes or move northward along the coast or in the ocean may put these species into competition with other species over food. Some diseases that are more prevalent in warm water may affect aquatic life. Changes in temperatures and seasons could affect the timing of reproduction and migration.

Coastlines of the United States and many other countries are highly populated. Climate change could affect coastal areas in a variety of ways. Coasts are sensitive to sea level rise, changes in the frequency and intensity of storms, increases in precipitation, and warmer ocean temperatures. In addition, rising atmospheric concentrations of CO_2 are causing the oceans to absorb more of the gas and become more acidic. The rising acidity could have significant impacts on coastal and marine ecosystems by leading to the loss of critical habitat (such as corals) and adversely affecting the health of species such as plankton, mollusks, and other shellfish.

Ecosystems are impacted as warming could force species to migrate to higher latitudes or higher elevations where temperatures are more conducive to their survival. Similarly, as sea level rises, saltwater intrusion into freshwater systems may force some species to relocate or die.

Energy demand and supply can be affected by climate change. Changes in temperature, precipitation, and sea level, and the frequency and severity of extreme events will affect how much energy is produced, delivered, and consumed. In warmer climates of industrialized nations, such as the United States, more electricity is used for air conditioning increasing summer peak demands requiring investments in new energy infrastructure. More energy would be required for summer cooling and there could be less demand for winter heating. Energy is needed to pump, transport, and treat drinking water and wastewater, and to operate power plants through cooling and, in the case of hydroelectric plants, electricity generation. Changes in precipitation, increased risk of drought, reduced snowpack, and changes in the timing of the snowmelt in spring will likely influence patterns of energy and water use.

Climate changes directly and indirectly affect the growth and productivity of forests: directly due to changes in atmospheric CO_2 and climate and indirectly through complex interactions in forest ecosystems. Increases in CO_2 concentrations may enable some trees to be more productive; however, warming temperatures may shift geographical ranges of some tree species and some species may die out if their current geographic range ceases to exist. Climate also affects the frequency and severity of forest disturbances such as drought, extreme precipitation and flooding, and increased wildfire risk.

Climate change is projected to increase the frequency and intensity of extreme weather events. These changes could increase the risk of delays, disruptions, damage, and failure across land-based, air, and marine transportation.

Water resources are important to both society and ecosystems. Water is needed to sustain human health and for agriculture, energy production, navigation, and manufacturing. Many of these uses put pressure on water resources, stresses that are likely to be exacerbated by climate change. Climate change is likely to increase water demand while shrinking water supplies.

Weather and climate play a significant role in people's health. Warmer average temperatures will likely lead to hotter days and more frequent and longer heat waves, which could increase the number of heat-related illnesses and deaths. Increases in the frequency or severity of extreme weather events such as storms could increase the risk of flooding, high winds, and other direct threats to people and property. Warmer temperatures could increase the concentrations of unhealthy air and water pollutants. Changes in temperature, precipitation patterns, and extreme events could enhance the spread of some diseases.

8.3 Greenhouse gas emissions

The U.S. Congress directed the EPA to collect annual greenhouse gas (GHG) information from the top emitting sectors in the U.S. Table 8.1 lists the Greenhouse Gas Reporting Program (GHGRP) sector classifications. The GHGRP does not represent total U.S. GHG emissions but provides facility level data for large sources of direct emissions, thus including the majority of U.S. GHG emissions [4]. One sector that contributes to greenhouse gas emissions – the transportation sector – is not covered by the reporting program but a summary of CO_2 emissions from transportation fuels is provided later in this section.

In 2013, 3.18 billion tons of CO_2 eq (see Chapter 2 for the definition of CO_2 equivalent) were reported by direct emitters [4]. The largest emitting sector was the power plant sector with 2.1 billion metric tons CO_2eq, followed by the petroleum and natural gas systems sector with 224 million metric tons CO_2eq and the petroleum refinery sector with 177 million metric tons CO_2eq. These data are shown in Figure 8.1 [4].

Table 8.2 and Figure 8.2 show the emissions trends from 2011 through 2013 [4]. Emissions increased by 0.62% from 2012 to 2013, which is a similar increase in

Table 8.1 GHGRP sector classifications

Carbon Dioxide Supply & Injection	Mining
Suppliers of CO_2	Underground coal mines
Injection of CO_2	Miscellaneous Combustion Sources
Geologic sequestration of CO_2	Stationary fuel combustion sources at facilities that are not part of any other sector
Chemicals	Natural Gas and Natural Gas Liquids Suppliers
Adipic acid production	Fractionators of natural gas liquids
Ammonia manufacturing	Local natural gas distribution companies
Hydrogen production	Petroleum & Nat. Gas Systems – Direct Emissions
Nitric acid production	Onshore production
Phosphoric acid production	Offshore production
Petrochemical production	Natural gas processing
Silicon carbide production	Natural gas transmission/compression
Titanium dioxide production	Natural gas distribution
Other chemicals production	Underground natural gas storage
Electrical Equipment	Liquefied natural gas storage
Electrical equipment manufacture & refurbishment	Liquefied natural gas import/export
Electrical transmission & distribution equipment use	Other petroleum & natural gas systems
Electronics Manufacturing	Petroleum Product Suppliers
Electronics manufacturing	Suppliers of coal-based liquid fuels
Fluorinated Chemicals	Suppliers of petroleum products
Fluorinated gas production	Power Plants
HCFC-22 production/HFC-23 destruction	Electricity generation
Industrial Gas Suppliers	Pulp & Paper
Suppliers of industrial GHGs	Chemical pulp & paper manufacturing
Imports and exports of equipment pre-charged with fluorinated GHGs or containing fluorinated GHGs in closed-cell foams	Other paper products
Metals	Refineries
Aluminum production	Petroleum refineries
Ferroalloy production	Waste
Iron & steel production	Municipal landfills
Lead production	Industrial waste landfills
Zinc production	Industrial wastewater treatment
Manganese production	Solid waste combustion
Other metals production	
Minerals	
Cement production	
Glass production	
Lime manufacturing	
Soda ash manufacturing	
Other minerals production	

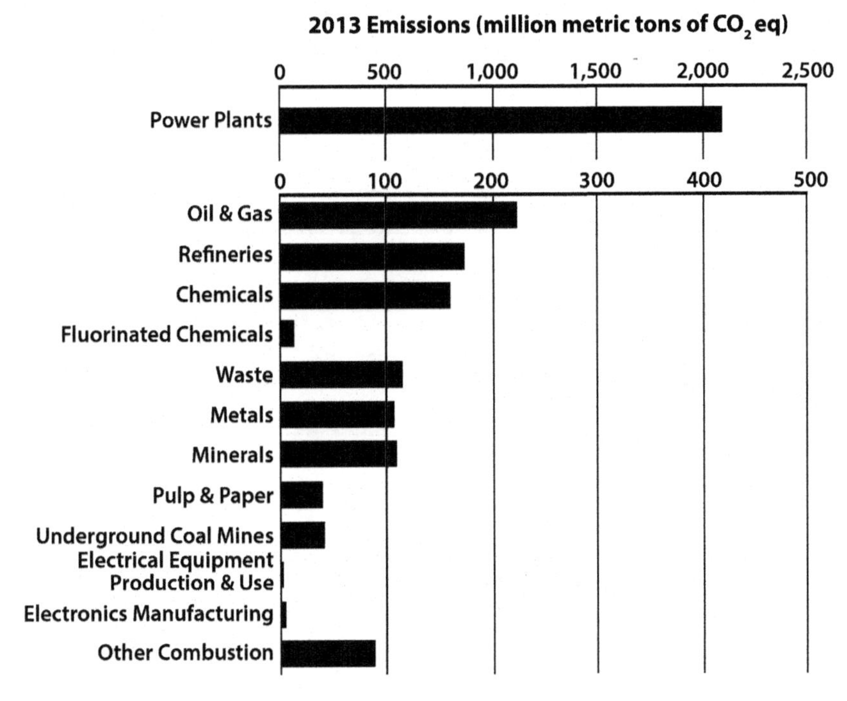

Figure 8.1 Direct greenhouse gas emissions reported by sector in 2013 (in million metric tons of CO_2eq).

Table 8.2 **Emission trends by sector for the period 2011–2013 (in million metric tons of CO_2eq)**

Sector	2011	2012	2013
Power Plants	2,221.3	2,088.1	2,100.9
Oil & Gas	223.1	226.4	224.1
Refineries	178.3	174.0	176.7
Chemicals	179.21	170.9	174.6
Fluorinated chemicals	15.7	12.4	13.5
Nonfluorinated chemicals	163.5	158.5	161.1
Waste	115.4	118.5	114.0
Minerals	103.2	107.5	111.3
Metals	112.8	106.9	106.8
Pulp & Paper	44.2	42.5	39.1
Other	136.8	130.0	136.9
Underground coal mines	37.1	34.0	41.5
Electrical equipment production & use	4.3	3.5	3.5
Electronics manufacturing	6.1	5.6	4.5
Other combustion	89.3	86.9	87.4

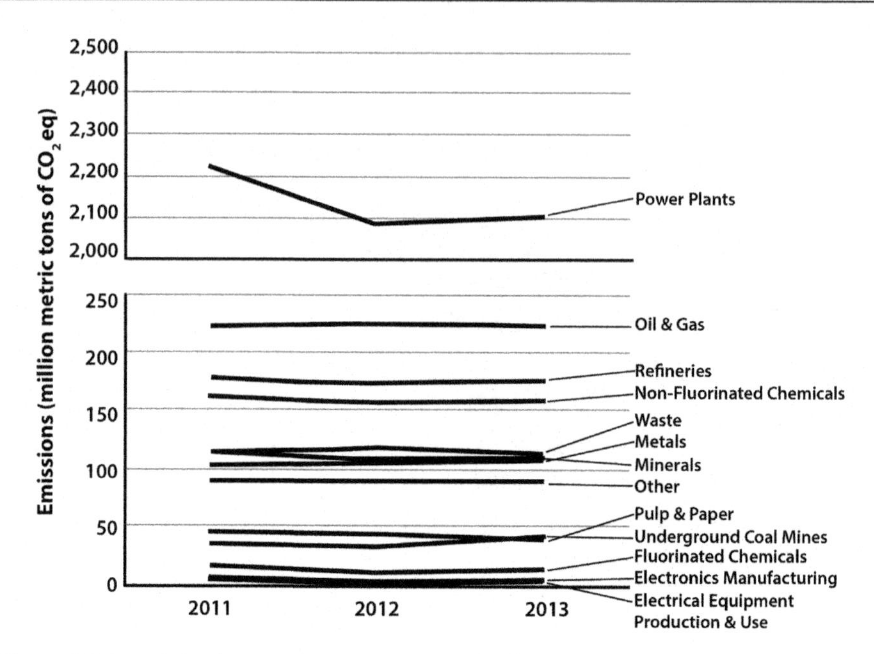

Figure 8.2 Trends in direct greenhouse gas emissions for the period 2011–2013.

power plant emissions. Over the past three reporting years, GHGRP reported emissions have declined by 3.9%, which is caused primarily by a 5.4% decline in reported emissions by power plants. Since 2010, emissions from power plants have decreased by 9.8% [4].

The geographic distribution of the GHG emissions are shown in Figure 8.3. Figure 8.3 shows the location and total reported emissions from GHGRP facilities in 2013 [4].

Of the greenhouse gases emitted, CO_2 is emitted in the largest quantities. The 2.9 billion metric tons of CO_2 reported for 2013 represent 91.4% of the GHGs reported in 2013 [4]. Methane emissions represent about 7% of the reported GHG emissions, N_2O represents about 0.8%, and fluorinated gases represent about 0.7%. This is illustrated in Figure 8.4.

Table 8.3 lists the primary sectors that emit each GHG and the source categories contributing most to the emissions [4]. The source categories in Table 8.3 account for 75% or more of the reported emissions of the corresponding GHG.

The emissions of CO_2 from fossil fuel-fired electric generating stations account for the majority of GHG emissions. Consequently, power plants have been targeted for reducing CO_2 emissions. EPA's Clean Power Plan was discussed in Chapter 2. The plan would require existing power plants to reduce CO_2 emissions by 30% below 2005 levels by 2030 although the specific reduction target for each state varies [7]. The CO_2 reduction targets for each state are based on that state's energy mix. Under the proposed rule, the benchmark year would be 2012, which means any improvements completed at coal-fired power plants during that year or before

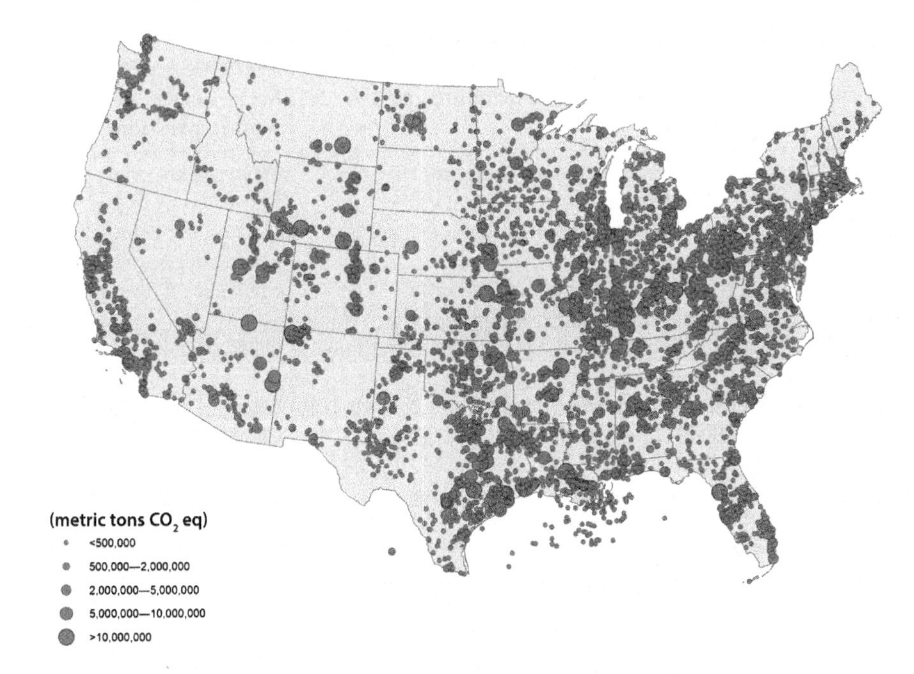

Figure 8.3 Location and total reported emissions from GHGRP facilities in 2013.

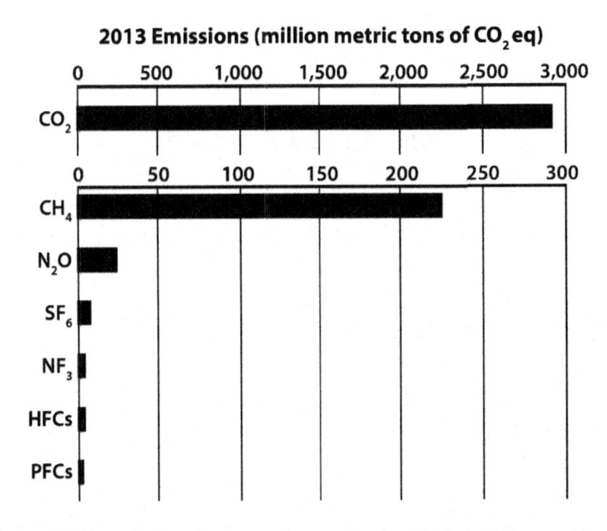

Figure 8.4 Direct GHG emissions by greenhouse gas in 2013 (million metric tons CO_2eq).

would not be recognized. Table 8.4 shows the proposed state reduction targets (lb/MWh) compared with historic emissions rates [modified from 7].

The U.S. EIA reported CO_2 emissions from energy consumption by major source for the period 1975 through 2013 [8]. Figure 8.5 shows the CO_2 emissions for

Table 8.3 **Largest sources of GHG emissions**

Greenhouse gas	Source categories contributing most to emissions	Sectors contributing most to emissions
CO_2	Electricity generation, Stationary combustion	Power plants
CH_4	Municipal landfills, Petroleum & natural gas systems	Waste, Petroleum & natural gas systems
N_2O	Nitric acid production, Electricity generation, Adipic acid production	Chemicals, Power plant
SF_6	SF_6 from electrical equipment, Magnesium production	Other, Metals
NF_3	Electronics manufacturers	Other
HFCs	HCFC-22 production and HFC-23 destruction	Chemicals
PFCs	Aluminum production, Electronics manufacturers	Metals, Other

petroleum, the largest major source, coal, and natural gas (the smallest major source). Data for coal, natural gas, and petroleum (various fuel sources and total petroleum) for the years 2011, 2012, and 2013 are listed in Table 8.5 for comparison to the GHG emissions reported by the EPA in Table 8.2.

8.4 CO_2 capture/reduced emissions technologies

Capture of CO_2 from electric power generation can be accomplished by three general methods: pre-combustion CO_2 capture, where CO_2 is separated from hydrogen (H_2) and other constituents in the syngas stream produced by the gasifier in an IGCC (integrated gasification combined cycle) power plant prior to H_2 combustion; oxy-fuel combustion, where the coal is combusted in an oxygen and CO_2-enriched environment; and post-combustion CO_2 capture, where coal is combusted normally in a boiler and CO_2 is removed from the flue gas. Figures 8.6–8.8 are generalized flow diagrams of the three processes, respectively, showing the various components for comparison. In addition, chemical looping is an advanced technology similar to oxy-fuel combustion in that it relies on combustion or gasification of coal in a nitrogen-free environment [9]. However, rather than using an air separation unit, chemical looping involves the use of a metal oxide or other compound as an oxygen carrier to transfer O_2 from air to the fuel. Figure 8.9 is a flow diagram of the chemical looping process. Biomass cofiring is a near-term, low-cost option (when it is compared to other capture technologies) for efficiently, and cleanly, converting biomass to electricity by feeding biomass as a partial substitute fuel in coal-fired boilers. Each of these technologies is discussed in the following sections.

Table 8.4 State CO_2 reduction targets under the proposed clean power plan

State	State	2012 Emission rate (lb/MWh)	2030 State goal (lb/MWh)	2030 Reduction target (%)	State	State	2012 Emission rate (lb/MWh)	2030 State goal (lb/MWh)	2030 Reduction target (%)
1	Alabama	1,444	1,059	27	26	Montana	2,245	1,771	21
2	Alaska	1,351	1,003	26	27	Nebraska	2,009	1,479	26
3	Arizona	1,453	702	52	28	Nevada	988	647	34
4	Arkansas	1,640	910	45	29	N. Hampshire	905	486	46
5	California	698	537	23	30	New Jersey	932	531	43
6	Colorado	1,714	1,108	35	31	New Mexico	1,586	1,048	34
7	Connecticut	765	540	29	32	New York	983	549	44
8	Delaware	1,234	841	32	33	North Carolina	1,646	992	40
9	Florida	1,200	740	38	34	North Dakota	1,994	1,783	11
10	Georgia	1,500	834	44	35	Ohio	1,850	1,338	28
11	Hawaii	1,540	1,306	15	36	Oklahoma	1,387	895	35
12	Idaho	339	228	15	37	Oregon	717	372	48
13	Illinois	1,895	1,271	33	38	Pennsylvania	1,540	1,052	32
14	Indiana	1,923	1,531	20	39	Rhode Island	907	782	14
15	Iowa	1,552	1,301	16	40	South Carolina	1,587	772	51
16	Kansas	1,940	1,499	23	41	South Dakota	1,135	741	35
17	Kentucky	2,158	1,763	18	42	Tennessee	1,903	1,163	39
18	Louisiana	1,466	883	40	43	Texas	1,298	791	39
19	Maine	437	378	14	44	Utah	1,813	1,322	27
20	Maryland	1,870	1,187	37	45	Virginia	1,297	810	38
21	Massachusetts	925	576	38	46	Washington	763	215	72
22	Michigan	1,696	1,161	32	47	West Virginia	2,019	1,620	20
23	Minnesota	1,470	873	41	48	Wisconsin	1,827	1,203	34
24	Mississippi	1,130	692	39	49	Wyoming	2,115	1,714	19
25	Missouri	1,963	1,544	21					

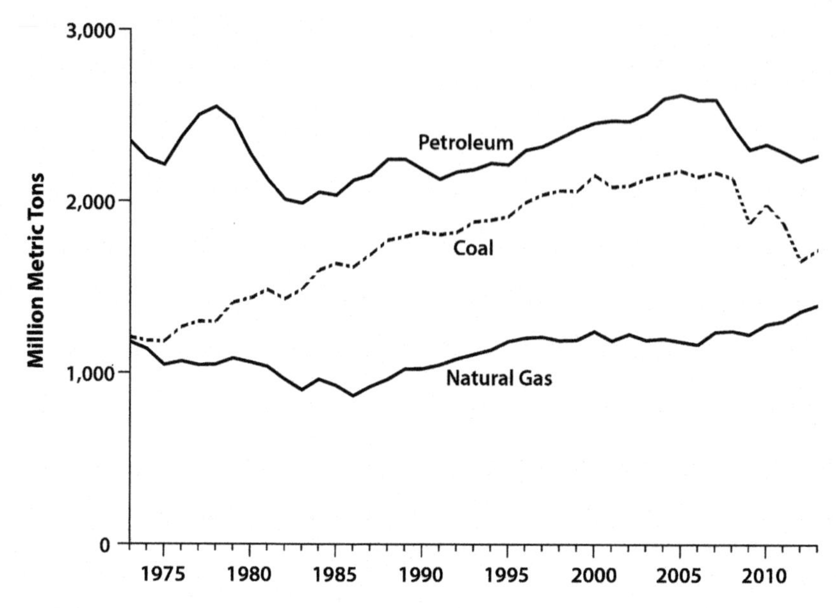

Figure 8.5 Carbon dioxide emissions from energy consumption by source for the period 1975 through 2013 (million metric tons of CO_2).

Table 8.5 Carbon dioxide emissions from energy consumption by source for the period 2011 through 2013 (million metric tons of CO_2)

	2011	2012	2013
Coal	1,876	1,657	1,722
Natural Gas	1,305	1,363	1,399
Petroleum (total)	2,291	2,240	2,272
Aviation gasoline	2	2	2
Distillate fuel oil	604	580	587
Jet fuel	209	206	210
Kerosene	2	1	1
Liquefied petroleum gas	78	81	88
Lubricants	10	9	10
Motor gasoline	1,112	1,106	1,123
Petroleum coke	78	78	76
Residual fuel oil	79	65	56
Other	117	113	119
Total coal, natural gas, & petroleum	5,483	5,272	5,405

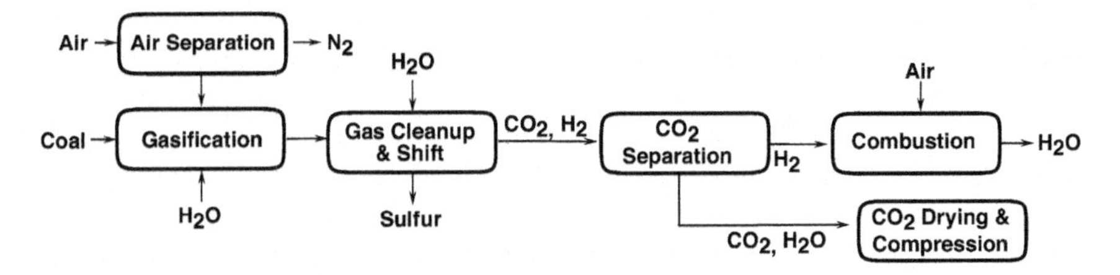

Figure 8.6 Flow diagram of pre-combustion CO_2 capture.

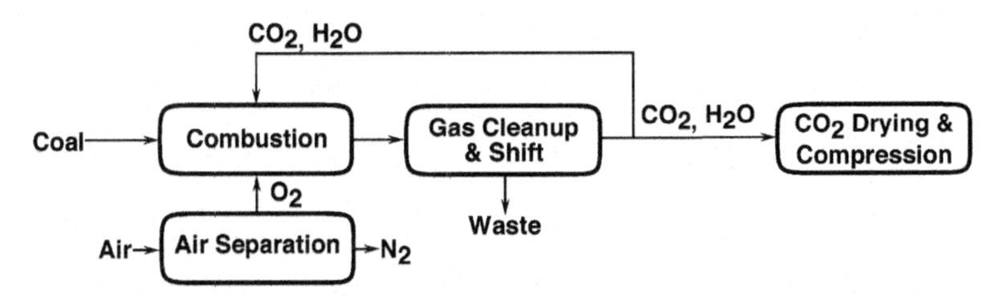

Figure 8.7 Flow diagram of oxy-fuel combustion.

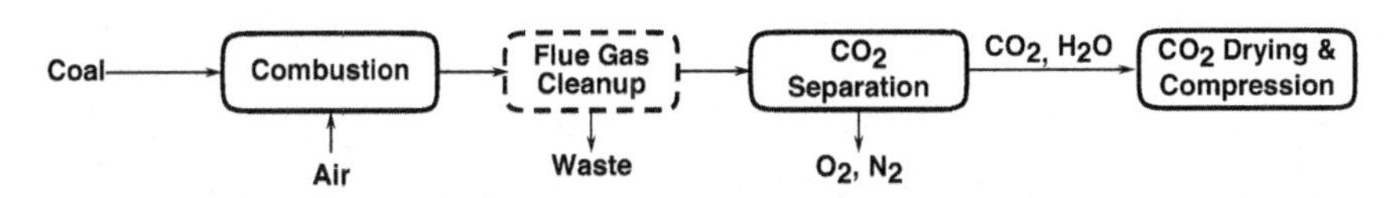

Figure 8.8 Flow diagram of post-combustion CO_2 capture.

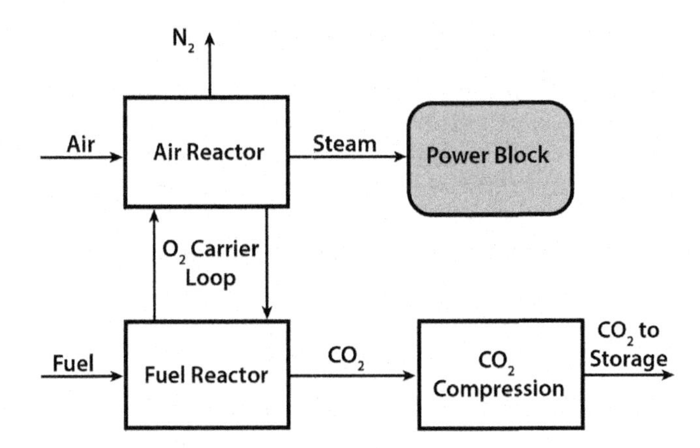

Figure 8.9 Flow diagram of a chemical looping process.

The various CO_2 capture technology types include absorption (physical and chemical), adsorption (physical, chemical, and mixed), membranes (organic and inorganic), cryogenics, mineralization, and others. The most common technologies used for pre-combustion CO_2 capture includes physical absorption, physical adsorption, and inorganic membranes. The most common technologies used for post-combustion CO_2 capture include chemical absorption, chemical and mixed adsorption, organic membranes, cryogenics, and others including mineralization and biological reduction. Adsorption technologies are used for oxy-fuel combustion applications. This chapter will primarily discuss absorption, adsorption, and membrane technologies.

8.4.1 Pre-combustion (IGCC)

Pre-combustion CO_2 capture is characterized by removing the carbon (as CO_2) from the fuel (in this case syngas) prior to utilization in a gas turbine. In this concept, coal is gasified through partial oxidation, using air or oxygen, to produce synthesis gas (syngas), which is composed of hydrogen (H_2), carbon monoxide (CO), and minor amounts of other constituents including methane (CH_4). The syngas then passes through gas clean-up stages and a shift reactor to convert the CO to CO_2 and increases the CO_2 and H_2 molar concentrations to approximately 40% and 55%, respectively. The CO_2 then has a high partial pressure and high chemical potential, which improves the driving force for various types of separation and capture technologies. After CO_2 removal, the hydrogen-rich syngas can be fired in a combustion turbine to produce electricity, similar to a natural gas combined cycle plant [10]. Additional electricity can be generated by extracting energy from the combustion turbine flue gas using a heat recovery steam generator, i.e. in an integrated gasification combined cycle (IGCC) power plant. The CO_2 that is removed is dried, compressed, and can be sequestered. For IGCC systems, the preferred separation method is physical absorption using a solvent, as this technology is commercially available. An IGCC process flowsheet with pre-combustion CO_2 capture and coal gas clean-up is shown in Figure 8.10 (modified from [11,12]).

There is an extremely wide variety of acid gas removal (AGR) systems on the market. These can be classified as chemical washes (which include all amines such as methyldiethanolamine (MDEA)) and physical washes such as Selexol or Rectisol. In addition it is possible to have a mixed characteristic solvent such as Sulfinol. All of these named processes have been used in IGCC or chemical plant gasification operations. Selection is based on requirements for high purity (Rectisol) versus low cost (MDEA) with Selexol and Sulfinol lying in between on both counts. All these processes have a long track record in industrial practice, all with high availability records. Advantages and disadvantages of pre-combustion capture are provided in Table 8.6 [13].

8.4.1.1 Chemical solvent processes

8.4.1.1.1 Amine processes

Figure 8.11 shows the flow sheet of a typical MDEA wash, although this flow sheet is representative for many other chemical washing processes [14]. The raw syngas

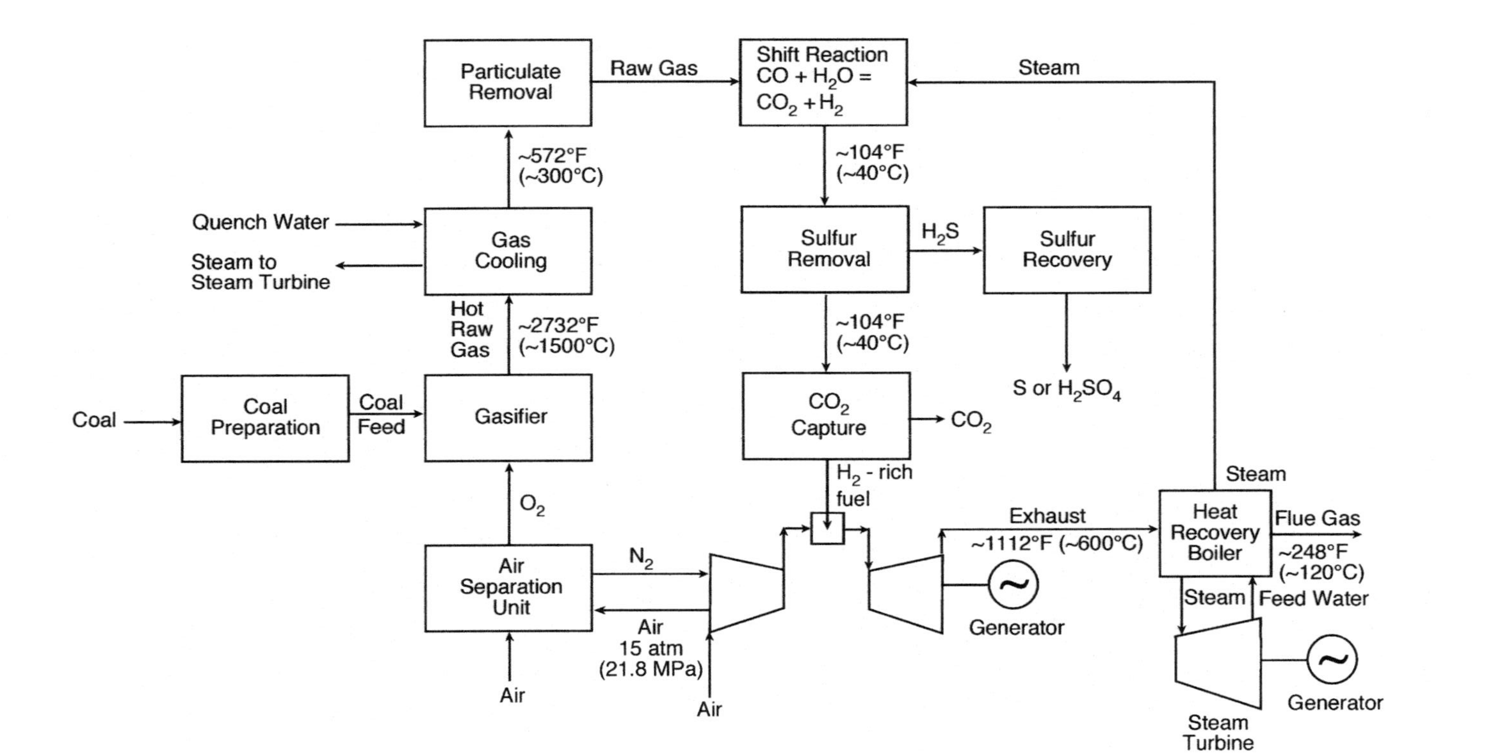

Figure 8.10 Simplified IGCC process flowsheet with pre-combustion CO_2 capture and coal gas clean-up.

Table 8.6 Advantages and disadvantages of pre-combustion capture

Advantages	Challenges/barriers to implementation
Technologies for pre-combustion capture of CO_2 via gasification are well established in the process industries	Lower (but still significant) energy loss compared to post-combustion capture
Capture using the water-gas shift reaction and removal of the CO_2 via AGR process is used widely	Capital costs of IGCC without capture are much higher than supercritical pulverized coal plant without capture
As smaller reaction volume is involved, at lower volumetric flow rates, elevated pressure, and higher component concentration, the CO_2 separation step consumes less energy than post-combustion capture	IGCC costs need reducing to compare more effectively
Syngas contains high concentration of CO_2 and is at high pressure resulting in: • high CO_2 partial pressure • increased driving force for separation • more technologies available for separation • potential for reduction in compression costs/ loads	Barriers to commercial application of gasification/ IGCC are common to pre-combustion capture: • availability • cost of equipment • extensive supporting systems requirements
Syngas produced as the first step of the process can be used to fuel a turbine cycle	Applicable mainly to new plants as relatively few coal gasification-based plants are in operation
Lower water use compared to post-combustion capture	

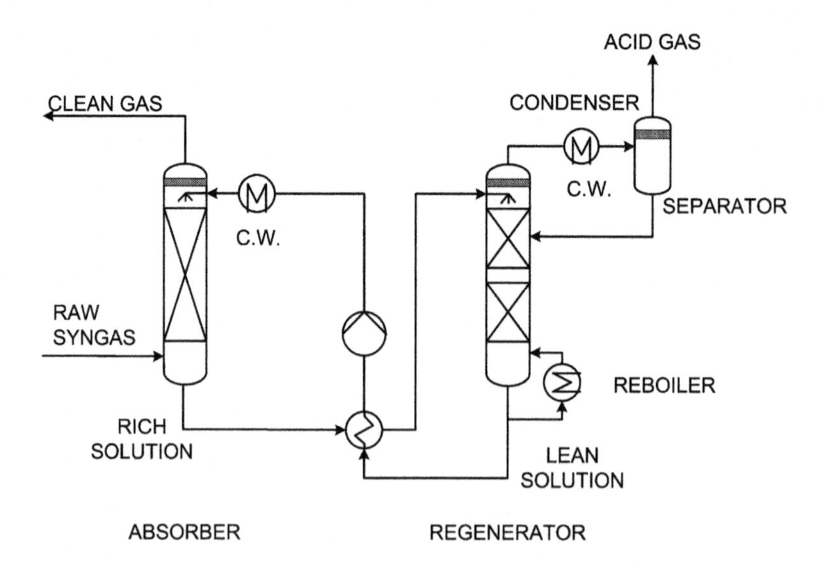

Figure 8.11 Typical MDEA Flowsheet.

is contacted in a wash column with lean MDEA solution, which absorbs the H_2S and some of the CO_2. MDEA is to some extent selective in that the bonding of the amine with H_2S takes places faster than with CO_2 and advantage can be taken of this in the design. The rich solution is preheated by heat exchange with the lean solution and enters the regenerator. Reboiling breaks the chemical bond and the acid gas components discharged at the top of the regenerator are cooled to condense the water, which is recycled. A discussion of amine chemistry of CO_2 capture is provided later in the chapter in the post-combustion section.

8.4.1.2 Physical solvent processes

8.4.1.2.1 Physical washes

The important characteristics for any successful physical solvent are:

- Good solubility for CO_2, H_2S and COS in the operating range, preferably with significantly better absorption for H_2S and COS compared with CO_2 if selectivity is an important issue for the application of interest.
- Low viscosity at the lower end of the operating temperature range. Although lowering the operating temperature increases the solubility, the viscosity governs, in effect, the practical limit to lowering the operating temperature.
- A high boiling point reduces vapor losses when operating at ambient or near ambient temperatures.

Table 8.7 lists the advantages and disadvantages of physical solvents [13].

8.4.1.2.2 Selexol

The Selexol process uses dimethyl ethers of polyethylene glycol (DMPEG) to absorb CO_2 and acid gases from syngas at relatively high pressure (usually between 290 and $\approx$1740 psig) [13]. The physical properties of DMPEG are listed in Table 8.8 [14]. The typical operating temperature range is 15 to 100°F. The ability to operate in this temperature range offers substantially reduced costs by eliminating or minimizing refrigeration duty. The acid gases are released using a pressure swing or steam stripping. Selexol has a number of properties that make its use attractive in commercial applications [13]. These include:

- a very low vapor pressure that limits its losses
- a low viscosity that avoids large pressure drop
- no heat of reaction and small heat of solution
- chemically inert, thermally stable, and no oxidation degradation
- nontoxic for environmental compatibility and worker safety
- noncorrosive to carbon steel construction due to its non-aqueous nature and inert chemical characteristics
- nonfoaming for operational stability
- high solubility for HCN and NH_3 allows removal without solvent degradation
- high solubility for nickel and iron carbonyls allows for their removal from syngas
- low heat requirements for regeneration as the solvent can be partially regenerated by a simple pressure let-down

Table 8.7 Advantages and disadvantages of physical solvents

Advantages	Disadvantages
Low utility consumption	CO_2 pressure is lost during flash recovery
CO_2 recovery does not require heat to reverse a chemical reaction	Necessary to cool syngas for CO_2 capture, then heat and rehumidify for firing in gas turbine
Common solvents also have high H_2S solubility, allowing for combined CO_2/H_2S removal	Low solubility can require circulation of large volumes of solvent, thereby increasing energy for pumping.
Rectisol uses inexpensive, easily available methanol	
Methanol is noncorrosive so carbon steel can be used for most plant equipment	Some H_2 may be lost with captured CO_2
Refrigeration costs can be high	Refrigeration is often required for the lean Selexol solution
Selexol has a higher capacity to absorb gases than amines	More economical at high pressures
Selexol can remove H_2S and organic sulphur compounds	Hydrocarbons are co-absorbed in Selexol, resulting in reduced product revenue and often requiring recycle compression
Both provide simultaneous dehydration of the gas stream	
Both can remove CO_2 and various contaminants in a single process	

Table 8.8 Properties of physical solvents

Process		Selexol	Rectisol
Solvent		DMPEG	Methanol
Formula		$CH_3O(C_2H_4O)_nCH_3$	CH_3OH
Mol. weight	lb/lb mol	178 to 442	32
Boiling point at 760 Torr	°F	415 to 870	147
Melting point	°F	-4 to -20	-137
Viscosity	cP	4.7 at 86°F	0.85 at 5°F
	cP	5.8 at 77°F	1.4 at -22°F
	cP	8.3 at 59°F	2.4 at -58°F
Specific weight	kg/m^3	1.031	790
Selectivity at working temperature	(H_2S:CO_2)	1:9	1:9.5

The Selexol flowsheet in Figure 8.12 exhibits the typical characteristics of most physical absorption systems [14]. The intermediate flash allows co-absorbed syngas components (H_2 and CO) to be recovered and recompressed back into the main stream. For other applications such as H_2S concentration in the acid gas or separate CO_2 recovery using staged flashing, techniques not shown here may be applied.

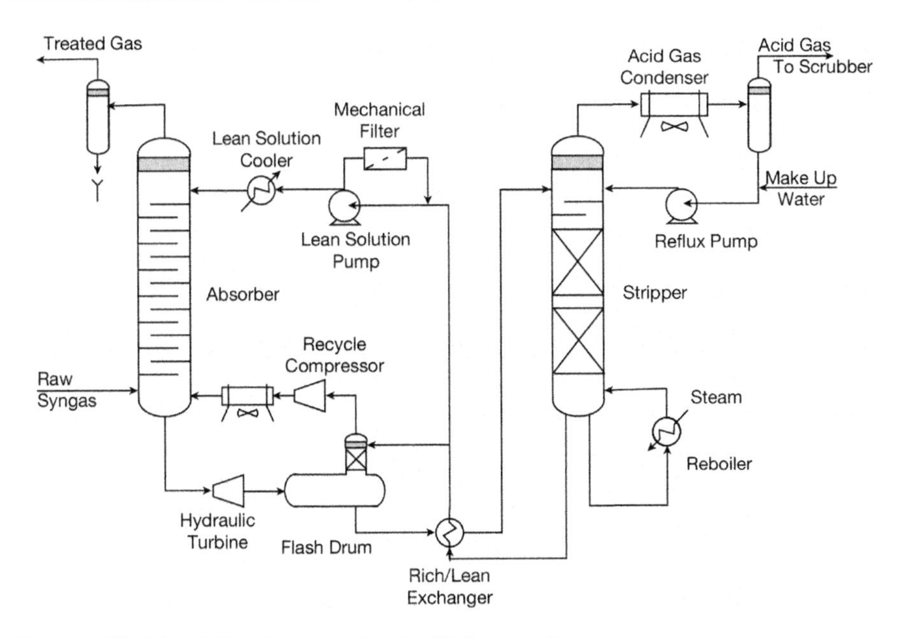

Figure 8.12 Selexol flowsheet for selective H_2S removal.

8.4.1.2.3 Rectisol

The Rectisol process, which uses cold methanol as solvent, was originally developed to provide a treatment for gas from the Lurgi moving-bed gasifier, which in addition to H_2S and CO_2 contains hydrocarbons, ammonia, hydrogen cyanide, and other impurities. Figure 8.13 contains a flowsheet of the Rectisol process [14].

In the typical operating range of -20 to -75°F, the Henry's law absorption coefficients of methanol are extremely high and the process can achieve gas purities unmatched by other processes. This has made it a standard solution in chemical applications such as ammonia, methanol or methanation, where the synthesis catalysts require sulfur removal to less than 0.1 ppm. This performance has a price, however, in that the refrigeration duty required for operation at these temperatures involves considerable capital and operating expense.

Methanol as a solvent exhibits considerable selectivity, as can be seen in Table 8.8 [14]. This allows substantial flexibility in the flowsheeting of the Rectisol process and both standard (nonselective) and selective variants of the process are regularly applied according to circumstances.

As a physical wash, which uses at least in part flash regeneration, part of the CO_2 can be recovered under an intermediate pressure. Typically, with a raw gas pressure of 710 psig, about 60 to 75% of the CO_2 would be recoverable at 40 to 60 psig. Where CO_2 recovery is desired, whether for urea production in an ammonia application or for sequestration, this can provide significant compression savings.

Following the solvent circuit is an intermediate H_2S flash from which co-absorbed hydrogen and carbon monoxide are recovered and recompressed back into the raw

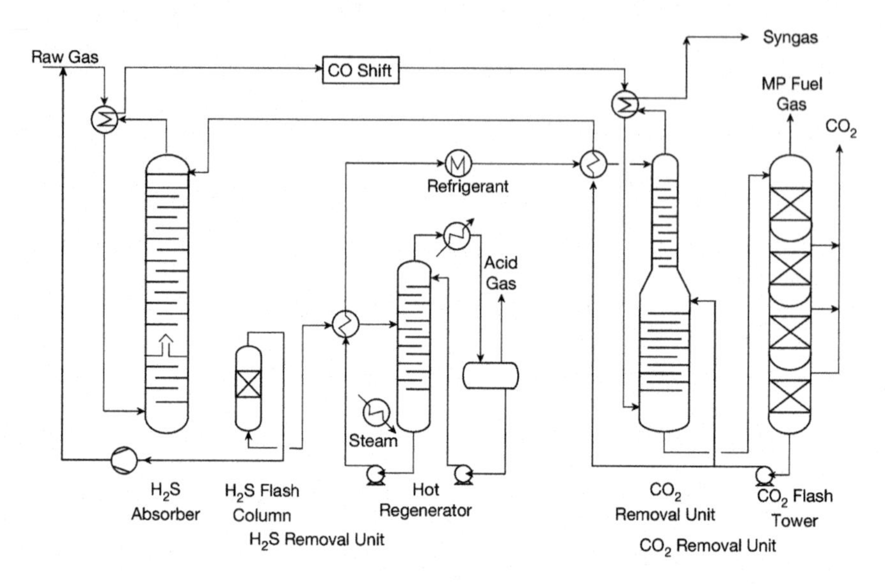

Figure 8.13 Flowsheet of selective Rectisol process.

gas. The flashed methanol is then reheated before entering the Hot Regenerator. Here the acid gas is driven out of the methanol by reboiling and a Claus gas with an H_2S content of 25–30% (depending on the sulfur content of the feedstock) is recovered. Minor adaptations are possible to increase the H_2S content if desired.

Water entering the Rectisol unit with the syngas must be removed and an additional small water-methanol distillation column is included in the process to cope with this. Typically the refrigerant is supplied at between -20 and -40°F. Depending on application, different refrigerants can be used. In an ammonia plant, ammonia is used and the refrigeration system is integrated with that of the synthesis. In a refinery environment, propane or propylene may be the media of choice.

The Rectisol technology is capable of removing not only conventional acid gas components but also, for example, HCN and hydrocarbons, and metallic mercury. Advantages and disadvantages of the Rectisol process are given in Table 8.9 [13].

8.4.2 Oxy-fuel combustion

Oxy-fuel combustion is one of the three main routes being pursued towards cost-effective, technically viable carbon capture. A schematic diagram of the most common oxy-fuel process, which involves the combustion of pulverized coal in pure oxygen (>95%) mixed with recycled flue gas is shown in Figure 8.14 (modified from [11,15]).

Oxy-fuel combustion is not technically a capture technology but rather is a process in which coal combustion occurs in an oxygen-enriched (i.e. nitrogen-depleted) environment, thereby producing a flue gas comprised mainly of CO_2 (up to 89 vol. percent) and water. The water is easily separated and the CO_2 is ready for sequestration.

Table 8.9 **Advantages and disadvantages of the rectisol process**

Advantages	Disadvantages
Selectively for H_2S over CO_2 is high – only slightly less than Selexol	Complex process scheme
Solubilities of H_2S and COS higher than in Selexol	Need to refrigerate solvent
Allows for deep sulfur removal to <0.1 ppmv H_2S + COS	Leads to high capital and operating costs
High selectivity for H_2S combined with ability to remove COS	Relatively high vaporization losses of the solvent even at low temperatures due to the appreciable vapor pressure of methanol
Also absorbs HCN, NH_3, and iron- and nickel carbonyls	
Applicability to deep cleaning of syngas for catalytical conversion (sensitive to contaminants) for such products as NH_3, H_2, and Fischer-Tropsch liquids	

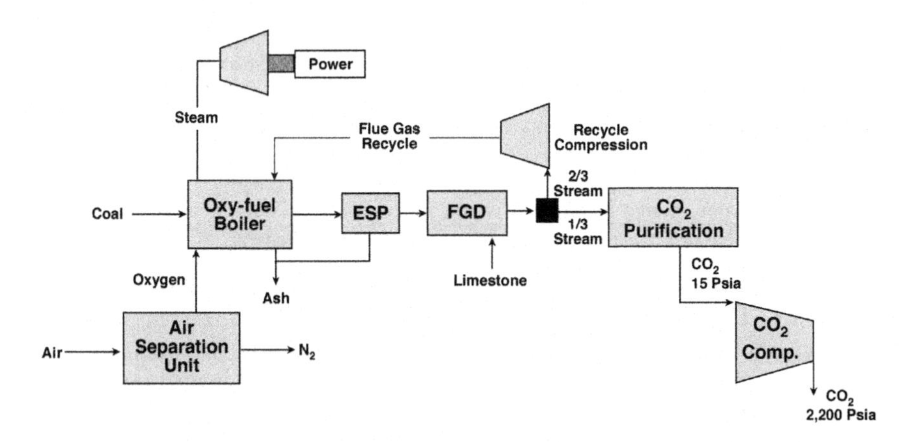

Figure 8.14 Schematic diagram of a pulverized coal oxy-fuel combustion system.

The most common oxy-fuel process involves the combustion of pulverized coal in an atmosphere of nearly pure oxygen (greater than 95% and up to 99%) mixed with recycled flue gas. Almost pure oxygen for combustion is produced in a cryogenic air separation unit (ASU) and is mixed with recycled flue gas (approximately ⅔ of the flue gas flow from the boiler) prior to combustion in the boiler to maintain combustion conditions similar to an air-fired configuration. Mixing the recycled flue gas with the pure oxygen is necessary to provide proper heat transfer characteristics because materials of construction currently available cannot withstand the high temperature resulting from coal combustion in pure oxygen. Research has shown that the oxygen/recirculated flue gas flow must be maintained such that the pulverized coal oxy-combustion flame has heat transfer characteristics similar to that of an air-fired system. This requires an oxygen level of about 30–35 vol. percent in the gas entering the boiler [16]. Power plants that are retrofitted to oxy-combustion would likely operate within the range of oxygen levels that will maintain combustion conditions near those of air-fired systems. New oxy-combustion plants or repowered plants, however, will not necessarily be constrained and smaller boiler equipment will be able to be used, which will increase efficiency. The critical parameter of oxygen content may be adjusted to increase radiative heat transfer at the expense of convective heat transfer zones further from the flame. This is in contrast to air-fired systems where approximately half of the heat transfer occurs in the boiler and the balance of the heat transfer is convective in the backpasses. In both retrofit and new systems, the changes inherent in oxy-combustion affect many parameters including flame behavior, heat and mass transfer, combustion gas chemistry and behavior, char burnout, and slag development, chemistry, and deposition.

Oxy-fuel combustion can essentially produce a near-zero emissions power plant. Oxy-fuel combustion offers a number of advantages over some competing carbon capture technologies [13]:

- The boiler and air pollution control equipment utilize conventional designs, materials of construction and arrangements
- Oxy-fuel systems look and operate in a similar manner to conventional power plants
- The oxy-fuel process can utilize all ranks of coal
- It should be easier and less complex to repower or retrofit an existing power plant compared to post-combustion control
- New chemicals or waste streams are introduced into the plant process
- There is no major change to the plant water process; for low-rank coals there may be a positive water balance from condensation of water from the flue gas

A major advantage of the technology is that it produces a flue gas that consists primarily of CO_2 (>80% by volume) and water. The water can be removed easily by condensation allowing the CO_2 to be purified relatively inexpensively [13]. Roughly three times more oxygen is needed for oxy-fuel systems than for an IGCC plant of comparable size, so the ASU adds significantly to the cost [10]. The main advantages and disadvantages of oxy-fuel combustion are summarized in Table 8.10.

8.4.3 Post-combustion

Technologies for capturing CO_2 from emissions streams have been used for many years to produce a pure stream of CO_2 from natural gas or industrial processing for use in the food processing and chemical industries. Methods currently used for CO_2 separation include:

- Physical and chemical solvents, particularly monoethanolamine (MEA);
- Various types of membranes;

Table 8.10 Advantages and disadvantages of oxy-fuel combustion

Advantages	Challenges/barriers to implementation
Oxy-combustion power plants should be able to deploy conventional, well-developed, high efficiency steam cycles without the need to remove significant quantities of steam from the cycle for CO_2 capture	Not possible to develop subscale oxy-combustion technology at existing power plants as it requires commitment of the entire power plant
Extra equipment consists mainly of conventional equipment and heat exchangers. The boiler and air pollution control devices utilize conventional designs, materials of construction, and arrangements, all well-established in industry	Energy penalty results from power needed for ASU air compression and CO_2 compression in the CO_2 purification unit will reduce net plant output by up to 25% compared to an air-fired power plant of same capacity without CO_2 capture
Very low emissions of conventional pollutants, which are usually achievable at relatively low cost	Currently, little geological or regulatory consensus on what CO_2 purity needed for compression, transportation, and storage. If purity requirements are lower, oxy-combustion costs could be reduced
On a cost per ton CO_2 captured basis, it should be possible to achieve 98 + % CO_2 capture at an incrementally lower cost than achieving a baseline 90% capture	Need to reduce overall costs although this issue is common to all capture technologies
Oxy-combustion with CO_2 capture should be competitive or slightly more cost effective than pre- and post-combustion capture	Technology needs proving through the integrated operation on a larger scale and under various operating conditions
Should be easier and less complex to repower or retrofit into an existing power plant than post-combustion capture	Installation must be air tight to avoid in-leakage. If overpressurized, there is risk of CO_2 leakage
Any oxy-fuel power plant will look and operate in a similar manner to a conventional power plant	
No on-site chemical operations are required. Waste streams are largely unchanged.	

- Adsorption onto solids; and
- Cryogenic separation.

These methods can be used on a range of industrial processes; however, their use for removing CO_2 from high volume, low CO_2 concentration flue gases, such as those produced by coal-fired power plants, is more problematic. The high capital

Table 8.11 Technology options for post-combustion capture

Chemical absorption		Physical absorption	Adsorption	Alternative approaches		
Amines	Alternative Solvents			Membranes	Cryogenic	Others
• primary • secondary • tertiary • sterically hindered	• ammonia • alkali-compound • aminosalt	• Rectisol • Selexol • press. water	• PSA • TSA • Solid sorbents	• gas absorption membrane	• distillation • frosting	• ionic liquid

costs for installing post-combustion separation systems to process the large volume of flue gas is a major impediment to post-combustion capture of CO_2. In addition, a large amount of energy is required to release the CO_2 from solvents or solid adsorbents after separation. There are some major technical and cost challenges that need to be overcome before retrofit of existing power plants with post-combustion capture systems becomes an effective mitigation option. Capture technologies will likely improve in the coming years, thereby improving economics. This section will discuss the status of technologies that are currently available for CO_2 capture as well as some key ones that are under research and development.

Post-combustion CO_2 capture technologies including state-of-the-art amine- and ammonia-based scrubbing systems, commercially available capture systems, and near-term technologies will be discussed in this section along with emerging technologies such as the use of adsorbents, membranes, and metal organic frameworks. Table 8.11 lists a number of possibilities to perform post-combustion CO_2 capture [10,11]. Note, however, that some of the technologies in Table 8.11 are more suited for pre-combustion applications (e.g. physical absorption such as Rectisol and Selexol) due to concentrated CO_2. The main advantages and disadvantages of post-combustion capture are summarized in Table 8.12 [13].

Post-combustion CO_2 capture mainly applies to coal-fired power plants but may also be applied to gas-fired combustion turbines. In a typical coal-fired power plant (see Figure 8.15; modified from [15]), fuel is burned with air in a boiler to produce steam, which drives a turbine to generate electricity. The boiler exhaust gas, i.e. flue gas, consists of mostly nitrogen, CO_2, O_2, moisture, and trace impurities. Separating the CO_2 from this gas stream is challenging because the CO_2 is present in dilute concentrations (i.e. 13 to 15 volume percent in coal-fired systems and 3–4 volume percent in natural gas-fired systems) and at low pressure (15 to 25 psia). In addition, trace impurities (particulate matter, sulfur dioxide, and nitrogen oxides) in the flue gas can degrade sorbents and reduce the effectiveness of some CO_2 capture processes. Also, compressing the captured or separated CO_2 from atmospheric pressure to pipeline pressure (approximately 2,000 psia) represents a large auxiliary power load on the overall power plant system. In spite of these difficulties, post-combustion capture has the greatest near-term potential for reducing CO_2 emissions because it can be retrofitted to existing units [9,15]. Retrofitting will primarily be

Table 8.12 Advantages and disadvantages of post-combustion capture

Advantages	Challenges/barriers to implementation
Applicable to the majority of new and existing coal-fired power plants. Retrofit technology option	Flue gas is dilute in CO_2 and at ambient pressure; low CO_2 partial pressure
Enables the continued use of well-established pulverized coal technology that is used worldwide for both retrofit and new construction	Amine-based processes are commercially available but at relatively small scale. Used mainly in nonpower industrial applications. Considerable scale-up is required
Extensive research & development to improve sorbents and capture equipment have been performed. This is leading to reduced energy penalty	Energy penalty – current amine technologies result in a lost of net power output of $\approx$30% and a reduction of 11 percentage points in efficiency
Future improvements and developments of pulverized coal systems (such as ultra-supercritical materials) will increase plant efficiency and reduce CO_2 emissions	Steam extraction for solvent regeneration reduces flow to low-pressure turbine; this impacts on efficiency and turn-down capability
	Significantly higher performance or circulation volume required for high capture levels
	CO_2 produced at low pressure compared to storage requirements
	Most sorbents need very pure flue gas to minimize sorbent usage and cost
	Considerable water requirement; nearly double per net MWh for water-cooled plants

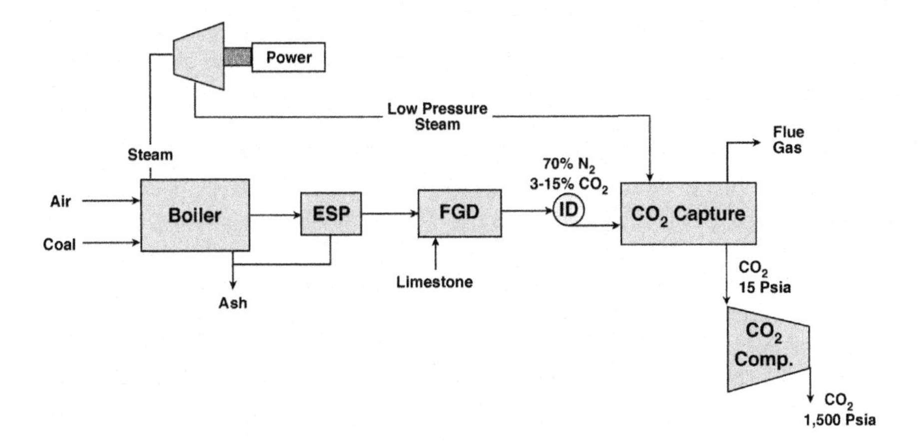

Figure 8.15 Block diagram illustrating a post-combustion system.

of interest in power plants with advanced efficiencies to compensate for the large efficiency loss from post-combustion capture. For this reason, capture-ready fossil fuel plants, retrofitted with a post-combustion capture process some time after the start of initial operation, could play an important role in the near future.

During the lifetime of these plants it is realistic to expect that post-combustion capture will become sufficiently mature. Some of the options for post-combustion CO_2 capture are discussed below. These range from state-of-the art amine-based systems to emerging technologies under development. Of the technologies presented in this section, chemical absorption technologies using amine or alternative solvents scrubbing are the nearest to commercial use in utility boilers. The others presented in this section are emerging technologies and at various stages of development. A general ordering of the technologies from nearest to commercialization in utility boilers to longer lead times until commercialization are: amine solvents, aqueous ammonia, chilled ammonia $<$ solid sorbents, membranes $<$ ionic liquids, metal organic frameworks.

8.4.3.1 Amine-based liquid solvent systems

Because the partial pressure of CO_2 in the flue gas of fossil fuel-fired power plants is low, technologies driven by high CO_2 partial pressure differentials, such as physical solvents or membranes, are not efficiently applicable for post-combustion capture [11]. At partial pressures typical of coal-fired power plants, the possible loading (in weight percent CO_2 per unit solvent) is very limited for physical solvents (e.g. ≈ 1 compared to ≈ 9 for chemical solvents), resulting in large solvent circulation rates and ultimately in large amounts of heat required for solvent regeneration. Only chemical solvents show an absorption capacity large enough to be applicable for CO_2 capture for CO_2 concentrations typical of coal-fired power plants. Amine-based absorption systems were developed over 60 years ago, primarily to remove acid gas impurities such as H_2S and CO_2 from natural gas streams [13]. Later the process was adapted to capture CO_2 from flue gas streams for subsequent use in chemical manufacture, food and beverage production, and enhanced oil recovery.

A system for post-combustion capture of CO_2 by chemical absorption is shown in Figure 8.16. The flue gas is usually cooled before entering the absorber column at the bottom. As the flue gas rises in the column, the CO_2 is absorbed by a solvent in counter-current flow. The CO_2-free gas is vented to the atmosphere. The absorbed CO_2 is stripped from the rich solution, which occurs in a separate regeneration column. At the bottom of the absorber the CO_2-rich solvent is collected and pumped through a heat exchanger where it is preheated before entering the regenerator. In the regenerator, the CO_2-rich solvent flows downwards. While flowing downwards, the temperature increases, thereby releasing CO_2, which rises to the top of the column and is removed.

The solvent most commonly used for CO_2 capture is amine based and is monoethanolamine (MEA). Amines are organic compounds with a functional group that contains nitrogen as the key atom. Structurally amines resemble ammonia, where one or more hydrogen atoms are replaced by an organic substituent. Primary amines

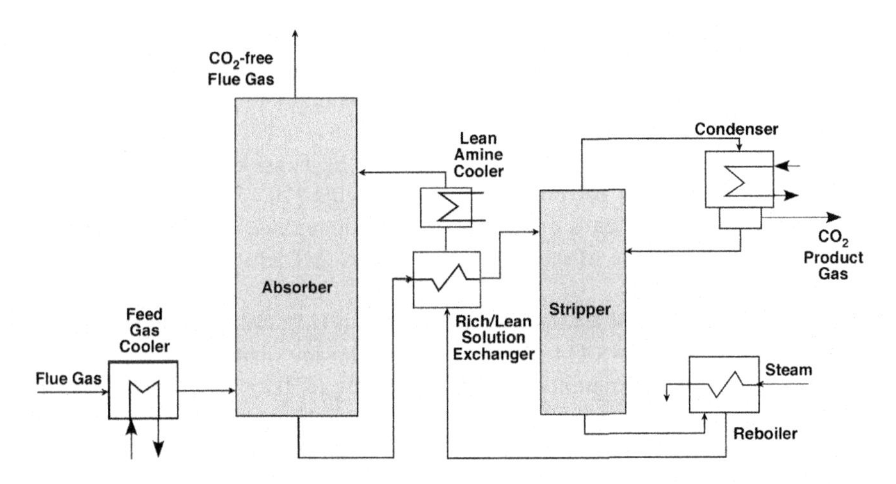

Figure 8.16 Schematic diagram of a post-combustion capture system using chemical absorption.

arise when one of the three hydrogen atoms in ammonia is replaced by an organic substituent, secondary amines have two hydrogen atoms replaced, and tertiary amines have all three hydrogen atoms substituted. The types of amine solvents include [17]:

- simple alkanolamines;
- primary – monoethanolamine (MEA) – $(C_2H_4OH)NH_2$;
- secondary – methylmonoethanolamine (MMEA), diethanolamine (DEA) – $(C_2H_4OH)_2NH$;
- tertiary – dimethylmonoethanolamine (DMMEA), methyldiethanolamine (MDEA);
- hindered amines;
- mildly hindered primary – alamine (ALA);
- moderately hindered – aminomethylpropoanol (AMP); and
- cyclic diamines – piperazine.

CO_2 solvent extraction is based on the reaction of a weak alkanolamine base with CO_2, which is a weak acid to produce a water-soluble salt. Amines react with CO_2 to form carbamate and bicarbonate via reactions (8.1) and (8.2), respectively:

$$2RNH_2 + CO_2 \leftrightarrow RNH_3^+ + RNHCOO^- \tag{8.1}$$

$$RNH_2 + H_2O + CO_2 \leftrightarrow RNH_3^+ + HCO_3^- \tag{8.2}$$

Chemical absorption with amines has been used for CO_2 separation in several industrial applications and in slipstream tests on coal-fired boilers; however, no amine-based CO_2 absorption system has been fully adapted to the specific conditions for complete flue gas treatment in a large coal-fired utility boiler, which would produce around 700 metric tons CO_2 per hour for a state-of-the-art 1000 MW_e plant. Amine, specifically MEA, based systems are the leading technology for retrofit applications since they are applicable for low-CO_2 partial pressure streams and provide recovery rates of up to 98% and product purity >99 vol. percent. Advantages and disadvantages of amine scrubbing are summarized in Table 8.13 [13].

There are a number of technologies that can be considered as commercially available capture systems for coal-fired power plants [10,13,18]. Most use either aqueous pure amines or blends of amines. Most commercially available systems are generally based on the use of amines such as MEA, MDEA, or sterically hindered amines.

The U.S. Department of Energy (DOE) adopted a U.S. National Aeronautics and Space Administration (NASA) method of describing the maturity of a technology, which consists of nine technology readiness levels (TRLs) [19]. DOE's view of technology development stages is shown in Figure 8.17 and ranges from the concept stage on paper (TRL of 1) to the full-scale commercial deployment (TRL of 9) [19,20]. Post-combustion absorption technologies are considered to be at TRLs of up to 7, the same level as oxy-fuel combustion. Absorption technologies are considered the most developed. Other post-combustion capture technologies are considered less developed; e.g. TRL levels for adsorption technologies are up to 4 while for membrane capture technologies the TRL level is up to 5. In contrast, pre-combustion systems are considered to be higher at TRL levels of up to 9 [10,13].

Amine-based absorption capture systems that are considered commercially available include [10,13]:

- Kerr-McGee/ABB Lummus Process (ABB/Lummus and Alstom/Dow Chemical)
- Fluor Econamine FG + SM Process
- Kansai Electric Power Company/Mitsubishi Heavy Industries KM-CDR Process®
- HTC Purenergy Process
- Aker Clean Carbon Just Catch Process
- Cansolv CO_2 Capture System
- Hitachi Process
- Toshiba Corporation Process
- Babcock & Wilcox OptiCap® Process

8.4.3.2 Aqueous ammonia process

Ammonia-based wet scrubbing is similar in operation to amine systems [10,13,18]. The process scheme is similar to the configuration shown in Figure 8.16. Ammonia-based systems operate efficiently at lower temperatures than those required for conventional MEA-based scrubbing. The lower temperatures also minimize ammonia volatility and the potential for ammonia slip. The main difference with that of the amine system is the precipitation of crystalline solid salts at the very low absorber temperatures of 32 to 50°F in the Chilled Ammonia Process (CAP) [18], and the relatively high vapor pressure of ammonia, which requires regeneration of the solvent at very high pressures to minimize ammonia slip.

The CO_2 transfer capacities of aqueous ammonia solutions can be higher than those of MEA solutions [13J]. In addition, the energy requirements for liquid mass

Table 8.13 Advantages and disadvantages of amine scrubbing systems

Advantages	Disadvantages
Well established in different industries; Can be retrofitted to some existing power plants in suitable locations; has been proven on a few small-scale coal-fired power plants	Need to treat large volumes of flue as with limited CO_2 content
Applicable to low CO_2-partial pressures	Process consumes considerable energy; retrofitting power plants with low thermal efficiency would result in efficiency losses rendering the plant uneconomic
MEA commonly used; solvents are inexpensive	Relatively low boiling point of MEA may result in solvent carryover into the CO_2 removal and regeneration steps
Recovery rates of up to 98% can be achieved	Presence of oxygen in the flue gas can increase corrosion and solvent degradation; uninhibited alkanolamines such as MEA and DEA can be oxidized to give carboxylic acids and heat-stable amine salts
Product purity >99% can be achieved	Concentrations of CO_2 and NO_x in the gas stream combine with the amine to form nonregenerable heat-stable salts
May offer flexibility if system can be switched between capture and no capture	Hot flue gases cause solvent degradation and decrease absorber efficiency; inlet flue gas limit of 120°F for MEA-based solvents
Amines offer potential cost reductions if the technology proves to be analogous to other technologies	SO_x in the flue gas reacts irreversibly with MEA-based solvents to produce nonreclaimable corrosive salts
There is a strong research base that should lead to improved solvents and processes	Because of its high reactivity, MEA can react with COS and CS_2, thereby degrading the solvent
	Fly ash in the absorption solvent may cause foaming in the absorber and stripper, scaling and plugging of equipment, erosion, corrosion, and increased solvent loss through chemical degradation and physical association with the waste sludge

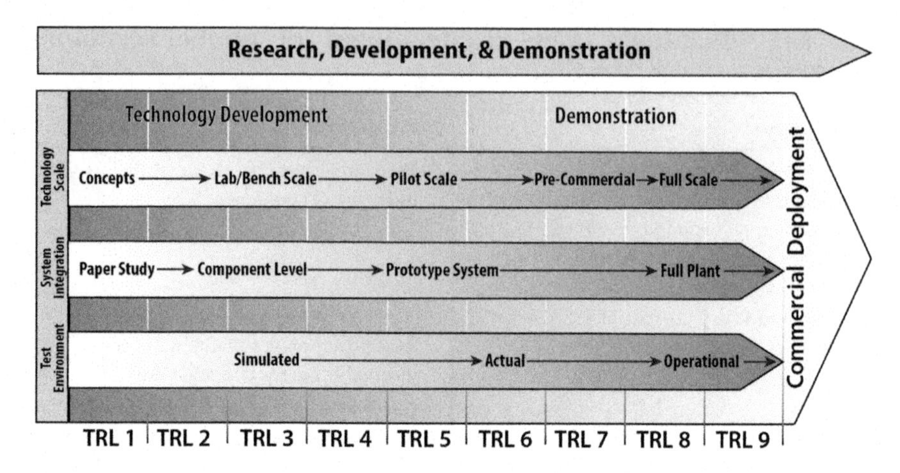

Figure 8.17 DOE's view of technology development stages and their corresponding TRLs.

circulation of ammonia solution is around half that required for MEA solution for an equal weight of CO_2 carried. The thermal energy required to regenerate CO_2 from the rich solution is substantially less for an ammonia solution than for MEA. Absorbent degradation and potential corrosion problems are also less.

In the absorber column, CO_2 from the flue gas is absorbed by reacting CO_2 with ammonium carbonate to form ammonium bicarbonate:

$$(NH_4)_2CO_3 + CO_2 + H_2O \leftrightarrow 2NH_4HCO_3 \tag{8.3}$$

Ammonium bicarbonate partly precipitates in the absorber as a crystalline product [6]. In doing so, the absorption process is promoted and the CO_2 capacity increased. The CO_2-rich slurry, consisting mainly of ammonium bicarbonate, is pumped through a heat exchanger to the high-pressure regenerator.

If the absorption/desorption cycle is controlled in such a way that it is primarily cycling between ammonium bicarbonate and ammonium carbonate, the regeneration energy can be reduced to approximately 30% of the necessary energy of a MEA system. In addition, reaction (8.3) has a significantly lower heat of reaction than amine-based systems, resulting in energy savings. Ammonia-based absorption has a number of other advantages over amine-based systems, such as tolerance to oxygen in the flue gas and potential for regeneration at high pressure. It also offers multipollutant control by reacting with sulfur and nitrogen oxides to form fertilizer (ammonium sulfate and ammonium nitrate) as a salable byproduct.

Disadvantages of the process include ammonia's higher volatility than MEA, which often results in ammonia slip into the exit gas and the requirement of the flue gas being cooled to the 60–80°F range to enhance CO_2 absorptivity of the ammonia compounds. Also, ammonia is consumed through the irreversible formation of ammonium sulfates and nitrates as well as removal of HCl and HF.

In the CAP process developed by Alstom, the absorber is kept at a significantly lower temperature, i.e., 32 to 50°F. This process uses the same ammonium carbonate/ammonium bicarbonate absorption chemistry as the aqueous system but differs in that no fertilizer is produced and a slurry of aqueous ammonium carbonate and ammonium bicarbonate and solid ammonium bicarbonate is circulated to capture CO_2 [15]. The main advantages of this process configuration are water condensation, smaller volume flow, less power duty for the blower, and less ammonia slip in the absorber [11]. Technical hurdles include cooling the flue gas and absorber to maintain operating temperatures below 50°F, mitigating the ammonia slip during absorption and regeneration, achieving 90% removal efficiencies in a single stage, and avoiding fouling of heat transfer and other equipment by ammonium bicarbonate deposition [15].

Ammonia-based solutions offer possibilities for developing absorption processes based on less corrosive and more stable solvents. At the same time, since ammonia is a toxic gas, prevention of ammonia slip to the atmosphere is a necessity [18]. Advantages and disadvantages of ammonia-based scrubbing are summarized in Table 8.14 [13].

Ammonia-based absorption capture systems that are considered commercially available include [10,13]:

- Alstom Chilled Ammonia Capture (CAP) Process
- PowerSpan ECO_2™ Ammonia-Based Capture Process

Table 8.14 Advantages and disadvantages of ammonia-based scrubbing

Advantages	Disadvantages
Lower heat of regeneration than amines, reducing the energy consumption associated with solution regeneration	Ammonium bicarbonate decomposes at 140°F, so temperatures in the absorber must be kept below this
Higher net CO_2 transfer than amines	Ammonia is more volatile than, for example, MEA, and can produce an ammonia slip into the exit gas
Higher loading capacity than amines	Because of lower reaction rate and CO_2 loading than amines, may require a larger absorber
Fewer corrosion issues than amines	May be issues of precipitation
Does not degrade in a flue gas environment, minimizing absorbent make-up	Cooling energy required for Chilled Ammonia Process
Lower cost than amines	
Stripping steam is not required	
More tolerant to pollutants such as SO_2	
Offers possibility of multipollutant control	

8.4.3.3 Alkali carbonate-based systems

Carbonate systems are based on the ability of a soluble carbonate to react with CO_2 to form a bicarbonate, which when heated releases CO_2 and reverts back to a carbonate. A major advantage of carbonates over amine-based systems is the significantly lower energy required for regeneration.

There are several commercial processes (e.g. Benfield™ and Catacarb™) for the removal of CO_2 from high-pressure gases that utilize aqueous potassium carbonate (K_2CO_3) and these are in industries producing natural gas and hydrogen for ammonia synthesis [11]. These processes are attractive because they are not expensive, use nonvolatile compounds, and are nontoxic. Currently, however, there are no plants using potassium carbonate for CO_2 removal from flue gas but this option is being explored. In fact, work has been conducted using sodium carbonate solutions as well, since sodium carbonate solutions are nonhazardous and nonvolatile, have low corrosion rates, have no problem with the solvent causing fouling in the piping system, are inexpensive and available in large quantities, have a low binding energy with CO_2, and provide the possibility for multipollutant control [21,22]. During absorption, the CO_2 becomes chemically bound according to the following reactions:

$$M_2CO_3 + H_2O + CO_2 \leftrightarrow 2MHCO_3 \tag{8.4}$$

where M is potassium (K) or sodium (Na).

The main drawback associated with the Na_2CO_3 system is the slower absorption rate of CO_2 in comparison to amine solutions, thereby leading to tall absorption columns [23]. Studies have been conducted studying rate-increasing additives/catalysts [21]. Most of the rate-increasing additives have been shown to increase the rate of absorption but also increase the energy needed for regeneration in the stripping column.

8.4.3.4 Solid sorbents

Solid particles can be used to capture CO_2 from flue gas through chemical absorption, physical adsorption, or a combination of the two. Amines and other chemicals, such as sodium carbonate, can be immobilized on the surface of solid supports to create a sorbent that reacts with CO_2. A number of patents have been published on soda-lime, activated carbon, zeolites, molecular sieves, alkali metal oxides, silver oxide, lithium oxide, lithium silicate, carbonates, silica gel, alumina, amine solid sorbents, metal organic frameworks (MOFs), and others [24]. Solid sorbents that physiosorb the CO_2 onto the surface include carbonaceous materials (activated carbon, carbon nanotubes, graphene), zeolites, ordered mesoporous silica, and MOFs [25,26]. Potential advantages of solid sorbents include [9,25,26]:

- ease of material handling because coal-fired plants are experienced with solids handling
- safe for local environment
- high CO_2 capacity on a per mass or volume basis than similar wet scrubbing chemicals

- lower regeneration capacity
- multipollutant control capabilities
- chemical sites provide larger capacities and fast kinetics, enabling capture from streams with low CO_2 partial pressure
- lower heating requirements than wet scrubbing in many cases
- dry process requires less sensible heat than wet scrubbing process

Challenges include [24]:

- heat required to reverse chemical reaction
- heat management in solid systems is difficult, which can limit capacity and/or create operational issues when absorption reaction is exothermic
- pressure drop can be large in flue gas applications
- sorbent attrition

Possible configurations for contacting the flue gas with the solid particles included fixed, moving, and fluidized beds. A number of solids can be used to react with CO_2 to form stable compounds at one set of operating conditions and then be regenerated to liberate the adsorbed CO_2 at another set of conditions. In the adsorption phase, the CO_2 is adsorbed by a molecular sieve, activated carbon, or zeolites. In the regeneration phase, the CO_2 is released in a second process stage using pressure swing adsorption (PSA) or temperature swing adsorption (TSA), although less attention is being given to TSA due to the longer cycle times needed for heating during regeneration.

Chemical sorbents that react with CO_2 in the flue gas include a high surface area support with an immobilized amine or other reactant on the surface. Examples of commonly used supports are alumina or silica, while common reactants include chemicals such as sodium carbonate (Na_2CO_3) or amines such as polyethylenimine.

An example of a polyethylenimine (PEI) sorbent for CO_2 capture is the novel nanoporous-supported polymer sorbent, called a molecular basket sorbent (MBS) under development by Penn State [27–30]. In this concept, a CO_2-philic polymer sorbent, PEI, is loaded onto the surface of a porous material, such as mesoporous silica and carbonaceous materials including MCM-41, SBA-15, and others, to increase the accessible sorption sites per weight/volume unit of sorbent, and to improve the mass transfer rate in the sorption/ desorption process by increasing the gas-PEI interface. The advantages of the MBS concept include:

- high sorption capacity and selectivity for CO_2;
- higher sorption-desorption rate;
- good regenerability and stability in the sorption-desorption cycles;
- low energy consumption;
- special functionality;
- promoting effect of moisture in the gas on sorption capacity; and
- no or reduced corrosion.

Supported amines share the benefits of liquid amine solvents but require less energy to regenerate because there is no water solution. The amine solvent can be supported by a number of different materials. Such sorbents have been shown to have high CO_2 carrying capacities compared to other solid sorbents [10].

Research is underway investigating a dry, inexpensive, regenerable, supported sorbent, sodium carbonate (Na_2CO_3), which reacts with CO_2 and water to form sodium bicarbonate ($NaHCO_3$) [15] via the equation:

$$Na_2CO_3 + H_2O + CO_2 \leftrightarrow 2NaHCO_3 \tag{8.5}$$

A temperature swing is used to regenerate the sorbent and produce a pure CO_2/water stream that can be separated during cooling and compression. The pure CO_2 stream can then be sequestered.

Carbon-based adsorbents such as activated carbon are attractive because they are relatively inexpensive and have large surface areas that can readily adsorb CO_2. They can also provide a support material for amines or other solid sorbents. Activated carbons possess several advantages including high thermal stability, favorable adsorption kinetics, wide range of starting materials for production of activated carbons (i.e. lower raw material costs), large adsorption capacity at elevated temperatures, and desorption can easily be accomplished by the pressure swing approach [22]. Disadvantages include low CO_2 capacity at mild conditions, wide variety of starting materials results in a wide variety of poor characteristics between adsorbents, and the presence of NO_x, SO_x, and H_2O negatively impacts performance.

MOFs and zeolites are crystalline sorbents that are receiving attention for post-combustion capture. MOFs are a new class of hybrid material built from metal ions with well-defined coordination geometry and organic bridging liquids [10,24]. They are extended structures with carefully sized cavities that can adsorb CO_2. High storage capacity is possible and the heat required for recovery of the adsorbed CO_2 is low. Although MOFs possess potential for CO_2 capture, some challenges inhibit their use including costly starting materials, lack of experimental data examining the impact of multiple adsorption/desorption cycles on performance, and the lack of experimental data describing the effects of pressure or temperature swing on regeneration of the MOF [22]. In addition, MOFs are negatively impacted by the presence of NO_x, SO_x, and H_2O, and they have low CO_2 selectivity in CO_2/N_2 gas streams. Advantages for using MOFs as adsorbents for CO_2 include high thermal stability, adjustable chemical functionality, extra high porosity, high adsorption capacity at elevated pressures (510 psig), and easily tunable pore characteristics [22,26].

Zeolites are highly ordered porous alumino-silicate materials that have high selectivity, but low carrying capacity for CO_2 and are subject to performance degradation in the presence of water. Zeolites possess several advantages for CO_2 adsorption including favorable adsorption kinetics, high adsorption capacity at mild operating conditions (32 to 212°F and 1.45 to 14.5 psig), and selectivity against CO and N_2 is good [18,22]. Disadvantages include the presence of impurities (NO_x, SO_x, and H_2O) significantly impacts performance and CO_2 has been shown to chemisorb to the zeolite surface.

8.4.3.5 Membranes

Membranes refer to a barrier or medium that are porous materials that can be used to selectively separate CO_2 from other components of a gas stream [6,9,10,31].

Based on the membrane material, a membrane can be organic (e.g. polymeric membranes) or inorganic (e.g. metallic, ceramic, and zeolitic membranes) [9]. The driving force for this separation process is a pressure differential across a membrane, which can be created by either compressing the gas on one side of the material or by creating a vacuum on the opposite side. Figure 8.18 is a simplified process schematic for a post-combustion gas separation membrane [modified from 9].

Membranes have been used for gas purification in a number of industrial applications since the 1980s [10]. Membrane properties such as permeability, selectivity, chemical/ thermal/ mechanical stability, and packing density all influence the performance of a membrane system. Two important physical parameters of a membrane are its selectivity and permeability. For post-combustion CO_2 capture, the selectivity to CO_2 over N_2 (the main constituent of flue gas) determines the purity of the captured CO_2 stream. The permeability of a membrane reflects the amount of a given substance that can be transported for a given pressure difference, thereby determining the membrane surface required and consequently the capital cost and footprint of the CO_2 capture system.

There are two basic membrane module design configurations that can be used for post-combustion CO_2 capture applications: hollow-fiber and spiral-wound [9]. Hollow-fiber modules are constructed using numerous small diameter (100 to 250 μm), hollow-fiber membranes packed into a module shell. Spiral-wound membrane modules are constructed of large membrane sheets that are wound around a collection pipe.

The major disadvantage in using conventional polymeric membranes for post-combustion CO_2 capture is the potentially large surface area required because of the large flue gas volume that needs to be processed coupled with the low concentration and partial pressure of CO_2 in the flue gas [9]. Another potential disadvantage of membrane technology for power plant applications is that although 90% CO_2 separation is technically achievable in a single-step process, a high level of

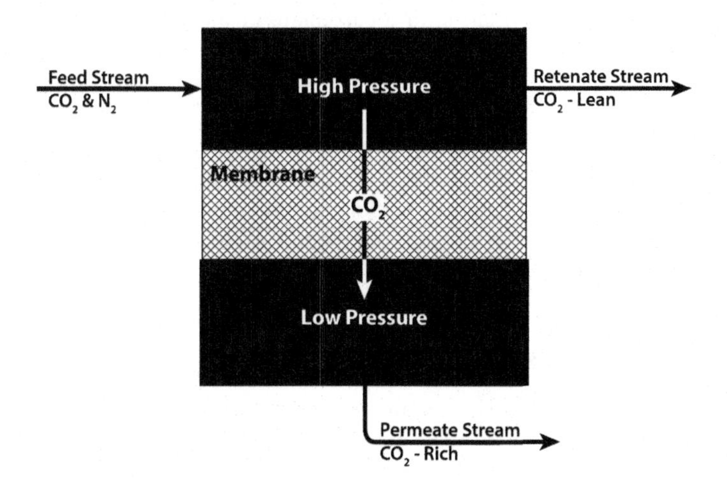

Figure 8.18 Membrane process schematic.

CO_2 purity will require a multistep process. Advantages to post-combustion membrane technologies include [9,24]:

- No steam load
- No chemicals
- Simple and modular design
- Unit operation vs. complex process

Challenges for using membranes for post-combustion CO_2 control include [9,24]:

- Membranes tend to be more suitable for high-pressure processes such as IGCC
- There is a trade-off between recovery rate and product purity (i.e. difficult to meet both high recovery rate and high purity)
- Requires high selectivity (due to CO_2 concentration and low pressure ratio)
- Requires good pretreatment
- Bad economy of scale
- Multiple stages and recycle streams may be required

8.4.3.6 Others

Other technologies considered for CO_2 capture include cryogenics, mineralization, and reduction technologies. The use of cryogenics for removing CO_2 has been considered for nearly 30 years [11]. Three general methods being investigated include: 1) compressing the flue gas to approximately 1100 psia with water used for cooling/condensing and removing the CO_2; 2) compressing the flue gas to 250 to 350 psia at 10° to 70°F, dehydrating the feed stream with activated alumina or a silica gel dryer, and distilling the condensate in a stripping column; and 3) cooling the flue gas stream to condense CO_2 at atmospheric pressure. The latter method is also known as frosting or antisublimation (i.e. phase change of a gas to a solid) on a cold surface. For removal of 90% of the CO_2 from a coal-fired flue gas, the sublimation temperature lies in the range of -103° to -122°F.

Mineralization is the formation of a carbonate or bicarbonate solid from CO_2. This type of process leaves the carbon in the same oxidation state as the carbon in CO_2, i.e. fully oxidized. The solid is then disposed or supplied as a product for beneficial reuse.

Reduction technologies involve the chemical transformation of the oxidized carbon to a reduced state through the input of energy during the application of chemical, photochemical, electrochemical, and/or biological processes [32]. This concept incorporates the CO_2 into the organic compound such as polycarbonate plastic, a fuel, or some other desired product.

8.4.4 Chemical looping

Chemical looping enables the production of a concentrated CO_2 stream similar to oxy-combustion but without the need for a separate ASU. Chemical looping can be applied to coal combustion, where it is known as chemical looping combustion (CLC), or to coal gasification, where it is known as chemical looping gasification

Table 8.15 Technical advantages and challenges for chemical looping technologies

Advantages	Challenges
CO_2 and H_2O kept separate from the rest of the flue gases ASU is not required and CO_2 separation takes place during combustion	Underdeveloped technology still conceptual and bench scale Attrition-resistant metal oxide carriers required during multiple cycles Reliable solids transport systems Providing efficient heat integration to the process Ash separation is problematic

(CLG) [9,10]. Table 8.15 summarizes the technical advantages and challenges for chemical looping technologies [9].

CLC involves the use of a metal oxide or other compound as an O_2 carrier to transfer O_2 from the combustion air to the fuel, avoiding direct contact between fuel and combustion air. The products of combustion, concentrated CO_2 and H_2O, are kept separate from the rest of the flue gases. This stream can be purified, compressed, and sent to storage. Combustion is divided into separate oxidation and reduction reactions, which are carried out in two separate reactors. The oxygen carrier releases the O_2 in a reducing atmosphere and the O_2 reacts with the fuel. The oxygen carrier is then recycled back to the oxidation chamber where the carrier is regenerated by contact with carrier. Figure 8.19 is a schematic diagram of the two-reactor CLC process [9]. The oxygen carrier is usually a solid, metal-based compound with chemical composition M_xO_{y-1}. The overall chemical reactions in the two reactors are expressed as:

$$\text{Oxidizer: } M_xO_{y-1} + 1/2O_2 \rightarrow M_xO_y \tag{8.6}$$

$$\text{Reducer: } C_nH_{2m} + (2n+m)M_xO_y \rightarrow nCO_2 + mH_2O + (2n+m)M_xO_{y-1} \tag{8.7}$$

$$\text{Net reaction: } C_nH_m + 1/2(2n+m)O_2 \rightarrow nCO_2 + mH_2O + \text{heat} \tag{8.8}$$

CLC offers the following advantages [9]:

- Avoids the large investment costs and parasitic power associated with either cryogenic ASU or ion transport membranes used for oxy-combustion
- Captures CO_2 at high temperature without additional energy, thus eliminating the thermodynamic penalty normally associated with CO_2 capture
- Involves small equipment and low capital cost because of the fast reactions
- Requires conventional material of construction and fabrication techniques

A chemical looping process can also be integrated into gasification and the water-gas shift (WGS) reaction [9]. In a CLG system, two or three solid particle loops are utilized to provide the O_2 for gasification and to capture CO_2. A loop,

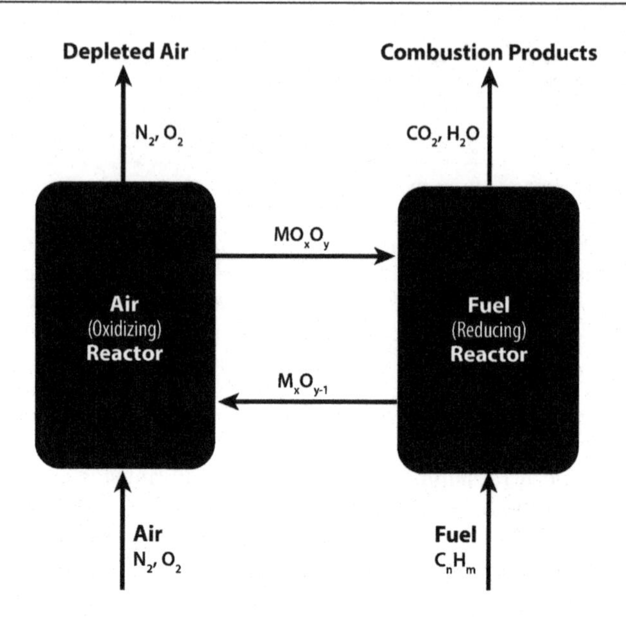

Figure 8.19 Schematic diagram of a two-reactor CLC process.

similar to that of CLC, is used to gasify the coal and produce syngas. A second loop is used in a WGS reactor where steam reacts with CO and converts it to H_2 and CO_2. The circulating solid absorbs the CO_2, which provides a greater driving force for the WGS reaction. The CO_2 is then released in a calcination step that produces nearly pure CO_2 for compression and storage. Figure 8.20 is a schematic diagram of a two-loop CLG process [9].

For the WGS reaction, the chemical looping process uses a solid carbon carrier (instead of an oxygen carrier) to separate CO_2 from the WGS reactor. An example is the iron oxide-based syngas chemical looping process, which requires a three-reactor configuration to accomplish WGS [9]. In the first reactor, syngas is burned by Fe_2O_3:

$$Fe_2O_3 + 3CO \rightarrow 2Fe + 3CO_2 \quad (8.9)$$

$$Fe_2O_3 + 3H_2 \rightarrow 2Fe + 3H_2O \quad (8.10)$$

In the second reactor, Fe is oxidized by steam to produce H_2:

$$3Fe + 4H_2O \rightarrow Fe_3O_4 + 4H_2 \quad (8.11)$$

In the third reactor, Fe_3O_4 is further oxidized to Fe_2O_3 to complete the cycle:

$$2Fe_3O_4 + 1/2O_2 \rightarrow 3Fe_2O_3 \quad (8.12)$$

The overall reaction converts syngas to H_2 with a small fraction of the syngas lost in the process.

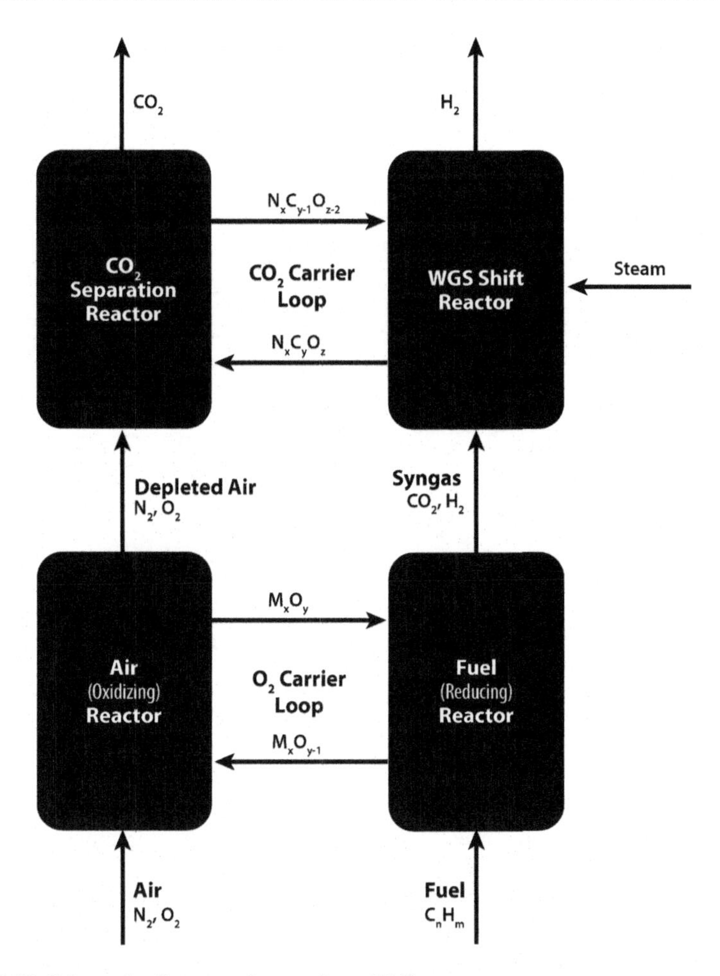

Figure 8.20 Schematic diagram of a two-loop CLG process.

8.4.5 Biomass cofiring

Cofiring is a near term, low-cost option (when compared to other capture technologies) for efficiently and cleanly converting biomass to electricity by feeding biomass as a partial substitute fuel in coal-fired boilers. This concept has been demonstrated successfully in over 150 installations worldwide and has resulted in some commercially operating units, using a variety of feedstocks in all boiler types commonly used by electric utilities including pulverized coal, cyclone, stoker, and FBC boilers. Common feedstocks include woody and herbaceous materials, energy crops, and agricultural and construction residues [33,34]. Extensive demonstrations and tests have confirmed that biomass energy can provide as much as 15 to 20% of the total energy input with only fuel feed system and burner modifications.

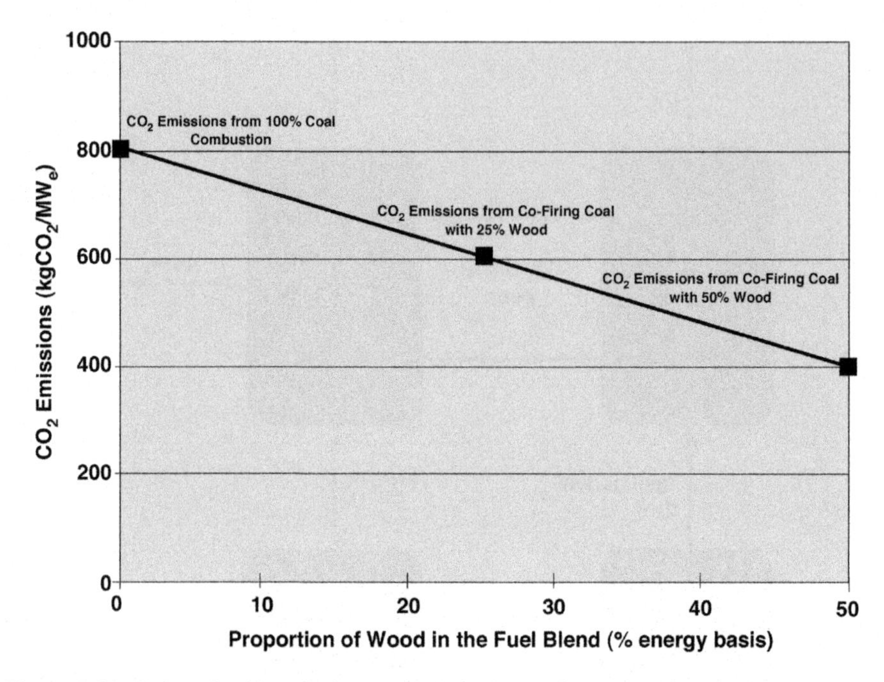

Figure 8.21 Carbon dioxide emissions reduction when cofiring wood in a coal-fired boiler.

Cofiring biomass with coal offers several environmental benefits including reductions in SO_2, NO_x, and greenhouse gas emissions. Carbon dioxide emissions from coal-fired power plants are essentially reduced by the percentage of biomass cofired on a heat input basis (see Figure 8.21) [35] because the biomass is considered carbon neutral, i.e. the CO_2 emitted from burning the biomass was removed from the atmosphere by the biomass-based material. When cofiring biomass at rates of 5% and 15% by heat input, greenhouse gas emissions have been reduced by 5.4% and 18.2%, on a CO_2-equivalent basis, respectively [36]. In studies comparing carbon emissions when producing biofuels and electricity from biomass, it has been shown that carbon displacement is significantly less when producing liquid fuels [37]. Thus, from the standpoint of reducing carbon emissions, it is better to use biomass to produce electricity. This would especially be the case if carbon were captured and sequestered from the biomass.

In addition to the environmental benefits, there are other benefits as well. For example, it has been shown that total energy consumption is lowered when cofiring biomass in coal-fired boilers by as much as 3.5% and 12.4% when cofiring 5% and 15% biomass, by heat input, respectively [36].

Although biomass cofiring is considered a proven technology, there are technical issues that must be considered when cofiring biomass with coal. Specifically, this includes fuel handling, storage, and preparation, NO_x formation, carbon conversion, and ash effects such as deposition, corrosion, impacts on selective catalytic reduction and other downstream processes, and loss of sales in selling fly ash for

Figure 8.22 The existing U.S. CO_2 pipeline infrastructure.

beneficial uses. These issues are dependent upon fuel characteristics, operating conditions, and boiler design. Proper combinations of coal and biomass and operating conditions can minimize or eliminate most impacts for most fuels.

8.5 Transport of CO_2

It is anticipated that most early carbon capture and storage (CCS) projects will be located near a geologic storage site (CO_2 storage is discussed in the following section). However, as the scale of CCS activities grow, it will become necessary to transport large volumes of CO_2 from the point of capture to the storage site. The primary method for CO_2 transport will be pipelines. Pipelines currently operate as a mature technology and are the most common method for transporting CO_2. Gaseous CO_2 is typically compressed to pressures above 1160 psi so it remains a supercritical fluid (i.e. avoids two-phase flow and increases the density) thereby making it easier and less costly to transport [38]. The energy required for compression corresponds to a loss in power plant efficiency by about 2 to 4 percentage points.

The first long-distance CO_2 pipeline came into operation in the early 1970s [38]. In the United States, 99 million metric tons (Mt) per year of CO_2 are transported over 4,100 miles of pipe for enhanced oil recovery (EOR) [39,40]. The existing infrastructure for CO_2 transport is shown in Figure 8.22 [41]. For comparison, the existing U.S. natural gas pipeline network transports 455 Mt per year of natural gas over 305,000 miles of pipe.

CO_2 can also be transported as a liquid in ships, over the road, and by rail in tankers that carry CO_2 in insulated tanks at a temperature well below ambient, and at much lower pressures [38]. In some situations, transport of CO_2 by ship may be more economical if the CO_2 has to be moved over large distances especially for distances greater than 600 miles and for less than a few million metric tons [39]. CO_2 can be transported by ship much in the same way liquefied petroleum gases currently are (typically at 101 psi pressure). Road and rail tankers are also technically feasible options. These systems transport CO_2 at a temperature of -4°F and at 290 psi pressure. Compared to pipelines and ships, road and rail tankers are uneconomical except on a very small scale and hence are unlikely to be relevant to large-scale CCS [38].

8.6 CO_2 storage

CO_2 (carbon) storage is defined as the placement of CO_2 into a repository such that it will remain stored or sequestered permanently. This is the final step in carbon management. At present, the world is releasing about 37 billion metric tons (37 Gt) per year of CO_2 into the environment [42]. Storage volumes could amount to several thousand Gt of CO_2 over the century; consequently, many different technologies are being considered for long-term CO_2 storage. These include storing the CO_2 in geologic formations, in the ocean, terrestrially, and through mineral sequestration. These various options are discussed in this section.

8.6.1 Geologic storage

Underground storage of CO_2 in geological formations is the most developed disposal option for CO_2 because it has the advantages of existing experience and low cost. Geologic sequestration involves the injection of CO_2 into underground reservoirs that have the ability to securely contain it over long periods of time. The geologic formations primarily considered for CO_2 storage are deep saline or depleted oil and gas reservoirs, which are layers of porous rock capped by a layer or multiple layers of nonporous rock above them, and unmineable coal beds. These are depicted in Figure 8.23. In each of these cases, CO_2 would be injected in a supercritical state (i.e. where CO_2 is dense like a liquid, fluid like a gas) below ground into a porous rock formation that holds or previously held fluids [39,43]. When CO_2 is injected at depths greater than 2600 feet in a typical reservoir, the pressure keeps the injected CO_2 in a supercritical state and it is less likely to migrate out of the geological formation.

The United States Geological Survey has performed an assessment of storage capacity for the United States and estimated that there is geologic capacity to sequester approximately 3,000 Gt of CO_2 on- or near-shore [44]. This amount is more than 500 times the 2013 annual U.S. energy-related CO_2 emissions of 5.4 Gt [8]. Figure 8.24 shows eight regions that were evaluated and assessed by the USGS [44].

In addition, two potential storage options include basalt formations and organic-rich shale basins. The use of these two formations are under study [43].

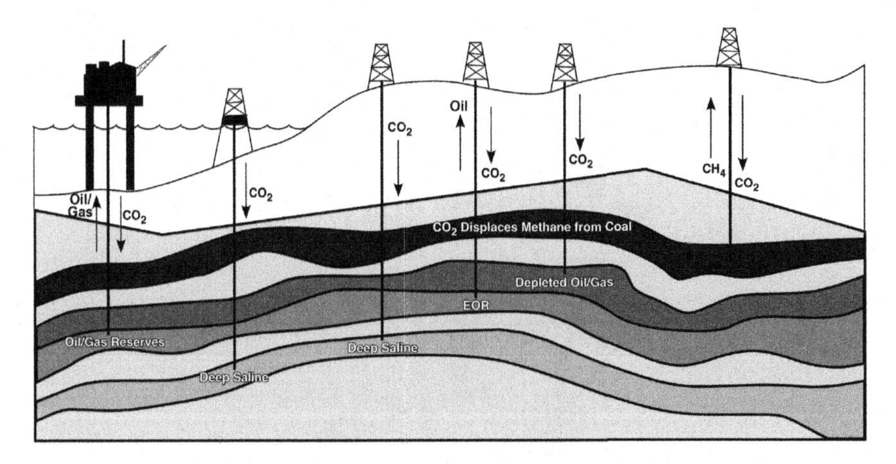

Figure 8.23 Depiction of primary geologic storage options.

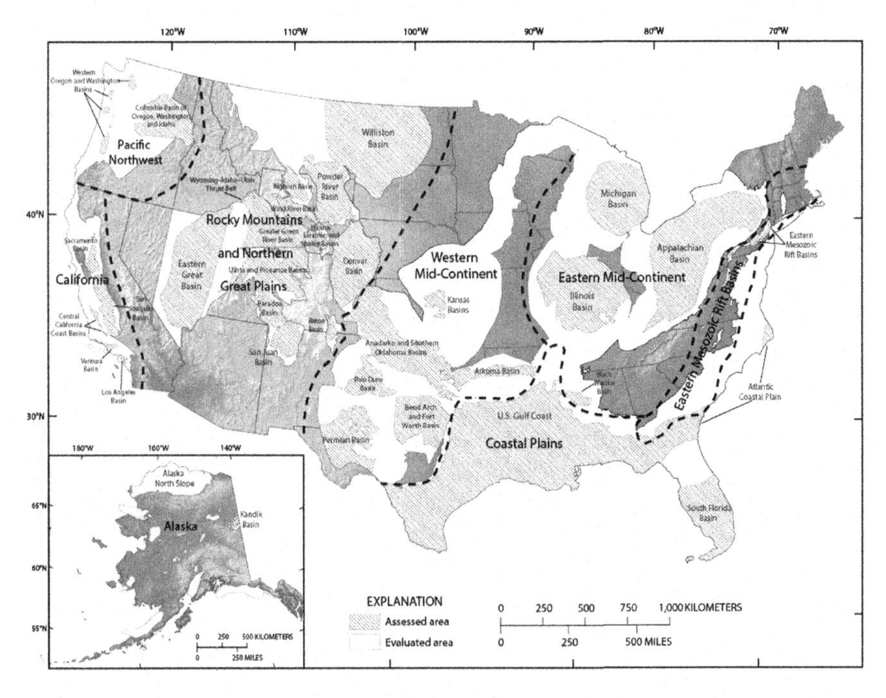

Figure 8.24 Eight regions evaluated and assessed by the USGS for geologic storage capacity.

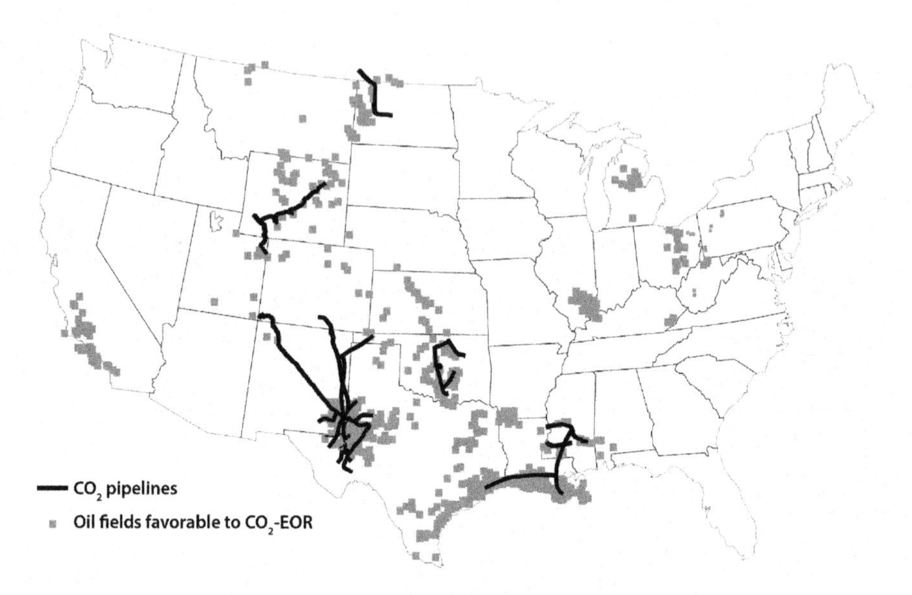

Figure 8.25 Location of oil-bearing formations in relation to existing CO_2 pipelines in the lower 48 states.

8.6.1.1 *Oil and gas reservoirs*

Oil and gas reservoirs are porous rock formations (usually sandstones or carbonates) containing hydrocarbons such as crude oil and/or natural gas that have been physically trapped. Oil and gas reservoirs are ideal because they have been shown to be impermeable over millions of years and have conditions suitable for CO_2 storage. This, coupled with the extensive exploration and removal of oil and gas during which the composition of the strata have become precisely known, makes depleted oil and gas reservoirs prime candidates for CO_2 storage. Depleted oil and gas fields provide an estimated sequestration capacity of 226 Gt of CO_2 [45,43]. In addition, the injection of CO_2 into active reservoirs for EOR is an applicable method for CO_2 storage. EOR capacity has been estimated at 70 to 200 Gt of CO_2 [46].

Depleted or abandoned oil and gas fields are considered prime candidates for CO_2 storage for several reasons [39]:

- oil and gas originally trapped did not escape for millions of years, demonstrating the structural integrity of the reservoirs
- extensive studies for oil and gas location and production have characterized the geology of the reservoir
- computer models have been developed to understand how hydrocarbons move in the reservoir, and the models can be applied to predicting how CO_2 could move
- infrastructure and wells from oil and gas extraction may be in place and might be used for handling CO_2 storage. An example of this is shown in Figure 8.25 where the oil-bearing formations favorable to CO_2-EOR are located in relation to existing CO_2 pipelines [45].

However, the biggest problem for safely storing CO_2 in these structures is posed by old abandoned drill holes, which in some cases may be present in oil and gas fields in large numbers. These drill holes need to be filled to ensure CO_2 is not readmitted into the atmosphere.

8.6.1.2 Unmineable coal

Similar to EOR, high-pressure CO_2 can also be used for enhanced methane recovery for coal that is considered unmineable because of geologic, technological, and economic factors (i.e. too deep, too thin, or lacking the internal continuity to be economically mined with today's technologies) provided the permeability is sufficient [39,43]. Coal preferentially adsorbs CO_2 over methane, which is naturally found in coal seams, at a ratio of 2 to 13 times [43]. Methane is typically recovered from coal seams by dewatering and depressurization; this can leave significant quantities of methane trapped in the coal seam. The process of inject and storing CO_2 in unmineable coal seams to enhance methane recovery is called enhanced coalbed methane (ECBM) recovery. Approximately 56 to 114 Gt of potential CO_2 storage in unmineable coal seams has been identified in the U.S. [45,43].

While CO_2 injection is known to displace methane, a greater understanding of the displacement mechanism is needed to optimize CO_2 storage and to understand the problems of coal swelling and permeability [47]. Coal has been shown to swell when it adsorbs CO_2 and in an underground formation, swelling can cause a sharp drop in permeability, which restricts the flow of CO_2 into the formation and impedes the recovery of coalbed methane.

8.6.1.3 Deep saline formations

Saline aquifers are highly porous sedimentary rocks, which are saturated with a strong saline solution (i.e. brine). Compared to coal seams or oil and gas reservoirs, saline formations are more common and offer the added benefits of greater proximity, higher CO_2 storage capacity, and fewer existing well penetrations [48]. Deep saline reservoirs are more widespread in the U.S. than oil and gas reservoirs and thus have greater probability of being close to large point sources of CO_2 and saline reservoirs have potentially the largest reservoir capacity of the three main types of geologic formations [39]. However, much less is known about saline formations because they lack the characterization experience that industry has acquired through resource recovery from oil and gas reservoirs and coal seams, which is one reason for the large range in estimated lower and upper CO_2 storage capacities. Therefore, there is more uncertainty regarding the suitability of saline formations for CO_2 storage. Storage capacities for CO_2 in saline formations, as estimated by DOE, range from a low of 2,100 Gt to a high of 20,043 Gt [43,45].

8.6.1.4 Basalt formations

Basalt formations, which are formations of solidified lava, have a unique chemical makeup that could potentially convert all of the injected CO_2 to a solid mineral form.

The U.S DOE is investigating these geologic formations because they are a relatively large potential storage resource that is widely geographically distributed [43]. These formations have a unique chemical composition that could potentially convert all of the injected CO_2 to a solid mineral form. Before basalt formations can be considered as a viable CO_2 storage option, a number of questions relating to the basic geology, the CO_2-trapping mechanisms and their kinetics, and monitoring and modeling tools need to be addressed.

8.6.1.5 Shale basins

Organic-rich shales are another geologic storage option. Shale is a sedimentary rock and is characterized by thin horizontal layers with very low permeability in the vertical direction. Many shales contain 1–5% organic material and this provides an adsorption substrate for CO_2 storage, similar to CO_2 storage in coal seams. Similar to questions regarding basalt formations, before organic-rich shale basins can be considered as a viable CO_2 storage option, a number of questions relating to the basic geology, the CO_2-trapping mechanisms and their kinetics, and monitoring and modeling tools need to be addressed [43].

8.6.1.6 Uncertainties in geologic storage

The uncertainty in geologic storage lies in the long-term stability of reservoirs. Generally, the risk of technical plants (e.g. separation equipment, compressors, pipelines) is considered to be low or manageable with the usual technical means and controls. The primary issue is long-term safety and long-term leakage rate, which ultimately will define the capacity of these technologies. Storage times under discussion vary from 1,000 to 10,000 years. Some important processes that could compromise the safety and permanence CO_2 storage are:

- geochemical processes such as the dissolution of carbonate rocks through acidic CO_2-water mixtures
- pressure-induced processes such as the expansion of existing small fissures in the seal rock through the overpressure of CO_2 injection
- leakage through existing drill holes
- leakage via undiscovered migration paths in the seal rock
- lateral expansion of the formation of water, which is displaced by the injected CO_2

8.6.2 Ocean storage

A potential CO_2 storage option is to inject captured CO_2 directly into the deep ocean where most of it would be isolated from the atmosphere for centuries. Estimates for ocean storage capacity in excess of 100,000 billion GtC (i.e. 360,000 Gt CO_2) have been given; however, such numbers are only feasible if alkalinity (e.g. NaOH) were added to the ocean to neutralize the carbonic acid produced because an addition of 1,000 Gt of carbon is projected to change ocean chemistry [46].

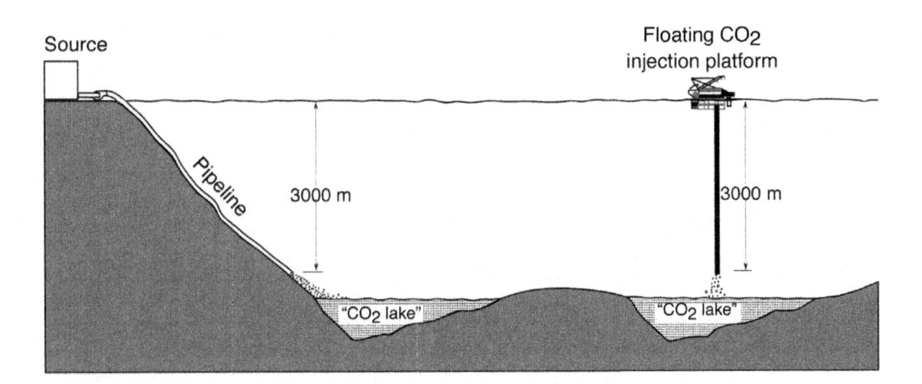

Figure 8.26 Methods of ocean storage.

A number of CO_2 injection methods have been investigated but deep ocean (at depths greater than 1,000 to 3,000 meters) injection is the option that would offer long-term (e.g. centuries) sequestration. This can be achieved by transporting CO_2 via pipelines or ships to an ocean storage site where it is injected into the water column of the ocean or at the sea floor (see Figure 8.26; modified from [38,49]). Carbon dioxide injected at depths shallower than 500 meters typically would be released as a gas and would rise to the surface [39]. At depths below 500 meters, CO_2 can exist as a liquid in the ocean, although it is less dense than seawater. After injection below 500 meters, CO_2 would also rise but an estimated 90% would dissolve in the first 200 meters. Below 3,000 meters in depth, CO_2 is a liquid and is denser than seawater. The injected CO_2 would sink and dissolve in the water column or possibly form a CO_2 pool or lake on the sea bottom. In addition, CO_2 will react with seawater to form a solid clathrate, which is an ice-like cage structure with approximately six water molecules per CO_2 molecule [46].

Ocean storage research has been performed since 1977; however, the technology has not yet been deployed or demonstrated at the pilot scale [39]. There is much resistance to this option due to environmental concerns related to long-term chronic issues of altering the ocean chemistry as well as on the local effects of low pH (which can be as low as 4) and its effect on marine organisms. Ocean acidification is a concern today even before it is being used for CO_2 storage. The ocean is an open system and it is difficult, if not impossible, to monitor the distribution of stored carbon to confirm residence times of CO_2. It is most unlikely that ocean storage will be a CO_2 storage option in the foreseeable future.

8.6.3 Terrestrial storage

Terrestrial carbon sequestration is defined as the removal of CO_2 from the atmosphere by soil and plants, both on land and in aquatic environments such as wetlands and tidal marshes, and/or the prevention of CO_2 net emissions from terrestrial ecosystems into the atmosphere [48,50]. This is considered a low-cost option as a near-term bridging technology (like biomass cofiring) until more resource-intensive

Table 8.16 Carbon dioxide sequestration rates for four biomes

Biome	Sequestration rate (metric tons/acre/year)
Cropland	0.2–0.6
Forest	0.05–3.9
Grassland	0.12–1.0
Swamp/Floodplain/Wetland	2.23–3.71

and technically complex geologic sequestration activities are brought on-line and can offer other benefits such as habitat and/or water quality improvements. Terrestrial sequestration efforts include tree plantings, no-till farming, wetlands restoration, land management on grasslands and grazing lands, fire management efforts, and forest preservation. It has been reported that terrestrial uptake of CO_2 offsets approximately one-third of global anthropogenic CO_2 emissions. This uptake is expected to decrease as forests in northeast U.S. mature. Ranges of CO_2 sequestration rates for four biomes are shown in Table 8.16 [50].

8.6.4 Mineral carbonation

Mineral carbonation is the fixation of CO_2 using alkaline and alkaline-earth oxides, such as magnesium oxide (MgO) and calcium oxide (CaO), which are present as naturally occurring silicate rocks such as serpentine and olivine [38]. Chemical reactions between these materials and dissolved CO_2 produce magnesium carbonate ($MgCO_3$) and calcium carbonate ($CaCO_3$). The process involves leaching out the Mg or Ca from the rock into the aqueous phase and reacting them with dissolved CO_2 to form the carbonate. Mineral carbonation produces silica and carbonates that are stable over long time scales and therefore can be disposed in areas such as silicate mines or reused in construction processes. The main challenge of this storage method is the slow dissolution kinetics, large energy requirements associated with mineral processing, and large volumes of minerals needed for the process and products produced. Minerals needed for reacting with CO_2 range from 1.6 to 3.7 metric tons per metric ton of CO_2 capture producing 2.6 to 4.7 metric tons of materials per metric ton of CO_2 sequestered [39]. This process is still in the research and experimental stage [39].

A commercial process would require mining, crushing, and milling of the mineral-bearing ores and their transport to a processing plant receiving a CO_2 stream from a power plant. The carbonation energy would be 30 to 50% of the power plant output. Factoring in the additional energy required to capture the CO_2 at the power plant, a CCS system with mineral carbonation would require 60 to 180% more energy input per kilowatt-hour than a reference electricity producing plant without capture or mineral carbonation [38].

Very early stages of investigation by the U.S. DOE are exploring the possibility of using basalts or their potential to react with CO_2 and form solid carbonates in situ. Instead of mining, crushing, and milling the reactant minerals, CO_2 would be injected directly into the basalt formations and would react with the rock over time and at depth to form solid carbonate minerals [39].

8.7 Economics of CO_2 capture and storage

The costs associated with CO_2 carbon and storage have been the focus of many studies and there is a wide range in costs reported. In addition, the studies have been performed over several years so the economics reported are based on a range of years. Costs reported in this section are in U.S. dollars.

The total cost for applying carbon capture and storage to a coal-fired power plant contains three main components – capture, transportation, and storage. This section will summarize the cost estimates for CO_2 capture (i.e. the cost of physically capturing the CO_2 at the plant and compressing it to transportation pressure), transportation (the cost of transporting the CO_2 from the plant to the storage site), and storage (the cost of injecting and storing the CO_2, including, in some cases, insurance, monitoring, and verification costs [51]). In general, the capture costs dominate the CCS.

Many studies combine capture and storage as a single cost and, in some cases, include transportation costs as well. Consequently, it is not always possible to cleanly summarize capture, transportation, and storage costs as separate costs. Therefore, this section will begin by discussing the economics of capture, storage, and transportation from several studies. These studies explore costs for: pulverized coal, IGCC, natural gas combined cycle (NGCC), oxy-fuel combustion, and fluidized-bed systems; firing of high- and low-rank coals; pre- and post-capture systems; regional transportation and storage in the U.S.; and comparisons of current and future (advanced) capture technologies. This will be followed by a discussion of the economics of biomass cofiring, and a generalized discussion of geologic, ocean, and mineral carbonization costs.

8.7.1 *Carbon capture and storage*

8.7.1.1 *Comparison of first-of-a-kind and* N*th-of-a-kind plants*

WorleyParsons, along with Schlumberger, Electric Power Research Institute and Baker & McKenzie, performed a detailed analysis of the capture, transport, and storage costs for power plants as part of a global comprehensive study [52]. They determined that the cost for CCS for power generation, based on the use of commercially available technology, ranged from \$57–\$107 per metric ton of CO_2 avoided or \$42–\$90 per metric ton of CO_2 captured. The lowest cost of CO_2 avoided was \$57 per metric ton of CO_2 for oxy-fuel combustion technology, while the highest CO_2 avoided cost was for NGCC with post-combustion capture. The lowest cost for captured CO_2 was \$39 and \$42 per metric ton of CO_2 for IGCC and

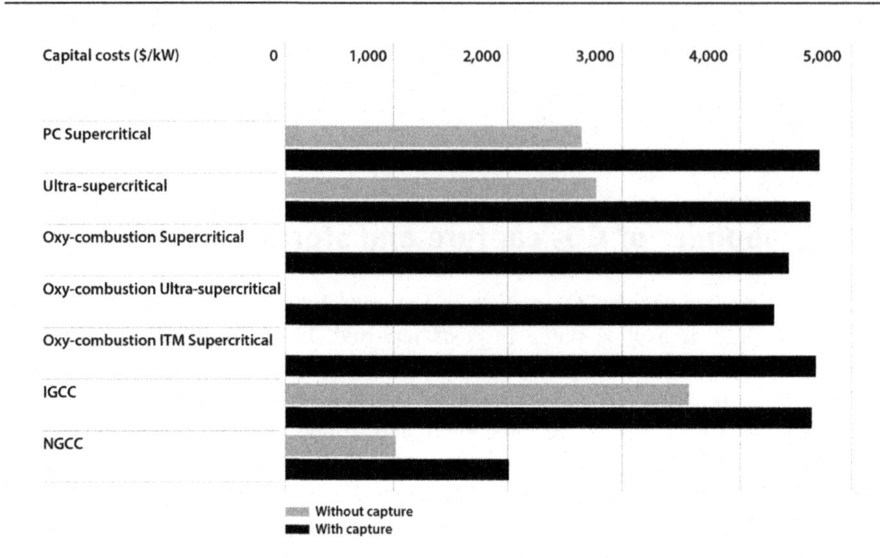

Figure 8.27 Installed (capital) costs ($/kW) for 550 MW net generation and FOAK CO_2 capture facility.

oxy-fuel combustion technologies, respectively. The highest CO_2 capture cost was $90 per ton of CO_2 for NGCC technologies.

Figure 8.27 shows the installed (capital) costs for a 550 MW net generation facility (for eight technologies) in the U.S. with and without CO_2 capture [52]. Costs used for the CO_2 capture facility were those for a first-of-a-kind (FOAK) facility. Table 8.17 lists a breakdown of levelized cost of electricity (LCOE) in the U.S. for the various technologies with and without CO_2 capture along with the cost of CO_2 avoided and cost of CO_2 captured for FOAK and *n*th-of-a-kind (NOAK) CO_2 capture plants [from 52].

8.7.1.2 Comparison of retrofit vs. greenfield costs

The U.S. DOE has performed several CCS studies. In one exploring the economic feasibility of CO_2 capture retrofits using an amine system for the U.S. power fleet, DOE compared costs for retrofitting a pulverized coal plant and a NGCC to costs associated with a new build (i.e. greenfield plant) [53]. Figure 8.28 shows the cost of electricity (COE) for greenfield and retrofitted plants for NGCC and pulverized coal with and without CO_2 capture and Figure 8.29 shows the cost of CO_2 capture for NGCC and pulverized coal plants for both greenfield and retrofit applications [53].

8.7.1.3 Technology comparison utilizing bituminous coal

The U.S. DOE performed an assessment of the cost and performance for pulverized coal combustion (PC), IGCC, and NGCC, all with and without CO_2 capture and sequestration assuming that the plants used current technologies [54]. The coal used in this study was an Illinois #6 bituminous coal and the power plants, 550 MW

Table 8.17 **Breakdown of LCOE for technologies with and without CCS for FOAK and NOAK plants; and cost of CO_2 avoided and captured for FOAK and NOAK plants**

	Pulverized coal (supercritical)	Pulverized coal (ultra-supercritical)	Oxy-fuel combustion (supercritical)	Oxy-fuel combustion (ultra-supercritical)	Oxy-fuel combustion; ITM (supercritical)	IGCC	NGCC
LCOE without capture ($/MWh)							
Generation	76	73	76	76	76	90	88
Transportation	0	0	0	0	0	0	0
Storage	0	0	0	0	0	0	0
Total	76	73	NA	76	76	90	88
LCOE with capture for FOAK ($/MWh)							
Generation and capture	124	113	114	107	116	116	115
Transportation	1	1	1	1	1	1	1
Storage	6	6	6	6	6	6	6
Total	131	120	121	114	123	123	122
LCOE with capture for NOAK ($/MWh)							
Generation and capture	122	109	112	104	114	113	114
Transportation	1	1	1	1	1	1	1
Storage	6	6	6	6	6	6	6
Total	129	116	119	111	121	120	121

(Continued)

Table 8.17 (Continued)

	Pulverized coal (supercritical)	Pulverized coal (ultra-supercritical)	Oxy-fuel combustion (supercritical)	Oxy-fuel combustion (ultra-supercritical)	Oxy-fuel combustion; ITM (supercritical)	IGCC	NGCC
FOAK							
Cost of CO_2 avoided ($/metric ton CO_2)	81	62	57	47	59	67	107
Cost of CO_2 captured ($/metric ton CO_2)	53	55	42	43	47	39	90
FOAK							
Cost of CO_2 avoided ($/metric ton CO_2)	78	57	54	44	57	63	103
Cost of CO_2 captured ($/metric ton CO_2)	52	52	41	42	45	38	87

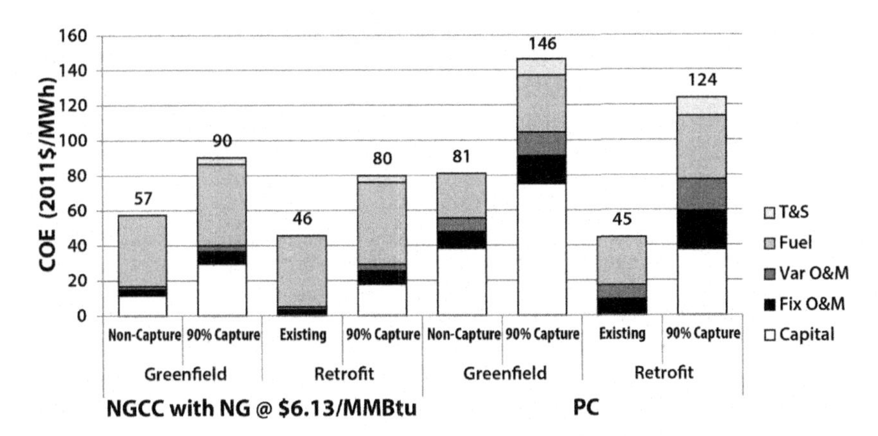

Figure 8.28 Cost of electricity for NGCC and pulverized coal plants with and without an amine scrubbing system.

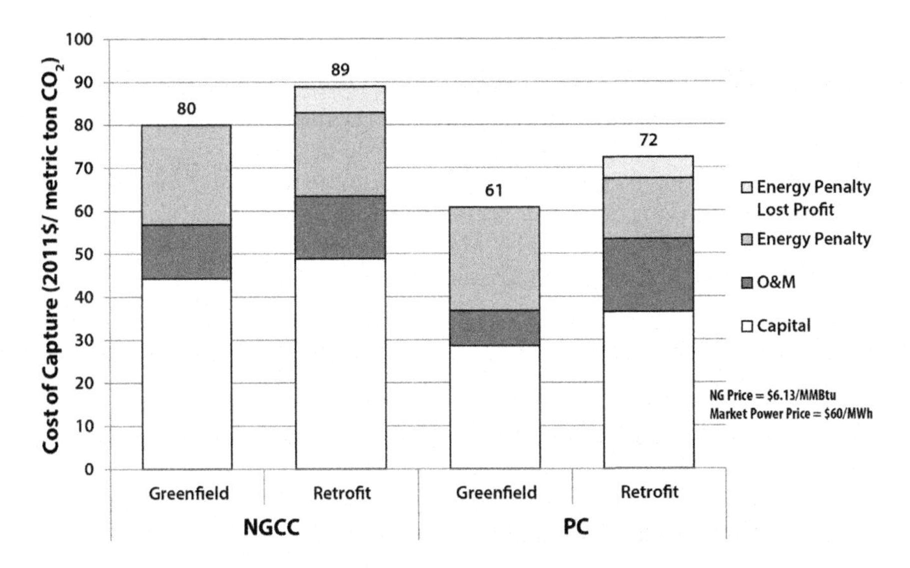

Figure 8.29 Cost of CO_2 capture for NGCC and pulverized coal plants.

nominal net plant output, were new construction. The following are highlights of the study [54]:

- Total overnight cost (TOC), which is defined later, for the noncapture plants are as follows: NGCC – $718/kW; PC – $2,010/kW (average); IGCC – $2,505/kW (average). With CO_2 capture, capital costs are: NGCC – $1,497/kW; PC – $3,590/kW (average); IGCC – $3,568/kW (average).
- At fuel costs of $1.64/MM Btu of coal and $6.56/MM Btu of natural gas, the COE for the noncapture plants is: NGCC – 59 mills/kWh; PC – 59 mills/kWh (average); IGCC – 77 mills/kWh (average).

- When CCS is integrated into these new power plants, the resultant COE, including the cost of CO_2 transportation, storage, and monitoring, is: NGCC – 86 mills/kWh; PC – 108 mills/kWh (average); IGCC – 112 mills/kWh (average). The cost of transporting CO_2 50 miles for storage in a geologic formation with over 30 years of monitoring is estimated to add about 3 to 6 mills/kWh, which represents less than 5.5% of the COE for each CO_2 capture case.
- A sensitivity analysis on natural gas prices shows that at a coal price of $1.64/MM Btu, the average COE for IGCC with capture equals that of NGCC with CO_2 capture at a natural gas price of $9.80/MM Btu. The average COE for PC with capture equals that of NGCC with capture at a natural gas price of $9.25/MM Btu.

Twelve power plant configurations were analyzed in the DOE study and their descriptions are given in Table 8.18 [54]. The list includes six IGCC cases utilizing General Electric Energy (GEE), ConocoPhillips (CoP), and Shell Global Solutions (Shell) gasifiers each with and without CO_2 capture; four PC cases, two subcritical and two supercritical, each with and without CO_2 capture; and two NGCC plants with and without CO_2 capture.

Table 8.19 summarizes the cost, performance, and CO_2 emissions for all cases [54]. Note that the nominal net plant output for this study was set at 550 MW. The actual net output varied between technologies because the combustion turbines in the IGCC and NGCC cases are manufactured in discrete sizes, but the boilers and steam turbines in the PC cases are readily available in a wide range of capacities. Consequently, all of the PC cases have a net output of 550 MW, but the IGCC cases have net outputs ranging from 497 to 629 MW. Likewise, the two NGCC cases have a net output of 555 and 474 MW. Carbon capture of 90% was used. The economic terms used in Table 8.19 are defined as [54]:

- Total plant cost (TPC) comprises the Base Erected Costs (BEC; which includes process equipment, supporting facilities, and direct and indirect labor) plus the cost of services provided by the engineering, procurement, and construction (EPC) contractor and project and process contingencies.
- The total overnight costs (TOC) comprise the TPC plus owner's costs, which are preproduction costs, inventory capital, financing costs, and other owner's costs.
- Total as spent capital (TASC) is sum of TOC plus escalation during capital expenditure period and interest on debt during capital expenditure period.

Some of the results from Table 8.19 are shown in Figures 8.30–8.32, which illustrate the COE by cost component, first year CO_2 avoided costs, and CO_2 emissions normalized by net output, respectively [54]. In both capture and noncapture cases, NGCC plants have the lowest COE, followed by PC and IGCC. Carbon dioxide avoided costs for IGCC plants using analogous noncapture plants as reference are substantially less than for PC and NGCC because the IGCC CO_2 removal is accomplished prior to combustion.

8.7.1.4 Technology comparison utilizing low-rank coals

In a manner similar to the U.S. DOE power plant study using bituminous coal, DOE performed an assessment of the cost and performance for IGCC, PC,

Table 8.18 Descriptions of the power plant configurations

Unit cycle	Steam cycle, psig/°F°/F	Combustion turbine	Gasifier/boiler technology	Oxidant	H_2S separation/removal	Sulfur removal/recovery	CO_2 separation
IGCC	1800/1050/1050	2× Advanced F Class	GEE Radiant	95% O_2	Selexol	Claus Plant	–
IGCC	1800/1000/1000	2× Advanced F Class	GEE Radiant	95% O_2	Selexol	Claus Plant	Selexol 2nd Stage
IGCC	1800/1050/1050	2× Advanced F Class	CoP E-Gas	95% O_2	Refrigerated MDEA	Claus Plant	–
IGCC	1800/1000/1000	2× Advanced F Class	CoP E-Gas	95% O_2	Selexol	Claus Plant	Selexol 2nd Stage
IGCC	1800/1050/1050	2× Advanced F Class	Shell	95% O_2	Sulfinol-M	Claus Plant	–
IGCC	1800/1000/1000	2× Advanced F Class	Shell	95% O_2	Selexol	Claus Plant	Selexol 2nd Stage
PC	2400/1050/1050	–	Subcritical PC	Air	–	Wet FGD/gypsum	–
PC	2400/1050/1050	–	Subcritical PC	Air	–	Wet FGD/gypsum	Amine absorber
PC	3500/1100/1100	–	Supercritical PC	Air	–	Wet FGD/gypsum	–
PC	3500/1100/1100	–	Supercritical PC	Air	–	Wet FGD/gypsum	Amine absorber
NGCC	2400/1050/1050	2× Advanced F Class	HRSG	Air	–	–	–
NGCC	2400/1050/1050	2× Advanced F Class	HRSG	Air	–	–	Amine absorber

Table 8.19 **Summary of cost, performance, and CO_2 emissions**

	Integrated gasification combined cycle						Pulverized coal boiler				NGCC	
	GEE		CoP E-Gas		Shell		Subcritical		Supercritical			
CO_2 Capture	0%	90%	0%	90%	0%	90%	0%	90%	0%	90%	0%	90%
Gross Power Output (kWe)	747,800	734,000	738,200	703,700	737,000	673,400	582,600	672,700	580,400	662,800	564,700	511,000
Net Power Output (kWe)	622,050	543,250	625,060	513,610	628,980	496,860	550,020	549,960	549,990	549,970	555,080	473,570
Net Plant HHV Efficiency (%)	39.0	32.6	39.7	31.0	42.1	31.2	36.8	26.2	39.3	28.4	50.2	42.8
CO_2 (lb/MM Btu)	197	20	199	20	197	20	204	20	204	20	118	12
CO_2 (lb/ MWh_{gross})	1,434	152	1,448	158	1,361	161	1,783	217	1,675	203	790	87
CO_2 (lb/ MWh_{net})	1,723	206	1,710	217	1,595	218	1,888	266	1,768	244	804	94
Total Plant Cost ($/kW)	1,987	2,711	1,913	2,817	2,217	3,181	1,622	2,942	1,647	2,913	584	1,226
Total Overnight Cost ($/kW)	2,447	3,334	2,351	3,466	2,716	3,904	1,996	3,610	2,024	3,570	718	1,497
Total As Spent Capital ($/kW)	2,789	3,801	2,680	3,952	3,097	4,451	2,264	4,115	2,296	4,070	771	1,614
COE (mills/kWh)	76.3	105.6	74.0	110.3	81.3	119.4	59.4	109.6	58.9	106.5	58.9	85.9
LCO (mills/kWh)	96.7	133.9	93.8	139.9	103.1	151.4	75.3	139.0	74.7	135.2	74.7	108.9

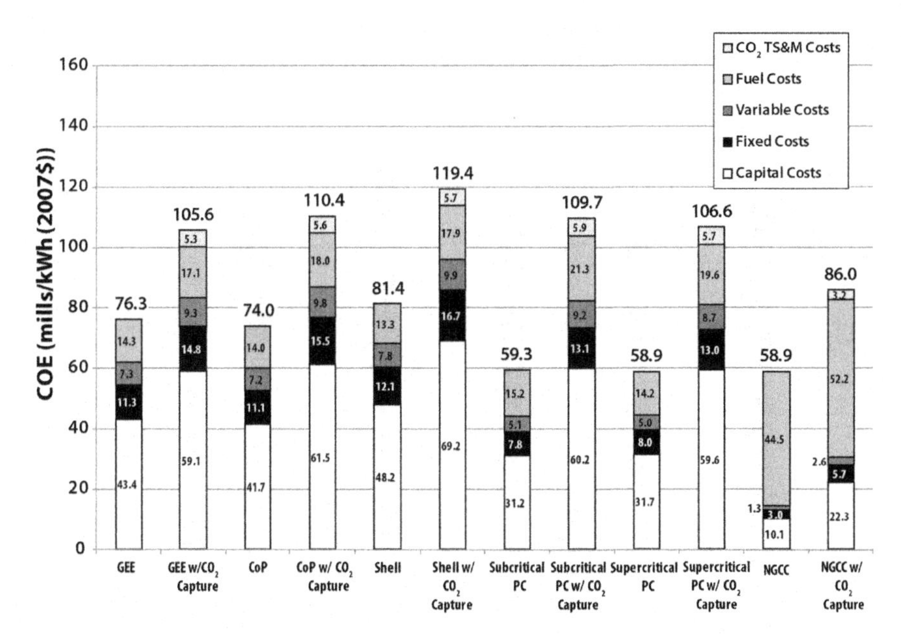

Figure 8.30 COE by cost component.

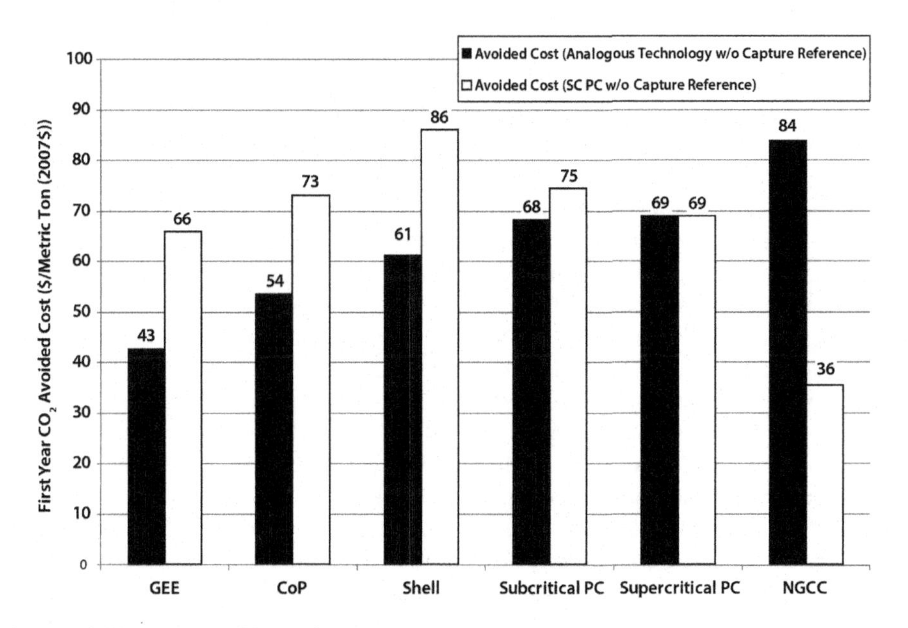

Figure 8.31 First year CO_2 avoided costs.

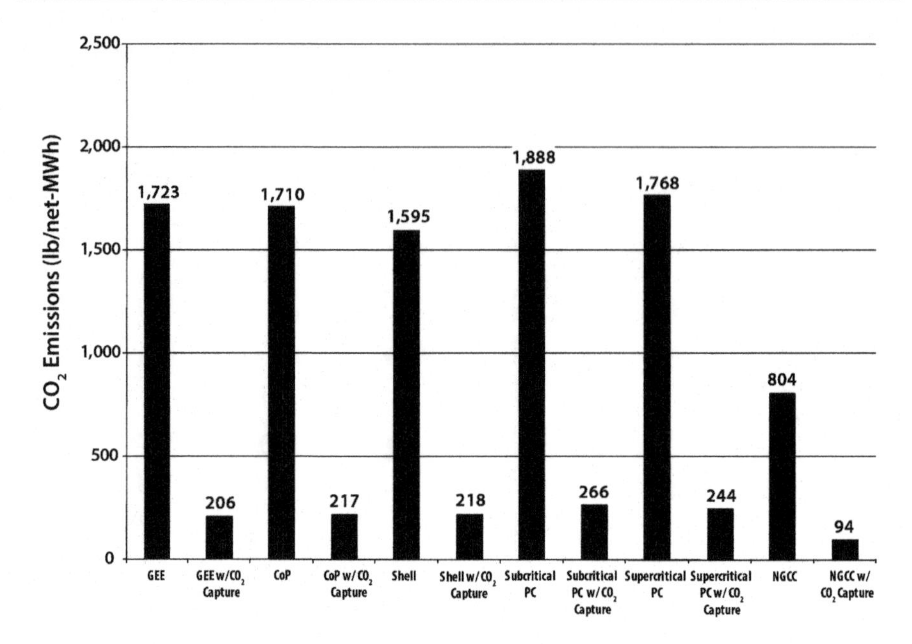

Figure 8.32 CO_2 emissions normalized by net output.

circulating fluidized-bed combustion (CFBC), and NGCC plants with and without CCS operating using low-rank coal at sites in Montana and North Dakota [55]. The Montana coal plants used Powder River Basin (PRB) subbituminous coal and the minemouth North Dakota coal plants used North Dakota lignites (NDL).

Twenty-eight power plant configurations were analyzed in the DOE study and their descriptions are given in Table 8.20 [55]. This study had 28 cases, one for each technology configuration, using either PRB or lignite at one of two site conditions, with and without CO_2 capture. This included configurations for 12 oxygen-blown IGCC plants based on the Shell Coal Gasification Process (Shell), Transport Integration Gasification (TRIG) gasifier, Siemens Fuel Gasifier (SFG), and ConocoPhillips (CoP) E-Gas gasifiers. Twelve combustion power plant configurations were analyzed including supercritical (SC) and ultra-supercritical (USC) PC plants and SC CFB plants. NGCC cases were also analyzed at each site.

Tables 8.21–8.23 summarize the cost, performance, and CO_2 emissions for the IGCC, coal combustion, and NGCC cases, respectively [55]. The nominal net plant output for this study was set at 550 MW. The actual net output varied between technologies because the combustion turbines in the IGCC and NGCC cases are manufactured in discrete sizes, but the boilers and steam turbines in the PC cases are readily available in a wide range of capacities. Consequently, all of the PC cases have a net output of 550 MW, but the IGCC cases have net outputs ranging from 445 to 617 MW. Likewise, the two NGCC cases have net outputs ranging from 435 to 547 MW. Carbon capture of 90% was used, except in one TRIG case carbon capture of 83% was used.

Some of the results from Table 8.19 are shown in Figures 8.33–8.38, which illustrate the COE by cost component, first year CO_2 avoided costs, and CO_2

Case	Gasifier/Boiler	Fuel	Steam cycle, psig/°F/°F	Sulfur removal	CO_2 separation
S1A	Shell	PRB	1800/1050/1050	Sulfinol-M	–
S1B	Shell	PRB	1800/1000/1000	Selexol	Selexol 2nd stage
L1A	Shell	NDL	1800/1050/1050	Sulfinol-M	–
L1B	Shell	NDL	1800/1000/1000	Selexol	Selexol 2nd stage
S2A	TRIG	PRB	1800/1050/1050	Sulfinol-M	–
S2B	TRIG	PRB	1800/1000/1000	Selexol	Selexol 2nd stage
S3A	Siemens SFG	PRB	1800/1050/1050	Sulfinol-M	–
S3B	Siemens SFG	PRB	1800/1000/1000	Selexol	Selexol 2nd stage
L3A	Siemens SFG	NDL	1800/1050/1050	Sulfinol-M	–
L3B	Siemens SFG	NDL	1800/1000/1000	Selexol	Selexol 2nd stage
S4A	CoP E-Gas	PRB	1800/1050/1050	MDEA	–
S4B	CoP E-Gas	PRB	1800/1000/1000	Selexol	Selexol 2nd stage
S12A	SC PC	PRB	3500/1100/1100	S.D. FGD[a]	–
S12B	SC PC	PRB	3500/1100/1100	S.D. FGD	Amine absorber
L12A	SC PC	NDL	3500/1100/1100	S.D. FGD	–
L12B	SC PC	NDL	3500/1100/1100	S.D. FGD	Amine absorber
S13A	USC PC	PRB	4000/1200/1200	S.D. FGD	–
S13B	USC PC	PRB	4000/1200/1200	S.D. FGD	Amine absorber
L13A	USC PC	NDL	4000/1200/1200	S.D. FGD	–
L13B	USC PC	NDL	4000/1200/1200	S.D. FGD	Amine absorber
S22A	SC CFB	PRB	3500/1100/1100	In-bed limestone	–
S22B	SC CFB	PRB	3500/1100/1100	In-bed limestone	Amine absorber
L22A	SC CFB	NDL	3500/1100/1100	In-bed limestone	–
L22B	SC CFB	NDL	3500/1100/1100	In-bed limestone	Amine absorber
S31A	NGCC	Nat. gas	2400/1050/1050	–	–
S31B	NGCC	Nat. gas	2400/1050/1050	–	Amine absorber
L31A	NGCC	Nat. gas	2400/1050/1050	–	–
L31B	NGCC	Nat. gas	2400/1050/1050	–	Amine absorber

[a]Spray drier flue gas desulfurization.

Table 8.21 **Summary of cost, performance, and CO_2 emissions for IGCC cases**

	Shell IGCC				TRIG IGCC		Siemens IGCC				CoP IGCC	
	S1A	L1A	S1B	L1B	S2A	S2B	S3A	L3A	S3B	L3B	S4A	S4B
CO_2 capture	0%	90%	0%	90%	0%	83%	0%	90%	0%	90%	0%	90%
Gross power output (kWe)	696,700	752,600	663,400	713,300	652,700	621,300	622,200	678,800	634,700	676,900	738,300	727,200
Net power output (kWe)	572,680	616,700	471,610	500,060	545,420	460,850	504,720	543,120	445,290	466,510	604,840	515,070
Net plant HHV efficiency (%)	42.0	41.8	32.1	31.7	39.9	31.8	37.9	37.6	30.6	30.0	36.7	30.4
CO_2 (lb/MM Btu)	214	219	22	22	211	36	214	219	22	22	213	22
CO_2 (lb/MWh_{gross})	1,426	1,461	165	170	1,507	287	1,563	1,585	172	175	1,620	174
CO_2 (lb/MWh_{net})	1,735	1,783	233	242	1,803	386	1,927	1,981	246	255	1,977	245
Total plant cost ($/kW)	2,506	2,539	3,480	3,584	2,236	3,019	2,610	2,656	3,533	3,626	2,265	3,144
Total overnight cost ($/kW)	3,056	3,094	4,253	4,378	2,728	3,691	3,185	3,239	4,318	4,430	2,771	3,851
Total as spent capital ($/kW)	3,484	3,527	4,849	4,991	3,110	4,208	3,631	3,692	4,922	5,050	3,159	4,390
COE (mills/kWh)	83.2	83.5	119.7	121.9	74.5	105.2	86.8	87.3	121.7	123.7	78.7	112.3
LCO (mills/kWh)	105.4	105.8	151.8	154.5	94.5	133.3	110.0	110.7	154.3	156.9	99.8	142.4

Table 8.22 **Summary of cost, performance, and CO_2 emissions for coal combustion cases**

	Supercritical PC boiler				Ultra-supercritical PC boiler				Supercritical CFB			
	S12A	**L12A**	**S12B**	**L12B**	**S13A**	**L13A**	**S13B**	**L13B**	**S22A**	**L22A**	**S22B**	**L22B**
CO_2 Capture	0%	0%	90%	90%	0%	0%	90%	90%	0%	0%	90%	90%
Gross Power Output (kWe)	582,700	584,700	673,000	683,900	581,500	583,200	665,400	675,200	578,400	578,700	664,000	672,900
Net Power Output (kWe)	550,040	550,060	550,060	550,050	550,070	550,030	5650,080	550,030	550,070	550,030	550,010	550,080
Net Plant HHV Efficiency (%)	38.7	37.5	27.0	25.5	39.9	38.8	28.7	27.2	38.9	38.0	27.3	25.9
CO_2 (lb/MM Btu)	215	219	21	22	215	219	21	22	213	219	21	22
CO_2 (lb/ MWh_{gross})	1,786	1,877	222	236	1,737	1,820	211	225	1,775	1,865	220	236
CO_2 (lb/ MWh_{net})	1,892	1,996	271	293	1,836	1,930	255	276	1,866	1,963	265	288
Total Overnight Cost ($/kW)	2,293	2,489	3,987	4,341	2,405	2,628	4,049	4,372	2,357	2,490	4,018	4,307
Total As Spent Capital ($/kW)	2,600	2,823	4,545	4,949	2,742	2,996	4,615	4,984	2,687	2,839	4,580	4,909
COE (mills/kWh)	57.8	62.2	107.5	116.4	62.2	67.3	107.7	115.4	61.5	64.6	108.0	115.2
LCO (mills/kWh)	73.3	78.8	136.3	147.5	78.8	85.3	136.5	146.3	78.0	81.9	136.9	146.0

Table 8.23 Summary of cost, performance, and CO_2 emissions for NGCC cases

	NGCC with advanced F class			
	S31A	L31A	S31B	L31B
CO_2 capture	0%	0%	90%	90%
Gross power output (kWe)	522,100	557,000	470,000	501,600
Net power output (kWe)	512,410	546,980	435,060	464,640
Net plant HHV efficiency (%)	50.5	50.6	42.9	43.0
CO_2 (lb/MM Btu)	118	118	12	12
CO_2 (lb/ MWh_{gross})	784	783	87	87
CO_2 (lb/ MWh_{net})	799	797	94	94
Total overnight cost ($/kW)	817	782	1,607	1,548
Total as spent capital ($/kW)	879	840	1,732	1,668
COE (mills/kWh)	64.4	63.6	92.9	91.4
LCO (mills/kWh)	81.7	80.6	117.8	115.8

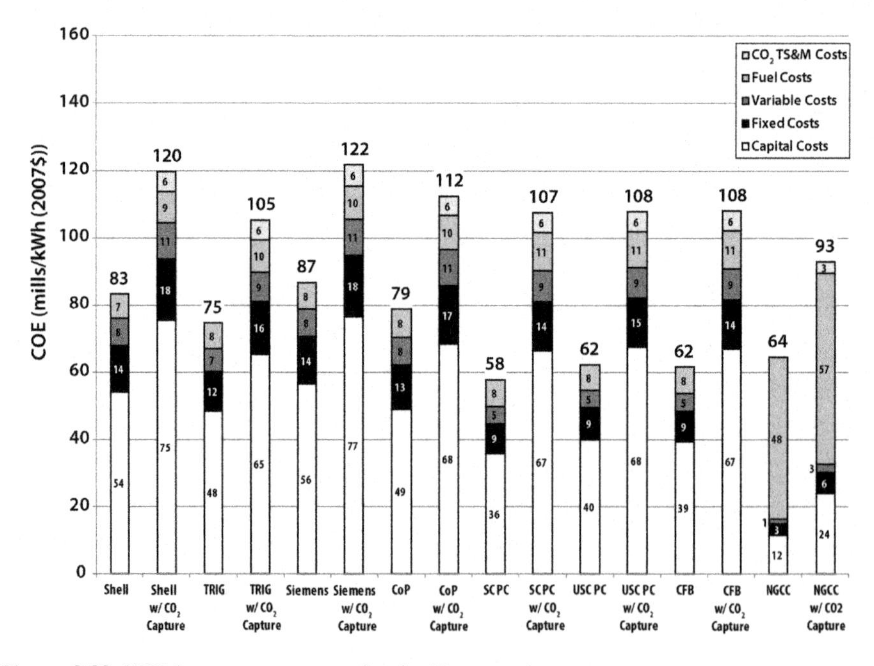

Figure 8.33 COE by cost component for the Montana site cases.

emissions normalized by net output, for both the Montana and North Dakota sites [55]. The COE results are shown in Figures 8.33 and 8.34 for the Montana and North Dakota sites, respectively. The following is observed [55]:

- In noncapture cases, at the Montana site, the coal combustion cases have the lowest COE (average 60.5 mills/kWh), followed by NGCC (64.4 mills/kWh) and IGCC (average

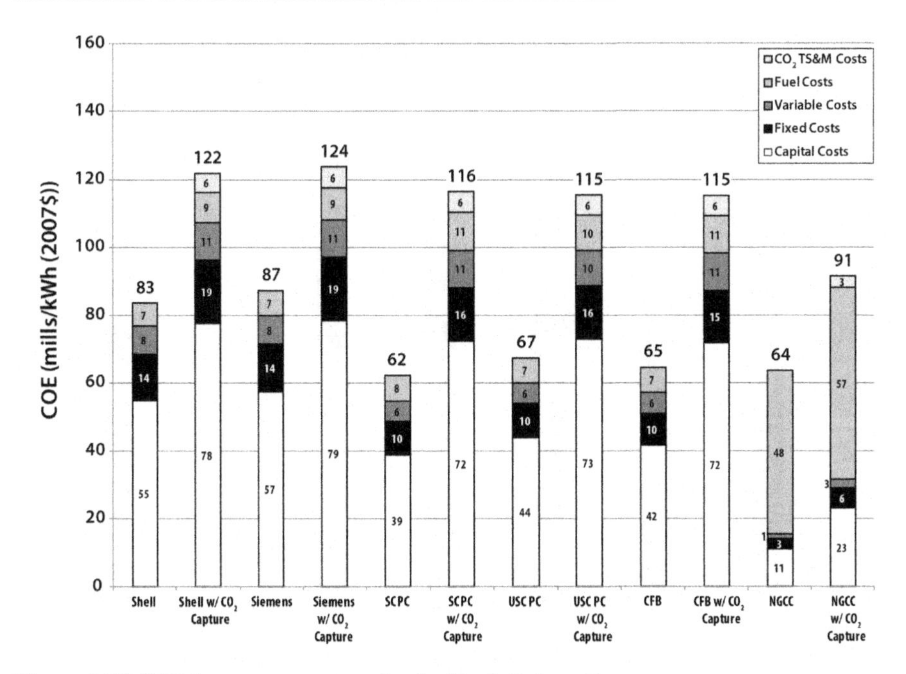

Figure 8.34 COE by cost component for the North Dakota site cases.

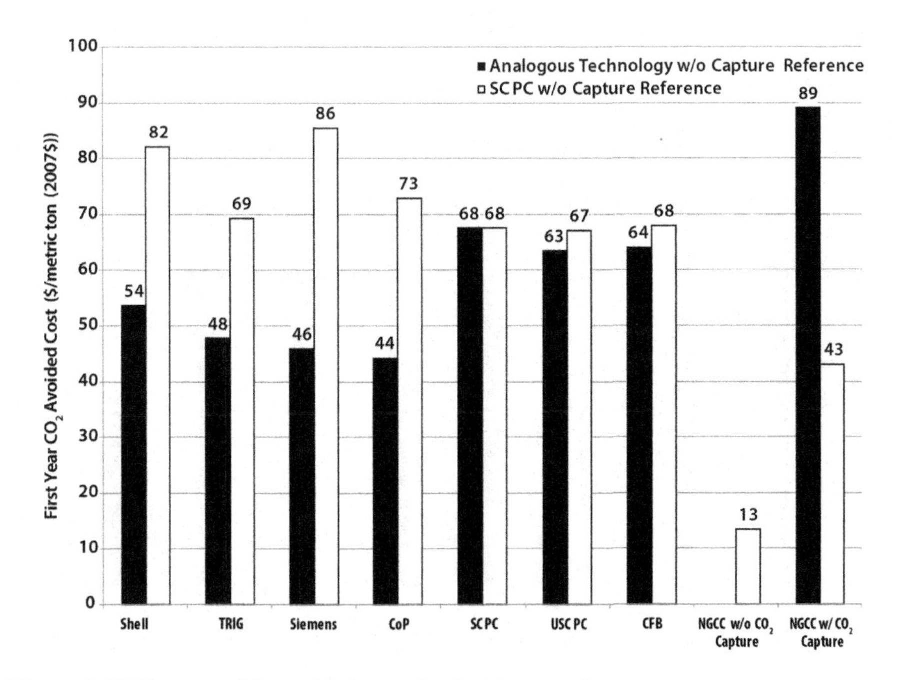

Figure 8.35 First year CO_2 avoided costs for the Montana site cases.

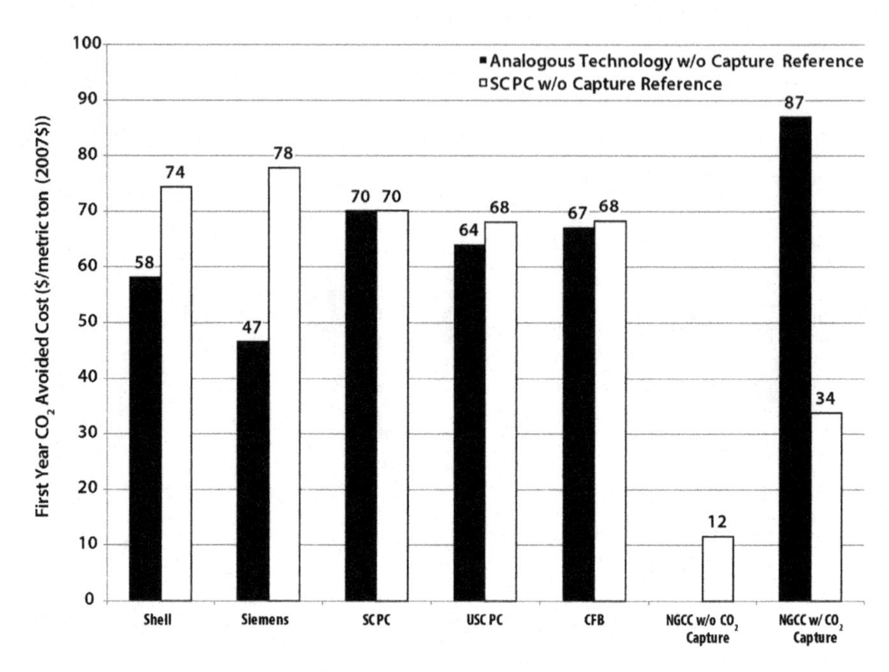

Figure 8.36 First year CO_2 avoided costs for the North Dakota site cases.

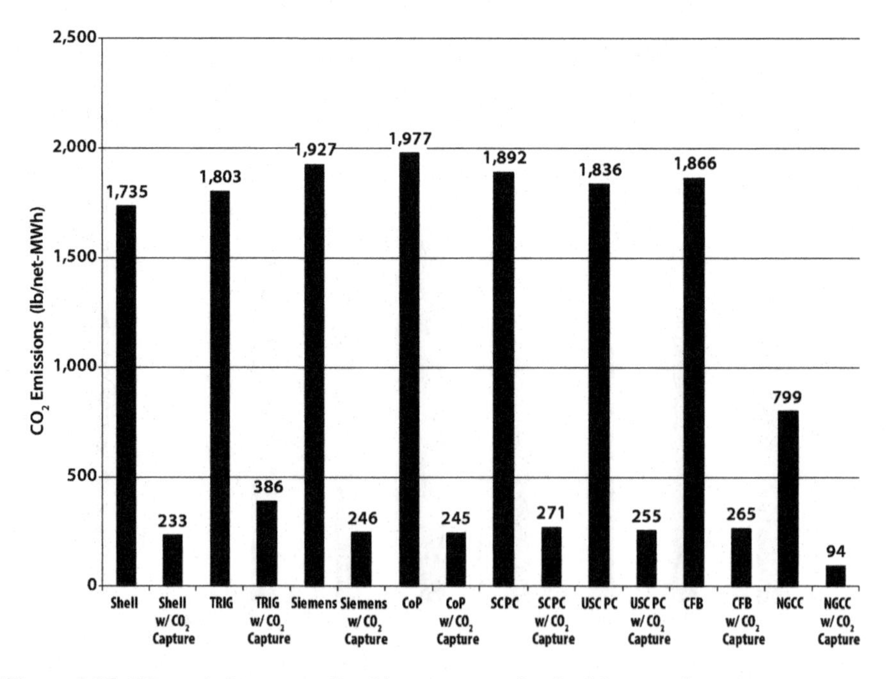

Figure 8.37 CO_2 emissions normalized by net output for the Montana site cases.

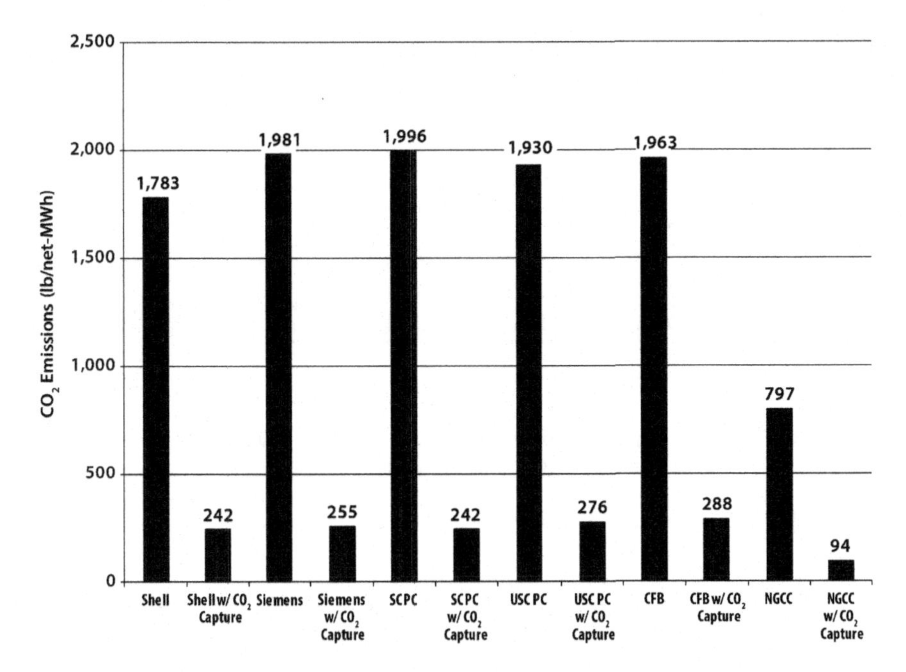

Figure 8.38 CO_2 emissions normalized by net output for the North Dakota site cases.

80.8 mills/kWh). At the North Dakota site, the SC PC plant has the lowest COE (61.5 mills/kWh, all coal combustion cases average 64.7 mills/kWh), followed by NGCC (63.6 mills/kWh) and IGCC (average 85.4 mills/kWh).

- In capture cases, at the Montana site, NGCC plants have the lowest COE (92.9 mills/kWh), followed by the coal combustion cases (average 107.7 mills/kWh) and IGCC (average 114.7 mills/kWh), although the TRIG case (105.2 mills/kWh), with an 83% CO_2 capture efficiency, is less expensive than the PC technologies. At the North Dakota site, NGCC plants have the lowest COE (91.4 mills/kWh), followed by the coal combustion cases (average 115.6 mills/kWh) and IGCC (average 122.8 mills/kWh).
- CO_2 transport, storage, and monitoring (TS&M) is estimated to add 3 to 6 mills/kWh to the COE, which is less than 6% of the total for all capture cases.

The COE with CO_2 removal includes the capture and compression as well as TS&M costs. The resulting avoided costs are shown in Figures 8.35 and 8.36 for the Montana site cases and North Dakota site cases, respectively [55]. The following is noted:

- CO_2 avoided costs for IGGC plants, using analogous noncapture plants as reference, are substantially less than for PC and NGCC because the IGCC CO_2 removal is accomplished prior to combustion and at elevated pressure.
- The CO_2 avoided costs for NGCC plants, using analogous noncapture as reference, are high in part because the NGCC noncapture configuration already emits a very low amount of carbon. As a result, the removal cost for NGCC plants is normalized by a smaller amount of CO_2 being captured.

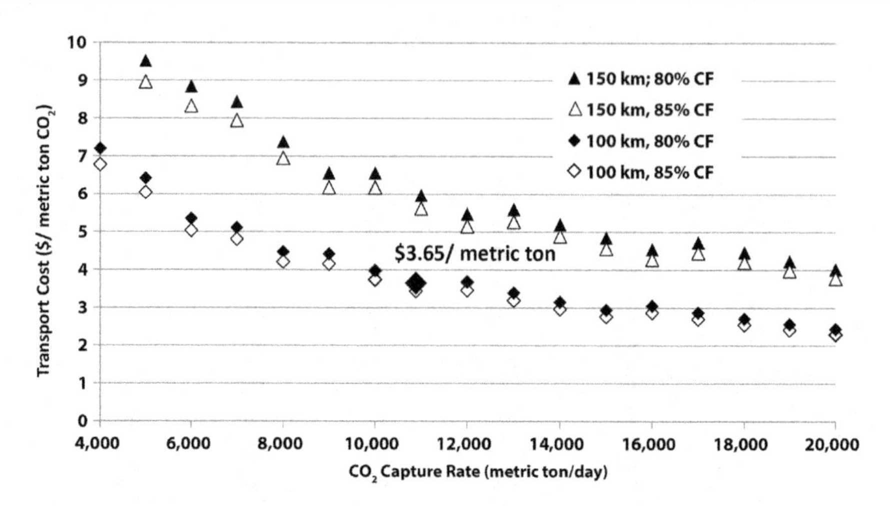

Figure 8.39 Sensitivity of transport costs to plant and distance assumptions.

- CO_2 avoided costs for NGCC plants without capture, using SC PC as reference, are substantially lower than all other technologies. The inherently low carbon intensity of the fuel makes even the noncapture NGCC configuration emit nearly 60% less CO_2 than the reference SC PC plant.

Carbon dioxide emissions, normalized by net output, are presented in Figures 8.37 and 8.38 for the Montana site cases and North Dakota site cases, respectively [55]. In cases with no CO_2 capture, NGCC emits 56% less CO_2 than the lowest PC case and 54% less CO_2 than the lowest IGCC case. The NGCC CO_2 emissions reflect the lower carbon intensity of natural gas relative to coal and the high cycle efficiency of NGCC relative to IGCC and PC. Controlled CO_2 emissions follow the same trend as the uncontrolled cases, i.e. the NGCC case emits less CO_2 than the IGCC cases, which emit less than the PC cases.

8.7.1.5 Transport and storage costs for four U.S. basins

A study was performed by the U.S. DOE to estimate the cost of CO_2 transport and storage in deep saline aquifers for plant locations used in energy system studies [56]. To account for variances in the geologic formations that comprise saline aquifers across the U.S., four locations were studied: the Midwest, Texas, North Dakota, and Montana. Transport and storage costs were based on a generic 62 mile (100 km) dedicated pipeline, a CO_2 flow rate of 10,900 metric tons/day or 3.2 million metric tons per year assuming an 80% capacity factor.

Transport of 10,900 metric tons per day of CO_2 a distance of 62 miles with 1,000 psi pressure drop requires a 16-inch diameter pipe and has estimated capital costs of $106 million and operating and maintenance costs of $0.53 million per year. Figure 8.39 shows the sensitivity of transport cost to distance, capacity factor (CF), and CO_2 capture rate for two pipeline lengths [56].

Costs for CO_2 storage were developed for four basins: Illinois, East Texas, Williston (North Dakota), and Powder River. The cost to store at least 25,000 million metric tons (Mt) is shown in Table 8.24 [56]. Combining storage and transport costs for the 25,000 Mt of cumulative storage resource potential for each basin results in the costs shown in Table 8.25.

8.7.1.6 Capture-ready plants

A study was performed by the U.S. DOE to evaluate options for new, CO_2 capture-ready supercritical pulverized coal-fired power plants to achieve an average 30-year CO_2 emission rate of 1,000 lb/MWh [57]. A CO_2 capture-ready plant is one that does not initially capture CO_2 emissions but fully intends to at some future date. Such a unit can minimize the future retrofit burden by incorporation of certain design considerations into the initial plant construction, which will take into account the installation of future CO_2 capture equipment, and minimizes the retrofit burden later.

The analysis found that the most cost-effective way for new supercritical units to achieve the emission rate is to build CO_2 capture-ready plants that seek to minimize the heat rate penalty, post-retrofit. The capital cost of this configuration is slightly higher but the benefit of maintaining a higher net power output far outweighs the increased capital expenditure [57]. If second-generation CO_2 capture is assumed, there could be a 20% capital cost savings realized over a noncapture-ready plant using today's amine scrubbing technology. Compared to nuclear generation, the capital cost savings of the CO_2 capture-ready plant could be substantial – when compared to recent nuclear estimates, a 50–60% capital cost savings is possible [57]. This is illustrated in Figure 8.40. The 30-year levelized COE is shown in Figure 8.41.

8.7.2 Economics of biomass cofiring

Biomass cofiring is technically feasible; however, costs are still a major barrier to increased coal and biomass cofiring. Cofiring economics depend on location, power plant type, and the availability of low-cost biomass fuels. Biomass is not economically competitive for electricity production in the current energy market except in niche applications where fuel costs are low or negative such as integrated pulp and paper facilities [58]. Biomass is generally more expensive than coal. Delivered costs for dedicated energy crops can be significantly higher [37].

A typical cofiring installation includes modifications to the fuel handling and storage systems, possibly burner modifications, and sometimes boiler modifications or separate biomass feed injection into the boiler. Costs can increase significantly if the biomass needs to be dried, size needs to be reduced, or pellets produced, if the boiler requires a separate feeder. Retrofit costs can range from \$100 to \$300/kW of biomass generation in pulverized coal boilers. Cyclone boilers offer the lowest cost opportunities, as low as \$50/kW [59].

Table 8.24 **Storage resource potential for four basins**

Basin	25,000 Mt storage resource potential				50,000 Mt storage resource potential			75,000 Mt storage resource potential		
	$/metric ton to 25,000 Mt (2011$)	Storage resource pot. at this $/t (Mt)	% of next 100 years captured emissions	GW storage resource potential	$/metric ton to 50,000 Mt (2011$)	Storage resource pot. at this $/t (Mt)	% of next 100 years captured emissions	$/metric ton to 75,000 Mt (2011$)	Storage resource pot. at this $/t (Mt)	% of next 100 years captured emissions
Illinois	5.98	36,931	9.68	185	6.16	66,961	17.56	6.26	95,968	25.16
E. Texas	6.34	28,860	7.57	144	9.36	56,021	14.69	10.28	78,991	20.71
Williston	11.67	38,649	10.13	194	11.67	69,879	18.32	11.81	100,308	26.30
P. River	19.84	43,667	11.45	214	20.44	75,910	19.90	20.44	75,910	19.90
Total		148,107	38.83	743		268,771	70.47		351,177	92.08

Table 8.25 **Total transport and storage costs developed by the U.S. DOE (2011$)**

Plant location	Basin	Transport ($/metric ton)	Storage ($/metric ton)	Total T&S ($/metric ton)
Midwest	Illinois	3.65	5.75	9.40
Texas	East Texas	3.65	6.06	9.71
North Dakota	Williston	3.65	10.96	14.60
Montana	Powder River	3.65	17.86	21.51

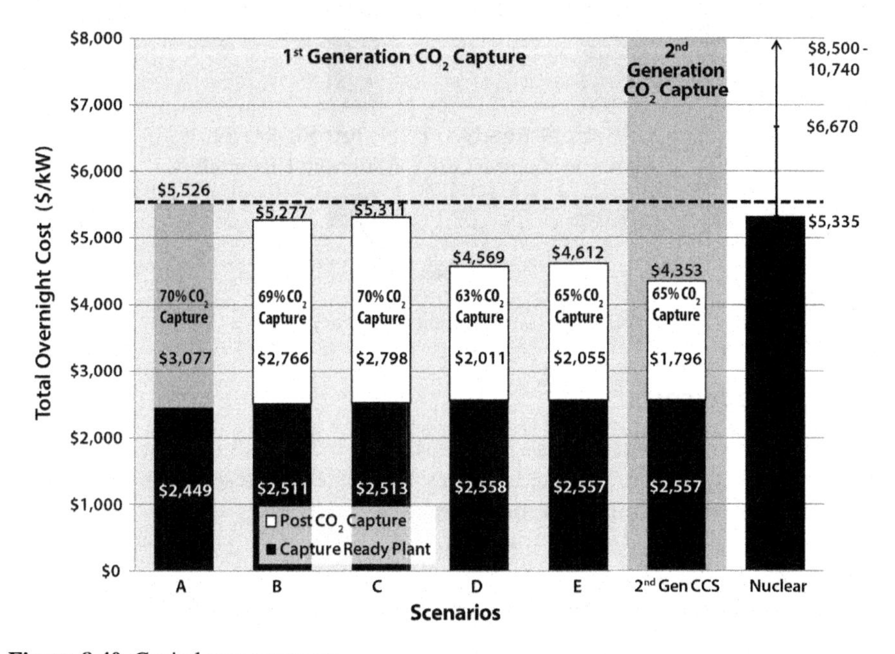

Figure 8.40 Capital cost summary.

Fuel supply is the most important cost factor. Costs for biomass depend on many factors such as climate, closeness to population centers, and the presence of industries that handle and dispose of wood. Low price, low shipping cost, and dependable supply are critical. Usually the cost of biomass must be equal to or less than the cost of coal, on heat basis, for cofiring to be economical. This, however, could change if carbon management legislation becomes enacted in the U.S., for example.

Although coal and biomass cofiring is generally not cost competitive in the current energy market, cofiring offers benefits that could result in more widespread application under certain policy scenarios. This is especially true as a near-term CO_2 mitigation strategy.

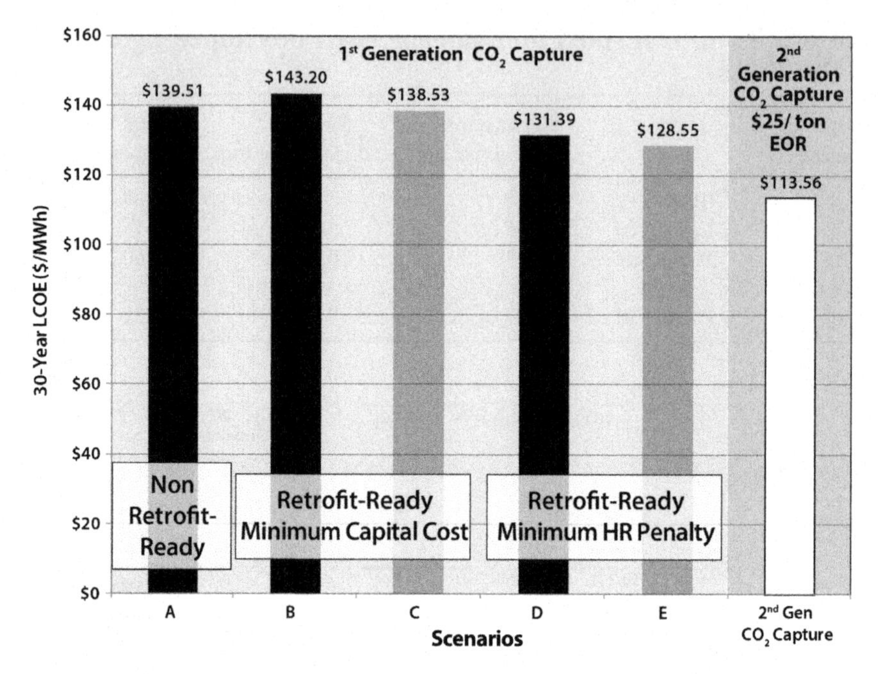

Figure 8.41 Thirty-year levelized cost of electricity summary.

8.7.3 Storage costs

The determination of storage costs is very site dependent and is a function of the type of technology being considered. Ranges of storage are presented in this section for geologic, ocean, and mineralization storage.

8.7.3.1 Geologic storage costs

The technologies and equipment used for geologic storage are widely used in the oil and gas industries, so cost estimates for this option have a fairly high degree of confidence for storage capacity in the lower range of technical potential [38]. However, there is a wide range and variability of costs due to site-specific factors such as onshore versus offshore, reservoir depth, geologic characteristics of the storage formation, type of reservoir (aquifer, coal seam, oil or gas reservoir), and injection method used [38,60].

Estimates for the cost for storage in saline formations and depleted oil and gas fields are between $0.5 and $8.0 per metric ton of CO_2 injected [38]. In the study presented earlier, storage costs in four saline formations in the U.S. ranged from $5.8 to $17.9 per metric ton of CO2 injected [57]. In addition, monitoring of costs of $0.1 to $0.3 per metric ton of CO_2 injected has been estimated. The lowest storage costs are for onshore, shallow, high permeability reservoirs, and/or storage sites where wells and infrastructure from existing oil and gas fields may be used [38].

8.7.3.2 Ocean storage costs

Although there is no experience with ocean storage, attempts have been made to estimate the cost of CO_2 storage projects that release CO_2 on the sea floor or in the deep ocean. Estimates range from $1 to $31 per metric ton CO_2 injected [38,46]. Some specific estimates reported by the Intergovernmental Panel on Climate Change (IPCC) for ocean storage at depths deeper than 3000 m (approximately 2 miles) using a fixed pipeline are $6 per metric ton of CO_2 injected $31 per metric ton of CO_2 injected for distances offshore of 100 km (62 miles) and 500 km (310 miles), respectively [38]. Similarly, the costs using a moving ship/platform are estimated to be $12 to $14 per metric ton CO_2 injected and $13 to $16 per metric ton CO_2 injected for distances offshore of 100 km (62 miles) and 500 km (310 miles), respectively.

8.7.3.3 Mineral carbonization costs

Although the cost of minerals, mineral preparation, and tailing disposal is well known and relatively inexpensive, the chemical reaction process is still too expensive for practical implementation. In situ mineral carbonation costs about $17 per metric ton of CO_2 while injection into saline formations tends to be less than that (see Table 8.25) [56,61]. Ex situ mineral carbonation ranges from $50 to $300 per metric ton of CO_2 mineralized [61]. To be more acceptable, costs below $30 per metric ton are required. In addition, for each metric ton of CO_2 mineralized, 1.6 to 3.7 metric tons of silicates needs to be mined and 2.6 to 4.7 metric tons of materials to be disposed for each metric ton of CO_2 stored as carbonates [38]. This will required a large operation with an environmental impact similar to a large-scale surface mining operation. Serpentine also often contains chrysotile, a natural form of asbestos; therefore, monitoring and mitigation measures used in the mining industry need to be implemented. It should be noted, however, that the products of mineral carbonation are chrysotile-free.

8.8 Monitoring, verification, accounting, and assessment

In addition to the major challenge of reducing CCS costs, another important challenge is the long-term fate or permanence of the stored CO_2. To ensure that carbon sequestration represents an effective pathway for CO_2 management, permanence must be confirmed at a high level of accuracy. This is especially true for geologic and ocean storage; mineral carbonization is expected to ensure long-term storage. The success of CCS is dependent upon the ability to measure the amount of CO_2 stored at a particular site, the ability to confirm that the stored CO_2 is not harming the host ecosystem, and the ability to effectively mitigate any impacts associated with a CO_2 leakage. Of the storage techniques discussed, ocean storage is the most problematic with this issue.

References

[1] U.S. Department of Energy. Energy information administration, energy and the environment, <http://www.eia.gov/energyexplained/index.cfm?page = environment_home>; last reviewed August 15, 2014.

[2] U.S. Department of Environmental Protection. Greenhouse gas emissions, <http://www.epa.gov/climatechange/ghgemissions/>; last updated March 18, 2014.

[3] U.S. Department of Environmental Protection. Climate change impacts and adapting to change, <http://www.epa.gov/climatechange/impacts-adaptation/>, last updated March 18, 2014.

[4] U.S. Department of Environmental Protection. Greenhouse gas reporting program, <http://www.epa.gov/ghgreporting/ghgdata/reported/index.html>, Last updated September 30, 2014.

[5] IPPC (Intergovernmental Panel on Climate Change). Climate change 2014: impacts, adaption, and vulnerability part A: global and sectoral aspects, working group II, contribution to the fifth assessment report of the intergovernmental panel on climate change; 2014.

[6] IPPC (Intergovernmental Panel on Climate Change). Climate change 2014: synthesis report; 2014.

[7] Power Engineering. EPA's clean power plan, <www.pwer-eng.com>; 2014. p. 20.

[8] U.S. Department of Energy. Energy information administration, monthly energy review, 2014.

[9] U.S. Department of Energy National Energy Technology Laboratory. DOE/NETL advanced carbon dioxide capture R&D program: technology update; 2013.

[10] Folger P. Carbon capture: a technology assessment, congressional research service report for congress; 2013.

[11] Kather A, Rafailidis S, Hersforf C, Klostermann M, Maschmann A, Mieske K, et al. Research and development needs for clean coal deployment. London: International Energy Association Clean Coal Centre; 2008.

[12] Henderson C. Clean coal technologies. London: International Energy Association Clean Coal Centre; 2003.

[13] Mills S. Coal-fired CCS demonstration plants, 2012. International Energy Agency, Clean Coal Centre; 2012.

[14] Miller BG, Tillman DA, editors. Combustion engineering issues for solid fuel systems. Oxford, United Kingdom: Elsevier; 2008.

[15] Figueroa JD, Fout T, Plasynski S, McIlvried H, Srivistava RD. Advances in CO_2 capture technology – The U.S. department of energy's carbon sequestration program. Int J Greenhouse Gas Control 2008;9–20.

[16] DOE (U.S. Department of Energy). Oxy-fuel combustion fact sheet, <www.netl.doe.gov>; 2008.

[17] Davidson RM. Post-combustion carbon capture from coal fired plants – solvent scrubbing. International Energy Association Clean Coal Centre; 2007.

[18] Herzog H, Meldon J, Hatton A. Advanced post-combustion CO_2 capture; 2009.

[19] U.S. Department of Energy. Technology readiness assessment guide; 2011.

[20] U.S. Department of Energy. Carbon capture program; 2013.

[21] Stolaroff JK. Carbonate solutions to carbon capture: a summary. Lawrence Livermore National Laboratory; 2013.

[22] Knuutila H, Svendsen HF, Anttila M. CO_2 capture from coal-fired power plants based on sodium carbonate slurry: a systems feasibility and sensitivity study. Int J Greenhouse Gas Control 2009;3:143–51.

[23] Spigarelli BP. A Novel Approach to carbon dioxide capture and storage, Masters Thesis, Michigan Technological University; 2013.
[24] Li B, Duan Y, Luebke D, Morreale B. Advances in CO_2 capture technology: a patent review. Appl Energy 2013;102:1439–47.
[25] Krutka HM, Sjostrom S, Bustard CJ, Durham M, Baldrey K, Stewart R. Summary of post-combustion CO_2 capture technologies for existing coal-fired power plants. Air & Waste Management Association Conference; 2008.
[26] Yu CH, Huang CH, Tan CS. A review of CO_2 capture by absorption and adsorption. Aerosol Air Qual Res 2012;12:745–69.
[27] Xu XC, Song CS, Andresen JM, Miller BG, Scaroni AW. Novel polyethyleneimine-modified mesoporous molecular sieve of MCM-41 type as adsorbent for CO_2 capture. Energy Fuels 2002;16:1463–9.
[28] Xu XC, Song CS, Andresen JM, Miller BG, Scaroni AW. Preparation and characterization of novel CO_2 "molecular basket" adsorbents based on polymer-modified mesoporous molecular sieve MCM-41. Microporous Mesoporous Mater 2003;62:29–45.
[29] Ma X, Wang X, Song C. Molecular basket sorbents for separation of CO_2 and H_2S from various gas streams. J Am Chem Soc 2009;131:5777–83.
[30] Wang X, Song C, Gaffney AM, Song R. New molecular basket sorbents for CO_2 capture based on mesoporous sponge-like TUD-1. Catal Today 2014;238:95–102.
[31] Luis P, Bruggen BV. The role of membranes in post-combustion CO_2 capture. Greenhouse Gas Sci Technol 2013;3:318–37.
[32] Cowan RM, Jensen MD, Pei P, Steadman EN, Harju JA. Current state of CO_2 capture technology development and application. Value-Added Report; 2011.
[33] Tillman DA, Harding NS. Fuels of opportunity: characteristics and uses in combustion systems. Elsevier; 2005.
[34] Miller BG, Tillman DA, editors. Combustion engineering systems for solid fuel systems. Elsevier; 2008.
[35] Patumsawad S. Co-firing biomass with coal for power generation. In: Proc. of the fourth biomass-Asia workshop "biomass: sources of renewable bioenergy and biomaterial"; 2007.
[36] Mann MK, Spath PL. A life cycle assessment of biomass cofiring in a coal-fired power plant. Clean Prod Processes 2001;3:81–91.
[37] Overend RP, Milbrandt A. Tackling climate change in the U.S.: potential carbon emissions reductions from biomass by 2030. In: Kutscher CF, editor. Tackling climate change in the U.S.: potential carbon emissions reductions from energy efficiency and renewable energy by 2030. American Solar Energy Society; 2007.
[38] Metz B, Davidson O, Coninck H, Loos M, Meyer L. Carbon dioxide capture and storage. Intergovernmental panel on climate change. UK: Cambridge University Press; 2005.
[39] Folger P. Carbon capture and sequestration (CCS): a primer. Congressional Research Service; 2013.
[40] Carbon Sequestration Leadership Forum. CO_2 transportation – is it safe and reliable?; 2011.
[41] CCSReg Project. Carbon capture and sequestration: framing the issues for regulation, Department of Engineering and Public Policy, Carnegie Mellon University; 2009.
[42] Global Carbon Project. Carbon budget 2014, <www.globalcarbonproject.org>; 2014.
[43] U.S. Department of Energy. National energy technology laboratory, CO_2 storage formation, <www.netl.doe.gov/research/coal/carbon-storage>, last updated; 2012.
[44] U.S. Geologic Survey. National assessment of geologic carbon dioxide storage resources – summary, Fact Sheet 2013–3020; 2013.

[45] U.S. Department of Energy. National energy technology laboratory, The United States 2012 Carbon Utilization and Storage Atlas, Fourth Edition; 2012.
[46] Sioshansi FP, editor. Generating electricity in a carbon-constrained world. Burlington, MA: Academic Press; 2010.
[47] U.S. Department of Energy. National Energy Technology Laboratory, Carbon storage technology program plan; 2013.
[48] U.S. Department of Energy. National Energy Technology Laboratory, Carbon sequestration technology roadmap and program plan; 2007.
[49] Sarv H. Large-scale CO_2 transportation and deep ocean sequestration, Phase I final report, Prepared for U.S. Department of Energy, National Energy Technology Laboratory, DE-AC26-98FT40412; 1999.
[50] Tribal Energy and Environmental Information Clearinghouse. Terrestrial sequestration of carbon dioxide, <http://teeic.indianaffairs.gov/er/carbon/apptech/terrapp/index.htm>; 2014 [accessed 17.01.14].
[51] Kessels J, Bakker S, Clemens A. Clean coal technologies for a carbon-constrained world. London: International Energy Association Clean Coal Centre; 2007.
[52] WorleyParsons and Schlumberger. Economic assessment of carbon capture and storage technologies, 2011 Update; 2011.
[53] Gerdes K. NETL studies on the economic feasibility of CO_2 capture retrofits for the U.S. Power Plant Fleet; 2014.
[54] U.S. Department of Energy. National energy technology laboratory, cost and performance baseline for fossil energy plants volume 1: bituminous coal and natural gas to electricity, Revision 2a; 2013.
[55] U.S. Department of Energy. National energy technology laboratory, cost and performance baseline for fossil energy plants volume 3: executive summary – low rank coal and natural gas to electricity, Revision 2a; 2011.
[56] U.S. Department of Energy. National energy technology laboratory, carbon dioxide transport and storage costs in NETL studies; 2013.
[57] U.S. Department of Energy. National energy technology laboratory, techno-economic analysis of CO_2 capture-ready coal-fired power plants; 2012.
[58] Robinson AL, Rhodes JS, Keith DW. Assessment of potential carbon dioxide reductions due to biomass-coal cofiring in the United States. Environ Sci Technol 2003;37 (22):5081–8.
[59] Hughes E. Biomass cofiring in the U.S. – status and prospects. In: Proc. of the third Chicago/Midwest renewable energy workshop; 2003.
[60] Kessels J, Bakker S, Clemens A. Clean coal technologies for a carbon-constrained World. London: International Energy Association Clean Coal Centre; 2007.
[61] Sanna A, Uibu M, Caramanna G, Kuusik R, Maroto-Valer MM. A review of mineral carbonation technologies to sequester CO_2. Chem Soc Rev 2014;43:8049–80.

Appendix A: Regional definitions

The country groupings used in this book are shown in Figure A.1 and defined as follows:

A.1 Organization for Economic Cooperation and Development (OECD) (18% of the 2013 world population):

- OECD Americas – Canada, Chile, Mexico, and United States.
- OECD Europe – Austria, Belgium, Czech Republic, Denmark, Estonia, Finland, France, Germany, Greece, Hungary, Iceland, Ireland, Italy, Luxembourg, the Netherlands, Norway, Poland, Portugal, Slovakia, Spain, Sweden, Switzerland, Turkey, and the United Kingdom.
- OECD Asia – Australia, Japan, New Zealand, South Korea.

A.2 Non-Organization for Economic Cooperation and Development (Non-OECD) (82% of the 2013 world population):

- Non-OECD Europe and Eurasia (5% of the 2013 world population) – Albania, Armenia, Azerbaijan, Belarus, Bosnia and Herzegovina, Bulgaria, Croatia, Cyprus, Georgia, Kazakhstan, Kyrgyzstan, Latvia, Lithuania, Macedonia, Malta, Moldova, Montenegro, Romania, Russia, Serbia, Tajikistan, Turkmenistan, Ukraine, and Uzbekistan.
- Non-OECD Asia (53% of the 2013 world population) – Afghanistan, American Samoa, Bangladesh, Bhutan, Brunei, Cambodia (Kampuchea), China, Cook Islands, Fiji, French Polynesia, Guam, Hong Kong, India, Indonesia, Kiribati, Laos, Macau, Malaysia, Maldives, Mongolia, Myanmar (Burma), Nauru, Nepal, New Caledonia, Niue, North Korea, Pakistan, Papua New Guinea, Philippines, Samoa, Singapore, Solomon Islands, Sri Lanka, Taiwan, Thailand, Timor-Leste (East Timor), Tonga, U.S. Pacific Islands, Vanuatu, Vietnam, and Wake Islands.
- Middle East (3% of the 2013 world population) – Bahrain, Iran, Iraq, Israel, Jordan, Kuwait, Lebanon, Oman, Qatar, Saudi Arabia, Syria, the United Arab Emirates, and Yemen.
- Africa (14% of the 2013 world population) – Algeria, Angola, Benin, Botswana, Burkina Faso, Burundi, Cameroon, Cape Verde, Central African Republic, Chad, Comoros, Congo (Brazzaville), Congo (Kinshasa), Côte dIvoire, Djibouti, Egypt, Equatorial Guinea, Eritrea, Ethiopia, Gabon, The Gambia, Ghana, Guinea, Guinea-Bissau, Kenya, Lesotho, Liberia, Libya, Madagascar, Malawi, Mali, Mauritania, Mauritius, Morocco, Mozambique, Namibia, Niger, Nigeria, Reunion, Rwanda, Sao Tome and Principe, Senegal, Seychelles, Sierra Leone, Somalia, South Africa, St. Helena, Sudan, Swaziland, Tanzania, Togo, Tunisia, Uganda, Western Sahara, Zambia, and Zimbabwe.
- Central and South America (7% of the 2013 world population) – Antarctica, Antigua and Barbuda, Argentina, Aruba, The Bahamas, Barbados, Belize, Bolivia, Brazil, British Virgin Islands, Cayman Islands, Colombia, Costa Rica, Cuba, Dominica, Dominican Republic, Ecuador, El Salvador, Falkland Islands, French Guiana, Grenada, Guadeloupe, Guatemala, Guyana, Haiti, Honduras, Jamaica, Martinique, Montserrat, Netherlands Antilles, Nicaragua, Panama, Paraguay, Peru, Puerto Rico, St. Kitts-Nevis, St. Lucia, St. Vincent/Grenadines, Suriname, Trinidad and Tobago, Turks and Caicos Islands, Uruguay, U.S. Virgin Islands, and Venezuela.

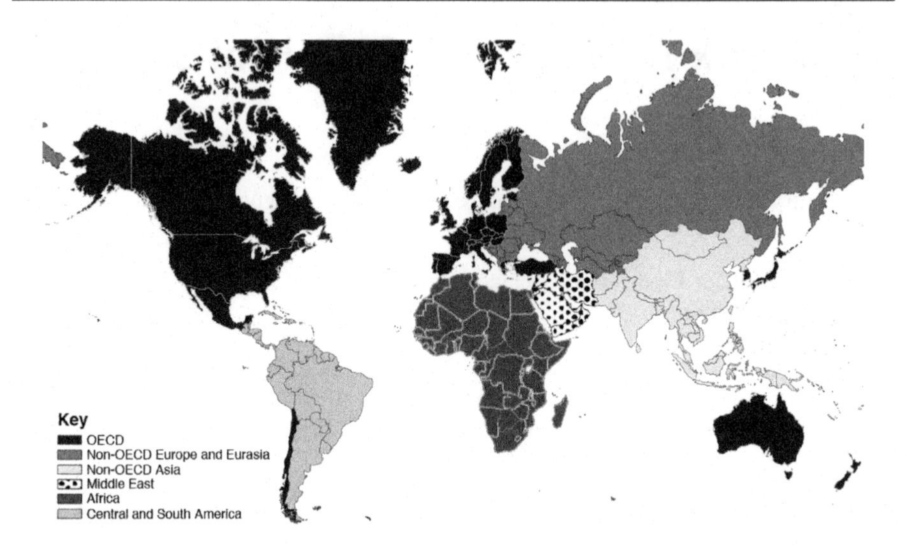

Figure A.1 Country groupings.

A.3 EU (European Union) – Austria, Belgium, Bulgaria, Cyprus, Czech Republic, Denmark, Estonia, Finland, France, Germany, Greece, Hungary, Ireland, Ireland, Italy, Latvia, Lithuania, Luxembourg, Malta, the Netherlands, Poland, Portugal, Romania, Slovakia, Slovenia, Spain, Sweden, and the United Kingdom.

A.4 OPEC (Organization of Petroleum Exporting Countries) – Algeria, Angola, Ecuador, Iran, Iraq, Kuwait, Libya, Nigeria, Qatar, Saudi Arabia, the United Arab Emirates, and Venezuela.

Appendix B: Fossil fuel-fired emission factors

B.1 Coal-fired emission factors

Bituminous Coal Combustion
Subbituminous Coal Combustion
Lignite Combustion
Anthracite Combustion

Table B.1 **Emission factors for SO_x, NO_x, and CO from bituminous and subbituminous coal combustion**[a]

Firing configuration	SO_x[b]		NO_x[c]		CO[d,e]	
	Emission factor (lb/ton)	Emission factor rating	Emission factor (lb/ton)	Emission factor rating	Emission factor (lb/ton)	Emission factor rating
PC, dry bottom, PC, dry bottom, wall-fired[f], bituminous pre-NSPS[g]	38S	A	22	A	0.5	A
PC, dry bottom, wall-fired[f], bituminous pre-NSPS[g] with low-NO_x burner	38S	A	11	A	0.5	A
PC, dry bottom, wall-fired[f], bituminous NSPS[g]	38S	A	12	A	0.5	A
PC, dry bottom, wall-fired[f], subbituminous pre-NSPS[g]	35S	A	12	C	0.5	A
PC, dry bottom, wall-fired[f], subbituminous NSPS[g]	35S	A	7.4	A	0.5	A
PC, dry bottom, cell burner fired, bituminous	38S	A	31	A	0.5	A
PC, dry bottom, cell burner fired, subbituminous	35S	A	14	E	0.5	A
PC, dry bottom, tangentially fired, bituminous, pre-NSPS[g]	38S	A	15	A	0.5	A
PC, dry bottom, tangentially fired, bituminous, pre-NSPS[g] with low-NO_x burner	38S	A	9.7	A	0.5	A
PC, dry bottom, tangentially fired, bituminous, NSPS[g]	38S		10	A	0.5	A
PC, dry bottom, tangentially fired, subbituminous, pre-NSPS[g]	35S	A	8.4	A	0.5	A
PC, dry bottom, tangentially fired, subbituminous, pre-NSPS[g]	35S	A	7.2	A	0.5	A

PC, wet bottom, wall-fired[f], bituminous, pre-NSPS[g]	38S	A	31	D	0.5	A
PC, wet bottom, tangentially fired, bituminous, NSPS[g]	38S	A	14	E	0.5	A
PC, wet bottom, wall-fired, subbituminous	35S	A	24	E	0.5	A
Cyclone furnace, bituminous	38S	A	33	A	0.5	A
Cyclone furnace, subbituminous	35S	A	17	C	0.5	A
Spreader stoker, bituminous	38S	B	11	B	5	A
Spreader stoker, subbituminous	35S	B	8.8	B	5	A
Overfeed stoker[h]	38S (35S)	B	7.5	A	6	B
Underfeed stoker	31S	B	9.5	A	11	B
Hand-fed units	31S	D	9.1	E	275	E
FBC, circulating bed	–[i]	E	5.0	D	18	E
FBC, bubbling bed	–[i]	E	15.2	D	18	D

[a]Source: AP-42. Factors represent uncontrolled emissions unless otherwise specified and should be applied to coal feed, as fired. Tons are short tons.
[b]Expressed as SO_2, including SO_2, SO_3, and gaseous sulfates. Factors in parentheses should be used to estimate gaseous SO_x emissions for subbituminous coal. In all cases, S is weight % sulfur content of coal as fired. Emission factor would be calculated by multiplying the weight percent sulfur in the coal by the numerical value preceding S. For example, if fuel is 1.2% sulfur, then $S = 1.2$. On average for bituminous coal, 95% of fuel sulfur is emitted as SO_2, and only about 0.7% of fuel sulfur is emitted as SO_3 and gaseous sulfate. An equally small percentage of fuel sulfur is emitted as particulate sulfate. Small quantities of sulfur are also retained in bottom ash. With subbituminous coal, about 10% more fuel sulfur is retained in the bottom ash and particulate because of the more alkaline nature of the coal ash. Conversion to gaseous sulfate appears about the same as for bituminous coal.
[c]Expressed as NO_2. Generally, 95 vol % or more of NO_x present in combustion exhaust will be in the form of NO, the rest NO_2. To express factors as NO, multiply factors by 0.66. All factors represent emissions at baseline operation (i.e. 60 to 110% load and no NO_x control measures).
[d]Nominal values achievable under normal operating conditions. Values 1 or 2 orders of magnitude higher can occur when combustion is not complete.
[e]Emission factors for CO_2 emissions from coal combustion should be calculated using lb CO_2/ton coal = 72.6C, where C is the weight % carbon content of the coal. For example, if carbon content is 85%, then C equals 85.
[f]Wall-fired includes front and rear wall-fired units, as well as opposed wall-fired units.
[g]Pre-NSPS boilers are not subject to any NSPS. NSPS boilers are subject to Subpart D or Subpart Da. Subpart D boilers are boilers constructed after August 17, 1971, and with a heat input rate greater than 250 million Btu per hour (MMBtu/h). Subpart Da boilers are boilers constructed after September 18, 1978, and with a heat input rate greater than 250 MMBtu/h.
[h]Includes traveling grate, vibrating grate, and chain grate stokers.
[i]SO_2 emission factors for fluidized bed combustion are a function of fuel sulfur content and calcium-to-sulfur ratio. For both bubbling bed and circulating bed design, use: lb SO_2/ton coal $= 39.6(S)(Ca/S)^{-1.9}$. In this equation, S is the weight percent sulfur in the fuel and Ca/S is the molar calcium-to-sulfur ratio in the bed. This equation may be used when the Ca/S is between 1.5 and 7. When no calcium-based sorbents are used and the bed material is inert with respect to sulfur capture the emission factor for underfeed stokers should be used to estimate the SO_2 emissions. In this case, the emission factor ratings are E for both bubbling and circulating units.

Table B.2 **Emission factors for CH_4, TNMOC, and N_2O from bituminous and subbituminous coal combustion**[a]

Firing configuration	CH_4[b]		TNMOC[b,c]		N_2O	
	Emission factor (lb/ton)	Emission factor rating	Emission factor (lb/ton)	Emission factor rating	Emission factor (lb/ton)	Emission factor rating
PC-fired, dry bottom, wall fired	0.04	B	0.06	B	0.03	B
PC-fired, dry bottom, tangentially fired	0.04	B	0.06	B	0.08	B
PC-fired, wet bottom	0.05	B	0.04	B	0.08	E
Cyclone furnace	0.01	B	0.11	B	0.09[c]	E
Spreader stoker	0.06	B	0.05	B	0.04	E
Spreader stoker, with multiple cyclones, and reinjection	0.06	B	0.05	B	0.04	E
Spreader stoker, with multiple cyclones, no reinjection	0.06	B	0.05	B	0.04	E
Overfeed stoker	0.06	B	0.05	B	0.04	E
Overfeed stoker, with multiple cyclones	0.06	B	0.05	B	0.04	E
Underfeed stoker	0.8	B	1.3	B	0.04	E
Underfeed stoker, with multiple cyclones	0.8	B	1.3	B	0.04	E
Hand-fed units	5	E	10	E	0.04	E
FBC, bubbling bed	0.06	E	0.05	E	3.5	B
FBC, circulating bed	0.06	E	0.05	E	3.5	B

[a]Source: AP-42. Tons are short tons. Factors represent uncontrolled emissions unless otherwise specified and should be applied to coal feed, as fired.
[b]Nominal values achievable under normal operating conditions; values 1 or 2 orders of magnitude higher can occur when combustion is not complete.
[c]TNMOC are expressed as C_2 to C_{16} alkane equivalents. Because of limited data, the effects of firing configuration on TNMOC emission factors could not be distinguished. As a result, all data were averaged collectively to develop a single average emission factor for pulverized coal units, cyclones, spreaders, and overfeed stokers.

Table B.3 **Uncontrolled emission factors for PM and PM_{10} from bituminous and subbituminous coal combustion**[a]

Firing configuration	Filterable PM[b]		Filterable PM_{10}	
	Emission factor (lb/ton)	Emission factor rating	Emission factor (lb/ton)	Emission factor rating
PC-fired, dry bottom, wall-fired	10A	A	2.3A	E
PC-fired, dry bottom, tangentially fired	10A	B	2.3A[c]	E
PC-fired, wet bottom	7A[d]	D	2.6A	E
Cyclone furnace	2A[d]	E	0.26A	E
Spreader stoker	66[e]	B	13.2	E
Spreader stoker, with multiple cyclones, and reinjection	17	B	12.4	E
Spreader stoker, with multiple cyclones, no reinjection	12	A	7.8	E
Overfeed stoker[f]	16[g]	C	6.0	E
Overfeed stoker, with multiple cyclones[f]	9	C	5.0	E
Underfeed stoker	15[h]	D	6.2	E
Underfeed stoker, with multiple cyclones	11	D	6.2[h]	E
Hand-fed units	15	E	6.2[i]	E
FBC, bubbling bed	—[j]	E	—[j]	E
FBC, circulating bed	—[j]	E	—[j]	E

[a]Source: AP-42. Factors represents uncontrolled emissions unless otherwise specified and should be applied to coal feed, as fired.Tons are short tons.
[b]Based on EPA Method 5 (front half catch). Where particulate is expressed in terms of coal ash content, the A factor is determined by multiplying weight % ash content of coal (as fired) by the numerical value preceding the A. For example, if coal with 8% ash is fired in a PC-fired, dry bottom unit, the PM emission factor would be 10×8, or 80 lb/ton.
[c]No data found; emission factor for PC-fired dry bottom boilers used.
[d]Uncontrolled particulate emissions, when no fly ash reinjection is employed. When control device is installed, and collected fly ash is reinjected to boiler, particulate from boiler reaching control equipment can increase up to a factor of 2.
[e]Accounts for fly ash settling in an economizer, air heater, or breaching upstream of control device or stack. (Particulate directly at boiler outlet typically will be twice this level). Factor should be applied even when fly ash is reinjected to boiler form air heater or economizer dust hoppers.
[f]Includes traveling grate, vibrating grate, and chain grate stokers.
[g]Accounts for fly ash settling in breaching or stack base. Particulate loadings directly at boiler outlet typically can be 50% higher.
[h]Accounts for fly ash settling in breaching downstream of boiler outlet.
[i]No data found; emission factor for underfeed stoker used.
[j]No data found; use emission factor for spreader stoker with multiple cyclones and reinjection.

Table B.4 **Condensable particulate matter emission factors for bituminous and subbituminous coal combustion**[a]

Firing configuration[b]	Controls[c]	CPM-TOT[d,e]		CPM-IOR[d,e]		CPM-ORG[d,e]	
		Emission factor (lb/MM Btu)	Emission factor rating	Emission factor (lb/MM Btu)	Emission factor rating	Emission factor (lb/MM Btu)	Emission factor rating
All pulverized coal-fired boilers	All PM controls (without FGD controls)	0.1S-0.03[f]	B	80% of CPM-TOT emission factor[e]	E	20% of CMP-TOT emission factor[e]	E
All pulverized coal-fired boilers	All PM controls combined with FGD controls	0.02	E	ND		ND	E
Spreader stoker, traveling grate overfeed stoker, underfeed stoker	All PM controls, or uncontrolled	0.04	C	80% of CPM-TOT emission factor[e]	E	20% of CPM-TOT emission factor[e]	E

[a]Source: AP-42. All condensable PM is assumed to be less than 1.0 μm in diameter.
[b]No data are available for cyclone boilers or for atmospheric fluidized bed combustion (AFBC) boilers. For cyclone boilers, use the factors provided for pulverized coal-fired boilers and applicable control devices. For AFBC boilers, use the factors provided for pulverized coal-fired boilers with PM and FGD controls.
[c]FGD = flue gas desulfurization.
[d]CPM-TOT, total condensable particulate matter; CPM-IOR, inorganic condensable particulate matter; CPM-ORG, organic condensable particulate matter; ND, no data.
[e]Factors should be multiplied by fuel rate on a heat input basis (MM Btu), as fired. To convert to lb/ton of bituminous coal, multiply by 26 MM Btu/ton. To convert to lb/ton of subbituminous coal, multiply by 20 MM Btu/ton.
[f]S = coal sulfur percent by weight, as fired. For example, if the sulfur percent is 1.04, then S = 1.04. If the coal sulfur percent is 0.4 or less, use a default emission factor of 0.01 lb/MM Btu rather than the emission equation.

Table B.5 Emission factors for trace elements, POM, and HCOH from uncontrolled bituminous and subbituminous coal combustion[a] (emission factor rating: E)[a]

Firing configuration	Emission factor, lb/10^{12} Btu									
	As	Be	Cd	Cr	Pb[b]	Mn	Hg	Ni	POM	HCOH
Pulverized coal, configuration unknown	ND	ND	ND	1,922	ND	ND	ND	ND	ND	112[c]
Pulverized coal, wet bottom	538	81	44–70	1,020–1,570	507	808–2,980	16	840–1,290	ND	ND
Pulverized coal, dry bottom	684	81	44.4	1,250–1,570	507	228–2,980	16	1,030–1,290	2.08	ND
Pulverized coal, dry bottom, tangential	ND	ND	ND	ND	ND	ND	ND	ND	2.4	ND
Cyclone furnace	115	<81	28	212–1,502	507	228–1,300	16	174–1,290	ND	ND
Stoker, configuration unknown	ND	73	ND	19–300	ND	2,170	16	775–1,290	ND	ND
Spreader stoker	264–542	ND	21–43	942–1,570	507	ND	ND	ND	ND	221[d]
Overfeed stoker, traveling grate	542–1,030	ND	43–82	ND	507	ND	ND	ND	ND	140[c]

[a]Source: AP-42. The emission factors in this table represent the ranges of factors reported in the literature. If only 1 data point was found, it is still reported in this table. To convert from lb/10^{12} Btu to pg/J, multiply by 0.43. ND = no data.
[b]Lead emission factors were taken directly from an EPA background document for support of the National Ambient Air Quality Standards.
[c]Based on 2 units; 133×10^6 Btu/h and 155×10^6 Btu/h.
[d]Based on 1 unit; 59×10^6 Btu/h.

Table B.6 Cumulative particle size distribution and size-specific emission factors for dry bottom boilers burning pulverized bituminous and subbituminous coal[a]

Particle size[b] (μm)	Cumulative mass % ≤ stated size					Cumulative emission factor[c] (lb/ton)				
	Uncontrolled	Controlled				Uncontrolled[d]	Controlled[e]			
		Multiple cyclones	Scrubber	ESP	Baghouse		Multiple cyclones[f]	Scrubber[g]	ESP[g]	Baghouse[f]
15	32	54	81	79	97	3.2A	1.08A	0.48A	0.064A	0.02A
10	23	29	71	67	92	2.3A	0.58A	0.42A	0.054A	0.02A
6	17	14	62	50	77	1.7A	0.28A	0.38A	0.024A	0.02A
2.5	6	3	51	29	53	0.6A	0.06A	0.3A	0.024A	0.01A
1.25	2	1	35	17	31	0.2A	0.02A	0.22A	0.01A	0.006A
1.00	2	1	31	14	25	0.2A	0.02A	0.18A	0.01A	0.006A
0.625	1	1	20	12	14	0.10A	0.02A	0.12A	0.01A	0.002A
TOTAL	100	100	100	100	100	10A	2A	0.6A	0.08A	0.02A

[a]Source: AP-42. Tons are short tons. To convert from lb/ton to kg/Mg, multiply by 0.5. Emission factors are lb of pollutant per ton of coal combusted, as fired. ESP = electrostatic precipitator.
[b]Expressed as aerodynamic equivalent diameter.
[c]A = coal ash weight percent as fired. For example, if coal ash weight is 8.2%, then A = 8.2.
[d]Emission factor rating = C.
[e]Estimated control efficiency for multiple cyclones is 80%; for scrubber, 94%; for ESP, 99.2%; and for baghouse, 99.8%.
[f]Emission factor rating = E.
[g]Emission factor rating = D.

Table B.7 Emission factors for polychlorinated dibenzo-p-dioxins and polychlorinated dibenzofurans from controlled bituminous and subbituminous coal combustion[a]

Controls	FGD-SDA with FF[b]		ESP or FF[c]	
Cogener	Emission factor[d] (lb/ton)	Emission factor rating	Emission factor (lb/ton)	Emission factor rating
2,3,7,8-TCDD	No data	–	1.43E-11	E
Total TCDD	3.93E-10	E	9.28E-11	D
Total PeCDD	7.06E-10	E	4.47E-11	D
Total HxCDD	3.00E-09	E	2.87E-11	D
Total HpCDD	1.00E-08	E	8.34E-11	D
Total OCDD	2.87E-08	E	4.16E-10	D
Total PCDD[e]	4.28E-08	E	6.66E-10	D
2,3,7,8-TCDF	No data	–	5.10E-11	D
Total TCDF	2.49E-09	E	4.04E-10	D
Total PeCDF	4.84E-09	E	3.53E-10	D
Total HxCDF	1.27E-08	E	1.92E-10	D
Total HpCDF	4.39E-08	E	7.678E-11	D
Total OCDF	1.37E-07	E	6.63E-11	D
Total PCDF[e]	2.01E-07	E	1.09E-09	D
Total PCDD/PCDF	2.44E-07	E	1.76E-09	D

[a]Source: AP-42.
[b]Flue gas desulfurization-spray dryer absorber with fabric filter.
[c]Electrostatic precipitator or fabric filter.
[d]Emission factor should be applied to coal feed, as fired. To convert from lb/ton to kg/Mg, multiply by 0.5. Emissions are lb of pollutant per tone of coal combusted.
[e]Total PCDD is the sum of Total TCDD through Total OCDD. Total PCDF is the sum of Total TCDF through Total OCDF.

Table B.8 Cumulative particle size distribution and size-specific emission factors for wet bottom boilers burning pulverized bituminous coal (emission factor rating: E)[a]

Particle size[b] (μm)	Cumulative mass % ≤ stated size			Cumulative emission factor[c] (lb/ton)		
	Uncontrolled	Controlled		Uncontrolled	Controlled[d]	
		Multiple cyclones	ESP		Multiple cyclones	ESP
15	40	99	83	2.8A	1.38A	0.046A
10	37	93	75	2.6A	1.3A	0.042A
6	33	84	63	2.32A	1.18A	0.036A
2.5	21	61	40	1.48A	0.86A	0.022A
1.25	6	31	17	0.42A	0.44A	0.01A
1.00	4	19	8	0.28A	0.26A	0.004A
0.625	2	–[e]	–[e]	0.14A	–[e]	–[e]
TOTAL	100	100	100	7.0A	1.4A	0.056A

[a]Source: AP-42. Tons are short tons. To convert from lb/ton to kg/Mg, multiply by 0.5. Emission factors are lb of pollutant per ton of coal combusted as fired. ESP = Electrostatic precipitator.
[b]Expressed as aerodynamic equivalent diameter.
[c]A = coal ash weight %, as fired. For example, if coal ash weight is 2.4%, then A = 2.4.
[d]Estimated control efficiency for multiple cyclones is 94%, and for ESPs, 99.2%.
[e]Insufficient data.

Table B.9 Cumulative size distribution and size-specific emission factors for cyclone furnaces burning bituminous coal (emission factor rating: E)[a]

Particle size[b] (μm)	Cumulative mass % ≤ stated size			Cumulative emission factor[c] (lb/ton)		
	Uncontrolled	Controlled		Uncontrolled	Controlled[d]	
		Multiple cyclones	ESP		Multiple cyclones	ESP
15	33	95	90	0.66A	0.114A	0.013A
10	13	94	68	0.26A	0.112A	0.011A
6	8	93	56	0.16A	0.112A	0.009A
2.5	5.5	92	36	0.11A[c]	0.11A	0.006A
1.25	5	85	22	0.10A[e]	0.10A	0.04A
1.00	5	82	17	0.10A[e]	0.10A	0.004A
0.625	0	—[e]	—[e]	0	—[f]	—[f]
TOTAL	100	100	100	2A	0.12A	0.016A

[a]Source: AP-42. Tons are short tons. To convert from lb/ton to kg/Mg, multiply by 0.5. Emission factors are lb of pollutant per ton of coal combusted as fired.
[b]Expressed as aerodynamic equivalent diameter.
[c]A = coal ash weight %, as fired. For example, if coal ash weight is 2.4%, then A = 2.4.
[d]Estimated control efficiency for multiple cyclones is 94%, and for ESPs, 99.2%.
[e]These values are estimates based on data from controlled source.
[f]Insufficient data.

Table B.10 Cumulative particle size distribution and size-specific emission factors for overfeed stokers burning bituminous coal[a]

Particle size[b] (μm)	Cumulative mass % ≤ stated size		Cumulative emission factor (lb/ton)			
	Uncontrolled	Multiple cyclones controlled	Uncontrolled		Multiple cyclones controlled[c]	
			Emission factor	Emission factor rating	Emission factor	Emission factor rating
15	49	60	7.8	C	5.4	E
10	37	55	6.0	C	5.0	E
6	24	49	3.8	C	4.4	E
2.5	14	43	2.2	C	3.8	E
1.25	13	39	2.0	C	3.6	E
1.00	12	39	2.0	C	3.6	E
0.625	—[d]	16	—[d]	C	1.4	E
TOTAL	100	100	16.0	C	9.0	E

[a]Source: AP-42. Tons are short tons. To convert from lb/ton to kg/Mg, multiply by 0.5. Emission factors are lb of pollutant per ton of coal combusted as fired. ESP = Electrostatic precipitator.
[b]Expressed as aerodynamic equivalent diameter.
[c]Estimated control efficiency for multiple cyclones is 80%.
[d]Insufficient data.

Table B.11 Cumulative particle size distribution and size-specific emission factors for underfeed stokers burning bituminous coal (emission factor rating: C)[a]

Particle size [b] (μm)	Cumulative mass % ≤ stated size	Uncontrolled cumulative emission factor[c] (lb/ton)
15	50	7.6
10	41	6.2
6	32	4.8
2.5	25	3.8
1.25	22	3.4
1.00	21	3.2
0.625	18	2.7
TOTAL	100	15.0

[a]Source: AP-42. Tons are short tons. To convert from lb/ton to kg/Mg, multiply by 0.5. Emission factors are lb of pollutant per ton of coal combusted, as fired.
[b]Expressed as aerodynamic equivalent diameter.
[c]May also be used for uncontrolled hand-fired units.

Table B.12 Cumulative particle size distribution and size-specific emission factors for spreader stokers burning bituminous coal[a]

Particle size[b] (μm)	Cumulative mass % ≤ stated size					Cumulative emission factor (lb/ton)				
	Uncontrolled	Controlled				Uncontrolled[e]	Controlled			
		Multiple cyclones[c]	Multiple cyclones[d]	ESP	Baghouse		Multiple cyclones[e,f]	Multiple cyclones[d,e]	ESP	Baghouse[e,g]
15	28	86	74	97	72	18.5	14.6	8.8	0.46	0.086
10	20	73	65	90	60	13.2	12	7.8	0.44	0.072
6	14	51	52	82	46	9.2	8.6	6.2	0.40	0.056
2.5	7	8	27	61	26	4.6	1.4	3.2	0.30	0.032
1.25	5	2	16	46	18	3.3	0.4	2.0	0.22	0.022
1.00	5	2	14	41	15	3.3	0.4	1.6	0.20	0.018
0.625	4	1	9	—[h]	7	2.6	0.2	1.0	—[h]	0.006
TOTAL	100	100	100	100	100	66.0	17.0	12.0	0.48	0.12

[a]Source: AP-42. Tons are short tons. To convert from lb/ton to kg/Mg, multiply by 0.5. Emissions are lb of pollutant per ton of coal combusted, as fired.
[b]Expressed as aerodynamic equivalent diameter.
[c]With flyash reinjection.
[d]Without flyash reinjection.
[e]Emission factor rating = C.
[f]Emission factor rating = E.
[g]Estimated control efficiency for ESP is 99.22%; and for baghouse, 99.8%.
[h]Insufficient data.

Table B.13 **Emission factors for SO_x, NO_x, CO, and CO_2 from uncontrolled lignite combustion (emission factor rating: C (except as noted))[a]**

Firing configuration	SO_x emission factor[b] (lb/ton)	NO_x emission factor (lb/ton)	CO emission factor (lb/ton)	CO_2 emission factor[e] (lb/ton)	TNMOC[g,h,I] emission factor (lb/ton)
Pulverized coal, dry bottom, tangential	30S	7.1[f]	ND	72.6C	0.04
Pulverized coal, dry bottom, wall fired[c], Pre-NSPS[d]	30S	13	0.25	72.6C	0,04
Pulverized coal, dry bottom, wall fired[c], NSPS[d]	30S	6.3	0.25	72.6C	0.04
Cyclone	30S	15	ND	72.6C	0.07
Spreader stoker	30S	5.8	ND	72.6C	0.03
Traveling grate overfeed stoker	30S	ND	ND	72.6C	0.03
Atmospheric fluidized bed combustor	10S[i]	3.6	0.15	72.6C	0.03

[a]Source: AP-42. Tons are short tons. To convert from lb/ton to kg/Mg, multiply by 0.5. To convert from lb/ton to lb/MM Btu, multiply by 0.0625. ND = no data.
[b]S = weight % sulfur content of lignite, wet basis. For example, if the sulfur content equals 3.4%, then S = 3.4. For high sodium ash ($Na_2O > 8\%$), use 22S. For low sodium ash ($Na_2O < 2\%$), use 34S. If ash sodium content is unknown, use 30S.
[c]Wall-fired includes front and rear wall-fired units, as well as opposed wall-fired units.
[d]Pre-NSPS boilers are not subject to an NSPS. NSPS boilers are subject to Subpart D or Subpart Da. Subpart D boilers are boilers constructed after August 17, 1971 and with a heat input greater than 250 million Btu per hour (MM Btu/h). Subpart Da boilers are boilers constructed after September 18, 1978 and with a heat input rate greater than 250 MM Btu/h.
[e]Emission Factor Rating: B. C = weight % carbon of lignite, as-fired basis. For example, if carbon content equals 63%, then C = 63. If the % C value is not known, a default CO_2 emission value of 4600 lb/ton may be used.
[f]Emission Factor Rating = A.
[g]TNMOC: Total nonmethane organic compounds. Emission factors were derived from bituminous coal data in the absence of lignite data assuming emissions are proportional to coal heating value. TNMOC are expressed as C_2 to C_{16} alkane equivalents. Because of limited data, the effects of firing configuration on TNMOC emission factors could not be distinguished. As a result, all data were averaged collectively to develop a single average emission factor for pulverized coal, cyclones, spreaders, and overfeed stokers.
[h]Nominal values achievable under normal operating conditions; values 1 or 2 orders of magnitude higher can occur when combustion is not complete.
[i]Using limestone bed material.

Table B.14 Emission factors for NO_x and CO from lignite combustion with NO_x controls[a]

Firing configuration	Control device	NO_x		CO	
		Emission factor (lb/ton)	Emission factor rating	Emission factor (lb/ton)	Emission factor rating
Subpart D boilers:[b] Pulverized coal, tangential-fired	Overfire Air	6.8	C	ND	NA
Pulverized coal, wall-fired	Overfire air and low NO_x burners	4.6	C	0.48	D
Subpart Da boilers:[b] Pulverized coal, tangential-fired	Overfire Air	6.0	C	0.1	D

[a]Source: AP-42. Tons are short tons. To convert from lb/ton to kg/Mg, multiply by 0.5. To convert from lb/ton to lb/MM Btu, multiply by 0.0625. ND = no data. NA = not applicable.
[b]Subpart D boilers are boilers constructed after August 17, 1971 and with a heat input rate greater than 250 million Btu per hour (MM Btu/h). Subpart Da boilers are boilers constructed after September 18, 1978 and with a heat input rate greater than 250 MM Btu/h.

Table B.15 Emission factors for POM from controlled lignite combustion (emission factor rating: E)[a]

Firing configuration	Control device	Emission factor (lb/10^{12} Btu)
		POM
Pulverized coal	High efficiency cold-side ESP	2.3
Pulverized dry bottom	Multi-cyclones	1.8–18[b]
	ESP	2.6[b]
Cyclone furnace	ESP	0.11[c]–1.6[b]
Spreader stoker	Multi-cyclones	15[c]

[a]Source: AP-42. To convert from lb/10^{12} Btu to pg/J, multiply by 0.43.
[b]Primarily trimethyl propenyl napthalene.
[c]Primarily biphenyl.

Table B.16 Emission factors for filterable PM and N_2O from uncontrolled lignite combustion (emission factor rating: E (except as noted))[a]

Firing configuration	Filterable PM emission factor [b] (lb/ton)	N_2O emission factor[c] (lb/ton)
Pulverized coal, dry bottom, tangential	6.5A	ND
Pulverized coal, dry bottom, wall fired	5.1A	ND
Cyclone	6.7A[c]	ND
Spreader stoker	8.0A	ND
Other stoker	3.4A	ND
Atmospheric fluidized bed combustor	ND	2.5

[a]Source: AP-42. Tons are short tons. To convert from lb/ton to kg/Mg, multiply by 0.5. To convert from lb/ton to lb/MM Btu, multiply by 0.0625. ND = no data.
[b]A = weight % ash content of lignite, wet basis. For example, if the ash content is 5%, then A = 5.
[c]Emission factor rating: C.

Table B.17 Emission factors for filterable PM emissions from controlled lignite combustion (emission factor rating: C (except as noted))[a]

Firing configuration	Control device	Filterable PM emission factor (lb/ton)
Subpart D boilers[b]	Baghouse	0.08A
	Wet Scrubber	0.05A
Subpart Da boilers[b]	Wet Scrubber	0.01A
Atmospheric fluidized bed combustor[c]	ESP	0.07A

[a]Source: AP-42. Tons are short tons. A = weight % ash content of lignite, wet basis. For example, if lignite is 2.3% ash, then A = 2.3. To convert from lb/ton to kg/Mg, multiply by 0.5. To convert from lb/ton to lb/MM Btu, multiply by 0.0625.
[b]Subpart D boilers are boilers constructed before August 17, 1971, and with a heat input rate greater than 250 million Btu per hour (MM Btu/h). Subpart Da boilers are boilers constructed after September 18, 1978, and with a heat input rate greater than 250 MM Btu/h.
[c]Emission factor rating: D.

Table B.18 Condensable particulate matter emission factors for lignite combustion[a]

Firing configuration[b]	Controls[c]	CPM – TOT[d,e]		CPM – IOR[d,e]		CPM - ORG[d,e]	
		lb/MM Btu	Rating	lb/MM Btu	Rating	lb/MM Btu	Rating
All pulverized coal-fired boilers	All PM controls (without FGD controls)	0.1S–0.03[f]	C	80% of CPM-TOT emission factor[e]	E	20% of CPM-TOT emission factor[e]	E
All pulverized coal-fired boilers	All PM controls combined with an FGD control	0.02[f]	E	ND		ND	
Traveling grate overfeed stoker, spreader stoker	All PM controls, or Uncontrolled	0.04	D	80% of CPM-TOT emission factor	E	20% of CPM-TOT emission factor	E

[a]Source: AP-42. All condensable PM is assumed to be less than 1.0 micron in diameter.
[b]No data are available for cyclone boilers. For cyclone boilers, use the factors provided for pulverized coal-fired boilers and applicable controls.
[c]FGD = flue gas desulfurization.
[d]CPM-TOT = total condensable particulate matter.
CPM-IOR = inorganic condensable particulate matter.
CPM-ORG = organic condensable particulate matter.
ND = no data.
[e]Factors should be multiplied by fuel rate on a heat input basis (MM Btu), as fired. To convert to lb/short ton of lignite, multiply by 16 MM Btu/short ton.
[f]S = coal sulfur percent by weight, as fired. For example, if the sulfur percent is 1.04, then S = 1.04. If the coal sulfur percent is 0.4 or less, use a default emission factor of 0.01 lb/MM Btu rather than the emission equation.

Table B.19 Emission factors for trace elements from uncontrolled lignite combustion (emission factor rating: E)[a]

Firing configuration	Emission factor (lb/10^{12}Btu)						
	As	Be	Cd	Cr	Mn	Hg	Ni
Pulverized, wet bottom	2,730	131	49–77	1,220–1,880	4,410–16,250	21	154–1,160
Pulverized, dry bottom	1,390	131	49	1,500–1,880	16,200	21	928–1,160
Cyclone furnace	235–632	131	31	253–1,880	3,760	21	157–1,160
Stoker configuration unknown	ND	118	ND	ND	11,800	21	ND
Spreader stoker	538–1,100	ND	23–47	1,130–1,880	ND	ND	696–1,160
Traveling grate (overfed) stoker	1,100–2,100	ND	47–90	ND	ND	ND	ND

[a]Source: AP-42. To convert from lb/10^{12} Btu to pg/J, multiply by 0.43. ND = no data.

Table B.20 Cumulative particle size distribution and size-specific emission factors for boilers firing pulverized lignite (emission factor rating: E)[a]

Particle size[b] (μm)	Cumulative mass % ≤ stated size		Cumulative emission[c] (lb/ton)	
	Uncontrolled	Multiple cyclone controlled	Uncontrolled	Multiple cyclone controlled[d]
15	51	77	3.4A	1.0A
10	35	67	2.3A	0.88A
6	26	57	1.7A	0.75A
2.5	10	27	0.66A	0.36A
1.25	7	16	0.47A	0.21A
1.00	6	14	0.40A	0.19A
0.625	3	8	0.19A	0.11A
TOTAL			6.6A	1.3A

[a]Source: AP-42. Tons are short tons. Based on tangential-fired units. For wall-fired units multiply emission factors in the table by 0.79.
[b]Expressed as aerodynamic equivalent diameter.
[c]A = weight % ash content of lignite, wet basis. For example, if lignite is 3.4% ash, then A = 3.4. To convert from lb/ton to kg/Mg, multiply by 0.5. To convert from lb/ton to lb/MM Btu, multiply by 0.0625.
[d]Estimated control efficiency for multiple cyclone is 80%, averaged over all particle sizes.

Table B.21 Cumulative particle size distribution and size-specific emission factors for lignite-fired spreader stokers (emission factor rating: E)[a]

Particle size[b] (μm)	Cumulative mass % ≤ stated size		Cumulative emission[c] (lb/ton)	
	Uncontrolled	Multiple cyclone controlled	Uncontrolled	Multiple cyclone controlled[d]
15	28	55	2.2A	0.88A
10	20	41	1.6A	0.66A
6	14	31	1.1A	0.50A
2.5	7	26	0.56A	0.42A
1.25	5	23	0.40A	0.37A
1.00	5	22	0.40A	0.35A
0.625	4	–[e]	0.33A	–[e]
TOTAL			8.0A	1.6A

[a]Source: AP-42. Tons are short tons.
[b]Expressed as aerodynamic equivalent diameter.
[c]A = weight % ash content of lignite, wet basis. For example, if lignite is 5% ash, then A = 5. To convert from lb/ton to kg/Mg, multiply by 0.5. To convert from lb/ton to lb/MM Btu, multiply by 0.0625.
[d]Estimated control efficiency for multiple cyclone is 80%.
[e]Insufficient data.

Table B.22 Default CO_2 emission factors for U.S. coals (emission factor rating: C)[a]

Coal type	Average %C[b]	Conversion factor[c]	Emission factor[d] lb/ton coal
Subbituminous	66.3	72.6	4810
High-volatile bituminous	75.9	72.6	5510
Medium-volatile bituminous	83.2	72.6	6040
Low-volatile bituminous	86.1	72.6	6250

[a]Source: AP-42. Tons are short tons. This table should be used only when an ultimate analysis is not available. If the ultimate analysis is available, CO_2 emissions should be calculated by multiplying the % carbon (%C) by 72.6. This resultant factor would receive a quality rating of B.
[b]Based on average carbon contents for each coal type (dry basis) based on extensive sampling of U.S. coals.
[c]Based on the following equation:

$$\frac{44 \text{ ton } CO_2}{12 \text{ ton C}} \times 0.99 \times 2000 \frac{\text{lb } CO_2}{\text{ton } CO_2} \times \frac{1}{100\%} = 72.6 \frac{\text{lb } CO_2}{\text{ton } \%C}$$

where 44 = molecular weight of CO_2; 12 = molecular weight of carbon; and 0.99 = fraction of fuel oxidized during combustion.
[d]To convert from lb/ton to kg/Mg, multiply by 0.5.

Table B.23 Emission factors for various organic compounds from controlled coal combustion[a]

Pollutant[b]	Emission factor (lb/ton)	Emission factor rating
Acetaldehyde	5.7E-04	C
Acetophenone	1.5E-05	D
Acrolein	2.9E-04	D
Benzene	1.3E-03	A
Benzyl chloride	7.0E-04	D
Bis(2-ethylhexyl)phthalate (DEHP)	7.3E-05	D
Bromoform	3.9E-05	E
Carbon disulfide	1.3E-04	D
2-Chloroacetophenone	7.0E-06	E
Chlorobenzene	2.2E-05	D
Chloroform	5.9E-05	D
Cumene	5.3E-06	E
Cyanide	2.5E-03	D
2,4-Dinitrotoluene	2.8E-07	D
Dimethyl sulfate	4.8E-05	E
Ethyl benzene	9.4E-05	D
Ethyl chloride	4.2E-05	D
Ethylene dichloride	4.0E-05	E
Ethylene dibromide	1.2E-06	E
Formaldehyde	2.4E-04	A
Hexane	6.7E-05	D
Isophorone	5.8E-04	D
Methyl bromide	1.6E-04	D
Methyl chloride	5.3E-04	D
Methyl ethyl ketone	3.9E-04	D
Methyl hydrazine	1.7E-04	E
Methyl methacrylate	2.0E-05	E
Methyl tert-butyl ether	3.5E-05	E
Methylene chloride	2.9E-04	D
Phenol	1.6E-05	D
Propionaldehyde	3.8E-04	D
Tetrachloroethylene	4.3E-05	D
Toluene	2.4E-04	A
1,1,1-Trichloroethane	2.0E-05	E
Styrene	2.5E-05	D
Xylenes	3.7E-05	C
Vinyl acetate	7.6E-06	E

[a]Source: AP-42. Tons are short tons. Factors were developed from emissions data from ten sites firing bituminous coal, eight sites firing subbituminous coal, and from one site firing lignite. The emission factors are applicable to boilers using both wet limestone scrubbers or spray dryers and an electrostatic precipitator (ESP) or fabric filter (FF). In addition, the factors apply to boilers utilizing only an ESP or FF.
[b]Pollutants sampled for but not detected in any sampling run include: carbon tetrachloride, 2 sites; 1,3-dichloropropyene, 2 sites; N-nitrosodimethylamine, 2 sites; ethylidene dichloride, 2 sites; hexachlorobutadiene, 2 sites; hexachloroethane, 1 site; propylene dichloride, 2 sites; 1,1,2,2-tetrachloro-ethane, 2 sites; 1,1,2-trichloroethane, 2 sites; vinyl chloride, 2 sites; and hexachlorobenzene, 2 sites.
[c]Emission factor should be applied to coal feed, as fired. To convert from lb/ton to kg/Mg, multiply by 0.5.

Table B.24 Emission factors for polynuclear aromatic hydrocarbons (PAH) from controlled coal combustion[a]

Pollutant	Emission factor[b] (lb/ton)	Emission factor rating
Biphenyl	1.7E-06	D
Acenaphthene	5.1E-07	B
Acenaphthylene	2.5E-07	B
Anthracene	2.1E-07	B
Benzo(a)anthracene	8.0E-08	B
Benzo(a)pyrene	3.8E-08	D
Benzo(b,j,k)fluoranthene	1.1E-07	B
Benzo(g,h,i)perylene	2.7E-08	D
Chrysene	1.0E-07	C
Fluoranthene	7.1E-07	B
Fluorene	9.1E-07	B
Indeno(1,2,3-cd)pyrene	6.1E-08	C
Naphthalene	1.3E-05	C
Phenanthrene	2.7E-06	B
Pyrene	3.3E-07	B
5-Methyl chrysene	2.2E-08	D

[a]Source: AP-42. Tons are short tons. Factors were developed from emissions data from six sites firing bituminous coal, four sites firing subbituminous coal, and from one site firing lignite. Factors apply to boilers utilizing both wet limestone scrubbers or spray dryers and an electrostatic precipitator (ESP) or fabric filter (FF). The factors apply to boilers utilizing only an ESP or FF.
[b]Emission factor should be applied to coal feed, as fired. To convert from lb/ton to lb/MM Btu, multiply by 0.0625. To convert from lb/ton to kg/Mg, multiply by 0.5. Emissions are lb of pollutant per ton of coal combusted.

Table B.25 Emission factors for hydrogen chloride (HCl) and hydrogen fluoride (HF) from coal combustion (emission factor rating: B)[a]

Firing configuration	HCl emission factor (lb/ton)	HF emission factor (lb/ton)
PC-fired	1.2	0.15
PC-fired, tangential	1.2	0.15
Cyclone furnace	1.2	0.15
Traveling grate (overfeed stoker)	1.2	0.15
Spreader stoker	1.2	0.15
FBC, Circulating bed	1.2	0.15

[a]Source: AP-42. Tons are short tons. The emission factors were developed from bituminous coal, subbituminous coal, and lignite emissions data. To convert from lb/ton to kg/Mg, multiply by 0.5. To convert from lb/ton to lb/MM Btu, multiply by 0.0625. The factors apply to both controlled and uncontrolled sources.

Table B.26 Emission factors for trace metals from controlled coal combustion[a]

Pollutant	Emission factor[b] (lb/ton)	Emission factor rating
Antimony	1.8E-05	A
Arsenic	4.1E-04	A
Beryllium	2.1E-05	A
Cadmium	5.1E-05	A
Chromium	2.6E-04	A
Chromium (VI)	7.9E-05	D
Cobalt	1.0E-04	A
Lead	4.2E-04	A
Magnesium	1.1E-02	A
Manganese	4.9E-04	A
Mercury	8.3E-05	A
Nickel	2.8E-04	A
Selenium	1.3E-03	A

[a]Source: AP-42. Tons are short tons. The emission factors were developed from emissions data at eleven facilities firing bituminous coal, fifteen facilities firing subbituminous coal, and from two facilities firing lignite. The factors apply to boilers utilizing either venturi scrubbers, spray dryer absorbers, or wet limestone scrubbers with an electrostatic precipitator (ESP) or fabric filter (FF). In addition, the factors apply to boilers using only an ESP, FF, or venturi scrubber. Firing configurations include pulverized coal-fired, dry bottom boilers; pulverized coal, dry bottom, tangentially fired boilers; cyclone boilers; and atmospheric fluidized bed combustors, circulating bed.
[b]Emission factor should be applied to coal feed, as fired. To convert from lb/ton to kg/Mg, multiply by 0.5.

Table B.27 Emission factor equations for trace elements from coal combustion[a] (emission factor equation rating: A)[b]

Pollutant	Emission equation ($lb/10^{12}Btu$)[c]
Antimony	$0.92\ [(C/A)PM]^{0.63}$
Arsenic	$3.1\ [(C/A)PM]^{0.85}$
Beryllium	$1.2\ [(C/A)PM]^{1.1}$
Cadmium	$3.3\ [(C/A)PM^{0.5}$
Chromium	$3.7\ [(C/A)PM]^{0.58}$
Cobalt	$1.7\ [(C/A)PM]^{0.69}$
Lead	$3.4\ [(C/A)PM]^{0.80}$
Manganese	$3.8\ [(C/A)PM]^{0.60}$
Nickel	$4.4\ [(C/A)PM]^{0.48}$

[a]Source: AP-42. The equations were developed from emissions data from bituminous coal combustion, subbituminous coal combustion, and from lignite combustion. The equations may be used to generate factors for both controlled and uncontrolled boilers. The emission factor equations are applicable to all typical firing configurations for electric generation (utility), industrial, and commercial/industrial boilers for bituminous coal, subbituminous coal, and lignite.
[b]AP-42 criteria for rating emission factors were used to rate the equations.
[c]The factors produced by the equations should be applied to heat input. To convert from $lb/10^{12}$ Btu to kg/joules, multiply by 4.31×10^{-16}. C = concentration of metal in the coal, parts per million by weight (ppmwt); A = weight fraction of ash in the coal. For example, 10% ash is 0.1 ash fraction; PM = site-specific emission factor for total particulate matter, $lb/10^6$ Btu.

Table B.28 Emission factors for SO_x and NO_x compounds from uncontrolled anthracite coal combustors[a]

Source category	SO_x		NO_x	
	Emission factor (lb/ton)	Emission factor rating	Emission factor (lb/ton)	Emission factor rating
Stoker-fired boilers	39S[b]	B	9.0	C
FBC boilers[c]	2.9	E	1.8	E
Pulverized coal boilers	39S[b]	B	18	B

[a]Source: AP-42. Tons are short tons. Units are lb of pollutant/ton of coal burned. To convert from lb/ton to kg/Mg, multiply by 0.5.
[b]S = weight percent sulfur. For example, if the sulfur content is 3.4%, then S = 3.4.
[c]FBC boilers burning culm fuel; all other sources burning anthracite coal.

Table B.29 Emission factors for CO and carbon dioxide (CO_2) from uncontrolled anthracite coal combustors[a]

Source category	CO		CO_2	
	Emission factor (lb/ton)	Emission factor rating	Emission factor (lb/ton)	Emission factor rating
Stoker-fired boilers	0.6	B	5,680	C
FBC boilers[b]	0.6	E	ND	NA

[a]Source: AP-42. Tons are short tons. Units are lb of pollutant/ton of coal burned. To convert from lb/ton to kg/Mg, multiply by 0.5. ND = no data. NA = not applicable.
[b]FBC boilers burning culm fuel; all other sources burning anthracite coal.

Table B.30 Emission factors for speciated organic compounds from anthracite coal combustors (emission factor rating: E)[a]

Pollutant	Stoker-fired boilers emission factor (lb/ton)
Acenaphthene	ND
Acenaphthylene	ND
Anthrene	ND
Anthracene	ND
Benzo(a)anthracene	ND
Benzo(a)pyrene	ND
Benzo(e)pyrene	ND
Benzo(g,h,i)perylene	ND
Benzo(k)fluoranthrene	ND
Biphenyl	2.5 E-02
Chrysene	ND
Coronene	ND
Fluoranthrene	ND
Fluorene	ND
Indeno(123-cd)perylene	ND
Naphthalene	1.3 E-01
Perylene	ND
Phenanthrene	6.8 E-03
Pyrene	ND

[a]Source: AP-42. Tons are short tons. Units are lb of pollutant/ton of coal burned. To convert from lb/ton to kg/Mg, multiply by 0.5. ND = no data.

Table B.31 Emission factors for TOC and methane (CH_4) from anthracite coal combustors (emission factor rating: E)[a]

Source category	TOC emission factor (lb/ton)	CH_4 emission factor (lb/ton)
Stoker fired boilers	0.30	ND

[a]Source: AP-42. Tons are short tons. Units are lb of pollutant/ton of coal burned. To convert from lb/ton to kg/Mg, multiply by 0.5. ND = no data.

Table B.32 Emission factors for speciated metals from anthracite coal combustion in stoker fired boilers (emission factor rating: E)[a]

Pollutant	Emission factor range (lb/ton)	Average emission factor (lb/ton)
Arsenic	BDL – 2.4 E-04	1.9 E-04
Antimony	BDL	BDL
Beryllium	3.0 E-05 – 5.4 E-04	3.1 E-04
Cadmium	4.5 E-05 – 1.1 E-04	7.1 E-05
Chromium	5.9 E-03 – 4.9 E-02	2.8 E-02
Manganese	9.8 E-04 – 5.3 E-03	3.6 E-03
Mercury	8.7 E-05 – 1.7 E-04	1.3 E-04
Nickel	7.8 E-03 – 3.5 E-02	2.6 E-02
Selenium	4.7 E-04 – 2.1 E-03	1.3 E-03

[a]Source: AP-42. Tons are short tons. Units are lb of pollutant/ton of coal burned. To convert from lb/ton to kg/Mg, multiply by 0.5. BDL = below detection limit.

Table B.33 Emission factors for PM and lead (Pb) from uncontrolled anthracite coal combustors[a]

Source category	Filterable PM		Condensable PM		Pb	
	Emission factor (lb/ton)	Emission factor rating	Emission factor (lb/ton)	Emission factor rating	Emission factor (lb/ton)	Emission factor rating
Stoker-fired boilers	0.8A[b]	C	0.08A[b]	C	8.9E-03	E
Hand-fired units	10	B	ND	NA	ND	NA

[a]Source: AP-42. Tons are short tons. Units are lb of pollutant/ton of coal burned. To convert from lb/ton to kg/Mg, multiply by 0.5. ND = no data. NA = not applicable.
[b]A = ash content of fuel, weight %. For example, if the ash content is 5%, then A = 5.

Table B.34 Cumulative particle size distribution and size-specific emission factors for dry bottom boilers burning pulverized anthracite coal (emission factor rating: D)[a]

Particle size[b] (μm)	Cumulative mass % ≤ stated size			Cumulative emission factor as fired[c] (lb/ton)		
	Uncontrolled	Controlled[d]		Uncontrolled	Controlled[d]	
		Multiple cyclone	Baghouse		Multiple cyclone	Baghouse
15	32	63	79	3.2A[e]	1.26A	0.016A
10	23	55	67	2.3A	1.10A	0.013A
6	17	46	51	1.7A	0.92A	0.010A
2.5	6	24	32	0.6A	0.48A	0.006A
1.25	2	13	21	0.2A	0.26A	0.004A
1.00	2	10	18	0.2A	0.20A	0.004A
0.625	1	7	—[f]	0.1A	0.14A	—[f]
TOTAL	100	100	100	10A	2A	0.02A

[a]Source: AP-42. Tons are short tons.
[b]Expressed as aerodynamic equivalent diameter.
[c]Units are lb of pollutant/ton of coal burned. To convert from lb/ton to kg/Mg, multiply by 0.5.
[d]Estimated control efficiency for multiple cyclone is 80%; for baghouse, 99.8%.
[e]A = coal ash weight %, as fired. For example, if ash content is 5%, then A = 5.
[f]Insufficient data.

B.2 Fuel oil emission factors

Table B.35 **Criteria pollutant emission factors for fuel oil combustion**[a]

Firing configuration	SO_2[b]		SO_3[b]		NO_x[c]		CO[d]		Filterable PM[e]	
	Emission factor ($lb/10^3$ gal)	Emission factor rating	Emission factor ($lb/10^3$ gal)	Emission factor rating	Emission factor ($lb/10^3$ gal)	Emission factor rating	Emission factor ($lb/10^3$ gal)	Emission factor rating	Emission factor ($lb/10^3$ gal)	Emission factor rating
Boilers >100 MMBtu/h										
No.6 FO, normal firing	157S	A	5.7S	C	47	A	5	A	9.19(S) + 3.22	A
No.6 FO, normal firing, low NO_x burner	157S	A	5.7S	C	40	B	5	A	9.19(S) + 3.22	A
No.6 FO, tangential firing	157S	A	5.7S	C	32	A	5	A	9.19(S) + 3.22	A
No.6 FO, tangential firing, low NO_x burner	157S	A	5.7S	C	26	E	5	A	9.19(S) + 3.22	A
No.5 FO, normal firing	157S	A	5.7S	C	47	B	5	A	10	B
No.5 FO, tangential firing	157S	A	5.7S	C	32	B	5	A	10	B
No.4 FO, normal firing	150S	A	5.7S	C	47	B	5	A	7	B
No.4 FO, tangential firing	150S	A	5.7S	C	32	B	5	A	7	B
No.2 FO fired	142S	A	5.7S	C	24	D	5	A	2	A
No.2 FO fired, LNB/ FGR	142S	A	5.7S	A	10	D	5	A	2	A

(Continued)

Table B.35 (Continued)

Firing configuration	SO_2 [b]		SO_3 [b]		NO_x [c]		CO [d]		Filterable PM [e]	
	Emission factor (lb/10^3 gal)	Emission factor rating	Emission factor (lb/10^3 gal)	Emission factor rating	Emission factor (lb/10^3 gal)	Emission factor rating	Emission factor (lb/10^3 gal)	Emission factor rating	Emission factor (lb/10^3 gal)	Emission factor rating
Boilers <100 MMBtu/h										
No.6 FO fired	157S	A	2S	A	55	A	5	A	9.19(S) + 3.22	B
No.5 FO fired	157S	A	2S	A	55	A	5	A	10	A
No.4 FO fired	150S	A	2S	A	20	A	5	A	7	B
Distillate oil fired	142S	A	2S	A	20	A	5	A	2	A
Residential furnace	142S	A	2S	A	18	A	5	A	0.4	B

[a]Source: AP-42. To convert from lb/10^3 gal to kb/10^3 L, multiply by 0.120.
[b]S indicates that the weight % of sulfur in the oil should be multiplied by the value given. For example, if the fuel is 1% sulfur, then S = 1.
[c]Expressed as NO_2. Test results indicate that at least 95% by weight of NO_x is NO for all boiler types except residential furnaces, where about 75% is NO. For utility vertical fired boilers use 105 lb/10^3 gal at full load and normal (>15%) excess air. Nitrogen oxides emissions from residual oil combustion is the weight % nitrogen in the oil. For example, if the fuel is 1% nitrogen, then N = 1.
[d]CO emissions may increase by a factor of 10 to 100 if the unit is improperly operated or not well maintained.
[e]Filterable PM is the particulate collected on or prior to the filter of an EPA method 5 (or equivalent) sampling train. Particulate emission factors for residual oil combustion are, on average, a function of fuel oil sulfur content where S is the weight % sulfur in oil. For example, if the fuel is 1% sulfur, then S = 1.

Table B.36 Condensable particulate matter emission factors for oil combustion[a]

Firing configuration[b]	Controls	CPM-TOT[c,d]		CPM-IOR[c,d]		CPM-ORG[c,d]	
		Emission factor ($lb/10^3$ gal)	Emission factor rating	Emission factor ($lb/10^3$ gal)	Emission factor rating	Emission factor ($lb/10^3$ gal)	Emission factor rating
No. 2 oil fired	All controls, or uncontrolled	1.3	D	65% of CPM-TOT emission factor[c]	D	35% of CMP-TOT emission factor[c]	D
No. 6 oil fired	All controls, or uncontrolled	1.5	D	85% of CPM-TOT emission factor[d]	E	15% CPM-TOT emission factor[d]	E

[a]Source: AP-42. All condensable PM is assumed to be less than 1.0 μm in diameter.
[b]No data are available for Nos. 4 and 5 oil. For Nos. 4 and 5 oil, use the factors provided for No. 6 oil.
[c]CPM-TOT, total condensable particulate matter; CPM-IOR, inorganic condensable particulate matter; CPM-ORG, organic condensable particulate matter; ND, no data.
[d]To convert to lb/MM Btu of No. 2 oil, divide by 140 $MMBtu/10^3$ gal. To convert to lb/MM Btu of No. 6 oil, divide by 150 MM $Btu/10^3$ gal.

Table B.37 Emission factors for total organic compounds (TOC), methane, and non-methane TOC (NMTOC) from uncontrolled fuel oil combustion (emission factor rating: A)[a]

Firing configuration	TOC[b] emission factor (lb/10^3 gal)	Methane[b] emission factor (lb/10^3 gal)	NMTOC[b] emission factor (lb/10^3 gal)
Utility Boilers			
No. 6 oil fired, normal firing	1.04	0.28	0.76
No. 6 oil fired, tangential firing	1.04	0.28	0.76
No. 5 oil fired, normal firing	1.04	0.28	0.76
No. 5 oil fired, tangential firing	1.04	0.28	0.76
No. 4 oil fired, normal firing	1.04	0.28	0.76
No. 4 oil fired, tangential firing	1.04	0.28	0.76
Industrial Boilers			
No. 6 oil fired	1.28	1.00	0.28
No. 5 oil fired	1.28	1.00	0.28
Distillate oil fired	0.252	0.052	0.2
No. 4 oil fired	0.252	0.052	0.2
Commercial/Institutional/Residential Combustors			
No. 6 oil fired	1.605	0.475	1.13
No. 5 oil fired	1.605	0.475	1.13
Distillate oil fired	0.556	0.216	0.34
No. 4 oil fired	0.556	0.216	0.34
Residential Furnace	2.493	1.78	0.713

[a]From AP-42.
[b]To convert from lb/10^3 gal to kb/10^3L, multiply by 0.12.

Table B.38 Cumulative particle size distribution and size-specific emission factors for utility boilers firing residual oil[a]

Particle size[b] (μm)	Cumulative mass % stated size			Cumulative emission factor, lb/10^3 gal					
	Uncontrolled	Controlled		Uncontrolled[c]		ESP controlled[d]		Scrubber controlled[e]	
		ESP	Scrubber	Emission factor	Emission factor rating	Emission factor	Emission factor rating	Emission factor	Emission factor rating
15	80	75	100	6.7A	C	0.05A	E	0.50A	D
10	71	63	100	5.9A	C	0.042A	E	0.50A	D
6	58	52	100	4.8A	C	0.035A	E	0.50A	D
2.5	52	41	97	4.3A	C	0.028A	E	0.48A	D
1.25	43	31	91	3.6A	C	0.021A	E	0.46A	D
1.00	39	25	84	3.3A	C	0.018A	E	0.42A	D
0.625	20	20	64	1.7A	C	0.007A	E	0.32A	D
Total	100	100	100	8.3A	C	0.067A	E	0.50A	D

[a]Source: AP-42. To convert from lb/10^3 gal to kg/m^3, multiply by 0.120. ESP = electrostatic precipitator.
[b]Expressed as aerodynamic equivalent diameter.
[c]Particulate emission factors for residual oil combustion without emission controls are, on average, a function of fuel oil grade and sulfur content where S is the weight % of sulfur in the fuel oil. For example, if the fuel is 1.00% sulfur, then S = 1.
No. 6 oil: A = 1.12(S) + 0.37.
No. 5 oil A = 1.2.
No. 4 oil: A = 0.84.
[d]Estimated control efficiency for ESP is 99.2%.
[e]Estimated control efficiency for scrubber is 94%.

Table B.39 Cumulative particle size distribution and size-specific factors for industrial boilers firing residual oil[a]

Particle size[b] (μm)	Cumulative mass %, stated size		Cumulative emission factor[c] ($lb/10^3$ gal)			
	Uncontrolled	Multiple cyclone controlled	Uncontrolled		Multiple cyclone controlled[d]	
			Emission factor	Emission factor rating	Emission factor	Emission factor rating
15	91	100	7.59A	D	1.67A	E
10	86	95	7.17A	D	1.58A	E
6	77	72	6.42A	D	1.17A	E
2.5	56	22	4.67A	D	0.33A	E
1.25	39	21	3.25A	D	0.33A	E
1.00	36	21	3.00A	D	0.33A	E
0.625	30	—[e]	2.50A	D	—[e]	NA
Total	100	100	8.34A	D	1.67A	E

[a]Source: AP-42. To convert from $lb/10^3$ gal to kg/m^3, multiply by 0.120. NA = not applicable.
[b]Expressed as aerodynamic equivalent diameter.
[c]Particulate emission factors for residual oil combustion without emission controls are, on average, a function of fuel oil grade and sulfur content where S is the weight % of sulfur in the fuel oil. For example, if the fuel is 1.00% sulfur, then S = 1.
No. 6 oil: A = 1.12(S) + 0.37.
No. 5 oil A = 1.2.
No. 4 oil: A = 0.84.
[d]Estimated control efficiency for multiple cyclone is 80%.
[e]Insufficient data.

Table B.40 Cumulative particle size distribution and size-specific emission factors for uncontrolled industrial boilers firing distillate oil (emission factor rating: E)[a]

Particle size[b] (μm)	Cumulative mass %, stated size	Cumulative emission factor (lb/10^3 gal)
15	68	1.33
10	50	1.00
6	30	0.58
2.5	12	0.25
1.25	9	0.17
1.00	8	0.17
0.625	2	0.04
Total	100	2.00

[a]Source: AP-42. To convert from lb/10^3 gal to kg/10^3 L, multiply by 0.12.
[b]Expressed as aerodynamic diameter.

Table B.41 Cumulative particle size distribution and size-specific emission factors uncontrolled commercial boilers burning residual or distillate oil (emission factor rating: D)[a]

Particle size[b] (μm)	Cumulative mass %, stated size		Cumulative emission factor[c] (lb/10^3 gal)	
	Residual oil	Distillate oil	Residual oil	Distillate oil
15	78	60	6.50A	1.17
10	62	55	5.17A	1.08
6	44	49	3.67A	1.00
2.5	23	42	1.92A	0.83
1.25	16	38	1.33A	0.75
1.00	14	37	1.17A	0.75
0.625	13	35	1.08A	0.67
Total	100	100	8.34A	2.00

[a]Source: AP-41. To convert from lb/10^3 gal to kg/10^3 L, multiply by 0.12.
[b]Expressed as aerodynamic diameter.
[c]Particulate emission factors for residual oil combustion without emission controls are, on average, a function of fuel oil grade and sulfur content where S is the weight % of sulfur in the fuel. For example, if the fuel is 1.0% sulfur, then S = 1.
No. 6 oil: A = 1.12(S) + 0.37.
No. 5 oil: A = 1.2.
No. 4 oil: A = 0.84.
No. 2 oil: A = 0.24.

Table B.42 Emission factors for nitrous oxide (N_2O), polycyclic organic matter (POM), and formaldehyde (HCOH) from fuel oil combustion (emission factor rating: E)[a]

Firing configuration	Emission factor (lb/10^3 gal)		
	N_2O[b]	POM	HCOH
Utility/industrial/commercial boilers	0.53	0.0011–0.0013[c]	0.024–0.061
No. 6 oil fired	0.26	0.033[d]	0.035–0.061
Distillate oil fired	0.05	ND	ND

[a]Source: AP-41. To convert from lb/10^3 gal to kg/10^3 L, multiply by 0.12. ND = no data.
[b]Emission factor rating = B.
[c]Particulate and gaseous POM.
[d]Particulate POM only.

Table B.43 Emission factors for speciated organic compounds from fuel oil combustion[a]

Organic compound	Average emission factor[b] (lb/10^3 gal)	Emission factor rating
Benzene	2.14E-04	C
Ethylbenzene	6.36E-05[c]	E
Formaldehyde[d]	3.30E-02	C
Naphthalene	1.13E-03	C
1,1,1-Trichlorethane	2.36E-04[d]	E
Toluene	6.20E-03	D
o-Xylene	1.09E-04	E
Acenaphthene	2.11E-05	C
Acenaphthylene	2.53E-07	D
Anthracene	1.22E-06	C
Benz(a)anthracene	4.01E-06	C
Benzo(b,k)fluoranthrene	1.48E-06	C
Benzo(g,h,i)perylene	2.26E-06	C
Chrysene	2.38E-06	C
Dibenzo(a,h)anthracene	1.67E-06	D
Fluoranthene	4.84E-06	C
Fluorene	4.47E-06	C
Indo(1,2,3-cd)pyrene	2.14E-06	C
Phananthrene	1.05E-05	C
Pyrene	4.25E-06	C
OCDD	3.10E-09	E

[a]Source: AP-41. Data are for residual oil fired boilers.
[b]To convert from lb/10^3 gal to kg/10^3 L, multiply by 0.12.
[c]Based on data from one source test.
[d]The formaldehyde number is based on data from utilities using No. 6 oil.

Table B.44 Emission factors for trace elements from distillate fuel oil combustion sources[a]

Firing configuration	Emission factor (lb/10^{12} Btu)										
	As	Be	Cd	r	Cu	Pb	Hg	Mn	Ni	Se	Zn
Distillate oil	4	3	3	3	6	9	3	6	3	15	4

[a]Source: AP-41. Data are for distillate oil fired boilers. To convert from lb/10^{12} But to pg/J, multiply by 0.43.

Table B.45 Emission factors for metals from uncontrolled No. 6 fuel oil combustion[a]

Metal	Average emission factor[b] (lb/10^3 gal)	Emission factor rating
Antimony	5.25E-03	E
Arsenic	1.32E-03	C
Barium	2.57E-03	D
Beryllium	2.78E-05	C
Cadmium	3.98E-04	C
Chloride	3.47E-01	D
Chromium	8.45E-04	C
Chromium IV	2.48E-04	C
Cobalt	6.02E-03	D
Copper	1.76E-03	C
Fluoride	3.73E-02	D
Lead	1.51E-03	C
Manganese	3.00E-03	C
Mercury	1.13E-04	C
Molybdenum	7.87E-04	D
Nickel	8.45E-02	C
Phosphorus	9.46E-03	D
Selenium	6.83E-04	C
Vanadium	3.18E-02	D
Zinc	2.91E-02	D

[a]Source: AP-41. Data are for residual oil-fired boilers.
[b]18 of 19 sources were uncontrolled and 1 source was controlled with low efficiency ESP. To convert from lb/10^3 gal to kg/10^3 L, multiply by 0.12.

Table B.46 Default CO_2 emission factors for liquid fuels (emission factor rating: B)[a]

Fuel type	% Carbon	Density (lb/gal)	Emission factor (lb/10^3 gal)
No. 1 (kerosene)	86.25	6.88	21,500
No. 2	87.25	7.05	22,300
Low sulfur No. 6	87.26	7.88	25,000
High sulfur No. 6	85.14	7.88	24,400

[a]Source: AP-42. Based on 99% conversion of fuel carbon to CO_2. To convert from lb/gal to gram/cm^3, multiply by 0.12.

B.3 Natural gas combustion emission factors

Table B.47 Emission factors for nitrogen oxides (NO_x) and carbon monoxide (CO) from natural gas combustion[a]

Combustor type (heat input)	NO_x[b]		CO	
	Emission factor ($lb/10^6$ scf)	Emission factor rating	Emission factor ($lb/10^6$ scf)	Emission factor rating
Large Wall-Fired Boilers (>100 MM Btu/h)				
Uncontrolled (Pre-NSPS)[c]	280	A	84	B
Uncontrolled (Post-NSPS)[c]	190	A	84	B
Controlled – Low-NO_x burners	140	A	84	B
Controlled – Flue gas recirculation (FGR)	100	D	84	B
Small Boilers (<100 MM Btu/h)				
Uncontrolled	100	B	84	B
Controlled – Low-NO_x burners	50	D	84	B
Controlled – Low-NO_x burners/FGR	32	C	84	B
Tangential-Fired Boilers				
Uncontrolled	170	A	24	C
Controlled – Flue gas recirculation	76	D	98	C
Residential Furnaces				
Uncontrolled	94	B	40	B

[a]Source: AP-42. Units are in pounds per million standard cubic feet of natural gas fired. To convert from $lb/10^6$ scf to $kg/10^6m^3$, multiply by 16. Emission factors are based on an average natural gas higher heating value of 1,020 Btu/scf. To convert from $lb/10^6$ scf to lb/MM Btu, divide by 1,020.

[b]Expressed as NO_2. For large and small wall-fired boilers with selective noncatalytic reduction (SNCR) control, apply a 24% reduction to the appropriate NO_x emission factor. For tangential-fired boilers with SNCR control, apply a 13% reduction to the appropriate NO_x emission factor.

[c]NSPS = New Source Performance Standard as defined in 40 CFR 60 Subparts D and Db. Post-NSPS units are boilers with greater than 250 MM Btu/h of heat input that commenced construction, modification, or reconstruction after August 17, 1971, and units with heat input capacities between 100 and 250 MM Btu/h that commenced construction, modification, or reconstruction after June 19, 1984.

Table B.48 Emission factors criteria pollutants and greenhouse gases from natural gas combustion[a]

Pollutant	Emission factor ($lb/10^6$ scf)	Emission factor rating
CO_2[b]	120,000	A
Lead	0.0005	D
N_2O (uncontrolled)	2.2	E
N_2O (Controlled – Low-NO_x burner)	0.64	E
PM (Total)[c]	7.6	D
PM (Condensable)[c]	5.7	D
PM (Filterable)[c]	1.9	B
SO_2[d]	0.6	A
TOC	11	B
Methane	2.3	B
VOC	5.5	C

[a]Source: AP-42. Units are in pounds of pollutant per million standard cubic feet of natural gas fired. Data are for all natural gas combustion sources. To convert from $lb/10^6$ scf to $kg/10^6 m^3$, multiply by 16. To convert from $lb/10^6$ scf to lb/MM Btu, divide by 1,020. TOC = Total Organic Compounds. VOC = Volatile Organic Compounds.
[b]Based on approximately 100% conversion of fuel carbon to CO_2. CO_2 in $lb/10^6$ scf = (3.67)(CON)(C)(D), where CON = fractional conversion of fuel carbon to CO_2, C = carbon content of fuel by weight (0.76), and D = density of fuel, 4.2×10^4 $lb/10^6$ scf.
[c]All PM (total, condensable, and filterable) is assumed to be less than 1.0 μm in diameter. Therefore, the PM emission factors may be used to estimate PM_{10}, $PM_{2.5}$ or PM_1 emissions. Total PM is the sum of the filterable PM and condensable PM. Condensable PM is the particulate matter collected using EPA Method 202 (or equivalent) sampling train.
[d]Based on 100% conversion of fuel sulfur to SO_2. Assumes sulfur content in natural gas of 2,000 grains/10^6 scf. The SO_2 emission factor can be converted to other natural gas sulfur contents by multiplying the SO_2 emission factor by the ratio of the site-specific sulfur content (grains/10^6 scf) to 2,000 grains/10^6 scf.

Table B.49 Emission factors for speciated organic compounds from natural gas combustion[a]

Pollutant	Emission factor ($lb/10^6$ scf)	Emission factor rating
2-Methylnaphthalene[b,c]	2.4E-05	D
3-Methylchloranthrene[b,c]	<1.8E-06	E
7,12-Dimethylbenz(a)anthracene[b,c]	<1.6E-05	E
Acenaphthene[b,c]	<1.8E-06	E
Acenaphthylene[b,c]	<1.8E-06	E
Anthracene[b,c]	<2.4E-06	E
Benz(a)anthracene[b,c]	<1.8E-06	E
Benzene[b]	2.1E-03	B
Bemzo(a)pyrene[b,c]	<1.2E-06	E
Benzo(b)fluoranthene[b,c]	<1.8E-06	E
Benzo(g,h,i)perylene[b,c]	<1.2E-06	E
Benzo(k)fluoranthene[b,c]	<1.8E-06	E
Butane	2.1	E
Chrysene[b,c]	<1.8E-06	E
Dibenzo(a,h)anthracene[b,c]	<1.2E-06	E
Dichlorbenzene	1.2E-03	E
Ethane	3.1	E
Fluoranthene[b,c]	3.0E-06	E
Fluorene[b,c]	2.8E-06	E
Formaldehyde[b]	7.5E-02	B
Hexane[b]	1.8	E
Indeno(1,2,3-cd)pyrene[b,c]	<1.8E-06	E
Naphthalene[b]	6.1E-04	E
Pentane	2.6	E
Phenanthrene[b,c]	1.75E-05	D
Propane	1.6	E
Pyrene[b,c]	5.0E-06	E
Toluene[b]	3.4E-03	C

[a]Source: AP-42. Units are in pounds of pollutant per million standard cubic feet of natural gas fired. Data are for all natural gas combustion sources. To convert from $lb/10^6$ scf to $kg/10^6 m^3$, multiply by 16. To convert from $lb/10^6$ scf to lb/MM Btu, divide by 1,020. Emission factors preceded with a less than symbol are based on method detection limits.

[b]Hazardous Air Pollutant (HAP) as defined by Section 112(b) of the Clean Air Act.

[c]HAP because it is Polycyclic Organic Matter (POM). POM is a HAP as defined by Section 112(b) of the Clean Air Act.

Table B.50 Emission factors for metals from natural gas combustion[a]

Pollutant	Emission factor (lb/ 10^6 scf)	Emission factor rating
Arsenic[b]	2.0E-04	E
Barium	4.4E-03	D
Beryllium[b]	<1.2E-05	E
Cadmium[b]	1.1E-03	D
Chromium[b]	1.4E-03	D
Cobalt[b]	8.4E-05	D
Copper	8.5E-04	C
Manganese[b]	3.8E-04	D
Mercury[b]	2.6E-04	D
Molybdenum	1.1E-03	D
Nickel[b]	2.1E-03	C
Selenium[b]	<2.4E-05	E
Vanadium	2.3E-03	D
Zinc	2.9E-02	E

[a]Source: AP-42. Units are in pounds of pollutant per million standard cubic feet of natural gas fired. Data are for all natural gas combustion sources. To convert from lb/ 10^6 scf to kg/$10^6 m^3$, multiply by 16. To convert from lb/10^6 scf to lb/MM Btu, divide by 1,020. Emission factors preceded with a less than symbol are based on method detection limits.

[b]Hazardous Air Pollutant (HAP) as defined by Section 112(b) of the Clean Air Act.

B.4 Liquefied petroleum gas combustion emission factors

Table B.51 Emission factors for LPG combustion (emission factor rating: E)[a]

Pollutant	Butane emission factor ($lb/10^3$ gal)		Propane emission factor ($lb/10^3$ gal)	
	Industrial boilers[b]	Commercial boilers[c]	Industrial boilers[b]	Commercial boilers[c]
PM, Filterable[d]	0.2	0.2	0.2	0.2
PM, Condensable	0.6	0.6	0.5	0.5
PM, Total	0.8	0.8	0.7	0.7
SO_2[e]	0.09S	0.09S	0.10S	0.10S
NO_x[f]	15	15	13	13
N_2O	0.9	0.9	0.9	0.9
CO_2[g,h]	14,300	14,300	12,500	12,500
CO	8.4	8.4	7.8	7.8
TOC	1.1	1.1	1.0	1.0
CH_4	0.2	0.2	0.2	0.2

[a]Source: AP-42. Assumes PM, CO, and TOC emissions are the same, on a heat input basis, as for natural gas combustion. Use heat contents of 91.5×10^6 Btu/10^3 gallons for propane, 102×10^6 Btu/10^3 gallons for butane, $1,020 \times 10^6$ Btu/10^6 scf for methane when calculating equivalent heat input basis. For example, the equation for converting from methane's emission factors to propane's emission factors is as follows: lb pollutant/10^3 gallons of propane = (lb pollutant/10^6 ft^3 methane) (91.5×10^6 Btu/10^3 gallons of propane) / ($1,020 \times 10^6$ Btu/10^6 scf of methane). The NO_x emission factors have been multiplied by a correction factor of 1.5, which is the approximate ratio of propane/butane NO_x emissions to natural gas NO_x emissions. To convert from lb/10^3 gal to kg/10^3L, multiply by 0.12.
[b]Heat input capacities generally between 10 and 100 MM Btu/h.
[c]Heat input capacities generally between 0.3 and 10 MM Btu/h.
[d]Filterable particulate matter (PM) is PM collected prior to the filter of an EPA Method 5 (or equivalent) sampling train. For natural gas, a fuel with similar combustion characteristics, all PM is less than 10 μm in aerodynamic equivalent diameter (PM_{10}).
[e]S equals the sulfur content expressed in gr/100 ft^3 of gas vapor. For example, if the butane sulfur content is 0.18 gr/100ft^3, the emission factor would be $(0.09 \times 0.18) = 0.016$ lb of SO_2/10^3 gal butane burned.
[f]Expressed as NO_2.
[g]Assuming 99.5% conversion of fuel carbon to CO_2.
[h]Emission factor rating = C.

B.5 Stationary gas turbines emission factors

Table B.52 Emission factors for nitrogen oxides (NO_x) and carbon monoxide (CO) from stationary gas turbines[a]

Turbine type	Nitrogen oxides		Carbon monoxide	
	lb/MMBtu[b] (Fuel input)	Emission factor rating	lb/MMBtu[b] (Fuel input)	Emission factor rating
Natural Gas Fired				
Uncontrolled	3.2E-01	A	8.2E-02	A
Water-steam injection	1.3E-01	A	3.0E-02	A
Lean-Premix	9.9E-02	D	1.5E-02	D
Distillate Oil Fired[c]				
Uncontrolled	8.8E-01	C	3.3E-03	C
Water-steam injection	2.4E-01	B	7.6E-02	C
Landfill Gas Fired				
Uncontrolled	1.4E-01	A	4.4E-01	A
Digestor Gas Fired				
Uncontrolled	1.6E-01	D	7.7E-02	D

[a]Source: AP-42. Factors are derived from units operating at high loads (≥80% load). The emission factors may be converted to other natural gas heating values by multiplying the given emission factor by the ratio of the specified heating value to the average heating value.

[b]Emission factors for natural gas-fired turbines are based on an average natural gas heating value (HHV) of 1,020 Btu/scf at 60°F. To convert from lb/MM Btu to lb/10^6 scf multiply by 1,020. Emission factors for distillate oil-fired turbines are based on an average distillate oil heating value of 139 MM Btu/10^3 gallons. To convert from lb/MM Btu to lb/10^3 gallons, multiply by 139. Emission factors for landfill gas-fired turbines are based on an average landfill gas heating value of 400 Btu/scf at 60°F. To convert from lb/MM Btu to lb/10^6 scf, multiply by 400. Emission factors for digestor gas-fired turbines are based on an average digestor gas heating value of 600 Btu/scf at 60°F. To convert from lb/MM Btu to lb/10^6scf, multiply by 600.

[c]It is recognized that the uncontrolled emission factor for CO is lower than the water-steam injection and lean-premix emission factors, which is contrary to expectation. EPA could not identify the reason for this except that the data sets used for developing the factors are different.

Table B.53 Emission factors for criteria pollutants and greenhouse gases from stationary gas turbines[a]

Pollutant	Natural gas-fired turbines		Distillate oil-fired turbines	
	lb/MMBtu[b] (Fuel input)	Emission factor rating	lb/MMBtu[c] (Fuel input)	Emission factor rating
CO_2[d]	110	A	157	A
N2O	0.003	E	ND	NA
Lead	ND	NA	1.45E-05	C
SO_2	0.94S[e]	B	1.01S[e]	B
Methane	8.6E-03	C	ND	NA
VOC	2.1E-03	D	4.1E-04[f]	E
TOC[g]	1.1E-02	B	4.0E-03[h]	C
PM (condensables)	4.7E-03[h]	C	7.2E-03[h]	C
PM (filterable)	1.9E-03[h]	C	4.3E-03[h]	C
PM (total)	6.6E-03[h]	C	1.2E-02[h]	C

[a]Source: AP-42. Factors are derived from units operating at high loads (≥80 percent load). ND = No Data, NA = No Applicable.
[b]Emission factors for natural gas-fired turbines are based on an average natural gas heating value (HHV) of 1,020 Btu/scf at 60°F. To convert from lb/MM Btu to lb/10^6 scf multiply by 1,020. Similarly, these emission factors can be converted to other natural gas heating values.
[c]Emission factors for distillate oil-fired turbines are based on an average distillate oil heating value of 139 MM Btu/10^3 gallons. To convert from lb/MM Btu to lb/10^3 gallons, multiply by 139.
[d]Based on approximately 99.5% conversion of fuel carbon to CO_2 for natural gas and 99% conversion of fuel carbon to CO_2 for distillate oil. CO_2 (Natural Gas) in units lb/MM Btu = (0.0036 scf/Btu)(%CON)(C)(D), where % CON = weight percent conversion of fuel carbon to CO_2, C = carbon content of fuel by weight, and D = density of fuel. For natural gas, C is assumed at 75%, and D is assumed at 4.1×10^4 lb/10^6 scf. For distillate oil, CO_2 (Distillate Oil) in units lb/MM Btu = (26.4 gal/MM Btu) (%CON)(C)(D), where C is assumed 87%, and D is assumed at 6.9 lb/gallon.
[e]All sulfur in the fuel is assumed to be converted to SO_2. S = percent fuel in fuel. For example, if sulfur content in the fuel is 3.4%, then S = 3.4. If S is not available, use 3.4×10^{-3} lb/MM Btu for natural gas turbines, and 3.3×10^{-2} lb/MM Btu for distillate oil turbines.
[f]VOC emissions are assumed equal to the sum of organic emissions.
[g]Pollutant references as THC in the gathered emission tests. It is assumed as TOC because it is based on EPA Test Method 25A.
[h]Emission factors are based on combustion turbines using water-steam injection.

Table B.54 Emission factors for hazardous air pollutants from natural gas-fired stationary gas turbines[a]

Pollutant	Emission factor (lb/MM Btu)[b]	Emission factor rating
1,3-Butadiene[c]	<4.3E-07	D
Acetaldehyde	4.0E-05	C
Acrolein	6.4E-06	C
Benzene[d]	1.2E-05	A
Ethylbenzene	3.2E-05	C
Formaldehyde[e]	7.1E-04	A
Naphthalene	1.3E-06	C
PAH	2.2E-06	C
Propylene Oxide[c]	<2.9E-05	D
Toluene	1.3E-04	C
Xylenes	6.4E-05	C

[a]Source: AP-42. Factors are derived from units operating at high loads (≥80% load).
[b]Emission factors for natural gas-fired turbines are based on an average natural gas heating value (HHV) of 1,020 Btu/scf at 60°F. To convert from lb/MM Btu to lb/10^6 scf multiply by 1,020. These emission factors can be converted to other natural gas heating values by multiplying the given emission factor by the ratio of the specified heating value to this heating value.
[c]Compound was not detected. The presented emission value is based on one-half of the detection limit.
[d]Benzene with SCONOX catalyst is 9.1×10^{-7}, rating of D.
[e]Formaldehyde with SCONOX catalyst is 2.0×10^{-5}, rating of D.

Table B.55 Emission factors for hazardous air pollutants from distillate-fired stationary gas turbines[a]

Pollutant	Emission factor (lb/MM Btu) [b]	Emission factor rating
1,3-Butadiene[c]	<1.6E-05	D
Benzene	5.5E-05	C
Formaldehyde	2.8E-04	B
Naphthalene	3.5E-05	C
PAH	4.0E-05	C

[a]Source: AP-42. Factors are derived from units operating at high loads (≥80% load).
[b]Emission factors based on an average distillate oil heating value (HHV) of 139 MM Btu/10^3 gallons. To convert from lb/MM Btu to lb/10^3 gallons, multiply by 139.
[c]Compound was not detected. The presented emission value is based on one-half of the detection limit.

Table B.56 Emission factors for metallic hazardous air pollutants from distillate oil-fired stationary gas turbines[a]

Pollutant	Emission factor (lb/10^6 scf)[b]	Emission factor rating
Arsenic[c]	<1.1E-05	D
Beryllium[c]	<3.1E-07	D
Cadmium	4.8E-06	D
Chromium	1.1E-05	D
Lead	1.4E-05	D
Manganese	7.9E-04	D
Mercury	1.2E-06	D
Nickel[c]	<4.6E-06	D
Selenium[c]	<2.5E-05	D

[a]Source: AP-42. Factors are derived from units operating at high loads ($\geq$80% load).
[b]Emission factors based on an average distillate oil heating value (HHV) of 139 MM Btu/10^3 gallons. To convert from lb/MM Btu to lb/10^3 gallons, multiply by 139.
[c]Compound was not detected. The presented emission value is based on one-half of the detection limit.

Appendix C: Original list of hazardous air pollutants

CAS number	Chemical name
75070	Acetaldehyde
60355	Acetamide
75058	Acetonitrile
98862	Acetophenone
53963	2-Acetylaminofluorene
107028	Acrolein
79061	Acrylamide
79107	Acrylic acid
107131	Acrylonitrile
107051	Allyl chloride
92671	4-Aminobiphenyl
62533	Aniline
90040	o-Anisidine
1332214	Asbestos
71432	Benzene (including benzene from gasoline)
92875	Benzidine
98077	Benzotrichloride
100447	Benzyl chloride
92524	Biphenyl
117817	Bis(2-ethylhexyl)phthalate (DEHP)
542881	Bis(chloromethyl)ether
75252	Bromoform
106990	1,3-Butadiene
156627	Calcium cyanamide
105602	Caprolactam[a]
133062	Captan
63252	Carbaryl
75150	Carbon disulfide
56235	Carbon tetrachloride
463581	Carbonyl sulfide
120809	Catechol
133904	Chloramben
57749	Chlordane
7782505	Chlorine
79118	Chloroacetic acid
532274	2-Chloroacetophenone
108907	Chlorobenzene
510156	Chlorobenzilate

CAS number	Chemical name
67663	Chloroform
107302	Chloromethyl methyl ether
126998	Chloropene
1319773	Cresols/Cresylic acid (isomers and mixture)
95487	o-Cresol
108394	m-Cresol
106445	p-Cresol
98828	Cumene
94757	2,4-D, salts and esters
3547044	DDE
334883	Diazomethane
132649	Dibenzofurans
96128	1,2-Dibromo-3-chloropropane
84742	Dibutylphthalate
106467	1,4-Dichlorobenzene(p)
91941	3,3-Dichlorobenzidene
111444	Dichloroethyl ether (Bis(2-chloroethyl)ether)
542756	1,3-Dichloropropene
62737	Dichlorvos
111422	Diethanolamine
121697	N,N-Diethyl aniline (N,N-Dimethylaniline)
64675	Diethyl sulfate
119904	3,3-Dimethoxybenzidine
60117	Dimethyl aminoazobenzene
119937	3,3′-Dimethyl benzidine
79447	Dimethyl carbamoyl chloride
68122	Dimethyl formamide
57147	1,1-Dimethyl hydrazine
131113	Dimethyl phthalate
77781	Dimethyl sulfate
534521	4,6-Dinitro-o-cresol, and salts
51285	2,4-Dinitrophenol
121142	2,4-Dinitrotoluene
123911	1,4-Dioxane (1,4-Diethyleneoxide)
122667	1,2-Diphenylhydrazine
106898	Epichlorohydrin (l-Chloro-2,3-epoxypropane)
106887	1,2-Epoxybutane
140885	Ethyl acrylate
100414	Ethyl benzene
51796	Ethyl carbamate (Urethane)
75003	Ethyl chloride (Chloroethane)
106934	Ethylene dibromide (Dibromoethane)
107062	Ethylene dichloride (1,2-Dichloroethane)
107211	Ethylene glycol
151564	Ethylene imine (Aziridine)
75218	Ethylene oxide

CAS number	Chemical name
96457	Ethylene thiourea
75343	Ethylidene dichloride (1,1-Dichloroethane)
50000	Formaldehyde
76448	Heptachlor
118741	Hexachlorobenzene
87683	Hexachlorobutadiene
77474	Hexachlorocyclopentadiene
67721	Hexachloroethane
822060	Hexamethylene-1,6-diisocyanate
680319	Hexamethylphosphoramide
110543	Hexane
302012	Hydrazine
7647010	Hydrochloric acid
7664393	Hydrogen fluoride (Hydrofluoric acid)
7783064	Hydrogen sulfide[b]
123319	Hydroquinone
78591	Isophorone
58899	Lindane (all isomers)
108316	Maleic anhydride
67561	Methanol
72435	Methoxychlor
74839	Methyl bromide (Bromomethane)
74873	Methyl chloride (Chloromethane)
71556	Methyl chloroform (1,1,1-Trichloroethane)
78933	Methyl ethyl ketone (2-Butanone)[c]
60344	Methyl hydrazine
74884	Methyl iodide (Iodomethane)
108101	Methyl isobutyl ketone (Hexone)
624839	Methyl isocyanate
80626	Methyl methacrylate
1634044	Methyl tert butyl ether
101144	4,4-Methylene bis(2-chloroaniline)
75092	Methylene chloride (Dichloromethane)
101688	Methylene diphenyl diisocyanate (MDI)
101779	4,4-Methylenedianiline
91203	Naphthalene
98953	Nitrobenzene
92933	4-Nitrobiphenyl
100027	4-Nitrophenol
79469	2-Nitropropane
684935	N-Nitroso-N-methylurea
62759	N-Nitrosodimethylamine
59892	N-Nitrosomorpholine
56382	Parathion
82688	Pentachloronitrobenzene (Quintobenzene)
87865	Pentachlorophenol

CAS number	Chemical name
108952	Phenol
106503	p-Phenylenediamine
75445	Phosgene
7803512	Phosphine
7723140	Phosphorus
85449	Phthalic anhydride
1336363	Polychlorinated biphenyls (Aroclors)
1120714	1,3-Propane sultone
57578	beta-Propiolactone
123386	Propionaldehyde
114261	Propoxur (Baygon)
78875	Propylene dichloride (1,2-Dichloropropane)
75569	Propylene oxide
75558	1,2-Propylenimine (2-Methyl aziridine)
91225	Quinoline
106514	Quinone
100425	Styrene
96093	Styrene oxide
1746016	2,3,7,8-Tetrachlorodibenzo-p-dioxin
79345	1,1,2,2-Tetrachloroethane
127184	Tetrachloroethylene (Perchloroethylene)
7550450	Titanium tetrachloride
108883	Toluene
95807	2,4-Toluene diamine
584849	2,4-Toluene diisocyanate
95534	o-Toluidine
8001352	Toxaphene (chlorinated camphene)
120821	1,2,4-Trichlorobenzene
79005	1,1,2-Trichloroethane
79016	Trichloroethylene
95954	2,4,5-Trichlorophenol
88062	2,4,6-Trichlorophenol
121448	Triethylamine
1582098	Trifluralin
540841	2,2,4-Trimethylpentane
108054	Vinyl acetate
593602	Vinyl bromide
75014	Vinyl chloride
75354	Vinylidene chloride (1,1,-Dichloroethylene)
1330207	Xylenes (isomers and mixture)
95476	o-Xylenes
108383	m-Xylenes
106423	p-Xylenes
0	Antimony Compounds
0	Arsenic Compounds (inorganic including arsine)
0	Beryllium Compounds

CAS number	Chemical name
0	Cadmium Compounds
0	Chromium Compounds
0	Cobalt Compounds
0	Coke Oven Emissions
0	Cyanide Compounds[d]
0	Glycol ethers[e]
0	Lead Compounds
0	Manganese Compounds
0	Mercury Compounds
0	Fine Mineral Fibers[f]
0	Nickel Compounds
0	Polycylic Organic Matter[g]
0	Radionuclides (including radon)[h]
0	Selenium Compounds

Note: For all listings above which contain the word "compounds" and for glycol ethers, the following applies: Unless otherwise specified, these listings are defined as including any unique chemical substance that contains the named chemical (i.e. antimony, arsenic, etc.) as part of that chemical's infrastructure.
[a]Caprolactam was delisted on June 18, 1996.
[b]Hydrogen sulfide was inadvertently added to the Section 112(b) list of HAPs through a clerical error. A joint resolution was passed by Congress and approved by the President on December 4, 1991 removing hydrogen sulfide from Section 112(b). Hydrogen sulfide is included in Section 112(r) and is subject to accidental release provisions.
[c]Methylethylketone (2-Butanone) was delisted December 19, 2005.
[d]X′CN where X = H′ or any other group where a formal dissociation may occur. For example KCN or $Ca(CN)_2$.
[e]Includes mono- and di-ethers of ethylene glycol. Diethylene glycol, and triethylene glycol R-(OCH_2CH_2)n-OR′ where

n = 1, 2, or 3

R = alkyl or aryl groups

R′ = R, H, or groups which, when removed, yield glycol ethers with the structure: R-(OCH_2CH)n-OH. Polymers are excluded from the glycol category.

[f]Includes mineral fiber emissions from facilities manufacturing or processing glass, rock, or slag fibers (or other mineral derived fibers) of average diameter 1 micrometer or less.
[g]Includes organic compounds with more than one benzene ring, and which have a boiling point greater than or equal to 100°C.
[h]A type of atom which spontaneously undergoes radioactive decay.

Index

Note: Page numbers followed by "*b*," "*f*," and, "*t*" refer to boxes, figures, and tables, respectively.

T

Printed in the United States
By Bookmasters